MEDITATIONES ALGEBRAICÆ

An English Translation of the Work of

Edward Waring

By permission of the Master and Fellows,
Magdalene College, Cambridge.

MEDITATIONES ALGEBRAICÆ

An English Translation of the Work of Edward Waring

Edited and Translated from the Latin
by
Dennis Weeks

American Mathematical Society
Providence, Rhode Island

1991 *Mathematics Subject Classification.* Primary 01A75, 01A50; Secondary 05E05, 11-03, 11P05.

Library of Congress Cataloging-in-Publication Data

Waring, Edward, 1734–1798.
[Meditationes algebraicae. English]
Meditationes algebraicae: an English translation of the work of Edward Waring/edited and translated from the Latin by Dennis Weeks.
p. cm.
Translation of: Meditationes algebraicae. 3rd ed. 1782.
Includes bibliographical references and index.
ISBN 0-8218-0169-4
1. Algebra—Early works to 1800. 2. Algebra—History—18th century. I. Weeks, Dennis, 1943– . II. Title.
QA154.8.W3713 1991 91-17347
512–dc20 CIP

Printed in the United States of America.
This volume was printed directly from author-prepared copy.
The paper used in this book is acid-free and falls within the guidelines
established to ensure permanence and durability. ♾

10 9 8 7 6 5 4 3 2 1 95 95 94 93 92 91

This translation is dedicated to Shana,

one of the few living people to have checked out a circulating copy of an eighteenth-century book from a university library,

to Kerensa, who told me to go to the Math Library,

and to Dave Dellinger, who introduced me to the wonders of whitespace management.

Contents

Translator's foreword

Some time ago I attended a lecture by a prominent computer scientist, a man of Chinese origin, who began with the words, "Since there are very few Chinese people in the audience, I will give my talk in the universal language of Science: broken English." While his command of English was actually quite good, I let my mind wander a bit during his talk, to speculate about what would happen if he wrote a paper in "broken English" which might then be read by someone in China, who would translate it into Chinese. How closely would the Chinese version resemble what the author, who probably wrote a draft in Chinese and then translated it into English, intended to say?

Across an interval of two centuries, I now ask a similar question. In the 18th Century, the universal language of Science was Latin, and it does no injustice to those (essentially all) scientific writers whose native language was not Latin to note that they probably wrote "broken Latin" in many cases. For reasons of historical interest and personal curiosity I have chosen to translate from Latin to English a work by Edward Waring. Can I then represent accurately what would have been found in his English language draft (if one ever existed) by matching my broken Latin to his, with virtually no English language fragments to aid me?

The translator of a work by Waring may encounter more obstacles than mere Latin vocabulary issues, as noted by William Wales, who (acording to Nangle [1934]) reviewed an earlier volume by Waring:

> But our learned Author must excuse us, as we have been sufferers by it, for taking notice of the obscurity of his Latin, which is rendered yet more so by the careless manner in which it is pointed; and, above all, by the incorrect manner in which it is printed.... [Wales 1778, 92–93]

Another commentator is even more severe:

> Waring was one of the profoundest mathematicians of the 18th century; but the inelegance and obscurity of his writings prevented him from obtaining that reputation to which he was entitled. Except Emerson, there is scarcely any writer whose works are so revolting as those of Waring. The great elegance and admirable simplicity and order of all the writings of Euler contributed as much to give him the great celebrity which he enjoyed, as his inventive genius. [Thomson 1812, 306]

An anonymous entry in Chalmers's *General Biographical Dictionary* (vol. XXXI, pp. 152–155 [1817]) helps to explain the cause of the difficulty, which sounds very much like what is now diagnosed as dyslexia:

> His extreme short-sightedness too, joined to a certain want of order and method in his mind, which appeared remarkably even in his hand-writing, rendered his mathematical compositions so confused and embarrassed, that in manuscript they were often utterly inexplicable, a circumstance which may account for the numerous typographical errors in his publications.

His printer, John Archdeacon, was also criticized by William Ludlam for the poor quality of his mathematical typesetting:

> For my own part, I am sometimes forced to make types, which are commonly brass, of which I here send you a specimen (±a ±b ±c). It is called plus-minus ±. I printed my first tracts at Cambridge, when Archdeacon (not Bentham) was their printer. I was very sick of it; the University meanly provided with mathematical types, insomuch that they used daggers turned sideways for *plus's*. As to table-work of which I had a deal, they knew nothing of it; and many a brass rule was I forced to make myself.... I complained of this to Mr. Bowyer, and would have had him print my essay on Halley's quadrant; but he was too full of more important work. I remember I told him I had marked all Archdeacon's damaged letters; which were not a few,

> especially in the italic. To which the old gentleman replied, "I dont like you the better for that." [Nichols 1814, Vol. VIII, 414; reprinted in Bowes 1888, 320–321]

(Some further remarks by Sylvester, regarding Waring's writing style, are reproduced in the translator's notes for Chapter 2.) What all this means, for me, is that Waring may not always have written what he meant; I have taken this as a challenge to be the editor and proofreader he never had. However, I have exercised these roles with restraint: the final product must be Waring's book, not mine.

This translation is based on the final edition of the work. The first edition, by Waring's reckoning, was a book he published under the title *Miscellanea Analytica*. (Printed by Archdeacon's predecessor Joseph Bentham, it was much better laid out and had proportionally far fewer typographical errors than the later editions.) That work was the subject of some controversy. Waring had submitted a draft of part of it to the Royal Society in 1757, where it apparently lay dormant for two years. Then on December 20, 1759, John Colson died and the search began for a new Lucasian Professor of Mathematics at Cambridge (a position held at the time of this writing by Stephen Hawking). In support of his candidacy Waring printed a few copies of the work, which he "circulated in vindication of his scientific character"; but the "appointment of a young man, scarcely twenty-five years of age, and still only a bachelor of arts, to a chair which had been honoured by the names of Newton, Saunderson, and Barrow, gave great offence to the senior members of the university, by whom the talents and pretensions of the new professor were severely arraigned" [Nichols 1812, Vol. III, p. 717]. The severest arraignment was *Observations on the First Chapter of a Book Called Miscellanea Analytica*, printed anonymously by William S. Powell, of St. John's College (coincidentally, the former college of Henry Lucas, whose bequest founded the professorship), in January 1760. He cast aspersions on Waring's writing ability and style, and tried with some specious mathematics to refute Waring's results. Waring produced a signed *Reply* on January 25, and his friend John (later Sir John) Wilson printed anonymously the *Vindication of the Miscellanea Analytica* about the same time, both showing that Powell's mathematical arguments were very much out of step with recognized authorities such as Newton, just before Waring was awarded the professorship on January 28. (The required M.A. degree was conferred *per literis regius* a few days earlier.) Unable to accept defeat, on March 3 Powell printed an anonymous *Defence of the Observations...*, which insinuated that Wilson, not Waring, had written Waring's signed reply. Waring followed with a *Letter to Dr. Powell...* which cast doubts on his pretext of objectivity (his hidden agenda was to promote the candidacy of William Ludlam, another St. John's fellow). Unable to launch any more anonymous attacks, Powell never responded to Waring's letter, in effect conceding that his conduct had been improper and that he was mathematically inept.

The *Miscellanea Analytica* was finally published in 1762 with the aid of Thomas Holles, Duke of Newcastle and Chancellor of Cambridge, who may also have helped secure Waring's early M.A. The first half covered the theory of equations and topics in number theory, and the second half covered what was later called analytic geometry. Later Waring separated the work into two books. In 1770 he published what he called the second edition of the *Meditationes Algebraicæ*, and the *Proprietates Algebraicarum Curvarum* appeared in 1772. The latter was only slightly larger than the equivalent chapters of the *Miscellanea Analytica*, but the former had grown to about the size of the entire original volume. Finally in 1782 the basis of this translation, the "third edition" (but only the second named *Meditationes Algebraicæ*) appeared.

There were actually at least three printings of the third edition. The first is represented by an autographed copy donated by Waring to the British Museum's printed book collection (now the British Library). It contained two unnumbered pages of "Corrigenda" and twelve of addenda; evidently there was no earlier printing, as the addenda are mentioned in Waring's preface. Where I have occasion to refer to these pages I have numbered them as C.1, C.2, and A1.1 through A1.12 respectively. A second printing (little more than a new binding, as none of the text had changed) incorporated four more pages of addenda, which I will refer to as pages A2.1 through A2.4: the only instance I have seen of this issue was bequeathed by Isaac Milner to Queens' College, Cambridge, in 1820, and has since passed from there to the Australian National University, in Canberra. (In a quarto volume, the four added pages would have represented one sheet from the printer.) All the other volumes I have seen have an additional eight pages (two quarto sheets) of addenda, which I refer to as pages A3.1 through A3.8. (In the volume at the University of California, Berkeley, which was used for this translation, the A3 pages have actually been inserted before the A2 pages; I have seen the same ordering in other volumes, but the University of Illinois, Urbana, has one that is consistent with my hypothesis based on the British Library and ANU printings.) The addenda pages themselves, predictably, have their share of typographical errors, the most vexing of which incorrectly specify the page and line number to which their corrections or insertions apply. In one particular case the referent was identified as "Pag. 229" where a section numbered 2.2 was to be inserted. Page 229 contained parts of two short theorems, neither having been broken into sections, although their subject matter was related to the topic of the addendum. I finally concluded that Waring actually intended this remark to be inserted at the end of page 209. In other cases he seems to have used a page reference to the start of a Theorem or Problem, coupled with a line number referring to a specific place on a later page within the same article.

There is evidence that Waring was painfully aware of the profusion of typographical errors in his publications, and though many remained uncorrected it was not for want of effort on his part. In 1794 an anonymous review of a work by Englefield [1793] was published in the *Monthly Review*; Nangle[1955] attributes it to Waring on the strength of a marginal note in a volume now at Oxford's Bodeleian Library. Believing that it was written by Waring, it is difficult to read the following passage without feeling a pang of sympathy for its author:

> Here we must take the liberty of adding one to the errata which the author has collected with so much industry and care; — not that the error is of moment, being only Tab. 3. instead of Tab. 4. (p. 199): but it may serve to show how liable our best-founded hopes are to be frustrated*, and that we have perused the book with some attention.
>
> *Sir Henry, after mentioning in his preface the great pains which he had taken, in the first place to avoid errors, and in the second to detect them, adds, 'It is hoped that no error is left undetected.'

(Even here the printing process appears to have frustrated Waring's hopes: the review, as published, does not identify which entries in Englefield's table are in error, nor how they should be corrected!)

I have incorporated the addenda and corrigenda (both known and inferred) "seamlessly" into the work, as if a fourth edition had been published by Waring including the additional text where it belonged. As to the translation itself, I try to achieve a balance between the 18th Century and 20th Century styles. Assuming a modest editorial role, I have allowed myself some flexibility in these matters. Waring spelled out all of his formulas in lengthy detail; I have compressed many of his longest ones into more compact form using, e.g., "factorial" notation, except in cases where stringing the whole thing out makes the pattern of the numbers or letters more obvious. With respect to terminology, I have substituted more modern terms for those Waring used, when Waring's words now might cause confusion. But in cases where it is easy to understand an archaic term today, I tend to keep the old expressions. Thus you will find "symmetric" instead of "similarly involved" (*similiter involvuntur*), but I seldom substitute "complex" or "imaginary" for "impossible" roots (though "possible" has usually become "real"), and you will still see fourth, fifth, and sixth powers referred to as "biquadrates", "quadrato-cubes" or "sursolids", and "cubo-cubes" or "bicubics". I generally substitute "degree" (a modern term for "the number of Units in the highest Power" [Wilson 1760, 7]) for "dimensions" of an equation, although Waring may have been the first to insist on this equivalence: Saunderson, a standard text at the time, allowed that $x^6+ax^3+b=0$ could be said to have the dimensions of a quadratic. The preface, which deals largely with historical issues, keeps more of the original terminology, but I have inclined toward modernization of the text of the mathematical chapters of the book. (Very little of the work, and virtually none of the preface, has ever been translated into any modern language; a paraphrased summary of the topics in the 1770 edition in Hutton [1815], Wales's [1778] translation of a substantial part of the preface to the *Meditationes Analyticæ*, and a 1787 paper by Waring, are the most extensive English renditions of any of Waring's Latin, and I have used them informally as style guides.)

Waring divided each chapter into numbered "Problems" and "Theorems", a few of the latter being preceded by unnumbered Lemmas, and each chapter ending with a Scholium. Some of the longer theorems and problems were subdivided into numbered sections, "Examples", and "Corollaries". In some cases a theorem has two or more occurrences of "Corollary 1": clearly Waring had in mind a more hierarchical structure to the work, and one of the corollaries belonged to one subsection or example, the other to a different one. I have accordingly replaced Waring's generally flat numbering, which went to a single decimal at most (e.g. "2.2"), with as many points as necessary to make it clear what section, example, or corollary belongs to what part of the article.

The entire work contained only two footnotes, which I have incorporated as parentheses in the text. Numbers in [brackets] are mine, and refer to the Translator's Notes at the end of each chapter. In the chapter notes I have concentrated mainly on indications of what has happened since Waring's time, with occasional mention of the evolution of ideas prior to their mention by Waring.

In any historical document, the reader is entitled to some biographical information about obscure characters who appear in the text. Of the sixty-four authors cited by Waring, only Buckley and Wassenaer are covered in no works such as the *Dictionary of Scientific Biography*, the *Dictionary of National Biography*, Poggendorff's *Biographisches-Litterarisches Handwörterbuch*, or Wallis and Wallis's *Biobibliography of British Mathematics and its Applications*. What information we have about the two exceptions, I have summarized in chapter endnotes.

In 1923 Franz X. Mayer (1875–1936, also known as Pater Augustin), a Cistercian monk who taught secondary school mathematics in Switzerland, published the only critical review of the *Meditationes* ever written, and although the coverage is somewhat uneven I include it as an Appendix to this translation. It is some consolation to know that Mayer too was impeded by Waring's typographical errors. Most of the individuals mentioned by Waring got brief biographical footnotes in Mayer's thesis, but he was utterly defeated in trying to identify someone named Kiraast, to whom Waring attributed a result in the theory of integer divisibility. By a stroke of good fortune I have determined that Waring was actually referring to Georg Wolfgang Krafft (the elongated "s" used by 18th Century printers aliases easily to an "f", and when followed by a "t" the two are virtually indistinguishable — which may explain why the error persisted from the 1770 to the 1782 edition). All footnotes in the Appendix are Mayer's; I have translated literally from the German and made virtually no changes, even to his mention of the nonexistent Kiraast (although I have succumbed to the desire to add a few endnotes of my own).

About Waring himself, Scott's entry in the *Dictionary of Scientific Biography* and Mayer's comments (derived in large part from Courtney's entry in the *Dictionary of National Biography*) suffice, with two exceptions and a digression. First, published biographies disagree, by as much as two years, on the date of Waring's birth. Courtney and Mayer say 1734, Scott says "around 1735", and Poggendorff believes 1736; an obituary notice in the *Gentleman's Magazine* said that his death, in August 1798, was "in his 63rd year". The most definitive source, *Shropshire County Parish Registers* (vol. 21, Diocese of Litchfield, published [1950] by the Shropshire Parish Register Society) says he was born on July 23, 1734. He was supposedly 17 years old when he entered Cambridge on March 24, 1753 (*Alumni Cantab.* II, vol. 6, p. 352): this would have been the last day of 1752 in the Julian calendar, but England had changed to the Gregorian calendar the previous September, causing enough confusion to account perhaps for this rather surprising inconsistency with his actual date of birth.

I also take exception to Mayer's contention that Waring avoided infinities and infinitesimals because he did not understand them. Reference to the *Meditationes Analyticæ* leaves no doubt that Waring had full mastery of the calculus; it is reasonable to suppose that in the *Meditationes Algebraicæ* he simply chose to limit his discussion (aside from some comments in the Scholia) to purely algebraic matters.

The digression is that another famous Shropshire lad, Charles Darwin, had an ancestor named Waring, and while it has so far proved impossible to confirm, it is intriguing to speculate that Shropshire's greatest mathematician and its greatest biologist were not-too-distant cousins. (However, a few generations earlier the Darwin family came to Shropshire from Nottinghamshire, and I have found no evidence that Edward Waring had ancestors there: the only possible Waring relative outside of Shropshire or London that I can identify is Thomas Waring, of Suffolk, one of the subscribers to the *Miscellanea Analytica* in 1762.)

This volume has the two conventional items at the end: a bibliography and an index. Most bibliographic entries are works cited in the endnotes, although I have also included material explicitly cited by Waring. Parenthesized names refer to authors, now known, of items that were published anonymously or only identified by initials (which I show outside the parentheses). I represent periodical volume numbers in boldface, followed (optionally) by an "issue number" in parentheses. The index attempts to be a full citation index for all persons named in any context in this volume, either in Waring's original text or in Mayer's or my commentaries, with full names given wherever possible. I have not tried to make a topic index.

Printing the original editions of this work took approximately two years, according to Waring's remarks in his Preface. A handwritten draft of the English translation was easy enough to do, but preparing it for a photo-typesetter has been a labor of comparable difficulty. By using the LaTeX system I have been able to take the document through multiple cycles of reviewing and proofreading, a great improvement over the type of iterations Waring and his printers had to endure. Nevertheless I could write a book about all the things I needed to do to prepare this work for publication, which were not easy to accomplish on my computer. Still I find the final result reasonably attractive to look at, and I hope the reader will agree that the composition of formulas makes them intelligible. As to the text itself, some of it was very easy to translate; other parts (especially portions of the Preface) proved quite difficult. Witmer, in the preface to his excellent translation of Cardan's *Ars Magna*, modestly and graciously thanks the people who reviewed his work and made suggestions as to how to improve it. I wish I had even more eloquence to give full credit to the dedicated referees of the American Mathematical Society and some other anonymous collaborators, involved by one of the referees in the task of helping to ensure that this translation would be both literally correct and consistent with some of the grammatical subtleties of the Latin language. Having already reproduced the criticisms by Waring's contemporaries of his poor typography and style, and having asserted my own intent to remedy those errors in the translation, I must now acknowledge that any errors remaining in this volume are mine.

Santa Clara, California
August 1, 1991

Preface

To accord recognition to its inventors, I have compiled this brief account of the stages of growth of Algebra; and I hope it may also be of some benefit to those who strive to keep abreast of current advances in the state of this art. I have borrowed from the histories of the famous men Wallis and Montucla — the former seems sometimes too much biased toward our compatriot Harriot, the latter in favor of the French school, but such lapses are human. It is generally acknowledged that Montucla's splendid history surpasses all others both in style and in content of philosophical prudence, but serious students of mathematics may desire a fuller account of more extensive recent research.

There remains disagreement as to the foundations of the subject, which were laid down in antiquity, and likewise as to whose works were written first; all show proficiency in certain common arithmetic operations and the solution of linear and quadratic equations. Diophantus of Alexandria, fl. c. AD 360, is the oldest of the algebraists whose works have survived. He reduced many equations involving two unknown quantities by relating one unknown to the other; until recently his methods for finding rational solutions to these problems have rarely been surpassed in conciseness and elegance. The numerals 0, 1, 2, 3, etc., invented by the Indians, first passed to the Arabs and from there, between the years 1000 and 1100 or perhaps slightly earlier, came into use among the Europeans. According to Wallis decimal arithmetic originated with al-Zarqālī and Regiomontanus about AD 1400; according to Montucla also with Peuerbach: these men divided the circle into 60,000,000 parts. But it is not clear whether they invented decimal fractions. The decimal method was nearly completed by Petrus Ramus in his Arithmetic, published c. 1560. He shows how to extract square and cube roots (if integer operations terminate with a remainder) by computing a quantity in integers, such that it shall have an integer root with a decimal adjunction. This had already been done, according to Wallis, by our own William Buckley (an alumnus of King's College, Cambridge), whose arithmetical manuscript, written in verse, was appended to Seton's *Logic* (Cambridge, 1631). Buckley died around 1550 (during the reign of Edward VI, who died in 1553) [1].

Our countryman Briggs studied the nth differences between the numbers $1^n, 2^n, 3^n, 4^n, 5^n, \ldots$, discovering that they are constant and each difference is equal to $1 \cdot 2 \cdot 3 \cdot 4 \cdots \times (n-1)\,n$; from this he was able to interpolate logarithms between given quantities. In this work nothing about this subject is pursued; nevertheless a method is given to find a function of the quantities $x,\ y,\ z,\ \ldots$ which will take the values $S,\ S',\ S'',\ \ldots$ when $x, y, z, \ldots$ take the values respectively $\alpha,\ \beta,\ \gamma,\ \ldots$, or $\alpha',\ \beta',\ \gamma',\ \ldots$, etc.

Mohammed ben Musa and Thebit ben Korah, the earliest of the Arab algebraists, were the first, according to Cardan, to solve quadratic equations; it is claimed that Omar ben Ibrahim reduced cubic equations, but the Arabic manuscripts have not been examined very closely. [2] Leonardo of Pisa, writing around AD 1400 or earlier, according to Vossius (following Blancanus), was in his time the first author to write about algebra. The first printed book on algebra was written by Lucas de Burgo; he showed how to solve all quadratic equations and stated that they must have two roots. Cardan attributed to Scipio del Ferro the method of solving cubic equations, which Ferro kept secret; meanwhile Tartaglia independently derived the same result, and disclosed it to Cardan at the latter's request, accompanied by a pledge of secrecy. But Cardan forswore his oath and published Tartaglia's method, with improvements of his own, the main one being to remove the second term of the equation; he also showed that the irreducible case must have three negative or positive roots. Ludovico Ferrari developed the solution of the biquadratic equation, and Wallis contends in a letter to Morland (*Algebra*, ch. 53) that Bombelli derived his method from Ferrari.

With respect to notation, Buteo's *Logistica*, published in 1559, designated various unknown quantities by letters A, B, C, etc. (at the end of Book III, in what he called the *Regula quantitatis*). Vieta first substituted letters for known quantities, in the process of showing that the coefficient of the second term of an equation is the sum of the roots, and that all coefficients are similarly constituted, also showing how the roots of a given equation could be increased or decreased, or multiplied or divided, by given amounts. He even showed a not inelegant method of finding rational solutions of a quadratic equation. He gave verbal rules for expressing the binomial quantity $(a+b)^n$ to any chosen power, showing the literal products to be a^n, $a^{n-1}b$, $a^{n-2}b^2$, etc. He further showed that the coefficient of the second term must be a triangular number, that of the third a pyramidal number, and so on. Properties of triangular, pyramidal, and other figurate numbers were discussed by Pascal, providing the basis for Newton's formulation of the binomial theorem, *viz.* $(a+b)^n = a^n + na^{n-1}b + n \cdot \frac{n-1}{2} a^{n-2}b^2 + \ldots$.

Harriot substituted the notation aa, aaa, etc., and sometimes a^2, a^3, etc., for the quantities Aq, Ac, etc. He also rewrote equations as single expressions, set equal to zero, which allowed him to show how equations of higher degree could be constructed from linear equations, and how their coefficients would be composed. The same principle is useful in detecting impossible roots in cubic equations, and from it Newton was later able to develop many similar rules. After this Descartes used the last letters of the alphabet, x, y, z, ... for unknown quantities and the first, a, b, c, ... for known quantities; it was he who first asserted that there are as many positive roots as there are changes in sign, and all the rest negative, if all the roots of the equation are real. The proof of this rule, based on multiplying by $x-a$ and $x+a$, if all roots are real, was first given by Gua. Segner stated that there are at most as many positive roots as there are changes in sign from $+$ to $-$ and from $-$ to $+$, and at most as many negative roots as there are continuous sequences $++$ and $--$. In the first edition of this work a method was first published showing how to decide if all roots of a given equation are real, and thus whether the above rule is applicable or not. Here it is perhaps worth mentioning that in arbitrary equations of high degree, very seldom are all the roots real, hence it is very desirable to have a rule which shows first the number of impossible roots of a given equation, thereafter, if the number of impossible roots is given, to find the number of positive and negative roots. These two problems, which in equations of high degree require painstaking and laborious calculations, have hardly ever been solved; something about these problems is contained in this work. Descartes also presented a method, different from Ferrari's, for the solution of biquadratic equations. Still other methods for the same problem were given by Euler, Bezout, Vandermonde, etc., and yet another new method will be found here. But it should be observed that all methods for this problem are deduced from the same first principle, and so all are basically the same.

Descartes showed that each integer root must be a divisor of a, the last term of the given equation $x^n - px^{n-1} + qx^{n-2} - \ldots = 0$, and conversely that if a given equation could be divided by $x - a$, then a must be a root of the given equation. This method is more simply expressed, and in most cases computationally easier, than the method of Wassenaer, although many authors of algebras seem to ignore it; moreover it leads to a method for finding "amicable numbers". Descartes also showed how to find limits between which the roots of a given equation must lie, by incrementing or decrementing the roots by certain quantities chosen so that its negative coefficients would all become positive, or its positive coefficients negative.

A method of extracting the nth root of a binomial quantity $a + \sqrt{b}$, provided the root can be expressed in the form $(x + \sqrt{y})/\sqrt[n]{Q}$, was given by Schooten. First he sought a number N, derived from the given quantity $a + \sqrt{b}$, such that $N^2(a^2 - b)$ would be a quantity whose n^{th} root $(N^2(a^2 - b))^{1/n} = \pi$ could be extracted; then he would find a rational quantity v, greater than $(N(a+\sqrt{b}))^{1/n}$ such that the difference would be not more than $\frac{1}{2}$; then he would select the nearest integer z, either larger or smaller respectively than $v \pm \frac{\pi}{v}$. From this $(\frac{1}{2}z + \sqrt{\frac{1}{4}z^2 - \pi})/\sqrt[n]{N}$ will be the desired root. Newton follows the principle laid down by Schooten, *viz.* to multiply a given quantity $a + \sqrt{b}$ or $\sqrt{a} + \sqrt{b} = A + B$, where $A > B$, by N, such that the nth root $(N \times (A^2 - B^2))^{1/n} = \pi$ can be extracted; then let r be the integer nearest to $(N^{1/2}(A+B))^{1/n}$, let s be the quotient of $A\sqrt{N}$ by its greatest rational factor, and let t be the integer nearest to $(r + \frac{\pi}{r})/2s$. Then $(tx + \sqrt{t^2s^2 - \pi})/\sqrt[2n]{N}$ will be the required root. Euler takes $(x + y/\sqrt[2n]{N} = \sqrt[n]{A + B}$, whence $x^2 - y^2 = \sqrt[n]{(A^2 - B^2)N} = p^n$, where x, y, and p are integers; he shows that $x^2+y^2 = \frac{1}{2}\sqrt[n]{(A^2 + B^2)N} + \frac{1}{2}\sqrt[n]{(A^2 - B^2)N}$, and observes, by inspecting x^2 and y^2, that Newton's rule sometimes fails.

To my knowledge I was the first to prove Schooten's rational principle as given above. I also proved that $a^n + n \cdot \frac{n-1}{2} a^{n-2} b + n \cdot \frac{n-1}{2} \cdot \frac{n-2}{3} \cdot \frac{n-3}{4} a^{n-4} b^2 + \cdots$ and $na^{n-1} + n \cdot \frac{n-1}{2} \cdot \frac{n-2}{3} a^{n-3} b + \cdots$, where a and b are mutually prime, have no common divisors except powers of 2 up to 2^n. It follows that when $A + B = 2^n$, it is always possible to extract their root by Newton's rule; moreover both these rules are particular cases of a principle more general than that stated by Schooten.

Jacob van Wassenaer showed how to find linear factors of a given equation by substituting ..., −1, 0, 1, ... for the unknown quantity; Newton extended Wassenaer's method to quadratic, cubic, and higher divisors. In this work Descartes' method is similarly extended to divisors of higher degree, and also a method is given for finding integral values of two unknown quantities satisfying two equations simultaneously; the same method can be applied to more equations in more unknowns. Hudde introduced the following technique for equations with multiple unknowns: set one, two, or more of the unknown quantities equal to zero, or to some fixed quantity; if this results in an equation which is irreducible, the original equation must be irreducible also. He showed how to find a divisor of a given equation, when one or more of its terms are constrained to be zero or some given value; and that if the terms of the given equation are multiplied by successive numbers in an arithmetic progression, equal roots may be detected, if any are contained in the given equation. Given an equation $x^6 + qx^4 + rx^3 + sx^2 + tx + v = 0$, with a divisor $x^2 + yx + h = 0$, he showed how to find an equation whose root is y. The method commonly in use to eliminate an unknown from two or more equations in two or more unknowns is due to Erasmus Bartholinus (his Chapter 16, "Of a general method by which the second root is used to remove a radical sign from a proposed equation"). His methods of reducing equations to eliminate unknowns, and getting rid of radical signs, are the same as those that later appeared in Newton's *Arithmetica Universalis*.

So-called "continued fractions" were first discussed by Viscount Brounker, who also showed several ways of finding integral values for x or y in the equation $ax^2 + 1 = y^2$. A general rule for this, by means of series, will be presented in this book. Euler extended the method to the equation $ax^2 + bx + c = y^2$; Lagrange adds something about the case in which x and y cannot both take integral values. Wallis gives a method of reducing fractions and ratios to their lowest terms, which is also applicable to integer approximations; he further gives a method of approximating with integers the unknown quantities in $ax^2 + 1 = y^2$ and similar equations. Bachet stated that every number is the sum of four squares, which was first proved by Euler using rational numbers, then by Lagrange using integers. He was also the first to propose a general method of finding integer solutions to the equation $ax + by = P$, where a, b, and P are positive integers. Euler accomplished this by dividing the greater coefficient, a, by the lesser, b, and dividing the divisor by the remainder, and so on, continuing until a remainder was found which divided the number P; from these quotients he found a single pair of values for x and y. Simpson later accomplished the same thing from the remainders.

In the first edition of the present work, transmitted to the Royal Society in 1757, printed and circulated among friends in 1760, and published jointly with *Proprietates Algebraicarum Curvarum* in 1762, a method was first published for finding the general expression for the integer values of the individual unknown quantities $x, y, z, \ldots$ in the equation $ax+by+cz+dv+\cdots = A$, with three or more unknowns x, y, z, v, etc. contained in the equation, in terms of quantities $u, w, \ldots$, where $a, b, c, d, \ldots$ are integers and $u, w, \ldots$ are arbitrary quantities. In the second edition of this work, printed in 1768 and 1769, and published at the beginning of the year 1770, a further method was given of reducing two equations $ax + by + \cdots = P$ and $a'x + b'y + \cdots = P'$ to one (by multiplying both equations by m and n respectively, and choosing integral values of m and n to satisfy the equation $ma - na' = 1$; or finding general integer values of a quantity x in terms of the specified quantities $u, w, \ldots$ from a given equation, and substituting these in another or in the remaining terms), thereby eliminating an unknown quantity, and if the unknown quantities in the resulting equations are integers, then the eliminated quantity is also an integer. It is asserted that the quantity $ax + by$, where a and b are mutually prime and x and y are integers, can be used to construct every integer greater than $ab - a - b$. In the 1762 edition, I also showed a reduction of two or more, say m, equations, at least one of them having $m+1$ or more unknowns, by which if the unknown quantities in the resultant equations are rational we can conclude that the eliminated unknown is rational as well; this relies on the common method due to Bartholinus whereby two or more equations are reduced such that the final resultant is $Px = Q$, where P and Q are rational functions of the other unknowns, and the reduction is completed by substituting P/Q for x in the given equations. However, it may happen that the form $Px = Q$ cannot be attained, but instead the process ends with two or more resultant equations of the form $Ax^2 + Bx^2 = C$ and $Ax^2 + Bx^2 = D$, ..., or $Ax^3 + Bx^2 + Cx = D$ and $Ax^3 + Bx^2 + Cx = E$, ..., where A, B, C, D, E, ... denote arbitrary rational functions of the unknown quantities y, z, v, ... contained in the given equation; in such a case each value of each of the unknown quantities y, z, v, ... has two (or more) values of the unknown x corresponding to it, so by this method the quantity x cannot be eliminated. One or more unknown quantities may be found from the rational terms of the other and new assumed quantities: only if this is possible, these values are substituted and what is required is carried out, repeating the process as often as necessary.

Fermat showed that every integer is a triangular number, or can be written as the sum of two or three triangular numbers, and several other theorems of like kind, for instance letting m denote an integer he asserted that every prime number of the form $(4m + 1)$ is the sum of two squares. In the second edition of this work it is asserted that every number is either a cube, or the sum of $1, 2, 3, 4, \ldots 8$, or 9 cubes, also that every number which is greater than a particular number, is a sum of numbers of the form $pa^2 + qb^2 + rc^2 + sd^2$, where a, b, c, d are arbitrary numbers and p, q, r, s are mutually prime. I say that the sum of two cubes, $a^3 + b^3 = (a + b)(a^2 - ab + b^2)$, cannot be equal to a third cube, c^3, nor can $ae(a^2 - e^2)$ or $x \times (x + 1)/2$ be a square (unless $x = 1$). Euler gave proofs of these propositions and added similar ones, such as that $\alpha^3\beta x^4 \pm \beta^3\alpha y^4, \ldots$ are not squares, which he proved along the same lines, namely that if the result is a square, he deduces quantities less than those given, which will also be squares, and so on *ad infinitum*. Lagrange applied the same principle to finding from smaller given quantities the larger values of the unknown quantities of a given equation. Fermat likewise discovered that $a^n - a$ is always divisible by n, if n is prime. Euler proved this theorem, and added an elegant corollary: he asserted that the numbers N^{2m+1} and N^{2m+2} can be composed in exactly $m+1$ distinct ways from numbers of the form p^2+rq^2, where $N=a^2+rb^2$, and he based some further corollaries on this. In my second edition for the first time the actual construction of the $m+1$ different compositions was given.

Albert Girard gave formulas for the sums of the squares, cubes, and fourth powers of the individual roots. Newton gave a rule by which from the sum of an given equation's roots, the sum of their squares, the sum of their cubes, and so on to some particular degree, the sum of the next higher power of the roots could be found. In my first edition I gave the rule for a series which computes these sums directly from the coefficients of the given equation. The same rule was later independently discovered by Euler. The above sums are useful for finding approximations to the largest root of a given equation, when all its roots are real, or when the largest root of the equation is real. But for transformations, or reduction and final resolution, or for eliminating unknown quantities from given equations, without the aid of subsequent developments, they are not very serviceable.

Cramer represents the roots of the equation $y^n - py^{n-1} + qy^{n-2} - ry^{n-3} + \ldots = 0$ by α, β, γ, δ, $\ldots$, and substitutes these values for y in the equation $V - Ty + Sy^2 - \ldots = 0$, and multiplies the resulting equations by themselves. Then from the sums of all individual values of the form $\alpha^a\beta^b\gamma^c \cdots + \ldots$, involving quantities of lower degrees, he deduces from the individual values sums of the same expression in the next higher degree. In the first edition of this work is given an easy rule to compute the value of $\alpha^a\beta^b\gamma^c\delta^d \cdots + \ldots$ using the sums of integer powers of the roots; in the second edition I added a method of evaluating the same expression using the coefficients of the given equation. In the first edition a method is given of finding the sum of the values of any algebraic function of the roots of one or more given equations, which is further explained in the second and third editions. From this rule which I discovered, in the first edition I deduced a new transformation of any equation into another whose roots should have, with respect to the roots of the given equation, any chosen relation; *viz.*, first the number of roots (and in consequence the dimensions of the required equation) is deduced from the doctrine of combinations, secondly from the assigned relation (letting α, β, γ, δ, ... and A, B, Γ, Δ, ... represent the roots of the given equations) the required roots of the equation are specified and thence the sum of the roots and the sums of products of two, three, four or more roots can be determined with the help of the above rules (since in such sums the roots α, β, $\gamma, \delta, \ldots$ all appear symmetrically). Never before had anyone transformed the roots, except by reducing equations in order to eliminate unknowns by substitution, nor had anyone previously shown how such a transformation could lead to a method for determining the sum of the individual values of any algebraic function of the roots. No one before me ever showed how to evaluate an irrational or fractional function of the roots of a given equation. No one else has deduced a method of determining the sum of distinct values of quantities $\alpha^a\beta^b\gamma^c \cdots + \ldots$ from the sums of powers of the roots of a given equation, or a like rule based on the coefficients, except for Cramer's inductive method of progressing from known sums of lower degree to the next higher degree, which could with difficulty be used for this purpose.

In the same edition of this work the above method of transforming equations was applied to the investigation of properties of curves, which is often helpful since the properties of interest often depend only on the first terms of the required equation. Also for finding impossible roots, removing surds from the terms of a given expression, and solving equations, my methods supplemented by those of Tschirnhaus are needed. In the second edition was given a new method of finding the aforesaid sums from assumed given equations with known roots, and so the required sums; also of progressing from equations of lower degree to equations of higher degree, and these methods are often quite helpful.

Schooten assigned a $+$ or $-$ to each null term in a given equation, and said that the number of imaginary roots was at least as many as the maximum difference in the number of sign changes based on the alternate possible assignments to each missing term. Newton stated a rule by which one may often determine the exact number of real and impossible roots; Campbell and Maclaurin published extensions of his rule. In my second edition I gave perhaps the most general principle so far published, of finding symmetric functions of several quantities which will always be positive, when all the quantities are real, and easy further generalizations of such functions. These lead to a multitude of rules for finding the number of impossible roots of a given equation. There I generalized the above rules, multiplying a given equation by

$x - a$, to advance to a rule for determining the number of impossible roots at a higher degree. In the first edition in 1762 is given a rule by means of which it is seen that the probability of finding the true number of impossible roots by Newton's rule is about $2^{n-2} : 3^{n-2}$ where $2n$ is the degree of the given equation.

I transformed a given equation, in the first edition, into one whose roots are the squares of the roots of the given equation, and in the second edition, into another equation all of whose roots would be positive, if all roots of the given equation were real; whence from the progressions of sign ($+ +$ or $- -$) the number of impossible roots may often be determined. In the second edition, printed in 1769, it is observed that by increasing or decreasing the roots of a given equation by suitably chosen quantities, it is often easy to discover impossible roots which were otherwise undetected. But all these rules have great difficulty in finding the true number of impossible roots in equations of high degree, and proofs are still wanting. Indeed the commonly known proofs only show that the equation under consideration has some impossible roots (if the rule so claims), not that it has exactly as many as the given rule asserts. A full solution to this problem must be very difficult and will involve a great deal of work. We know that roots pass from possibility through equality to impossibility, therefore a general solution to this problem depends on transformations in which roots of a given equation become equal. This is itself a very laborious task, hence the general resolution to the broader problem will be even more laborious.

In 1762 I published a transformation of a given equation into another whose roots are the squares of the differences of the roots of the given equation, and there noted, "if all the signs of the terms of the resulting equation alternate $+ -$ and $- +$, the given equation has no impossible roots, otherwise it has some impossible roots." In the *Philosophical Transactions*, London 1764, examples are given of this transformation for equations of the fifth degree, where it is asserted that the given equation has two impossible roots if the final term of the resulting equation is negative; but if it is positive, then either none or four will be impossible. I also discussed the case where the final term is zero. For one who understands them well, it is not impossible to apply the same observations to equations of higher degree. In the same edition I also mentioned that given an equation $x^n - px^{n-1} + qx^{n-2} - \ldots = 0$, we may propose equations $v^{n-1} - \frac{n-1}{n} p v^{n-2} + \ldots = 0$ and $v^n - pv^{n-1} + \ldots = z$, and transforming these two equations into one in order to eliminate v, by means of the resulting equation in z we may often detect impossible roots: it is asserted that the given equation has no real roots if the signs of the terms of the equation with root z continuously alternate + − and − +. Also, the given equation may be transformed into another, whose roots are $1/(\alpha - \beta)$, $1/(\alpha - \gamma)$, $1/(\beta - \gamma)$, ..., where α, β, γ, ... are roots of the given equation. Then we may determine a number $1/A$ which is greater than the greatest root of the resulting equation; then A is less than the difference between any two roots. Hence A, $2A$, $3A$, $4A$, ..., 0, $-A$, ... are limits between successive roots, with which information the number of impossible roots of the given equation is easily determined. In the second edition is shown a method of finding, from the given number of impossible roots in a given equation, the number of impossible roots in an equation whose roots stand in a given relation to the roots of the given equation; examples of this rule are given, one of which is the case in which the roots of the resulting equation are the squares of the differences of any two roots of the given equation. From the method of finding the number of impossible roots of a given equation in one unknown, I derive a method of finding the limits between which the possible solutions to a given equation in two or more unknowns must lie. Clairaut (*Algebra*, 1760) gives a method of finding the number of impossible roots in a biquadratic equation.

To my knowledge, Maclaurin was the first to assert that the roots of the equation $nx^{n-1} - (n-1)px^{n-2} + (n-2)qx^{n-3} - \ldots = 0$ are limits of the roots of the equation $x^n - px^{n-1} + qx^{n-2} - \ldots = 0$. Milner proved that the equation $ax^n - (a+b)px^{n-1} + (a+2b)qx^{n-2} - (a+3b)rx^{n-3} + \ldots = 0$ does not always have a root between the least positive and least negative roots of $x^n - px^{n-1} + qx^{n-3} - rx^{n-4} + \ldots = 0$. Let A be the equation $x^n - px^{n-1} + qx^{n-2} - \ldots = 0$, whose roots are $\alpha > \beta > \gamma \ldots$, and let $\pi, \rho, \sigma, \ldots$ be the roots of $\mathrm{B} \equiv nx^{n-1} - (n-1)px^{n-2} + \ldots = 0$, lying between α and β, β and γ, etc.: then in this work it is shown that if h and m have the same sign, the roots of $h\mathrm{A} + m\mathrm{B} = 0$ lie between α and π, β and ρ, etc. respectively; but if h and m have opposite signs, then one root of the equation $h\mathrm{A} + m\mathrm{B} = 0$ is greater than α, while the others are located between π and β, ρ and γ, etc. Further, if h and m have the same sign, then two roots of the equation $h\mathrm{A} + m\mathrm{B}x$

$= 0$ lie between the least positive, θ, and the least negative, ι, roots of the given equation $A = 0$, and one of them will be positive, the other negative. If there is a change of sign from $+$ to $-$ or from $-$ to $+$ between the penultimate and final terms of the given equation, then the positive root will be between θ and 0, the negative between $-u$ and $-\iota$, otherwise the positive root will be between θ and u, and the negative between 0 and $-\iota$, where u is the root of the equation $nx^{n-1} - (n-1)px^{n-2} + \ldots = 0$ lying between the least positive and least negative roots of the given equation $x^n - px^{n-1} + \ldots = 0$, and u will be negative or positive according as there does or does not exist a change of sign between the penultimate and final terms of the given equation. More results similar to this are contained in this work.

Newton showed how to extract the square root of a quantity $a + \sqrt{b} + \sqrt{c} + \sqrt{d} + \ldots$, provided the root could be expressed by a quantity of the same type, and he reduced certain equations of degree $2n$ to equations of degree n. Similar reductions to degree $n/2$, provided $n/2$ is an integer, were given in the first edition of this work. But in this and similar reductions we must first seek denominators, which may be reciprocals of quantities in the reduced equations, which are consistent with the integer values of the coefficients of the given equation. In the same edition a method is added of deciding whether or not a given equation can be solved by extracting square or higher roots. Later Newton gave a rule for selecting the terms with which to begin the calculation, *viz.*, look for two quantities x and y which are similarly represented in the given equations, that we may transform them so as not to be dependent on either of them, but instead to be expressed in terms of a third, z, which will represent the equivalent relation with respect to both of them. In my first edition I gave the first proof of this proposition, based on showing that the number of permissible values of z is only half as great as the number of values of x or y; the argument is that, if a value of x is α, and the corresponding value of y is β, then since the quantities x and y are symmetric in the given equations, if x is β then y will be α, so if the values α and β, or β and α, are substituted respectively for x and y in the rational expression z, in which x and y are also symmetrically represented, the same value of z must result from each substitution, so that z has two equal values and must have half as many distinct values as the quantity y. Now if the number of values of the quantities x and y is not even, then one value of the quantity x will be equal to the corresponding value of y; hence if the equation whose root is x or y has m roots, then the number of roots of the equation in z, which are rational symmetric functions of the quantities x and y, will be $m/2$ or $(m+1)/2$; and likewise for higher degrees of symmetry.

The second edition treats this problem more generally, applying the same reasoning to n equations having n unknown quantities, which are symmetric in $m, r, s, \ldots$ of them. The same reasoning can thus be applied to the first part of Newton's rule, *viz.*, when there exists such an affinity between two terms, or a symmetry to all the other terms of the required equation, then it may happen that identical equations are produced, and they are of no use to us. But if identical equations involving x and y are produced, then let α be a value of x and π a value of y, and in the said equations for x write y and for y write x; then if α is a value of the quantity y it follows that π must be the corresponding value of the quantity x, whence it follows that the two different values α and π of the quantities x and y will yield only one value ($\alpha + \pi$ or $\pi + \alpha$) of the quantity ($z = x + y$) corresponding to the two of them. In consequence the number of values of the quantity x or y is twice the number of values (*exceptis excipiendis*) of the quantity z.

In this edition of this work is proved, that if we are given equations which are symmetric with respect to x and y, and assume a rational quantity $P = z$, in whose square x and y are symmetrically involved, then the equation whose root is z will have roots in pairs, one negative and one positive, of equal magnitude in each case, and consequently will be of a form which is half the degree of the equation whose root is x or y. Newton does not appear to have known that the degree of the equation whose root is z would be half the degree of the equation whose root is x or y; I was the first to assert and prove the subsequent proposition, that if two equations are given, which are symmetric in x and y, the equation whose roots are a rational symmetric function of x and y will be of only half the degree (*exceptis excipiendis*) of the equation whose root is x or y. In this work I add the following proposition: let $x = \phi(z)$, such that z is the same function of the quantity x, *viz.*, $z = \phi(x)$; then in any given equation $A = 0$ which is a function of x and y for x write its value $\phi(z)$, and in the resulting equation for z write x, and let the resulting equation be $B = 0$; then the values of the quantity x in the assumed equation $x = \phi(x)$ will be the same as the values of the quantity x in the assumed equations $A = 0$ and $B = 0$; likewise let v be a rational function of the quantities x and $\phi(x)$, in which x and $\phi(x)$ are symmetrically involved, then the equation (*exceptis excipiendis*) whose root is v will be of degree no greater than half the degree of the equation whose root is x or $\phi(x)$. More is adjoined about these matters.

Leibnitz or Bernoulli was the first to decompose the fraction $1/(x^n - px^{n-1} + qx^{n-2} - \dots)$ into fractions $p/(x-\alpha) + q/(x-\beta) + \dots$. Euler shows that

$$\frac{\alpha^r}{(\alpha-\beta)(\alpha-\gamma)(\alpha-\delta)\cdots} + \frac{\beta^r}{(\beta-\alpha)(\beta-\gamma)(\beta-\delta)\cdots} + \frac{\gamma^r}{(\gamma-\alpha)(\gamma-\beta)(\gamma-\delta)\cdots} + \cdots = 0$$

provided that r is an integer less than the number of letters α, β, γ, δ, etc. I generalized this proposition in the *Philosophical Transactions*, *viz.*, it is proved there, using new principles, when α^r, β^r, γ^r, ... are replaced by products, all of the same dimension, of two, three, or more different roots. In this edition of this work from the above Bernoulli/Leibnitz theorem I deduce the value of the aforesaid sum, when r is any number, whether positive or negative, leading to a series which expresses the said sum when r is a fraction.

DeMoivre achieved a solution of equations of the form $x^n - npx^{n-2} + n \cdot \frac{n-3}{2} p^2 x^{n-4} - \dots = A$, and a reduction of the equation $x^{2n} + px^{2n-1} + \dots + px + 1$ to the equation $x^n - ax^{n-1} + \dots = 0$. In my first edition I showed that the four roots of the equation $x^4 + 1 = 0$ can be written as $\pm\sqrt{\frac{1}{2}} \pm \sqrt{\frac{-1}{2}}$, the five roots of the equation $x^5 + 1 = 0$ as

$$-1, \quad \pm\frac{\sqrt{1\frac{1}{4}} + \frac{1}{2}}{2} \pm \frac{\sqrt{-2\frac{1}{2} \pm \sqrt{1\frac{1}{4}}}}{2},$$

and the six roots of the equation $x^6+1=0$ as $\pm\sqrt{-1}$, $(\pm\sqrt{3}\pm\sqrt{-1})/2$. I also showed how to reduce the equation $x^{2n} \pm x^{2n-1} + x^{2n-2} \pm x^{2n-3} + x^{2n-4} \pm \dots = 0$ to a quadratic with the aid of the equation $v^n \mp v^{n-1} - (n-1)v^{n-2} \pm (n-2)v^{n-3} + \frac{(n-2)(n-3)}{2} v^{n-4} \mp \dots = 0$, and showed that $p^n - np^{n-2} + \frac{n(n-3)}{2} p^{n-4} - \frac{n(n-4)(n-5)}{2 \cdot 3} p^{n-6} + \dots = (\sqrt{\frac{1}{4}p^2 - 1} + \frac{1}{2}p)^n + (-\sqrt{\frac{1}{4}p^2 - 1} + \frac{1}{2}p)^n = \alpha^n + 1/\alpha^n$, etc., where α and $1/\alpha$ are the roots of the equation $x^2 - px + 1 = 0$. In the second edition more is added on the reduction of recurrent equations.

Bezout set $y = (x+a)/(x+b)$ and substituted this expression for y in the equation $y^n + h = 0$, resulting in

$$x^n + n\frac{a+hb}{1+h}x^{n-1} + n\cdot\frac{n-1}{2}\cdot\frac{a^2+hb^2}{1+h}x^{n-2} + \ldots = 0;$$

I produced the series $x^n = a^{n-1}b \pm a^{n-2}b^2 + na^{n-2}bx + n\cdot\frac{n-3}{2}a^{n-3}bx^2 + n\cdot\frac{n-4}{2}\cdot\frac{n-5}{3}a^{n-4}bx^3 + \ldots$, whose resolution is $x = \sqrt[n]{a^{n-1}b} + \sqrt[n]{a^{n-2}b^2}$, in the *Philosophical Transactions*, London, and in this edition of this work I produce the identity $x^{2n} - 2b^npx^n - \frac{n}{1\cdot 2}\cdot 2nb^{n-1}a^2px^{n-1} - \frac{n\cdot(n^2-1)}{1\cdot 2\cdot 3\cdot 4}\times 2nb^{n-2}a^4px^{n-2} - \ldots = a^{2n}p - b^{2n}p^2$, and further $x^n - p(na^{n-3}bx^2 + n\cdot\frac{n-5}{2}a^{n-6}b^2x^4 + \ldots) \pm p^2(na^{(n-3)/2}b^{(n+1)/2}x - \ldots) = a^np + b^np^3$, whose resolution is $x = a\sqrt[n]{p} + b\sqrt[n]{p^3}$, and other similar results. These resolutions may easily be made even more general, e.g. one may construct equations whose roots shall be $x = a\sqrt[n]{p^m} + b\sqrt[n]{p^{m+n}}$.

Leseur asserted that the equation $x^n - px^{n-1} + \ldots = 0$ can be depressed to an equation $x^m - ax^{m-1} + \ldots = 0$ with the aid of an equation of degree $n\cdot(n-1)\cdots(n-m-1)/m!$ (the same principle was earlier applied by Sanderson to biquadratic equations) and hence was able to show that the equation $x^{4n+2} + \ldots = 0$ has at least one quadratic divisor. Using the same principle Euler proved that every quantity $x^n - px^{n-1} + \ldots$ is actually a product of linear and quadratic divisors. I did not see either Leseur's work, or Euler's, until after the publication of the *Miscellanea Analytica*, that is to say, the first edition of this work. Euler's method is not much different from mine, as I showed the biquadratic equation $x^2 - 2(a+b\sqrt{-1})x - c - d\sqrt{-1} = 0$ as two rational quadratics. In my second edition I showed that if a given equation, $x^n - px^{n-1} + qx^{n-2} - \ldots = 0$, of degree n, is reduced to an equation $x^m - ax^{m-1} + bx^{m-2} - \ldots = 0$, of degree m, then the equations defining a, b, c, ... will all be of the same degree, and corresponding values of the unknown quantities (a, b, c, ...) will contain the same irrationalities, unless two or more values of the original unknown quantity are equal to each other, and so on. There also I assert that if two equations involve two unknown quantities, say x and y, the corresponding values of the unknown quantities x and y will have the same irrationalities, unless two or more (say m) values of one of them (e.g. x) are equal to each other, in which case the same irrationalities will be found in the square (or some higher power) of the equation whose root is x, as exist in the corresponding values of y.

A method is given by Euler of reducing two equations, $y^n + ay^{n-1} + by^{n-2} + \ldots = 0$ and $y^m + Ay^{m-1} + By^{m-2} + \ldots = 0$, into one, in order to eliminate y. He multiplies the two said equations by the two quantities $Py^{m-1} + a'y^{m-2} + b'y^{m-3} + \ldots$ and $P'y^{n-1} + A'y^{n-2} + B'y^{n-3} + \ldots$, then adds the two products and imposes on the coefficients $P, a', b', \ldots, P', A', B', \ldots$ the condition that in the resulting equation all terms involving y must vanish. Similar methods may be applied to three or more equations having three or more unknown quantities. Bezout reduces two or more equations to one, in order to eliminate unknown quantities, by multiplying the given equations, say

$$\begin{aligned} &Ay^n + (a + bx + cz + \ldots)y^{n-1} + (d + ex + fz + gx^2 + hxz + \ldots)y^{n-2} + \ldots = 0,\\ &A'y^m + (a' + b'x + c'z + \ldots)y^{m-1} + \ldots = 0,\\ &A''y^r + (a'' + b''x + c''z + \ldots)y^{r-1} + \ldots = 0, \end{aligned}$$

etc., by quantities of the form

$$\begin{aligned} &P\, y^{nmr\cdots-n} + (p + q\,x + r\,z + \ldots)y^{nmr\cdots-n-1} + \ldots,\\ &P'\, y^{nmr\cdots-m} + (p' + q'\,x + r'\,z + \ldots)y^{nmr\cdots-m-1} + \ldots,\\ &P''y^{nmr\cdots-r} + (p'' + q''x + r''z + \ldots)y^{nmr\cdots-r-1} + \ldots, \end{aligned}$$

etc., respectively, and adding the resulting products, then values of the coefficients of these quantities are so chosen that all terms involving the quantities to be eliminated shall be equal to zero. In many cases he gives even simpler formulas for the multipliers, making the elimination process often much easier. He develops other elegant propositions from Taylor's principle of increments. However, the same proposition is easily deduced from a proposition given by me in the first edition, *viz.*, that the number of coefficients in an equation of degree n, $y^n + (a + bx + cz + \ldots)y^{n-1} + \ldots = 0$, with m unknowns, will be $\frac{(n+1)(n+2)(n+3)\cdots(n+m)}{m!} = S$, and from an elegant proposition of his, *viz.*, that if $N(h)^n$ denotes the number of terms in an equation of degree n containing h unknown quantities (x, y, z, v, $\ldots$), then the number of terms in an equation of degree n divisible by x^p or y^q or z^r or v^s or $\ldots$ will be equal to $(N(h)^{n-p} + N(h)^{n-q} + N(h)^{n-r} + N(h)^{n-s} + \ldots) - (N(h)^{n-p-q} + N(h)^{n-p-r} + N)h)^{n-p-s} + N(h)^{n-q-s} + \ldots) + (N(h)^{n-p-q-r} + N(h)^{n-p-q-s} + \ldots) - (N(h)^{n-p-q-r-s} + \ldots) + \ldots$. If there are h equations respectively of degrees $n, m, l, k, \ldots$, all involving the same unknown quantities x, y, z, $v, \ldots$, and if p, q, r, $s, \ldots$ p', q', r', s', $\ldots$, p'', q'', r'', $s'', \ldots$, etc., are the maximum degrees to which $x, y, z, v, \ldots$, appear in the equations of degree $n, m, l, k, \ldots$ respectively, then the equation whose root is x or y or z, etc., cannot be of degree higher than $n \times m \times l \times k \times \ldots - (n-p) \times (m-p') \times (l-p'') \times \ldots - (n-q) \times (m-q') \times (l-q'') \times \ldots - (n-r) \times (m-r') \times (l-r'') \times \ldots - \ldots = P$; or if the degrees of the quantities $x, y, \ldots$ do not exceed $a, a', a'', \ldots$ respectively in any of the said equations, then the equation whose root is x, etc., will be of degree no greater than $P + (n-a) \times (m-a') \times \cdots - (p+q-a) \times (p'-a') \times \cdots$. These propositions, which I first published in the second edition of this work, could perhaps more easily be derived directly from the principle of Problem 41, as I there indicated. In my first edition I showed how to reduce two equations to one, to eliminate unknown quantities, by means of infinite series, and in this edition that method is extended to the case of more equations in more unknowns. I also give a method of resolving this problem, which combines the methods of Cramer and Bezout. Lagrange resolves the same problem by means of fluxions and logarithms.

In the first edition I asserted, for the first time to my knowledge, that in two given equations of degrees n and m respectively, involving two unknown quantities x and y, if the terms in which the two unknowns appear to the maximum degree have a common divisor, then the resolvent equation will be of degree less than $n \times m$; in the same edition, in the section on properties of curves, I went further to show that if there are three equations of degree n, m, r having three unknown quantities x, y, z, the equation whose root is x or y or z will be of degree at most $n \times m \times r$. In my second edition I show how, given h equations of degree $n, m, r, s, \ldots$ respectively in h unknowns, by examining the terms of maximum degree in the given equations, to decide whether the equation whose root is x or y or z or $\ldots$ will be of degree $n \times m \times r \times s \times \cdots$, which method will be explained in this edition. In the same edition I asserted that if there are more equations than unknown quantities, and if they have a common divisor involving only one unknown quantity, say x, then there will be as many solutions to the given equations as there are distinct values of the quantity x, unless the said unknown quantity has two or more equal values. Likewise, if there are more equations than unknown quantities, and all unknown quantities are eliminated, there result equations involving only known quantities, which reveals in which cases equations of this sort are valid. I also give an example which contains three equations in as many unknown quantities, however one of them is set to an arbitrary value, leading to an equation expressing a relation among known quantities. In this edition is asserted that if there are n equations in m unknowns, and among them are r independent equations exhibiting relations among known quantities only, then $n - m + r$ of the unknown quantities may be assigned arbitrary values. I am not aware if any cases of these equations, of degree greater than 1, were previously treated by others.

Bezout gave a new method of finding values of a single unknown quantity in simple equations, and a new method of finding such values is given in this work. Euler gives a method of finding certain equations, whose solutions are known, *viz.* supposing $x = \sqrt[n]{\alpha}+ \sqrt[n]{\beta}+ \sqrt[n]{\gamma}+ \sqrt[n]{\delta}+\dots$ is the solution, where α, β, γ, δ, ... denote respectively the m distinct roots, not yet determined, of an equation of degree m. In the first edition is given a general method of finding an equation from its resolution, which requires only the elimination of irrational quantities from the given solution, in connection with which are given two new general methods of eliminating irrationalities. To this end, from a given solution you must find all the different values whose products together will be the required rational quantities, i.e., if the irrational quantity is $\sqrt[n]{A}$, then $\pi \sqrt[n]{A}$, $\rho \sqrt[n]{A}$, $\sigma \sqrt[n]{A}$, ... will be the distinct values of the irrational quantity $\sqrt[n]{A}$, where π, ρ, σ, τ, ... are the various roots of the equation $x^n - 1 = 0$, and so on. From this method of eliminating irrational quantities, which I was the first to discover and disclosed in the first edition of this work, the number of roots and in consequence the dimensionality of the resulting equation is immediately seen. From this a judgment as to the generality of the proposed resolution can be formed. A resolution may appear to be more general, where in the equation multiple unknown quantities of given dimensions are given, and assigned arbitrary values. In the general resloution of an equation of degree h at least h independent quantities of this sort will be required, whence Euler's resolution is less general since in his solution, with all irrational quantities eliminated, the unknown quantity x rises to degree n^m, and only contains m unknown quantities. It is also easy to see that a new irrational quantity $\sqrt[k]{\ }$ contained in the solution will increase the dimensionality of the equation whose root is x, in the ratio $k:1$, whence by introduction of several irrational quantities the degree of the resulting equation is greatly increased without the introduction of many independent unknown quantities. Whence I deduced that if in the solution the different powers $1, 2, 3, 4, \ldots, n-2, n-1$ are assumed, then $n-1$ independent unknown quantities can be introduced into it without introducing any new irrational quantity, leading to the solution $x = a\sqrt[n]{p} + b\sqrt[n]{p^2} + c\sqrt[n]{p^3} + \ldots + B\sqrt[n]{p^{n-3}} + C\sqrt[n]{p^{n-2}} + D\sqrt[n]{p^{n-1}}$, in which if the irrational quantities are eliminated, there results the equation $x^n - n(aD + bC + cB + \ldots)x^{n-1} + \ldots = 0$. This method was contained in the first edition, and at the beginning of the year 1763 I sent a copy of it to the renowned Euler, then in May 1770 copies of the second edition were sent to the great mathematicians D'Alembert, Bezout, and Montucla, as well as to Euler, Lagrange, Frisi, and others. Whether they were received, I do not know; none sent letters to me acknowledging receipt, except Frisi, whose own outstanding work I received in return, and whose friendship with me will remain cordial as long as I live. After my publication of this result, both Euler and Bezout have suggested the same solution, $x = a\sqrt[n]{p} + b\sqrt[n]{p^2} + c\sqrt[n]{p^3} + \ldots$. M. Bezout's method of clearing irrationalities is different from mine. He himself first applied the same principle to the reduction of equations, an approach which differs hardly at all from that which I published later in the *Philosophical Transactions*. In my second edition, printed in the years 1768 and 1769, and published in 1770, the above general solution is found from n given values α, β, γ, δ, $\ldots$, *viz.* $x = a\sqrt[n]{p} + b\sqrt[n]{p^2} + c\sqrt[n]{p^3} + \ldots$, whose roots must be the specified quantities α, β, γ, $\ldots$ (for the sake of easier computation we suppose $p = 1$, as Bezout did at first). Then $\alpha = \pi a + \pi^2 b + \pi^3 c + \ldots$, $\beta = \rho a + \rho^2 b + \rho^3 c + \ldots$, $\gamma = \sigma a + \sigma^2 b + \sigma^3 c + \ldots$, etc., where π, ρ, σ, $\ldots$ are roots of the equation $z^n - 1 = 0$. Then it is readily apparent that

$$\begin{aligned}
n \times a &= \pi^{n-1}\alpha + \rho^{n-1}\beta + \sigma^{n-1}\gamma + \tau^{n-1}\delta + \ldots \\
&= \pi^{n-1}(\pi a + \pi^2 b + \pi^3 c + \ldots) + \rho^{n-1}(\rho a + \rho^2 b + \rho^3 c + \ldots) \\
&\quad + \sigma^{n-1}(\sigma a + \sigma^2 b + \sigma^3 c + \ldots) + \ldots; \\
n \times b &= \pi^{n-2}\alpha + \rho^{n-2}\beta + \sigma^{n-2}\gamma + \tau^{n-2}\delta + \ldots \\
&= \pi^{n-2}(\pi a + \pi^2 b + \pi^3 c + \ldots) + \rho^{n-2}(\rho a + \rho^2 b + \rho^3 c + \ldots) \\
&\quad + \sigma^{n-2}(\sigma a + \sigma^2 b + \sigma^3 c + \ldots) + \ldots; \\
n \times c &= \pi^{n-3}\alpha + \rho^{n-3}\beta + \sigma^{n-3}\gamma + \tau^{n-3}\delta + \ldots \\
&= \pi^{n-3}(\pi a + \pi^2 b + \pi^3 c + \ldots) + \rho^{n-3}(\rho a + \rho^2 b + \rho^3 c + \ldots) \\
&\quad + \sigma^{n-3}(\sigma a + \sigma^2 b + \sigma^3 c + \ldots) + \ldots;
\end{aligned}$$

and so on; so $\alpha = \pi a + \rho b + \sigma c + \ldots = \frac{1}{n}(\pi(\pi^{n-1}\alpha + \rho^{n-1}\beta + \ldots) + \rho(\pi^{n-2}\alpha + \rho^{n-2}\beta + \ldots) + \sigma(\pi^{n-3}\alpha + \rho^{n-3}\beta + \ldots) + \ldots)$. Hence from this resolution, which I was the first to present, we can solve an equation whose general resolution becomes known by some other method. Many similar resolutions can be given, e.g.

(1) Let the resolution be

$$\begin{aligned} x &= \frac{a\sqrt[n]{p} + b\sqrt[n]{p^2} + c\sqrt[n]{p^3} + \ldots + H\sqrt[n]{p^{n-1}}}{\alpha\sqrt[n]{p} + \beta\sqrt[n]{p^2} + \gamma\sqrt[n]{p^3} + \ldots + \theta\sqrt[n]{p^{n-1}}} \\ &+ \frac{a'\sqrt[n]{p} + b'\sqrt[n]{p^2} + c'\sqrt[n]{p^3} + \ldots + H'\sqrt[n]{p^{n-1}}}{\alpha'\sqrt[n]{p} + \beta'\sqrt[n]{p^2} + \gamma'\sqrt[n]{p^3} + \ldots + \theta'\sqrt[n]{p^{n-1}}} \\ &+ \frac{a''\sqrt[n]{p} + b''\sqrt[n]{p^2} + c''\sqrt[n]{p^3} + \ldots + H''\sqrt[n]{p^{n-1}}}{\alpha''\sqrt[n]{p} + \beta''\sqrt[n]{p^2} + \gamma''\sqrt[n]{p^3} + \ldots + \theta''\sqrt[n]{p^{n-1}}} + \ldots; \end{aligned}$$

the equation whose root is x will be only of degree n, when the irrational quantity $\sqrt[n]{p}$ and all its powers are eliminated.

(2) Let $n = m \cdot r \cdots$ and assume $x = A + (a\sqrt[m]{p} + b\sqrt[m]{p^2} + c\sqrt[m]{p^3} + \ldots) + \sqrt[r]{q}(a'\sqrt[m]{p} + b'\sqrt[m]{p^2} + \ldots + H'\sqrt[m]{p^{m-1}}) + \sqrt[r]{q^2}(a''\sqrt[m]{p} + b''\sqrt[m]{p^2} + \ldots) + \ldots + \sqrt[r]{q^{r-1}}(a'^{(r-1)}\sqrt[m]{p} + b'^{(r-1)}\sqrt[m]{p^2} + c'^{(r-1)}\sqrt[m]{p^3} + \ldots) + (a'^{(r)}\sqrt[r]{q} + b'^{(r)}\sqrt[r]{q^2} + c'^{(r)}\sqrt[r]{q^3} + \ldots + H'^{(r)}\sqrt[r]{q^{r-1}})$. If this equation is reduced to eliminate the irrational quantities $\sqrt[m]{p}$ and $\sqrt[r]{q}$, there results an equation of degree nm, free of irrationalities, in the nm unknown quantities A, a, b, c, $\ldots$, a', b', c', $\ldots$, etc., which may be assigned arbitrary values, and the two quantities p and q, which have a connection with the others.

(3) Let $x = \dfrac{a\sqrt[m]{p} + b\sqrt[m]{p^2} + c\sqrt[m]{p}\sqrt[r]{q} + d\sqrt[r]{q} + \ldots}{\alpha\sqrt[m]{p} + \beta\sqrt[m]{p^2} + \gamma\sqrt[m]{p}\sqrt[r]{q} + \delta\sqrt[r]{q} + \ldots} + \ldots$, i.e. in general let $n = m \times r \times s \times t \times \ldots$, and let $x =$ any rational algebraic function of the quantities $\sqrt[m]{p}$, $\sqrt[r]{q}$, $\sqrt[s]{k}$, $\sqrt[t]{l}, \ldots$, in which the highest power of each quantity $\sqrt[m]{p}$, $\sqrt[r]{q}$, $\sqrt[s]{k}$, $\sqrt[t]{l}, \ldots$ does not exceed $m-1$, $r-1$, $s-1$, $t-1$, $\ldots$ respectively. Then the equation whose root is x will have only $m \times r \times s \times t \times \ldots = n$ dimensions.

All the preceding techniques are applicable to the reductions of equations, or to equations having many unknown quantities. By the same principle Euler's resolution, $x = \sqrt[n]{\alpha} + \sqrt[n]{\beta} + \sqrt[n]{\gamma} + \ldots$, where α, β, γ are the required roots of the equation, can be generalized to assume

$$
\begin{aligned}
x &= a(\sqrt[n]{\alpha} + \sqrt[n]{\beta} + \sqrt[n]{\gamma} + \ldots) \quad + b(\sqrt[n]{\alpha^2} + \sqrt[n]{\beta^2} + \sqrt[n]{\gamma^2} + \ldots) \\
&+ c(\sqrt[n]{\alpha\beta} + \sqrt[n]{\alpha\gamma} + \sqrt[n]{\beta\gamma} + \ldots) + d(\sqrt[n]{\alpha^3} + \sqrt[n]{\beta^3} + \sqrt[n]{\gamma^3} + \ldots) \\
&+ e(\sqrt[n]{\alpha^2\beta} + \sqrt[n]{\beta^2\alpha} + \sqrt[n]{\alpha^2\gamma} + \sqrt[n]{\gamma^2\alpha} + \sqrt[n]{\beta^2\gamma} + \sqrt[n]{\gamma^2\beta} + \ldots) \\
&+ f(\sqrt[n]{\alpha\beta\gamma} + \ldots) + \ldots
\end{aligned}
$$

in which resolution the unknown quantities $\alpha, \beta, \gamma, \delta, \ldots$ all enter symmetrically; or more generally still assume $x =$ any rational algebraic function of the unknown quantities $\alpha, \beta, \gamma, \delta, \ldots$, in which all of them enter symmetrically. Now if we reduce this resolution, in order to eliminate the irrational quantities, there results an equation which is symmetric in all the unknown quantities $\alpha, \beta, \gamma, \delta, \ldots$, whence we may find an equation whose roots are $\alpha, \beta, \gamma, \delta, \ldots$, which will be of no greater degree than if the resolution had been simply $x = \sqrt[n]{\alpha} + \sqrt[n]{\beta} + \sqrt[n]{\gamma} + \ldots$. But it contains more quantities which can be set to arbitrary values, and hence is a more general form of resolution. And if the quantities $\alpha, \beta, \gamma, \delta, \ldots$ do not enter symmetrically in the above function, the same can still be asserted: unless some resolution has a completely different general form, the above principles will suffice to produce an equation of which it is the solution.

M. Vandermonde assumes the required equation to have roots of the forms discussed before. Thus if $1, r', r'', \ldots$ or equivalently $\pi, \rho, \sigma, \ldots$ are roots of the equation $z^n - 1 = 0$, we know that $\pi, \pi^2, \pi^3, \ldots, \rho, \rho^2, \rho^3, \ldots$, etc. are all roots of the same equation. Hence if in place of $\pi, \rho, \sigma, \ldots$, or $\pi, \pi^2, \pi^3, \ldots, \rho, \rho^2, \rho^3, \ldots$, etc. we substitute $1, r', r'', \ldots$, and for $\alpha, \beta, \gamma, \delta, \ldots$ we substitute $a, b, c, d, \ldots$, we obtain

$$a = \frac{a+b+c+d+\ldots+ \quad (a+r'b+r''c+\ldots)+ \quad (a+r'^2b+r''^2c+\ldots)+\ldots}{n},$$

$$b = \frac{a+b+c+d+\ldots+r'^{n-1}(a+r'b+r''c+\ldots)+r'^{n-2}(a+r'^2b+r''^2c+\ldots)+\ldots}{n},$$

$$c = \frac{a+b+c+d+\ldots+r''^{n-1}(a+r'b+r''c+\ldots)+r''^{n-2}(a+r'^2b+r''^2c+\ldots)+\ldots}{n},$$

etc., from which with great brilliance he pointed out that we can rewrite the quantities $a + r'b + r''c + \ldots$, $a + r'^2b + r''^2c + \ldots$, etc. in the above resolution as $\sqrt[n]{(a + r'b + r''c + \ldots)^n}$, $\sqrt[n]{(a + r'^2b + r''^2c + \ldots)^n}$, etc. Expanding the powers, we have

$$
\begin{aligned}
(a+r'\,b+r''\,c+\ldots)^n &= a^n + na^{n-1}r'\,b + na^{n-1}r''\,c + \ldots + \ldots, \\
(a+r'^2b+r''^2c+\ldots)^n &= a^n + na^{n-1}r'^2b + na^{n-1}r''^2c + \ldots + \ldots, \text{ etc.}
\end{aligned}
$$

Now we must divide these quantities, if possible, into others which are symmetric with respect to a, b, c, d, etc.; or alternatively, into other quantities, certain powers of which are symmetric in the prescribed roots; or into other quantities, powers of which contain the original quantities, raised to the same powers, and some additional terms which likewise, on elevation to some power, contain the original quantities raised to the same power; and so on *ad infinitum*. Then from my theorem, given in 1762, for finding the sum of the terms $\alpha^a\beta^b\gamma^c \cdots + \alpha^b\beta^a\gamma^c \cdots + \ldots$ from the sum of powers of the roots of the given equation, or from similar rules that I have since published (with changes in notation as needed), the sum of any function which is symmetric in the roots $a, b, c, d, \ldots$, is easily determined. If in fact the given resolution can be expressed in terms of roots of the given quantities, then it can always be expressed by this rule.

Tschirnhaus proposed the equation $y = ax^{n-1} + bx^{n-2} + cx^{n-3} + \ldots$ for eliminating the nth term of the given equation $x^n - px^{n-1} + qx^{n-2} - rx^{n-3} + \ldots = 0$. From the resulting equation all intermediate terms are likewise made to vanish, resulting finally in an equation $y^n - H = 0$. He used this method to solve the cubic equation $x^3 + px^2 + qx + r = 0$, eliminating its second and third terms (px^2, qx). In my second edition I first used this method to eliminate the second and fourth terms of a given equation, thereby deriving a resolution of the biquadratic equation. There I also pointed out, first, that the resolution

$$\sqrt[3]{-\frac{1}{2}b + \sqrt{\frac{1}{4}b^2 + a^3/27}}\ (= A) + \sqrt[3]{-\frac{1}{2}b - \sqrt{\frac{1}{4}b^2 + a^3/27}}\ (= B)$$

has nine values, which are the resolutions of the three cubic equations $x^3 + ax + b = 0$, $x^3 - \dfrac{1+\sqrt{-3}}{2}ax + b = 0$, and $x^3 - \dfrac{1-\sqrt{-3}}{2} = 0$; second, that the three roots of the equation $x^3 + ax + b = 0$ are $A + B$, $\alpha A + B/\alpha$, and $\beta A + B/\beta$, where $1, \alpha, \beta$ are the roots of the equation $x^3 - 1 = 0$; and third, given an equation $x^3 + qx - r = 0$, whose roots are α, β, $-\alpha - \beta$, if we transform it into another, whose root is z such that $z^2 - zx = q = -\dfrac{\alpha^2 + \alpha\beta + \beta^2}{3}$, then z will have six values,

$$z = \frac{\alpha}{2} \pm \frac{(\alpha + 2\beta)\sqrt{-3}}{6},\ \frac{\beta}{2} \pm \frac{(\beta + 2\alpha)\sqrt{-3}}{6},\ \frac{-\alpha - \beta}{2} \pm \frac{(-\alpha - \beta)\sqrt{-3}}{6}.$$

Lagrange takes a similar approach, but in place of $\frac{\alpha}{2} + \frac{\alpha+2\beta}{6}\sqrt{-3} = \frac{3\alpha+(\beta-\gamma)\sqrt{-3}}{6}, \ldots$, where α, β, and $\gamma=-\alpha-\beta$ are the three roots of the given equation, he writes $\frac{1}{3}\left(\alpha - \frac{1+\sqrt{-3}}{2}\beta - \frac{1-\sqrt{-3}}{2}\gamma\right)$, etc. Considering these values together he shows that the roots of the equation $z^6 - rz^3 = \frac{q^3}{27}$ will be impossible, if the roots of the given equation are real. Here it is perhaps worth pointing out that the cube $6^3z^3 = (3\beta+(\beta+2\alpha)\sqrt{-3})^3$ or $(3\alpha+(\alpha+2\beta)\sqrt{-3})^3$ or $(-3\alpha-3\beta+(\alpha-\beta)\sqrt{-3})^3$ only has two distinct values, namely $-27(4\beta^2\alpha+4\alpha^2\beta)+(24\beta^3+36\beta^2\alpha-36\alpha^2\beta-24\beta^3)\sqrt{-3} = A + B\sqrt{-3}$ and $A - B\sqrt{3}$; the product of these two values, $(A+B\sqrt{-3})(A-B\sqrt{-3}) = A^2 + 3B^2$ is a quantity which is symmetric with respect to the pairs $(\alpha,\beta),(\alpha,-\alpha-\beta),(\beta,-\alpha-\beta)$ and the ternary $(\alpha,\ \beta,\ \gamma=-\alpha-\beta)$.

In the second edition of this work it is asserted that the resolution $x = \sqrt[3]{a+\sqrt{c}} + \frac{h}{\sqrt[3]{a+\sqrt{c}}}$ is a solution of an equation of $3 \times 2 = 6$ dimensions, and in general such a resolution of an equation of degree n will be a resolution of an equation of $1 \cdot 2 \cdot 3 \cdot 4 \cdots (n-1) \cdot n = N$ dimensions. And it may be added that if the degree of an equation with the above sort of roots does not exceed N, then these quantities necessarily have denominators which are functions of the coefficients of the given equation, and so on.

In the first edition I gave a resolution, shown below, of a biquadratic equation $x^4 + 2px^3 = qx^2 + rx + s$, in which the second term is not removed: to each side of the equation the quantity $(p^2+2n)x^2+2pnx+n^2$ is added, giving the equation $(x^2+px+n)^2 = (p^2+2n+q)x^2 + (2pn+r)x + s + n^2$. Now we require that $4(p^2+2n+q)(s+n^2) = (2pn+r)^2$, or equivalently $8n^3 + 4qn^2 + (8s-4rp)n + 4qs + rp^2s - r^2$ $(=A) = 0$, whence $x^2 + px + n = \sqrt{p^2+2n+q}x + \sqrt{s+n^2}$. Then in the second edition I proved that $\frac{\alpha\beta+\gamma\delta}{2}, \frac{\alpha\gamma+\beta\delta}{2}, \frac{\alpha\delta+\beta\gamma}{2}$ are the three values of the quantity n, where $\alpha, \beta, \gamma, \delta$ are the roots of the given equation, and $\frac{\alpha+\beta-\gamma-\delta}{2}, \frac{\gamma+\delta-\alpha-\beta}{2}, \frac{\alpha+\gamma-\beta-\delta}{2}, \frac{\beta+\delta-\alpha-\gamma}{2}, \frac{\alpha+\delta-\beta-\gamma}{2}$, and $\frac{\beta+\gamma-\alpha-\delta}{2}$, are the six values of the quantity $\sqrt{p^2+2n+q}$, and finally $\frac{\alpha\beta-\gamma\delta}{2}, \frac{\gamma\delta-\alpha\beta}{2}, \frac{\alpha\gamma-\beta\delta}{2}, \ldots$ are the six values of the quantity $\sqrt{s+n^2}$. The same result was later published by Lagrange. Using all three roots of the cubic ($A=0$), we have a resolution with 24 values, of which 12, as I showed in the second edition, are extraneous, while the other 12 are true, i.e. they are roots of the given biquadratic, each of them occurring three times.

Lagrange substitutes the roots π, ρ, σ, ... of the equation $y^n+h=0$, respectively for y in Tschirnhaus's equation $a+bx+cx^2+dx^3+\ldots=y$, where x is one of the roots α, β, γ, δ, ... of the given equation $x^n-px^{n-1}+qx^{n-2}-\ldots=0$, and since the value π of y is the same for all roots α, β, ... of the given equation, then $a+b\alpha+c\alpha^2+\ldots=\pi$ or $a+b\beta+c\beta^2+\ldots=\pi$, etc., and so for the other values ρ, σ, τ, etc. From this he easily deduces the number of different values which the coefficients a, b, c, ... may have. In the second edition the equation whose root is $a, \ldots$ is asserted to be of degree $1\cdot2\cdot3\cdot4\cdots(n-1)$, where the given resolution is $x=a(\sqrt[n]{1})+b(\sqrt[n]{1})^2+c(\sqrt[n]{1})^3+\ldots$. Supposing α, β, γ, δ, ... to be the roots of the given equation, and π, ρ, σ, τ, ... to be the roots of the equation $x^n-1=0$, then the assumed resolutions are $\alpha=\pi a+\pi^2b+\pi^3c+\ldots$, $\beta=\rho a+\rho^2b+\rho^3c+\ldots$, etc. But since none of π, ρ, σ, ... are lacking in one root, α, or another, β, but rather all appear equally in all roots, Lagrange was able to show the number of values that each of the individual coefficients a, b, $c, \ldots$ may have. He was the first to show that α, α^2, α^3, α^4, ..., $\alpha^n=1$ must be the n distinct roots of the equation $x^n-1=0$, where n is prime and $\alpha\neq1$. More rules of this sort are added in this edition of my work.

DeMoivre gives a method of generating the sum of a series of terms that are equal, or alternating, or cyclic over an interval of distance two, three, or more, $a+bx+cx^2+\ldots$ through division of unity by a rational multinomial expression $p+qx+rx^2+\ldots$. In 1757 I sent the first version of this work to the Royal Society of London, which Simpson read, then in 1758 he inserted in the *Philosophical Transactions* a short piece containing a rule that was in the work I had submitted, *viz.* let S be a given function of the quantity x, which is expanded into a series proceeding according to the dimensions of x, say $a+bx+cx^2+dx^3+\ldots$; in S we now substitute for x respectively αx, βx, γx, δx, etc., where α, β, γ, δ, ... are roots of the equation $x^n-1=0$, resulting in a total of n quantities A, B, C, D, etc., then $\frac{A+B+C+D+\cdots}{n}$ will be the sum of the first term and of those whose position is respectively n, $2n$, $3n$, etc. beyond the first term. Nevertheless at the end of his paper he says of the series $a+bx+cx^2+\ldots$ that he has given a solution of an example of this problem by a method which differs a little from another one where a general method is indicated. But I say that no one, before my submission to the Royal Society in 1757, had ever claimed to have devised the general method, and it was my notes that Simpson had read, in which the above method was contained. In the first edition the rule was extended, but here I show that:

(1) If a series with progressing degrees of the quantity x has the form $a+bx^{n/m}+cx^{2n/m}+dx^{3n/m}+\ldots$, and the sum of its terms is required, then for $x^{n/m}$ we substitute in turn $\alpha x^{n/m}$, $\beta x^{n/m}$, $\gamma x^{n/m}$, $\delta x^{n/m}, \ldots$ and specify the resulting quantities $a+b\alpha x^{n/m}+c\alpha^2x^{2n/m}+\ldots = A$, B, C, etc. Then $\frac{A+B+C+\cdots}{n}$ will be the required sum.

(2) If the given quantity contains irrationalities, then the same root π of the irrationality must be used in each of A, B, C, D, etc., e.g., let the irrational quantity be $\sqrt[r]{e+fx+gx^2+\ldots}$, and the roots of the equation $x^r-1=0$ are π, ρ, $\sigma, \ldots$. Then if in the case of the quantity $A=\pi\sqrt[r]{e+f\alpha x+g\alpha^2x^2+\ldots}$ we have formed a product with π, in the other quantities $B=\pi\sqrt[r]{e+f\beta x+g\beta^2x^2+\ldots}$, $C=\pi\sqrt[r]{e+f\gamma x+g\gamma^2x^2+\ldots}$, D, E, etc., the products must all be written combining the same root π with each value of $\sqrt[r]{e+\ldots}$.

(3) If a radicand $\sqrt[s]{ax^h+\ldots}$ begins with a term involving a power or a root of the quantity x, then it must be reduced in such a way that x no longer appears in the first term. For example, supposing the expression is $\sqrt[n]{ax^h+bx^{h+m}+\ldots}$, then it should be rewritten as $x^{h/n}\times\sqrt[n]{a+bx^m+\ldots}$. For another example, consider the quantity

$$\sqrt[n]{ax^r+bx^{r+s}+\cdots+\sqrt[m]{a'x^t+b'x^{t+v}+\cdots}+\cdots}\,.$$

This is first reduced to the form

$$x^{r/n}\times\sqrt[n]{a+bx^s+\cdots+x^{\frac{t}{m}-r}\sqrt[m]{a'+b'x^v+\cdots}+\cdots}\,;$$

now if $\frac{t}{m-r}$ is less than 0, then it must be rewritten again, as

$$x^{r/n}\times\sqrt[n]{a+bx^s+\cdots+x^{\frac{t}{m}-r}\sqrt[m]{a'+b'x^v+\cdots}+\cdots}$$

$$=x^{t/nm}\times\sqrt[n]{\sqrt[m]{a'+b'x^v+\cdots}+x^{r-\frac{t}{m}}(a+bx^s+\cdots)+\cdots}\,.$$

And so one must proceed reducing all the other roots, such that each radicand begins with a term that does not contain x or any power or root of x. All this reasoning may be applied to quantities which do not begin with the first term of the series, or which contain several quantities, say x, y, z, etc., or to the equations.

For finding integral or rational values of the unknown quantities in given equations, solutions are given which are in some cases general, in others very specific to the problem. In the case of specific solutions, particular values of some of the unknown quantities are assumed, which could have been chosen by no general reasoning process (such a general process would lead to a general solution) unless from previously known facts. The majority of cases of equations of all dimensions (except first degree equations) have no general solution, except by repeated trial and error. If there is a general solution, and the number of unknown quantities exceeds by one the number of equations, then the general values of the unknown quantities are expressed by a formula given in the first edition of this work, namely

$$Az^n + Bz^{n-1} + Cz^{n-2} + \dots,\ az^m + bz^{m-1} + cz^{m-2} + \dots, \text{ etc.}$$

or

$$\frac{Az^n + Bz^{n-1} + Cz^{n-2} + \dots}{Pz^m + Qz^{m-1} + Rz^{m-2} + \dots},\ \frac{az^r + bz^{r-1} + cz^{r-2} + \dots}{Pz^m + Qz^{m-1} + Rz^{m-2} + \dots}, \text{ etc.}$$

or (a special case of the preceding example)

$$\frac{Az^n + Bz^{n-1} + Cz^{n-2} + \dots}{Pz^m + Qz^{m-1} + Rz^{m-2} + \dots},\ \frac{az^r + bz^{r-2} + \dots}{pz^s + qz^{s-1} + \dots}, \text{ etc.}$$

and so on, in which n, m, r, $s, \dots$ denote integers, A, $B, \dots$, P, $Q, \dots$, a, $b, \dots$, p, $q, \dots$ denote rational quantities, and z is a quantity to be chosen at will. Now these quantities are to be substituted for their values in the given equations, and from equality, or the vanishing, of coefficients of corresponding terms of the resulting equations, the prescribed integral and rational quantities can be extracted. And if the number of unknown quantities exceeds by m the number of equations, then we are to assume general rational functions of rational quantities z, w, $u, \dots$ to be chosen at will, each containing all m of z, w, $u, \dots$, and for each unknown quantity we will substitute a different quantity of the form $\dfrac{az^n + (b + cw + du + \dots)z^{n-1} + \dots}{Pz^m + (Q + Rw + Su + \dots)z^{m-1} + \dots}$, etc. Then individual coefficients of corresponding terms of the resulting equations are equated to zero, from which one obtains rational values of the coefficients a, b, ... , P, Q, ... , etc., and integral values of the numbers n and m.

Now suppose that some integral or rational values corresponding to some of the unknown quantities are given, and a method is required of finding, consistent with them, more corresponding values. Then in the given equations, for the required quantities substitute general rational functions of the given quantities, or of the given quantities and others arbitrarily chosen; and by the method described above the required rational corresponding values will be found. If on the other hand irrational values of a given type are required for unknown quantities of the given equations, then for each of them assume a general irrational function of the specified type, and proceed as before. But often these general substitutions lead to extremely laborious calculations, and it will be easier to fall back on particular solutions. I cannot enumerate the many elegant solutions of problems of this sort that have been found by numerous diligent researchers.

In this preface I will add a little concerning what is known about the theory of divisibility and residues of numbers. Euler gives the following elegant property of numbers: if $S(a)$ denotes the sum of the divisors of the number a, then $S(n) = S(n-1) + S(n-2) - S(n-5) - S(n-7) + S(n-12) + S(n-15) - \dots$. He also showed that if α, β, γ, $\delta, \dots$ are the residues of the division of the squares $1, 4, 9, 16, \dots$ by a prime number, and π is a non-residue, then $\pi\alpha$, $\pi\beta$, $\pi\gamma, \dots$ are also non-residues, from which he deduced the following proposition: Let p be a prime number, and a^m the least power of the number a which has a remainder of 1 when divided by p; then if m is less than $p-1$, it cannot be greater than $(p-1)/2$. Similar propositions may be extended to some other cases.

In my second edition I gave a method of determining whether a given series can be expressed by a rational fraction, or not, and I added that there cannot be n prime numbers in an arithmetic progression, unless their common difference is divisible by $1 \cdot 2 \cdot 3 \cdot 4 \cdots n$. Several important properties of the divisors of numbers have been found by Euler, Lagrange, Beguelin, and others. In this edition I show that the number of residues of the division of numbers 1^m, 2^m, 3^m, $4^m, \dots$ by a prime number p will be $(p-1)/r$, where r is the greatest common divisor of m and $p-1$. Not until after this work was printed did I see Euler's proposition, that if $m \times s + 1 = p$, then s will be the number of residues of the above type, which is a special case of the proposition I have given. It is said that the ratio of the sum of the divisors of a number $1 \cdot 2 \cdot 3 \cdot 4 \cdots n$, to the number itself, is less than the ratio of any smaller number to its divisors. I prove that the sum $1^r + 2^r + 3^r + 4^r + \ldots + x^r$ is always divisible by x or $x+1$, provided x or $x+1$ is a prime number greater than $r+1$. In the same way I prove that the sum of products of the form $\alpha^a\beta^b\gamma^c\delta^d \cdots$, where α, β, γ, $\delta, \ldots$ represent different numbers $1, 2, 3, 4, \ldots, x$, and the sum $a+b+c+d+\ldots$ of the given indices is less than $x-1$ or x, will always be divisible by the prime number x or $x+1$. Even more generally, we can assert that $1 \cdot 2 \cdot 3 \cdot 4 \cdots (r+1) \times S$ is always divisible by $(2x+1) \times (x+1) \times x$, if r is even, and by $x^2 + (x+1)^2$, if r is odd, from which the proof of an elegant property of parabolas, similar to a property of Archimedes' conical parabola, and published by me in my *Proprietates Algibraicarum Curvarum*, can easily be deduced. A proof is also given of an elegant property of prime numbers which was discovered later, that the number $1 \cdot 2 \cdot 3 \cdot 4 \cdots (n-1) + 1$ is always divisible by n, if n is a prime number; the elegant proof of this theorem was first discovered by Lagrange.

After so much labor by so many workers in the field of Algebra, some might be astonished that I have produced another work, one with so many pages, on the subject. But the fault would be mine had I done so, if the pages were fewer and the first book that could compile all the discoveries of recent writers should emphasize none of them, nor any of my own discoveries which also perhaps deserve elaboration. But of this the erudite reader must be the judge. My duty indeed has been to serve faithfully, and to promote as much as I can, the cause of knowledge. I beg you then to accept this book, despite its tedious verbiage, for a little greater mastery of the subject that it seeks to convey.

Summary of Contents

Chapter 1

(1) The rule is derived which governs the series for finding the sum $\alpha^m + \beta^m + \gamma^m + \delta^m + \ldots$, where α, β, γ, $\delta, \ldots$ are the roots of the given equation $x^n - px^{n-1} + qx^{n-2} - rx^{n-3} + \ldots = 0$, in terms of its coefficients p, q, r, s, etc.

(2) To this is added the series which expresses the aggregate of several such sums.

(3) This is followed by a series for evaluating the sum of all quantities of the type $\alpha^a\beta^b\gamma^c\cdots + \alpha^b\gamma^a\beta^c\cdots + \ldots$, from given sums of the abovementioned type $\alpha^m + \beta^m + \gamma^m + \ldots$, for which series the rule is admirably simple.

(4) To this is added a method of finding the same sums in terms of the coefficients.

(5) Following this is a rule for finding the sum of all products of m individual roots at a time (given the sum, as well as the sums of the squares, cubes, etc., of the individual roots).

(6) I teach how to find the sum of any rational or irrational algebraic function of the roots of a given equation or equations. In the second and third editions this method is more extensively explained. From it it is possible to find the value of an infinite power series in the roots of the given equations, provided that this infinite power series can be expressed as an algebraic function of the roots.

(7) Given one, two or more equations, a method is presented for finding an equation whose roots are any algebraic function of the roots of the given equations, by finding first the expression for one such value, then finding the sum of the products, two at a time, then three at a time, etc., of the values of the abovementioned function, which will be respectively the second, third, etc., terms of the equation whose root is the prescribed function. This transformation, whose principle is quite noteworthy, was presented by me before it was published by anyone else.

(8) The equation is found whose roots are the mth powers of the roots of a given equation, and from this is derived a means of deleting surds from a given equation; moreover equations are shown whose roots are the sums or differences of pairs of roots of the given equation, or the squares of the said differences, and some others as well.

(9) A useful method is taught for computing, from the known roots of an assumed equation, the coefficients of the required equation, and further a method to find some of the higher terms of the equation once some of the lower terms are known.

Chapter 2

(1) Every algebraic equation whose roots are all imaginary is shown to be a sum of different squares.

(2) Every equation of even degree is proved to be the sum or difference of two real squares.

(3) Let equation (A) be $x^n - px^{n-1} + qx^{n-2} - \ldots = 0$, with roots $\alpha > \beta > \gamma > \ldots$, and let $\pi, \rho, \sigma, \ldots$ be the roots of the equation (B) $nx^{n-1} - (n-1)px^{n-2} + \ldots = 0$ situated respectively between α and β, β and γ, etc. Then if h and m have the same sign, the roots of the equation $hA + mB = 0$ will be between α and π, β and ρ, and so on; whereas if they have opposite signs one root will be greater than α, and the others will be placed between π and β, ρ and γ, etc.

(4) If h and m have the same sign, two roots of the equation $hA + mB = 0$ will lie between the least positive and least negative roots of the given equation $A = 0$, and one of them will be positive, the other negative; etc.

(5) Given two equations, of which some roots of each are limits of some roots of the other, a method is shown to deduce other pairs of equations, whose roots are likewise limits of each other.

(6) From the equation whose roots are $1/(\alpha-\beta), 1/(\alpha-\gamma), 1/(\beta-\gamma)$, etc., where α, β, γ, ... are roots of a given equation, is shown a method of finding limits of the roots of the given equation.

(7) If α, β, γ, δ, ... are any real quantities, then the sum of the individual values of the products $\alpha^{2a}\beta^{2b}\gamma^{2c}\cdots+\alpha^{2b}\beta^{2a}\gamma^{2c}\cdots+\ldots$ will have, to the sum of the values of $\alpha^p\beta^q\gamma^r\cdots+\alpha^q\beta^p\gamma^r\cdots+\ldots$, a ratio equal to or greater than the ratio of the number of terms in the first sum compared to the number of terms in the latter sum, provided that $2a > p$, etc.

(8) From a series of quantities which are always positive, when all the roots of a given equation are real, principles are given by which it is often possible to say how many impossible roots the given equation has.

(9) To this is adjoined a rule for finding impossible roots which always finds as many as Newton's rule finds, and often detects impossible roots when Newton's rule fails.

(10) A rule is given for finding impossible roots based on an equation whose roots are squares, or higher powers, of the roots of the given equation.

(11) A given equation is transformed into another whose roots are the squares of the differences of the given equation's roots, from which it is possible to determine whether or not all of the given equation's roots are real.

(12) Given the equation $x^n - px^{n-1} + qx^{n-2} - \ldots = 0$, I assume $nx^{n-1} - (n-1)px^{n-2} + \ldots = 0$ and $x^n - px^{n-1} + \ldots = w$, and find the equation whose root is w, from which it is often possible to detect impossible roots.

(13) If therc arc two equations, $Hx^n - ax^{n-1} + \ldots = 0$ and $Lx^m - px^{m-1} + \ldots = 0$, whose roots are α, β, γ, $\delta, \ldots$ and π, ρ, σ, $\tau, \ldots$ respectively, then on setting $Hx^n - ax^{n-1} + \ldots = 0$ and $Lx^m - px^{m-1} + \ldots = w$, and afterwards $Lx^m - px^{m-1} + \ldots = 0$ and $Hx^n - ax^{n-1} + \ldots = u$, I assert that the product of all the values of w will be, to the product of all the values of u, in the ratio $L^n : H^m$.

(14) Rules are given by which the number of real positive and negative roots may be found, for a given equation $x^n + Ax^m + B = 0$.

(15) To which is added a general method of determining the number of positive and negative roots contained in a biquadratic equation.

(16) Assuming a method exists to find the number of impossible roots in any given equation involving only one unknown quantity, and given an equation having multiple unknown quantities, a method is given to determine whether these unknown quantities can have real corresponding values.

(17) From a known number of real, positive and negative roots in a given equation, a method follows for determining the number of impossible, etc., roots in an equation whose roots have some assignable relation, e.g. the square of the differences, to the roots of the given equation.

(18) If the given equation is $x^3 + qx - r = 0$, the relation $z^2 - zx = q/3$ is assumed, whence it is shown that z has the six values $\frac{\alpha}{2} + \frac{(\alpha + 2\beta)\sqrt{-3}}{6}$, etc., where α, β (and γ) are roots of the

given equation, from which I conclude that if all the roots of the given equation are real, all the roots of the equation whose root is z must be imaginary.

(19) Let the equation be $x^4 + qx^2 - rx + s = 0$, and find the equation whose root z is $\alpha + \beta$, $\alpha + \gamma$, $\beta + \gamma$, etc., where α, β, γ, δ are the roots of the given equation; thence cases are identified in which real (etc.) roots of the resulting equation in z correspond to real (etc.) roots of the given equation.

Chapter 3

(1) A method is given of finding, among the roots of a given equation, those that have a given relation to each other.

(2) The remainder, upon division of $x^n - px^{n-1} + qx^{n-2} - \ldots$ by $(x-a)^m$, is discussed.

(3) The infinite equation $1 - \frac{a}{1-r}x + \frac{ra^2}{(1-r)(1-r^2)}x^2 - \ldots = 0$ is shown to have an infinite number of real roots.

(4) When there are more equations than unknowns, I assert that there are as many, and no more, values of all unknown quantities, as there are values of a single unknown x, unless two values of x are equal to each other, or other similar exceptions.

(5) A method is shown for determining whether given irrational quantities have real values.

(6) The equation $x^{2n} \pm x^{2n-1} + \ldots = 0$ is reduced to an equation $x^2 + vx + 1 = 0$ with the aid of the equation $v^n = v^{n-1} - (n-1)$ $v^{n-2} + \ldots = 0$, whence are revealed the four roots of the equation $x^4 + 1 = 0$ (which are $\sqrt{\frac{1}{2}} \mp \sqrt{-\frac{1}{2}}$), the five roots of the equation $x^5 + 1 = 0$, and the six roots of the equation $x^6 + 1 = 0$.

(7) Some new points about reciprocal polynomials are added.

(8) Irrational expressions are transformed into others which are equal to the given ones, but lie hidden under different forms of irrationalities.

(9) Remarks are added concerning the roots of the equation $x^n \pm 1 = 0$, and on the conjugate roots $a + \sqrt{-b^2}$, $a - \sqrt{-b^2}$.

(10) The conjugate roots of a given resolution are shown.

(11) If the equation $x^n - px^{n-1} + \ldots = 0$ can be reduced in degree to an equation $x^m - ax^{m-1} + bx^{m-2} - \ldots = 0$, then the equations whose roots are a, b, c, ... are shown to have the same dimensions; and given one coefficient, by the method of finding common divisors corresponding values of each remainder may be found from it, unless two or more values of the given coefficient are equal to each other; and all coefficients are shown to involve the same irrationalities.

(12) From the roots of the equation $P^6 + 2qP^4 + (q^2 - 4s)P^2 - r^2 = 0$, which are given, the roots of the equation $x^4 + qx^2 - rx + s = 0$ are found.

(13) A biquadratic equation of the form $x^2 - 2(a + b\sqrt{-1})x - c - d\sqrt{-1} = 0$ is separated into two equations having all real coefficients.

(14) The resolution of the biquadratic equation $x^4 + 2px^3 = qx^2 + rx + s$ is discussed, using the method of adding $(p^2 + 2n)x^2 + 2pnx + n^2$ to each side of the equation, and stipulating that $4(s + n^2) \times (p^2 + 2n + q) = (2pn + r)^2$; it is further shown that the values of n are respectively $\frac{\alpha\beta + \gamma\delta}{2}, \frac{\alpha\gamma + \beta\delta}{2}, \ldots$, of $\sqrt{q + p^2 + 2n}$ are respectively $\frac{\alpha + \beta - \gamma - \delta}{2}, \frac{\alpha + \gamma - \beta - \delta}{2}, \ldots$, of $\sqrt{s + n^2}$ are respectively $\frac{\alpha\beta - \gamma\delta}{2}, \frac{\alpha\gamma - \beta\delta}{2}, \ldots$, etc.

(15) Also a method is given by means of which extraneous roots enter into the resolution, which were not contained in the given equation.

(16) By a similar method certain equations of the fifth and sixth degree are solved; in which resolution there are three independent equations involving three unknown quantities, but, as I was the first to observe, one of these unknown quantities may be assigned an arbitrary value, causing one of the three equations now to involve only known quantities.

(17) An equation may be assumed for resolution of a given equation, but it is shown that in the assumed equation one must guard against the possibility that by equating terms of the assumed

and given equations, an equation would arise depending on other terms; of which an example is given.

(18) A resolution of a biquadratic equation, by eliminating the second and fourth terms, is presented, and generalized to a method of reducing an equation $x^n - px^{n-1} + qx^{n-2} - \ldots = 0$ to an equation $v^n + Q = 0$, which as I first showed, requires a reducing equation of degree $1 \cdot 2 \cdot 3 \cdots (n-1)$.

(19) I show a way to find equations whose degree may be reduced.

(20) Given any arbitrary irrational quantity, another is found, which multiplied by the first renders it rational; from which given an equation involving multiple irrationalities, I show the dimensions of the equation which will be free of all the abovementioned surds.

(21) It is observed that in multiplying irrational quantities, errors are easily introduced, unless care is taken with respect to the different radicals; e.g. if $\sqrt{-a^2}$ is multiplied by $\sqrt{-a^2}$, the product will be either $-a^2$ or $\sqrt{+a^4} = +a^2$ or $-a^2$, but $+a^2$ is not a true root; and if it is necessary to be concerned with different signs and roots in mathematical matters, it is even more so in physical science. [3]

(22) Given a quantity, expressed as a simple series of consecutive terms of successive dimensions in a literal x, it is shown how to find a quantity equal to the sum of terms separated by a distance of $n,\ 2n,\ 3n,\ \ldots$.

(23) Formulas are given for terms, which multiplied together will yield arbitrary given quantities.

(24) $x = \sqrt[3]{a+\sqrt{b}} + \sqrt[3]{a-\sqrt{b}}$ is shown to be the resolution of an equation of nine dimensions, i.e. of three different cubics, and similar principles are applied to other equations.

(25) The sum of all distinct quantities of the form $\alpha^a\beta^b\gamma^c\cdots$ is revealed, where $\alpha,\ \beta,\ \gamma,\ \ldots$ are roots of the equation $x^n - 1 = 0$.

(26) New methods are introduced for resolution of cubic and biquadratic equations, and also resolutions of equations of the form $x^{2n} - 2b^n px^n - \frac{p}{1 \cdot 2} \times 2nb^{n-1} - \ldots = a^{2n} - b^{2n}p^2$, etc.

(27) I discuss a resolution of equations, *viz.*, $x = H + a\sqrt[n]{p} + b\sqrt[n]{p^2} + \ldots + \sqrt[n]{p^{n-1}}$, and show how, from given roots, to construct a resolution of this type containing the prescribed roots.

(28) Letting $\alpha, \beta, \gamma, \delta, \ldots$ be the $n-1$ roots of the equation $v^{n-1} - pv^{n-2} + \ldots = 0$, the dimensions are determined to which the equation must ascend, one of whose roots is $x = \sqrt[n]{\alpha} + \sqrt[n]{\beta} + \sqrt[n]{\gamma} + \ldots$.

(29) From every general resolution of equations of higher dimensions, are deduced resolutions of all equations of lower degrees.

(30) The general resolution of an equation of degree n is shown to be equivalent to a resolution of an equation of degree at least $1 \cdot 2 \cdot 3 \cdots (n-1) \cdot n$.

(31) In investigating approximations to roots of a given equation, it is shown that the convergence of the approximation does not depend on the ratio of the assumed value to the true value, but on being much nearer to one value than to any others.

Chapter 4

(1) A method is given, which I believe is new, of reducing n simple equations to $n-1$, and so on.

(2) It is shown how to identify roots which arise extraneously in the equation resulting from the common method of reducing two or more equations to one, such that all but one of the unknowns are exterminated.

(3) Let there be two equations of degrees n and m having two unknowns, x and y; it is shown that the equation whose roots include x and y is of degree less than $n \times m$ if the terms of highest degree in the two equations have a common divisor.

(4) If there are three or more equations of degrees $n, m, o, \ldots$ respectively, involving unknowns $x, y, z, \ldots$, the conditions are shown under which the resolvent equation is of degree less than $n \times m \times o \times \ldots$.

(5) A proof is given concerning which terms of the several given equations are sufficient to determine the sum of the roots, the sum of the squares of the roots, etc.

(6) A method is given to transform $h \geq 2$ equations into one, so as to eliminate $h-1$ of the unknown quantities by substituting either finite or infinite series for their values in a given equation.

(7) Where there are h equations, of degrees n, m, l, $k, \ldots$ in h unknown quantities x, y, $z, \ldots$, and where a, b, c, $d, \ldots$ are the greatest degrees of the terms of each involving only the $h-1$ unknown quantities x, z, $v, \ldots$, it is proved that the degree of the equation whose root is y is no greater than the largest of the products $n \cdot b \cdot c \cdot d \cdots$, $m \cdot a \cdot c \cdot d \cdots$, etc.

(8) A new method is given of transforming two or more equations, having two or more unknown quantities, into one, in order to eliminate unknown quantities, by a combination of substitution and multiplication by assumed quantities.

(9) From given equations are deduced other equations, whose solutions with respect to all of their unknown quantities will be the same as the solutions with respect to the unknown quantities of the given equations.

(10) A method is also given of determining whether all roots of m equations are the same as all the roots of m other given equations.

(11) A principle is given, by which it is often possible to determine how many solutions with respect to an unknown quantity, in the first set of given equations, are not likewise solutions in the second set.

(12) In several cases the value of an algebraic quantity is determined.

(13) Given an equation of degree n in $h \geq 1$ unknown quantities x, y, z, etc., the constitution of its coefficients is found, and from the $(n+1) \cdot \frac{n+2}{2} \cdots \frac{n+h}{h} - 1$ corresponding values of the unknown quantities x, y, z, etc., the required constitution follows.

(14) The general equation of degree n in m unknown quantities is shown to have $(n+1) \cdot \frac{n+2}{2} \cdots \frac{n+m}{m} - 1$ coefficients which may take on arbitrarily assumed values.

(15) If there are n equations involving m unknown quantities, and they can be so reduced that there result r equations involving only known quantities, then, it is asserted, in the given equations $n - m + r$ unknown quantities may be assumed at will.

(16) Where there are two or more equations involving two or more unknown quantities, whose solutions with respect to these unknowns

are α, β, $\gamma, \ldots$, π, ρ, $\sigma, \ldots$, κ, λ, $\mu, \ldots$, etc. respectively, the symmetric sum of quantities of the form $\alpha^m \pi^a \kappa^l \cdots$ is derived.

(17) A rule is given, by which it is sometimes possible to deduce limits between which the real roots of a given set of equations must lie.

(18) Given an expression involving two or more quantities x, y, z, etc., which is to be reduced to a series in order of the dimensions with respect to x, y, $z, \ldots$, a method is given here of finding sums of terms of the form $Ay^{pn+d}z^{p'm+d'} \ldots$.

(19) Given the same conditions, a method is given for finding those roots of the given equation which have an algebraic relation to each other.

(20) From this follows a method of predicting how many roots corresponding to the unknown quantities will prove to be equal to 0.

(21) If there are two or more equations in exactly the same number of unknowns x, y, z, etc., I assert that the unknown quantity y will have the same number of impossible roots as will the quantities z or x, unless one of these quantities has two or more values which are found to be equal to each other.

(22) Given two or more equations in two or more unknowns, which have factors, the given equations can be reduced into others of lower dimensions, from which is easily determined the degree of the equation whose root is an unknown quantity in two or more equations having two or more unknown quantities.

(23) (1) Sets of two or more equations involving two or more unknowns are constructed, which can be solved; (2) the most general resolution so far known of two or more such equations in two or more unknowns is presented; and (3) by a method similar to a principle laid down in Chapter 3, the most general resolution so far known of the prescribed equations is deduced.

(24) Resolutions of certain types of equations are shown.

(25) If there are three or more equations, with terms of dimensions respectively n, $n-m$, $n-2m, \ldots$, n', $n'-m$, $n'-2m, \ldots$, etc., then I assert that the equation whose root is x in the above equations will be of the form $x^\alpha + Px^{\alpha-m} + Qx^{\alpha-2m} + \ldots = 0$, and so on.

(26) If there are two or more equations symmetrically involving the quantities x and y, then it is proved that the equation whose roots are a rational function, in which or in the square of which x and y are symmetrically involved, has a degree only half as large, or is of a form involving only half the dimensions, as that of the general equation whose root is x or y, of which case many examples are given.

(27) Similar propositions are given with respect to more quantities symmetrically involved among themselves.

(28) Given an equation $A = 0$, which involves two unknown quantities x and y, for x is written $1/x$, obtaining an equation $B = 0$, then from the two equations $A = 0$ and $B = 0$ is deduced an equation whose root is x; and I assert that the equation so derived will always yield the values 1 and -1 for the quantity x, and all its other roots will be contained in the recurrent equation. Many related propositions are added.

(29) If A and B are functions of x and y, a method is given of determining whether A is a function of B, or not.

(30) Let an equation be given $x^n - px^{n-1} + qx^{n-2} - \ldots = 0$, let $p' = P$, $pP - q' = Q$, $pQ - qP + r' = R$, etc., and let a, b, c, $d, \ldots$ be the sum of the given equation's roots, the squares of its roots, etc. Then I assert that $\dot{a} = P$, $\frac{1}{2}\dot{b} = Q$, $\frac{1}{3}\dot{c} = R$, etc., where in the fluxions we substitute p', q', etc. for $\dot{p}$, $\dot{q}$, etc.

(31) I prove that the ratio of quantities found by common methods of approximation, in two or more equations involving two or more unknown quantities, to the true values of the roots, depends on how close to one of the roots, rather than to the others, the initial estimate is; the same principle may be extended to cases where the estimate is closer to two, three, or more of the roots than it is to the rest of them.

(32) A method is given for finding approximations to impossible roots of the given equations.

(33) A method is discussed, which by incrementing coefficients leads to increments of the various roots, from which it is shown that all equations have roots that are real, or impossible roots of the form $\alpha + \beta\sqrt{-1}$.

Chapter 5

(1) A method is given, following Descartes' principle, to find divisors of two or more dimensions of a given equation, and several methods are given to prove that a given equation has no rational roots.

(2) Given two or more equations with two or more unknown quantities, it is shown that the corresponding values of all of these quantities will involve the same irrationalities, unless one of the unknowns has a multiple root.

(3) A method is given to find integral corresponding values of the unknown quantities contained in two or more equations.

(4) Given an equation whose root is x, I show how to determine if it can be solved by extraction of square, cube, or higher roots.

(5) Given arbitrary rational and integral functions of several quantities, an expression is found in which all the divisors of the given quantities are contained.

(6) I prove that the numbers $(a+b)^n$ and $(a-b)^n$, where a and b are mutually prime, have no common divisor except 2^m.

(7) A method is discussed for determining whether an equation of degree $4n$, etc., can be reduced to an equation of degree n by extraction of square roots.

(8) I prove that $\sqrt[2n+1]{\sqrt{e}+\sqrt{f}}$ cannot be a quantity of the form $\sqrt[2n+1]{g}$, if $\sqrt[2n+1]{e-f}$ is a rational quantity; to which are added new rules for finding its real and imaginary solutions; the rules of Newton and Schooten are explained; and it is shown that in several cases it is necessary to multiply the given binomial by 2^n.

(9) Similar rules are added for the extraction of the nth root of two or more irrational quantities of the form $\sqrt[m]{e}+\sqrt[m]{f}+\ldots$, which can easily be further generalized.

(10) A method is given of deciding whether a given infinite rational expression can be expressed as a rational function of arbitrary quantities.

(11) A rule is given which governs the terms of a continued fraction which approximates a given fraction, such that no better approximation exists unless it involves a fraction whose terms are larger than those of the given fraction.

(12) Given a simple equation involving two, three, or more unknowns, its integral solutions are found in terms of arbitrarily assumed values, etc.
(13) Equations in two or more unknown quantities are reduced to one, so that some unknowns are eliminated, and if the solutions of the resulting equation are rational, so are the corresponding values of the eliminated unknowns.
(14) I describe a method of constructing an equation in unknown quantities x, y, etc., whose corresponding values are $x = Az^n + Bz^{n-1} + \dots$, $y = az^n + bz^{n-1} + \dots$, etc.
(15) Given an equation having two or more unknown quantities, I describe a method of determining whether these quantities may be expressed as rational functions of a quantity z.
(16) Some material is added about general integral solutions to given equations, and a method is given, from given corresponding values of some of the quantities, to find infinities of values corresponding to the other unknowns in a given equation or equations.
(17) There follows a method of transforming numbers in the common notation into other notations.
(18) Where $a^x b^z c^y d^v \cdots = A$, it is added that the products $(x+1)(z+1)(y+1)\cdots$ and $\frac{a^{x+1}-1}{a-1} \times \frac{b^{z+1}-1}{b-1} \times \cdots$ are minimized, when $a^{x+1} = b^{z+1} = \dots$.
(19) I point out that the sum of the divisors of a number $1 \cdot 2 \cdot 3 \cdot 4 \cdots x$ has to the number itself a larger ratio than has the sum of the divisors of any lesser number to that lesser number.
(20) Several other properties of integers are added, but not proved, e.g. that every number is composed of nine cubes.
(21) Let $a^2 + rb^2 = N$, then N^{2m+1} or N^{2m+2} is composed in $m+1$ different ways of quantities of the same form as the others; the prescribed compositions are shown with proofs.
(22) Similar propositions are stated and proved with respect to $(a^2 + ab + b^2)^n$, e.g., that every number which is expressible as a^2+ab+b^2 can also be expressed in the form $a^2 - ab + b^2$.
(23) I show that the number of non-residues, on division of numbers of the form $A^h B^k C^l \cdots$ by a prime number, will be either 0, or n, or $2n, \dots$, where n is the number of residues.

(24) Let r be the greatest common divisor of the numbers m and $n-1$. Then $\frac{n-1}{r}+1$ is the number of distinct residues that will result from dividing the numbers 1^m, 2^m, 3^m, etc. by the prime number n.

(25) From the theory of non-residues, certain equations are proved to be susceptible of no integral solutions.

(26) If 0, 1, 2, 3, ... are the final digits of the roots, then they will also be the final digits of the 1st, 5th, 9th, 13th, etc., powers, whence it is easily proved that there cannot be four squares in arithmetic progression.

(27) It is proved that there cannot be three, five, seven, etc. prime numbers in an arithmetic progression, unless the common difference of the progression is divisible respectively by 3!, 5!, etc.

(28) If $1^r + 2^r + 3^r + 4^r + \ldots = S$, then I show that $1 \cdot 2 \cdot 3 \cdots r \cdot (r+1)$ is divisible by $x^2(x+1)^2$ if r is odd, and by $x(x+1)(2x+1)$ if r is an even number; further if α, β, γ, δ, etc. are prime numbers and $a+b+c+d+\ldots$ is less than x, then $\alpha^a\beta^b\gamma^c\delta^d\cdots$ is divisible by $x+1$ if $x+1$ is a prime number. From this also is given a proof of a most elegant theorem regarding prime numbers, discovered and communicated to me by my best friend, one most versatile in all aspects of mathematics, John Wilson, Esq.

(29) A theorem on equality of numbers, taken from my *Proprietates Algebraicarum Curvarum*, is given.

(30) Given m corresponding values for each of two, three, or more unknown quantities, I show a method of deriving a function of the quantities x, y, z, etc., which will have the prescribed solutions.

(31) I discuss the cases in which the exponential equation $x^a = a$ or $x^x = a$ has one positive or negative, or two real roots.

In the addenda are certain other propositions, which follow easily from the foregoing ones, and should be incorporated into this book.

I heartily pray and hope that very little need be added, either to provide a firmer foundation for these my labors, or to remedy defects in calculations that so easily arise in the intricacies of the subject matter of Algebra.

July 12, 1782.

TRANSLATOR'S NOTES

[1] Modern scholarship (e.g. Taylor [1954]) holds that Buckley lived until 1570 or 1571, thus unfortunately invalidating the Wallis-Waring argument in favor of his priority in this matter, since his manuscript contains no internal evidence by which it can be reliably dated to a specific year before Ramus's work appeared. There is little doubt, in any case, that Buckley obtained his results independently, whether before or after Ramus's publication.

It may seem surprising that Buckley published his rules "in verse". In fact, by the same manner Tartaglia delivered to Cardan his method of resolving cubic equations (see Mayer's thesis in the Appendix for more details). As pointed out by Boorstin [1983], use of verse was actually an aid to memorization, at a time when books were scarce and paper was expensive.

[2] Close examination of the Arab manuscripts happened about seventy years later; see e.g. Woepcke [1850].

[3] Several lines of commentary have been moved from here, where they originally appeared (making the summary lengthier than the material that it summarized), to the conclusion of Problem XXVI, something I believe Waring would have done if the time and expense of re-setting the type for the rest of Chapter 3 had not been so prohibitive.

Chapter 1

A method of finding an equation whose roots are any algebraic function of the roots of a given equation or equations.

Problem I

Let $\alpha,\ \beta,\ \gamma,\ \delta,\ \epsilon,\ \ldots$ *be the roots of a given equation* $\mathrm{x^n - px^{n-1} + qx^{n-2} - rx^{n-3} + sx^{n-4} - tx^{n-5} + vx^{n-6} - wx^{n-7} + zx^{n-8}} - \ldots = 0$*; to find the sum* $\alpha^{\mathrm{m}} + \beta^{\mathrm{m}} + \gamma^{\mathrm{m}} + \delta^{\mathrm{m}} + \epsilon^{\mathrm{m}} + \ldots$.

The required sum will be $\quad p^m \ - \ mqp^{m-2} \ + \ mrp^{m-3}$

$$+\left\{\begin{matrix} +ms \\ +\dfrac{m(m-3)}{2}q^2 \end{matrix}\right\}p^{m-4} + \left\{\begin{matrix} +mt \\ -m(m-4)qr \end{matrix}\right\}p^{m-5} + \left\{\begin{matrix} -mv \\ +m(m-5)\,qs \\ +\dfrac{m(m-5)}{2}r^2 \\ -\dfrac{m(m-5)(m-4)}{2\cdot 3}q^3 \end{matrix}\right\}p^{m-6}$$

$$+\left\{\begin{matrix} +mw \\ -m(m-6)qt \\ -m(m-6)rs \\ +\dfrac{m(m-6)(m-5)}{2}q^2r \end{matrix}\right\}p^{m-7} + \left\{\begin{matrix} -mz \\ +m(m-7)qv \\ +m(m-7)\,rt \\ -\dfrac{m(m-7)(m-6)}{2}q^2s \\ -\dfrac{m(m-7)(m-6)}{2\cdot 3}q\,r^2 \\ +\dfrac{m(m-7)(m-6)(m-5)}{2\cdot 3\cdot 4}q^4 \end{matrix}\right\}p^{m-8} + \ldots$$

It is a double rule which this series obeys: one, respecting the literal products, the other with regard to the unciæ, or numerical coefficients.

For the literal parts the rule is this: to form the product of the exponents of each individual coefficient of the given equation, with their respective powers, joining them symbolically to the quantity p, such that they sum up to u, where p^{m-u} is the power of p to which the products of these letters will be adjoined.

By the exponents of the coefficients of the given equation, I mean the numbers designating their distance from the first term of the equation, namely, the exponent of the coefficient q is 2; of r, 3; of s, 4; of t, 5; and so on—and in consequence, that the exponent of a power of the coefficient q, such as q^μ, shall be 2μ; the exponent of any power of r, say r^ν, shall be 3ν; the exponent of s^π shall be 4π; the exponent of t^ρ shall be 5ρ; and likewise for all the rest of them.

Every literal product which can be so composed from the sum u must be adjoined to the power p^{m-u}.

Of the numerical coefficients, or unciæ, the rule is this: for each literal product $p^{m-u}q^\mu r^\nu s^\pi t^\rho v^\sigma w^\tau \cdots$ the coefficient will be a fraction, whose numerator is

$m(m-u+1)(m-u+2)(m-u+3)(m-u+4)(m-u+5)(m-u+6)\cdots$ (in which the number of factors is equal to the sum of the indices, i.e. $\mu+\nu+\rho+\sigma+\tau+\ldots$), and whose denominator is a product of factorials of the indices μ, ν, π, ρ, etc. respectively. More explicitly, in our working example the coefficient of the literal product $p^{m-u}q^\mu r^\nu s^\pi t^\rho\cdots$ will be

$$\frac{m(m-u+1)(m-u+2)(m-u+3)(m-u+4)\cdots(m-u+\mu+\nu+\pi+\rho+\sigma+\ldots-1)}{\mu!\,\nu!\,\pi!\,\rho!\,\cdots}.$$

To the coefficient of the literal product is affixed a sign which will be positive or negative, when u is an even number, according as the sum $\mu+\nu+\pi+\rho+\sigma+\ldots$ is even or odd, and the opposite sign will be affixed when u is an odd number; more generally, the sign will be positive or negative according as the sum $u+\mu+\nu+\pi+\rho+\ldots$ is even or odd.

PROOF

Let $p=a$, $p^2-2q=b$, $pb-qa+3r=c$, $pc-qb+ra-4s=d$, $pd-qc+rb-sa+5t=e$, $pe-qd+rc-sb+ta-6v=f$, and so on. Then a will be the sum of the roots, b the sum of the squares of the individual roots, c the sum of their cubes, d the sum of the fourth powers, e the sum of the fifth powers, f the sum of the sixth powers, and likewise for the rest. (Newton, *Arith. Univ.*, page 251.)

In this rule it is easy enough to see that the coefficient of the literal product $p^{m-u}q^{\mu}r^{\nu}s^{\pi}t^{\rho}\cdots$, when computing the sum of $\alpha^m+\beta^m+\gamma^m+\delta^m+\ldots$, must be composed of the coefficients of the literal products

$$\begin{array}{llllll} p^{m-u-1} & q^{\mu} & r^{\nu} & s^{\pi} & t^{\rho} & \cdots, \\ p^{m-u} & q^{\mu-1} & r^{\nu} & s^{\pi} & t^{\rho} & \cdots, \\ p^{m-u} & q^{\mu} & r^{\nu-1} & s^{\pi} & t^{\rho} & \cdots, \\ p^{m-u} & q^{\mu} & r^{\nu} & s^{\pi-1} & t^{\rho} & \cdots, \\ p^{m-u} & q^{\mu} & r^{\nu} & s^{\pi} & t^{\rho-1} & \cdots, \text{ etc.} \end{array}$$

These coefficients themselves will be, respectively,

$$\begin{aligned} (m-1)\cdot(m-u)\cdot &\frac{(m-u+1)(m-u+2)\cdots(m-u+\mu+\nu+\pi+\rho+\ldots-2)}{\mu!\,\nu!\,\pi!\,\rho!\,\cdots} \\ \mu\cdot(m-2)\cdot &\frac{(m-u+1)(m-u+2)\cdots(m-u+\mu+\nu+\pi+\rho+\ldots-2)}{\mu!\,\nu!\,\pi!\,\rho!\,\cdots} \\ \nu\cdot(m-3)\cdot &\frac{(m-u+1)(m-u+2)\cdots(m-u+\mu+\nu+\pi+\rho+\ldots-2)}{\mu!\,\nu!\,\pi!\,\rho!\,\cdots} \\ \pi\cdot(m-4)\cdot &\frac{(m-u+1)(m-u+2)\cdots(m-u+\mu+\nu+\pi+\rho+\ldots-2)}{\mu!\,\nu!\,\pi!\,\rho!\,\cdots} \\ \rho\cdot(m-5)\cdot &\frac{(m-u+1)(m-u+2)\cdots(m-u+\mu+\nu+\pi+\rho+\ldots-2)}{\mu!\,\nu!\,\pi!\,\rho!\,\cdots} \end{aligned}$$

etc., and the sum of these numerical coefficients will be

$$\begin{aligned} &\Big(m^2-(u-1+\mu+\nu+\pi+\rho+\ldots)m+(u-2\mu-3\nu-4\pi-5\rho-\ldots)\Big) \\ &\times\frac{(m-u+1)(m-u+2)(m-u+3)\cdots(m-u+\mu+\nu+\pi+\rho+\ldots-2)}{\mu!\,\nu!\,\pi!\,\rho!\,\cdots}. \end{aligned}$$

In the first factor of this sum, for u substitute its value $2\mu+3\nu+4\pi+5\rho+\ldots$, whereupon the sum becomes

$$\begin{aligned} &m(m-u+\mu+\nu+\pi+\rho+\ldots-1) \\ &\times\frac{(m-u+1)(m-u+2)(m-u+3)\cdots(m-u+\mu+\nu+\pi+\rho+\ldots-2)}{\mu!\,\nu!\,\pi!\,\rho!\,\cdots} \end{aligned}$$

This is precisely the numerical coefficient which is to be substituted, in the given series, for the coefficient of the literal product $q^{\mu}r^{\nu}s^{\pi}t^{\rho}\cdots$ in the sum of $\alpha^m+\beta^m+\gamma^m+\delta^m+\ldots$; in consequence, if the series is correct for sums of lower powers, $\alpha^{m-1}+\beta^{m-1}+\gamma^{m-1}+\delta^{m-1}+\ldots$, $\alpha^{m-2}+\beta^{m-2}+\gamma^{m-2}+\delta^{m-2}+\ldots$, $\alpha^{m-3}+\beta^{m-3}+\gamma^{m-3}+\delta^{m-3}+\ldots$, $\alpha^{m-4}+\beta^{m-4}+\gamma^{m-4}+\delta^{m-4}+\ldots$, etc., then $\alpha^m+\beta^m+\gamma^m+\delta^m+\ldots$ is also correctly expressed by the next higher sum. [1]

Example 1. Let the equation be $x^4-2x^3-7x^2+20x-12=0$. In this example p, q, r, s, t are respectively 2, –7, –20, –12, 0. Whence the sum of the squares of each of the roots is $p^2 - 2q = 2^2 + 2 \times 7 = 18$; the sum of the cubes is $p^3 - 3pq + 3r = 2^3 + 3 \times 2 \times 7 - 3 \times 20 = -10$; the sum of the fourth powers is $p^4 - 4pq^2 + 4rp - 4s + 2q^2 = 2^4 + 4 \times 7 \times 2^2 - 4 \times 20 \times 2 + 4 \times 12 + 2 \times 7^2 = 114$; and the sum of the fifth powers is $p^5 - 5qp^3 + 5rp^2 + (5q^2 - 5s)p + 5t - 5qr = 2^5 + 5 \times 7 \times 2^3 - 5 \times 20 \times 2^2 + 5 \times 12 \times 2 + 5 \times 7^2 \times 2 - 5 \times 7 \times 20 = -178$.

It is perhaps worth mentioning here that the infinite series $p^m - mqp^{m-2} + mrp^{m-3} + (m \cdot \frac{m-3}{2} \cdot q^2 - ms)p^{m-4} + (mt - m(m-4)qr)p^{m-5} - \ldots$, multiplied by a similar series (also infinite), $p^n - nqp^{n-2} + nrp^{n-3} + (n \cdot \frac{n-3}{2} \cdot q^2 - ns)p^{n-4} + (nt - n(n-4)qr)p^{n-5} - \ldots$, generates a series

$$p^{n+m} - (n+m)qp^{n+m-2} + (n+m)r\,p^{n+m-3} +$$
$$\left((n+m)\tfrac{n+m-3}{2}q^2 - (n+m\qquad)s\right)p^{n+m-4} +$$
$$\left((n+m)t - (n+m)(n+m-4)qr\right)p^{n+m-5} + \ldots,$$

with the same rules governing the composition of all product terms.

Corollary 1. Let the equation be $x^n - Px^{n-m} + Qx^{n-2m} - Rx^{n-3m} + \ldots \mp Hx^{n-\alpha m} \pm Kx^{n-\alpha m-1} \mp \ldots = A$; then the sum of all a^{th} powers, where a is not divisible by m and $a < \alpha m$, is equal to 0.

Corollary 2. Let the equation be $x^n - px^{n-1} + qx^{n-2} - Rx^{n-3} + \ldots \pm hx^{n-m} \mp kx^{n-m-1} \pm \ldots = 0$, and let h be the first coefficient which is a function of some quantity z. Then the values of all power sums of index less than m will be the same, regardless of how z may vary.

Corollary 3. Given an equation whose resolution is known, it is often possible to construct the sums of the powers of its roots.

Example 2. Let the equation be $x^n - nax^{n-2} + n \cdot \frac{n-3}{2}a^2x^{n-4} - \ldots = -b$, whose resolution is $x = \sqrt[n]{-\frac{1}{2}b + \sqrt{\frac{1}{4}b^2 - a^n}}(= P) + \sqrt[n]{-\frac{1}{2}b - \sqrt{\frac{1}{4}b^2 - a^n}}$ $(= Q)$; and let $\alpha, \beta, \gamma, \delta, \ldots$ respectively be the roots of the equation $x^n - 1 = 0$. Now the roots of the original equation will be $\alpha P + Q/\alpha$, $\beta P + Q/\beta$, $\gamma P + Q/\gamma$, $\delta P + Q/\delta$, etc., whence the sum of the $(2r)^{\text{th}}$ powers of each of the roots (for $2r < n$) is

$$\alpha\left(\sqrt[n]{-\frac{1}{2}b + \sqrt{\frac{1}{4}b^2 - a^n}}\right)^{2r} + \frac{1}{\alpha}\left(\sqrt[n]{-\frac{1}{2}b - \sqrt{\frac{1}{4}b^2 - a^n}}\right)^{2r} =$$

$$(\alpha P+\tfrac{1}{\alpha}Q)^{2r}=\alpha^{2r}P^{2r}+2r\alpha^{2r-2}P^{2r-1}Q+\ldots+\tfrac{2r(2r-1)\cdots(r+1)}{r!}\underbrace{\alpha^r\cdot\tfrac{1}{\alpha^r}}_{=1}\underbrace{P^rQ^r}_{=a^r}+\ldots$$

$$(\beta P+\tfrac{1}{\beta}Q)^{2r}=\beta^{2r}P^{2r}+2r\beta^{2r-2}P^{2r-1}Q+\ldots+\tfrac{2r(2r-1)\cdots(r+1)}{r!}a^r+\ldots$$

$$(\gamma P+\tfrac{1}{\gamma}Q)^{2r}=\gamma^{2r}P^{2r}+2r\gamma^{2r-2}P^{2r-1}Q+\ldots+\tfrac{2r(2r-1)\cdots(r+1)}{r!}a^r+\ldots$$

$$(\delta P+\tfrac{1}{\delta}Q)^{2r}=\delta^{2r}P^{2r}+2r\delta^{2r-2}P^{2r-1}Q+\ldots+\tfrac{2r(2r-1)\cdots(r+1)}{r!}a^r+\ldots$$

etc., but, since $\alpha^{2r}+\beta^{2r}+\gamma^{2r}+\delta^{2r}+\ldots=0$, $\alpha^{2r-2}+\beta^{2r-2}+\gamma^{2r-2}+\delta^{2r-2}+\ldots=0$, etc., then I claim that the sum of the $(2r)^{\text{th}}$ powers of the roots of the given equation will be $\pi=\frac{2r(2r-1)(2r-2)\cdots(r+1)}{r!}\times na^r$.

For by this problem $\pi=(2n^r-2rn^{r-2}\cdot n\cdot\frac{n-3}{2}+2rn^{r-3}\cdot n\cdot\frac{n-4}{2}\cdot\frac{n-5}{3}-2rn^{r-4}\cdot n\cdot\frac{n-5}{2}\cdot\frac{n-6}{3}\cdot\frac{n-7}{4}+\ldots+2r\cdot\frac{r-3}{2}\cdot n^{r-4}\cdot(n\cdot\frac{n-3}{2})^2-\ldots)a^r=\frac{2r(2r-1)(2r-2)\cdots(r+1)}{r!}na^r$.

Corollary 4. Let the equation be $x^n-nax^{n-1}+\frac{n(n-1)}{2}a^2x^{n-2}-\frac{n(n-1)(n-2)}{2\cdot3}a^3x^{n-3}+\ldots=b$, a root of which is $a+\sqrt[n]{b}$ (hence the several roots are $a+\alpha\sqrt[n]{b}$, $a+\beta\sqrt[n]{b}$, $a+\gamma\sqrt[n]{b}$, $a+\delta\sqrt[n]{b}$, etc.) Then the sum of the rth powers (if $r<n$) of each of these roots will be

$$\begin{aligned}(a+\alpha\sqrt[n]{b})^r&=a^r+ra^{r-1}\alpha\sqrt[n]{b}+\ldots\\(a+\beta\sqrt[n]{b})^r&=a^r+ra^{r-1}\beta\sqrt[n]{b}+\ldots\\(a+\gamma\sqrt[n]{b})^r&=a^r+ra^{r-1}\gamma\sqrt[n]{b}+\ldots\\\text{etc.}&=\text{etc.}\end{aligned}$$

but since $\alpha+\beta+\gamma+\ldots=0$, $\alpha^2+\beta^2+\gamma^2+\ldots=0$, and so on, the required sum is $n\times a^r$.

Directly by the problem's formula, the required sum will be

$$\left(\begin{matrix}n^r-rn^{n-2}\frac{n(n-1)}{2}+rn^{n-3}\frac{n(n-1)(n-2)}{2\cdot3}\\-rn^{n-4}\frac{n(n-1)(n-2)(n-3)}{2\cdot3\cdot4}+\ldots+\frac{r(r-3)}{2}n^{n-4}\left(\frac{n(n-1)}{2}\right)^2-\ldots\end{matrix}\right)a^r$$

$=n\times a^r$: thus the sum, computed by two distinct methods, comes to the same value. [2]

Problem II

To find the aggregate of several consecutive terms of these series.

Terms of the series satisfying the preceding problem are to be added together to produce the new series.

Example 1. Let a be the sum of the roots, b the sum of the squares of the roots, c the sum of their cubes, d the sum of their fourth powers, e the sum of their fifth powers, f the sum of their sixth powers, etc. We are to find the sum of the first m terms of the series

$$a + b + c + d + e + \ldots + M.$$

By the preceding problem,

$$\begin{aligned}
a &= p \\
b &= p^2 - 2q \\
c &= p^3 - 3qp + 3r \\
d &= p^4 - 4qp^2 + 4rp - 4s + 2q^2 \\
e &= p^5 - 5qp^3 + 5rp^2 - 5sp + 5q^2p + 5t - 5qr \\
\text{etc.} &= \text{etc.} \quad \text{etc.} \\
M &= p^m - mqp^{m-2} + mrp^{m-3} - msp^{m-4} + \frac{m(m-3)}{2}q^2p^{m-4} + mtp^{m-5} \\
&\quad - m(m-4)qrp^{m-5} - \ldots
\end{aligned}$$

Now the sum of the terms in the first column is

$$p + p^2 + p^3 + p^4 + p^5 + \ldots + p^m = \frac{p^{m-1}}{p-1};$$

of the second,

$$2q + 3qp + 4qp^2 + 5qp^3 + \ldots + mqp^{m-2} = \frac{2 - p - (m+1)p^{m-1} + mp^m}{(1-p)^2}q;$$

of the third,

$$3r + 4rp + 5rp^2 + 6rp^3 + \ldots + mrp^{m-3} = \frac{3 - 2p - (m+1)p^{m-2} + mp^{m-1}}{(1-p)^2}r;$$

of the fourth,

$$4s + 5sp + 6sp^2 + 7sp^3 + \ldots + msp^{m-4} = \frac{4 - 3p - (m+1)p^{m-3} + mp^{m-2}}{(1-p)^2}s;$$

of the fifth,

$$2q^2 + 5q^2p + 9q^2p^2 + \ldots + \frac{m(m-3)}{2}q^2p^{m-4}$$

$$= \frac{4 - 2p - (m+1)(m-2)p^{m-3} + (2m^2 - 4m - 4)p^{m-2} - m(m-3)p^{m-1}}{2(1-p)^3}q^2;$$

etc. etc.

and in consequence the required sum will be

$$\frac{p^{m+1}-p}{p-1} - \frac{2-p-(m+1)p^{m-1}+mp^m}{(1-p)^2} \times q + \frac{3-2p-(m+1)p^{m-2}+mp^{m-1}}{(1-p)^2} \times r$$

$$- \frac{4-3p-(m+1)p^{m-3}+mp^{m-2}}{(1-p)^2} \times s + \frac{5-4p-(m+1)p^{m-4}+mp^{m-3}}{(1-p)^2} \times t$$

$$+ \frac{4-2p-(m+1)(m-2)p^{m-3}+(2m^2-4m-4)p^{m-2}-m(m-3)p^{m-1}}{2(1-p)^3} \times q^2$$

$$- \frac{5-3p-(m+1)(m-3)p^{m-4}+(2m^2-6m-5)p^{m-3}-m(m-4)p^{m-2}}{(1-p)^3} \times qr$$

$$+ \ldots.$$

The literal products will be the same as in the preceding series, aside from the quantity p and its powers.

And here is the rule which the coefficients obey: write $u = 2\mu+3\nu+4\pi+5\rho+6\sigma+\ldots$, and $l = 1+\mu+\nu+\pi+\rho+\sigma+\ldots$; then the coefficient of the literal product $q^{\mu}r^{\nu}s^{\pi}t^{\rho}v^{\sigma}\cdots$ will be equal to the product of $(l-2)!$, multiplied by a fraction whose denominator is $(1-p)^{l}\cdot\mu!\cdot\nu!\cdot\pi!\cdot\rho!\cdot\sigma!\cdots$, and whose numerator is

$$u \qquad -(u-l+1)\ p$$

$$-(m+1)\ \frac{m-u+2}{1}\cdot\frac{m-u+3}{2}\cdot\frac{m-u+4}{3}\cdot\frac{m-u+5}{4}\cdot\frac{m-u+6}{5}\cdot\ \cdots\frac{m-u+l-1}{l-2}\,p^{m-u+1}$$

$$\left.\begin{array}{l} +u\ \cdot\frac{m-u+1}{1}\cdot\ \frac{m-u+3}{1}\cdot\frac{m-u+4}{2}\cdot\frac{m-u+5}{3}\cdot\frac{m-u+6}{4}\cdot\frac{m-u+7}{5}\cdot\frac{m-u+8}{6}\cdots\frac{m-u+l}{l-2} \\ -(u-l+1)\frac{m-u}{1}\cdot\ \frac{m-u+2}{1}\cdot\frac{m-u+3}{2}\cdot\frac{m-u+4}{3}\cdot\frac{m-u+5}{4}\cdot\frac{m-u+6}{5}\cdot\frac{m-u+7}{6}\cdots\frac{m-u+l-1}{l-2}\end{array}\right\}p^{m-u+2}$$

$$\left.\begin{array}{l} -u\ \cdot\frac{m-u+1}{1}\cdot\frac{m-u+2}{2}\cdot\ \frac{m-u+4}{1}\cdot\frac{m-u+5}{2}\cdot\frac{m-u+6}{3}\cdot\frac{m-u+7}{4}\cdot\frac{m-u+8}{5}\cdots\frac{m-u+l}{l-3} \\ +(u-l+1)\frac{m-u}{1}\cdot\frac{m-u+1}{2}\cdot\ \frac{m-u+3}{1}\cdot\frac{m-u+4}{2}\cdot\frac{m-u+5}{3}\cdot\frac{m-u+6}{4}\cdot\frac{m-u+7}{5}\cdots\frac{m-u+l-1}{l-3}\end{array}\right\}p^{m-u+3}$$

$$\left.\begin{array}{l} +u\ \cdot\frac{m-u+1}{1}\cdot\frac{m-u+2}{2}\cdot\frac{m-u+3}{3}\cdot\ \frac{m-u+5}{1}\cdot\frac{m-u+6}{2}\cdot\frac{m-u+7}{3}\cdot\frac{m-u+8}{4}\cdots\frac{m-u+l}{l-4} \\ -(u-l+1)\frac{m-u}{1}\cdot\frac{m-u+1}{2}\cdot\frac{m-u+2}{3}\cdot\ \frac{m-u+4}{1}\cdot\frac{m-u+5}{2}\cdot\frac{m-u+6}{3}\cdot\frac{m-u+7}{4}\cdots\frac{m-u+l-1}{l-4}\end{array}\right\}p^{m-u+4}$$

$$\left.\begin{array}{l} -u\ \cdot\frac{m-u+1}{1}\cdot\frac{m-u+2}{2}\cdot\frac{m-u+3}{3}\cdot\frac{m-u+4}{4}\cdot\ \frac{m-u+6}{1}\cdot\frac{m-u+7}{2}\cdot\frac{m-u+8}{3}\cdots\frac{m-u+l}{l-5} \\ +(u-l+1)\frac{m-u}{1}\cdot\frac{m-u+1}{2}\cdot\frac{m-u+2}{3}\cdot\frac{m-u+3}{4}\cdot\ \frac{m-u+5}{1}\cdot\frac{m-u+6}{2}\cdot\frac{m-u+7}{3}\cdots\frac{m-u+l-1}{l-5}\end{array}\right\}p^{m-u+5}$$

$$\left.\begin{array}{l} +u\ \cdot\frac{m-u+1}{1}\cdot\frac{m-u+2}{2}\cdot\frac{m-u+3}{3}\cdot\frac{m-u+4}{4}\cdot\frac{m-u+5}{5}\cdot\ \frac{m-u+7}{1}\cdot\frac{m-u+8}{2}\cdots\frac{m-u+l}{l-6} \\ -(u-l+1)\frac{m-u}{1}\cdot\frac{m-u+1}{2}\cdot\frac{m-u+2}{3}\cdot\frac{m-u+3}{4}\cdot\frac{m-u+4}{5}\cdot\ \frac{m-u+6}{1}\cdot\frac{m-u+7}{2}\cdots\frac{m-u+l-1}{l-6}\end{array}\right\}p^{m-u+6}$$

$$\left.\begin{array}{l} -u\ \cdot\frac{m-u+1}{1}\cdot\frac{m-u+2}{2}\cdot\frac{m-u+3}{3}\cdot\frac{m-u+4}{4}\cdot\frac{m-u+5}{5}\cdot\frac{m-u+6}{6}\cdot\ \frac{m-u+8}{1}\cdots\frac{m-u+l}{l-7} \\ +(u-l+1)\frac{m-u}{1}\cdot\frac{m-u+1}{2}\cdot\frac{m-u+2}{3}\cdot\frac{m-u+3}{4}\cdot\frac{m-u+4}{5}\cdot\frac{m-u+5}{6}\cdot\ \frac{m-u+7}{1}\cdots\frac{m-u+l-1}{l-7}\end{array}\right\}p^{m-u+7}$$

and so on until the penultimate term, $p^{m-u+l-1}$; the final term will be

$$\pm\, m\cdot\frac{m-u+1}{1}\cdot\frac{m-u+2}{2}\cdot\frac{m-u+3}{3}\cdot\frac{m-u+4}{4}\cdots\frac{m-u+l-2}{l-2}p^{m-u+l}.$$

The derivation and proof of this problem are easy consequences of the method of finding the sum of the $m-u+1$ terms of the series emerging from division of the quantity $u-(u-l+1)$ by $(1-p)^{l}$.

Example 2. Given the equation $x^4-3x^3+x^2+3x-2=0$, to find the sum $a+b+c+d$.

The values of p, q, r, s, and m are 3, 1, -3, -2, and 4 respectively. The sum will be

$a+b+c+d=$

$$\begin{array}{lll} \frac{p-p^5}{1-p} & -q\frac{2-p-5p^3+4p^4}{(1-p)^2} & +q^2\frac{4-2p-10p+12p^2-4p^3}{2(1-p)^3} \\ =120 & =-47 & =2 \\ & +r\frac{3-2p-5p^2+4p^3}{(1-p)^2} & \\ & =-45 & \\ & -s\frac{4-3p-5p\ +4p^2}{(1-p)^2} & \\ & =8 & \end{array} = 38.$$

Corollary. When $p=1$, replace $\frac{p-p^{m+1}}{1-p}$ by m, $q\frac{2-p-(m+1)p^{m-1}+mp^m}{(1-p)^2}$ by $q(m+2)\frac{m-1}{2}$, etc. Then the desired sum becomes

$$m-(m+2)\frac{m-1}{2}q+(m+3)\frac{m-2}{2}r-(m+4)\frac{m-3}{2}s \quad +\dots$$
$$+\tfrac{1}{2}(2m+4)\frac{m-3}{2}\cdot\frac{m-2}{3}q^2 \quad -\dots$$

etc. The coefficient of the literal product $q^\mu r^\nu s^\pi t^\rho\cdots$ in this series is equal to the fraction

$$((l-1)m+u)\frac{(m-u+1)(m-u+2)(m-u+3)\cdots(m-u+l-1)}{l(l-1)\mu!\,\nu!\,\pi!\,\rho!\,\cdots}.$$

PROOF

Let S be the sum of the coefficients of the terms $q^\mu r^\nu s^\pi t^\rho\dots$, $pq^\mu r^\nu s^\pi t^\rho\dots$, $p^2q^\mu r^\nu s^\pi t^\rho\dots$, through $p^{m-u}q^\mu r^\nu s^\pi t^\rho\dots$, appearing in the sums of the uth, $(u+1)^{\text{th}}$, $(u+2)^{\text{th}}$, $\dots m^{\text{th}}$ powers respectively of the several roots. Then

$$S+(m+1)\frac{(m-u+2)(m-u+3)(m-u+4)\cdots(m-u+l-1)}{\mu!\,\nu!\,\pi!\,\rho!}$$

(which by Problem I is the coefficient of the τth term $p^{m-u+1}q^\mu r^\nu s^\pi t^\rho\cdots$ in the sum of the $(m+1)^{\text{th}}$ powers of the individual roots) will be equal to S', the next sum, in which the power m has been increased by 1. But by this corollary, S and S' will be, respectively,

$$\frac{((l-1)m+u)(m-u+1)(m-u+2)(m-u+3)\cdots(m-u+l-1)}{l(l-1)\mu!\,\nu!\,\pi!\,\rho!\,\cdots}$$

and $$\frac{((l-1)(m+1)+u)(m-u+2)(m-u+3)(m-u+4)\cdots(m-u+l-0)}{l(l-1)\mu!\,\nu!\,\pi!\,\rho!\,\cdots},$$

whence the difference $S'-S$ is

$$\Big(((l-1)(m+1)+u)(m-u+l)-((l-1)m+u)(m-u+1)\Big)$$
$$\times\frac{(m-u+2)(m-u+3)\cdots(m-u+l-1)}{l(l-1)\mu!\,\nu!\,\pi!\,\rho!\,\cdots}$$
$$= \quad l(l-1)(m+1)\frac{(m-u+2)(m-u+3)\cdots(m-u+l-1)}{l(l-1)\mu!\,\nu!\,\pi!\,\rho!\,\cdots}$$
$$= \quad (m+1)\frac{(m-u+2)(m-u+3)\cdots(m-u+l-1)}{\mu!\,\nu!\,\pi!\,\rho!\,\cdots}$$

which is precisely the coefficient of the τth term $p^{m-u+1}q^{\mu}r^{\nu}s^{\pi}t^{\rho}\cdots$ as stated above. Therefore if in the corollary S correctly designates the aforesaid sum up to the mth power, then S' indeed extends the cumulative sum to the next higher power, $m+1$.

Problem III

Let α, β, γ, δ, ϵ, *etc. be roots of a given equation* $\mathrm{x^n - px^{n-1} + qx^{n-2} - rx^{n-3}} + \ldots = 0$, *and let* a, b, c, d, e, f, *etc. be given indices,* m *in number* (m$\leq$n). *To find the sum of all quantities of the following form:*

$$\alpha^{\mathrm a}\beta^{\mathrm b}\gamma^{\mathrm c}\delta^{\mathrm d}\epsilon^{\mathrm e}\cdots+\alpha^{\mathrm b}\beta^{\mathrm a}\gamma^{\mathrm c}\delta^{\mathrm d}\epsilon^{\mathrm e}\cdots+\alpha^{\mathrm c}\beta^{\mathrm a}\gamma^{\mathrm b}\delta^{\mathrm d}\epsilon^{\mathrm e}\cdots+$$
$$\alpha^{\mathrm d}\beta^{\mathrm a}\gamma^{\mathrm b}\delta^{\mathrm c}\epsilon^{\mathrm e}\cdots+\alpha^{\mathrm e}\beta^{\mathrm a}\gamma^{\mathrm b}\delta^{\mathrm c}\epsilon^{\mathrm d}\cdots+\alpha^{\mathrm a}\beta^{\mathrm c}\gamma^{\mathrm b}\delta^{\mathrm d}\epsilon^{\mathrm e}\cdots+\ldots$$

in which the number of factors is m.

Case 1. Let us write Sa, Sb, Sc, etc. for the sums of the ath, b^{th}, c^{th}, etc. powers respectively of the individual roots, i.e.,

Sa	$=\alpha^a$	$+\beta^a$	$+\gamma^a$	$+\delta^a$	$+\epsilon^a$	$+\ldots$
Sb	$=\alpha^b$	$+\beta^b$	$+\gamma^b$	$+\delta^b$	$+\epsilon^b$	$+\ldots$
Sc	$=\alpha^c$	$+\beta^c$	$+\gamma^c$	$+\delta^c$	$+\epsilon^c$	$+\ldots$
Sd	$=\alpha^d$	$+\beta^d$	$+\gamma^d$	$+\delta^d$	$+\epsilon^d$	$+\ldots$
Se	$=\alpha^e$	$+\beta^e$	$+\gamma^e$	$+\delta^e$	$+\epsilon^e$	$+\ldots$
etc.	= etc.	etc.				
$S(a{+}b)$	$=\alpha^{a+b}$	$+\beta^{a+b}$	$+\gamma^{a+b}$	$+\delta^{a+b}$	$+\epsilon^{a+b}$	$+\ldots$
$S(a{+}c)$	$=\alpha^{a+c}$	$+\beta^{a+c}$	$+\gamma^{a+c}$	$+\delta^{a+c}$	$+\epsilon^{a+c}$	$+\ldots$
$S(b{+}c)$	$=\alpha^{b+c}$	$+\beta^{b+c}$	$+\gamma^{b+c}$	$+\delta^{b+c}$	$+\epsilon^{b+c}$	$+\ldots$
$S(a{+}d)$	$=\alpha^{a+d}$	$+\beta^{a+d}$	$+\gamma^{a+d}$	$+\delta^{a+d}$	$+\epsilon^{a+d}$	$+\ldots$
etc.	= etc.	etc.				

$$\begin{array}{lllllll}
S(a+b+c) & =\alpha^{a+b+c} & +\beta^{a+b+c} & +\gamma^{a+b+c} & +\delta^{a+b+c} & +\epsilon^{a+b+c} & +\dots \\
S(a+b+d) & =\alpha^{a+b+d} & +\beta^{a+b+d} & +\gamma^{a+b+d} & +\delta^{a+b+d} & +\epsilon^{a+b+d} & +\dots \\
S(a+b+e) & =\alpha^{a+b+e} & +\beta^{a+b+e} & +\gamma^{a+b+e} & +\delta^{a+b+e} & +\epsilon^{a+b+e} & +\dots \\
S(a+c+d) & =\alpha^{a+c+d} & +\beta^{a+c+d} & +\gamma^{a+c+d} & +\delta^{a+c+d} & +\epsilon^{a+c+d} & +\dots \\
S(b+c+d) & =\alpha^{b+c+d} & +\beta^{b+c+d} & +\gamma^{b+c+d} & +\delta^{b+c+d} & +\epsilon^{b+c+d} & +\dots \\
\text{etc.} & = \text{etc.} & \text{etc.} & & & & \\
S(a+b+c+d) & =\alpha^{a+b+c+d} & +\beta^{a+b+c+d} & +\gamma^{a+b+c+d} & +\delta^{a+b+c+d} & +\epsilon^{a+b+c+d} & +\dots \\
S(a+b+c+e) & =\alpha^{a+b+c+e} & +\beta^{a+b+c+e} & +\gamma^{a+b+c+e} & +\delta^{a+b+c+e} & +\epsilon^{a+b+c+e} & +\dots \\
S(a+b+d+e) & =\alpha^{a+b+d+e} & +\beta^{a+b+d+e} & +\gamma^{a+b+d+e} & +\delta^{a+b+d+e} & +\epsilon^{a+b+d+e} & +\dots \\
S(b+c+d+e) & =\alpha^{b+c+d+e} & +\beta^{b+c+d+e} & +\gamma^{b+c+d+e} & +\delta^{b+c+d+e} & +\epsilon^{b+c+d+e} & +\dots \\
\text{etc.} & = \text{etc.} & \text{etc.} & & & & \\
S(a+b+c+d+e) & =\alpha^{a+b+c+d+e} & +\beta^{a+b+c+d+e} & +\gamma^{a+b+c+d+e} & +\delta^{a+b+c+d+e} & & +\dots \\
\text{etc.} & = \text{etc.} & \text{etc.} & & & &
\end{array}$$

and likewise for all the rest. Also, write

$$A = Sa\,Sb\,Sc\,Sd\,Se\cdots$$

$$B = \frac{S(a+b)A}{Sa\,Sb} + \frac{S(a+c)A}{Sa\,Sb} + \frac{S(b+c)A}{Sb\,Sc} + \frac{S(a+d)A}{Sa\,Sd} + \dots$$

$$C = \frac{S(a+b+c)A}{Sa\,Sb\,Sc} + \frac{S(a+b+d)A}{Sa\,Sb\,Sd} + \frac{S(a+c+d)A}{Sa\,Sc\,Sd} + \frac{S(b+c+d)A}{Sb\,Sc\,Sd} + \frac{S(a+b+e)A}{Sa\,Sb\,Se} + \dots$$

$$\begin{cases} D = \dfrac{S(a+b+c+d)A}{Sa\,Sb\,Sc\,Sd} + \dfrac{S(a+b+c+e)A}{Sa\,Sb\,Sc\,Se} + \dfrac{S(a+b+d+e)A}{Sa\,Sb\,Sd\,Se} + \dfrac{S(b+c+d+e)A}{Sb\,Sc\,Sd\,Se} + \dots \\ BB = \dfrac{S(a+b)S(c+d)A}{Sa\,Sb\,Sc\,Sd} + \dfrac{S(a+c)S(b+d)A}{Sa\,Sc\,Sb\,Sd} + \dfrac{S(a+e)S(b+c)A}{Sa\,Se\,Sb\,Sc} + \dots \end{cases}$$

$$\begin{cases} E = \dfrac{S(a+b+c+d+e)A}{Sa\,Sb\,Sc\,Sd\,Se} + \dfrac{S(a+b+c+d+f)A}{Sa\,Sb\,Sc\,Sd\,Sf} + \dfrac{S(a+b+c+e+f)A}{Sa\,Sb\,Sc\,Se\,Sf} + \dots \\ BC = \dfrac{S(a+b)S(c+d+e)A}{Sa\,Sb\,Sc\,Sd\,Se} + \dfrac{S(a+c)S(b+d+e)A}{Sa\,Sc\,Sb\,Sd\,Se} + \dfrac{S(a+d)S(b+c+e)A}{Sa\,Sd\,Sb\,Sc\,Se} + \dots \end{cases}$$

$$\begin{cases} F = \dfrac{S(a+b+c+d+e+f)A}{Sa\,Sb\,Sc\,Sd\,Se\,Sf} + \dots \\ BD = \dfrac{S(a+b)S(c+d+e+f)A}{Sa\,Sb\,Sc\,Sd\,Se\,Sf} + \dfrac{S(a+c)S(b+d+e+f)A}{Sa\,Sc\,Sb\,Sd\,Se\,Sf} + \dots \\ CC = \dfrac{S(a+b+c)S(d+e+f)A}{Sa\,Sb\,Sc\,Sd\,Se\,Sf} + \dfrac{S(a+b+d)S(c+e+f)A}{Sa\,Sb\,Sd\,Sc\,Se\,Sf} + \dots \\ BBB = \dfrac{S(a+b)S(c+d)S(e+f)A}{Sa\,Sb\,Sc\,Sd\,Se\,Sf} + \dfrac{S(a+c)S(b+d)S(e+f)A}{Sa\,Sc\,Sb\,Sd\,Se\,Sf} + \dots \end{cases}$$

and so on.

(Observe that in the foregoing expressions, Sa, Sb, $Sc, \ldots$, $S(a+b)$, $S(a+c)$, $S(b+c), \ldots$, $S(a+b+c)$, $S(a+b+d)$, etc. designate quantities which are equal to each other when the parenthesized expressions are identical; the same is to be understood for the terms BB, BC, BD, CC, BBB, etc.)

The required sum is

$$\begin{array}{llllll}
A - B + 1\cdot 2C & -1\cdot 2\cdot 3D & +1\cdot 2\cdot 3\cdot 4E & -1\cdot 2\cdot 3\cdot 4\cdot 5F \\
 & +1\cdot 1\,BB & -1\cdot 1\cdot 2\,BC & +1\cdot 1\cdot 2\cdot 3\,BD \\
 & & & +1\cdot 2\cdot 1\cdot 2\,CC \\
 & & & -1\cdot 1\cdot 1\,BBB
\end{array}$$

$$\begin{array}{l}
+1\cdot 2\cdot 3\cdot 4\cdot 5\cdot 6G - \ldots \\
-1\cdot 1\cdot 2\cdot 3\cdot 4\,BE + \ldots \\
-1\cdot 2\cdot 1\cdot 2\cdot 3\,CD + \ldots \\
+1\cdot 1\cdot 1\cdot 2\,BBC - \ldots
\end{array}$$

Of the literal quantities this is the rule: exponents of the literal quantities of the terms subjoined to them complete the same sum, which sum in fact will be successively 1, 2, 3, 4, 5, 6, etc.

(By the exponents of letters I mean the numbers which correspond to their alphabetical order. Thus the exponent of the letter A is 1, of B is 2, of C is 3, and so on, which means that the exponent of the expression $BBB\cdots CCC\cdots DDD\cdots EEE\cdots FFF\cdots GGG\cdots$ etc., in which the numbers of letters B, C, D, E, F, $G, \ldots$ are μ, ν, π, ρ, σ, $\tau, \ldots$ respectively, will be $2\mu + 3\nu + 4\pi + 5\rho + 6\sigma + 7\tau + \ldots$.)

All terms whose literal quantities combine to form the same sum are added together.

The coefficient of the literal quantity $BBB\ldots CCC\ldots DDD\ldots EEE\ldots FFF\ldots GGG\ldots$ etc. (in which the letter B occurs μ times, C occurs ν times, D, E, F, $G, \ldots$ occur π, ρ, σ, $\tau, \ldots$ times respectively) is equal to the product (of powers of factorials) $1!^{\mu}\cdot 2!^{\nu}\cdot 3!^{\pi}\cdot 4!^{\rho}\cdot 5!^{\sigma}\cdot 6!^{\tau}\cdots$.

But if there are k indices equal to a, l indices equal to b, r' and s' equal to c and d respectively, and so on, then the above sum must be divided by the product $k!\cdot l!\cdot r'!\cdot s'!\cdots$.

The sign of the term is easily deduced from Problem I, i.e. it will be positive or negative according as $\mu + 2\nu + 3\pi + 4\rho + \ldots$ is even or odd.

Example 1.1. Let the equation be $x^7 - 6x^6 + 9x^5 + 10x^4 - 45x^3 + 54x^2 - 29x + 16 = 0$, and say its roots are α, β, γ, δ, ϵ, etc.; required is the sum $\alpha^2\beta^3 + \beta^2\alpha^3 + \alpha^2\gamma^3 + \gamma^2\alpha^3 + \beta^2\gamma^3 + \gamma^2\beta^3 + \alpha^2\delta^3 + \delta^2\alpha^3 + \beta^2\delta^3 + \delta^2\beta^3 + \gamma^2\delta^3 + \delta^2\gamma^3 + \ldots$.

As the problem instructs we first compute

$$S2 = \alpha^2+\beta^2+\gamma^2+\delta^2+\epsilon^2+\ldots = p^2 - 2q = 18$$
$$S3 = \alpha^3+\beta^3+\gamma^3+\delta^3+\epsilon^3+\ldots = p^3 - 3pq + 3r = 24$$
$$S(2{+}3) = \alpha^5+\beta^5+\gamma^5+\delta^5+\epsilon^5+\ldots = p^5-5p^3q+5p^2r+5pq^2-5ps-5qr+5t = 216$$

In this case we have $A = S2 \times S3 = 18 \times 24 = 432$, $B = S(2{+}3) = 216$, and it follows that the required sum will be $A - B = 432 - 216 = 216$.

Example 1.2. Let the equation be $x^6 - 5x^5 + 4x^4 + 10x^3 - 11x^2 - 5x + 6 = 0$, whose roots are α, β, γ, δ, ϵ, ζ, to find the sum of the quantities $\alpha^2\beta^3\gamma^4 + \alpha^2\gamma^3\beta^4 + \beta^2\alpha^3\gamma^4 + \beta^2\gamma^3\alpha^4 + \gamma^2\alpha^3\beta^4 + \gamma^2\beta^3\alpha^4 + \alpha^2\beta^3\delta^4 + \delta^2\alpha^3\beta^4 + \delta^2\beta^3\alpha^4 + \ldots$. In this problem we must compute, first,

$$S2 = \alpha^2+\beta^2+\gamma^2+\delta^2+\epsilon^2+\zeta^2 = 17,$$
$$S3 = \alpha^3+\beta^3+\gamma^3+\delta^3+\epsilon^3+\zeta^3 = 35,$$
$$S4 = \alpha^4+\beta^4+\gamma^4+\delta^4+\epsilon^4+\zeta^4 = 101;$$

then

$$S(2+3) = \alpha^5+\beta^5+\gamma^5+\delta^5+\epsilon^5+\zeta^5 = 275,$$
$$S(2+4) = \alpha^6+\beta^6+\gamma^6+\delta^6+\epsilon^6+\zeta^6 = 797,$$
$$S(3+4) = \alpha^7+\beta^7+\gamma^7+\delta^7+\epsilon^7+\zeta^7 = 7315;$$

and lastly,

$$S(2+3+4) = \alpha^9+\beta^9+\gamma^9+\delta^9+\epsilon^9+\zeta^9 = 20195.$$

In consequence of these results we will have

$$A = S2 \times S3 \times S4 = 17 \times 35 \times 101 = 60095,$$
$$B = S(2{+}3) \times S4 + S(2{+}4) \times S3 + S(3{+}4) \times S2$$
$$= 275 \times 101 + 797 \times 35 + 2315 \times 17 = 95025,$$
$$C = S(2{+}3{+}4) = 20195;$$

and so by this problem the required sum will be $A - B + 2C = 5460$.

Case 2. With the same letters denoting the same quantities, now let $S_a = P$, $PS_b - S_{a+b} = Q$, $QS_c - (S_{c+a}S_b + S_{c+b}S_a) + 1 \cdot 2S_{a+b+c} = R$,

$$RS_d - (S_{d+a}S_bS_c + S_{d+b}S_aS_c + S_{d+c}S_aS_b) + 1 \cdot 2(S_{d+a+b}S_c + S_{d+a+c}S_b + S_{d+b+c}S_a) - 1 \cdot 2 \cdot 3S_{d+a+b+c} + 1 \cdot 1(S_{d+a}S_{b+c} + S_{d+b}S_{a+c} + S_{d+c}S_{a+b}) = T,$$

$$TS_e - (S_{e+a}S_bS_cS_d + S_{e+b}S_aS_cS_D + S_{e+c}S_aS_bS_d + S_{e+d}S_aS_bS_c)$$
$$+1 \cdot 2(S_{e+a+b}S_cS_d + S_{e+a+c}S_bS_d + S_{e+b+c}S_aS_d + S_{e+a+d}S_bS_c + S_{e+b+d}S_aS_c + S_{e+c+d}S_aS_b)$$
$$-1 \cdot 2 \cdot 3(S_{e+a+b+c}S_d + S_{e+a+b+d}S_c + S_{e+a+c+d}S_b + S_{e+b+c+d}S_a)$$

$$+1\cdot 1(S_{e+a}S_{b+c}S_d + S_{e+a}S_{b+d}S_c + S_{e+a}S_{c+d}S_b + S_{e+b}S_{a+c}S_d + S_{e+b}S_{a+d}S_c$$
$$+S_{e+b}S_{c+d}S_a + S_{e+c}S_{a+b}S_d + S_{e+c}S_{a+d}S_b + S_{e+c}S_{b+d}S_a + S_{e+d}S_{a+b}S_c$$
$$+S_{e+d}S_{a+c}S_b + S_{e+d}S_{b+c}S_a)$$
$$+1\cdot 2\cdot 3\cdot 4\cdot S_{e+a+b+c+d}$$
$$-1\cdot 1\cdot 2(S_{e+a}S_{b+c+d} + S_{e+b}S_{a+c+d} + S_{e+c}S_{a+b+d} + S_{e+d}S_{a+b+c}$$
$$+S_{e+a+b}S_{c+d} + S_{e+a+c}S_{b+d} + S_{e+b+c}S_{a+d} + S_{e+a+d}S_{b+c}$$
$$+S_{e+b+d}S_{a+c} + S_{e+c+d}S_{a+b}) \qquad = V,$$

and so on; then will be

$$P = \alpha^a + \beta^a + \gamma^a + \delta^a + \dots$$
$$Q = \alpha^a\beta^b + \beta^a\alpha^b + \alpha^a\gamma^b + \gamma^a\alpha^b + \beta^a\gamma^b + \dots$$
$$R = \alpha^a\beta^b\gamma^c + \alpha^a\beta^c\gamma^b + \alpha^b\beta^a\gamma^c + \dots$$
$$T = \alpha^a\beta^b\gamma^c\delta^d + \alpha^b\beta^a\gamma^c\delta^d + \alpha^a\beta^b\gamma^d\delta^c + \dots$$
$$V = \alpha^a\beta^b\gamma^c\delta^d\epsilon^e + \dots$$

and so for the rest.

The law governing this series is the same as the previous one, and by noting that the first terms PS_b, QS_c, RS_d, TS_e, etc., of this series are equal to the sum of the individual terms of the previous series, multiplied respectively by the sums S_b, S_c, S_d, S_e, etc. respectively, it is easily derived therefrom.

Case 3. It is often easier to derive the required sum from the coefficients of the given equation. Here follows a solution of this type to illustrate the method.

Let the index a be less than b, b less than c, c less than d, d less than e, etc., and let the number of indices be v, and let the given equation be $x^n - px^{n-1} + qx^{n-2} - rx^{n-3} + \dots \pm Px^{n-v+3} \mp Qx^{n-v+2} \pm Rx^{n-v+1} \mp Sx^{n-v} \pm \dots = 0$.

The first step of the solution is to find $S^a \times R^{b-a} \times Q^{c-b} \times P^{d-c} \cdots$; among these terms one particular product will be the required sum, while the remaining terms must be identified by the same method and then discarded.

Example 3.1. Given the equation $x^4 - px^3 + qx^2 - rx + s = 0$, whose roots are α, β, γ, δ, to find the symmetric sum $\alpha\beta^2\gamma^3 + \alpha\gamma^2\beta^3 + \beta\alpha^2\gamma^3 + \beta\gamma^2\alpha^2 + \gamma\alpha^2\beta^3 + \gamma\beta^2\alpha^3 + \alpha\delta^2\beta^3 + \beta\alpha^2\delta^3 + \dots$.

Here the indices a, b, c, are respectively 1, 2, 3; the number of indices is $v = 3$, thus $n - v = 4 - 3 = 1$; moreover $S = r$, $R = q$, $Q = p$, and $S^a \times R^{b-a} \times Q^{c-b} = rqp$. The product, resulting from multiplication of the factors $r \times q \times p = (\alpha\beta\gamma + \alpha\beta\delta + \alpha\gamma\delta + \beta\gamma\delta) \times (\alpha\beta + \alpha\gamma + \beta\gamma + \alpha\delta +$

$\beta\delta+\gamma\delta)\times(\alpha+\beta+\gamma+\delta)$, will be $\alpha\beta^2\gamma^3+\alpha\gamma^2\beta^3+\beta\alpha^2\gamma^3+\ldots+3(\alpha^2\beta^2\gamma^2+\alpha^2\beta^2\delta^2+\alpha^2\gamma^2\delta^2+\beta^2\gamma^2\delta^2)+8\alpha\beta\gamma\delta(\alpha\beta+\alpha\gamma+\beta\gamma+\alpha\delta+\beta\delta+\gamma\delta)(=8sq)+3\alpha\beta\gamma\delta(\alpha^2+\beta^2+\gamma^2+\delta^2)(=3sp^2-6sq)$. The first of the products, $\alpha\beta^2\gamma^3+\alpha\gamma^2\beta^3+\beta\alpha^2\gamma^3+\ldots$, is the required sum. The second product, $\alpha^2\beta^2\gamma^2+\alpha^2\beta^2\delta^2+\alpha^2\gamma^2\delta^2+\beta^2\gamma^2\delta^2$, must be evaluated by the same method: in this case the indices a, b, c are all equal to 2, and there are $v=3$ of them, whence $n-v=4-3=1$, and $S=r$, finally $S^a\times Q^{b-a}\times R^{c-b}=S^2Q^0R^0=S^2=r^2=(\alpha\beta\gamma+\alpha\beta\delta+\alpha\gamma\delta+\beta\gamma\delta)^2=\alpha^2\beta^2\gamma^2+\alpha^2\beta^2\delta^2+\alpha^2\gamma^2\delta^2+\beta^2\gamma^2\delta^2+2\alpha\beta\gamma\delta(\alpha\beta+\alpha\gamma+\alpha\delta+\beta\gamma+\beta\delta+\gamma\delta)$ $(=2sq)$, so we have $\alpha^2\beta^2\gamma^2+\alpha^2\beta^2\delta^2+\alpha^2\gamma^2\delta^2+\beta^2\gamma^2\delta^2=r^2-2qs$. Therefore the required sum $\alpha\beta^2\gamma^3+\alpha\gamma^2\beta^3+\beta\alpha^2\gamma^3+\beta\gamma^2\alpha^3+\ldots$ is $rqp-3(r^2-2qs)-8sq-3sp^2+6sq=rqp+rqs-3r^2-3sp^2$.

Example 3.2. Given an equation $x^n-px^{m-1}+qx^{n-2}-\ldots\mp Ox^{n-r+5}\pm Px^{n-r+4}\mp Qx^{n-r+3}\pm Rx^{n-r+2}\mp Sx^{n-r+1}\pm Tx^{n-r}\mp\ldots\pm ax^{n-m-r}\mp bx^{n-m-r-1}\pm cx^{n-m-r-2}\mp dx^{n-m-r-3}\pm ex^{n-m-r-4}\mp fx^{n-m-r-5}\pm\ldots=0$, whose roots are α, β, γ, δ, ϵ, ζ, η, θ, etc., to find the symmetric sum of quantities of the form $\alpha^2\beta^2\gamma^2\delta^2\cdots\epsilon\zeta\eta\cdots+\alpha^2\beta^2\gamma^2\epsilon^2\cdots\zeta\eta\theta\cdots+\alpha^2\beta^2\gamma^2\zeta^2\cdots\delta\eta\theta\cdots+\beta^2\gamma^2\delta^2\epsilon^2\cdots\zeta\eta\theta\cdots+$etc., in which the number of letters appearing to the second power is r, and the number of letters appearing to the first power only is m.

We begin by multiplying the coefficient T, of the term involving x^{n-r}, which is equal to $\alpha\beta\gamma\delta\epsilon\cdots+\alpha\beta\gamma\delta\zeta\cdots+\alpha\gamma\delta\zeta\eta\cdots+\beta\gamma\delta\epsilon\zeta\cdots+\ldots$ (in which each term is the product of r letters) by the coefficient a, of the term involving x^{n-m-r} in the given equation, which is $\alpha\beta\gamma\delta\epsilon\zeta\cdots+\alpha\beta\gamma\delta\epsilon\theta\cdots+\alpha\beta\gamma\epsilon\zeta\theta\cdots+\beta\gamma\delta\epsilon\zeta\theta\cdots+\ldots$ (in which each term is the product of $m+4r$ letters.) The product will be

$\alpha^2\beta^2\gamma^2\delta^2\cdots\epsilon\zeta\eta\cdots+\alpha^2\beta^2\gamma^2\epsilon^2\cdots\zeta\eta\theta\cdots+\alpha^2\beta^2\gamma^2\zeta^2\cdots\delta\eta\theta\cdots$
$+\beta^2\gamma^2\delta^2\epsilon^2\cdots\zeta\eta\theta\cdots+\ldots$ (the required sum)
$+(m+2)A+\frac{(m+3)(m+4)}{2!}B+\frac{(m+4)(m+5)(m+6)}{3!}C+\frac{(m+5)(m+6)(m+7)(m+8)}{4!}D+\frac{(m+6)(m+7)(m+8)(m+9)(m+10)}{5!}E+\ldots$, to r terms (A, B, C, D, E, etc.) where

$$\begin{aligned}
A = {} & \alpha^2\beta^2\gamma^2\cdots \text{ (to } r-1 \text{ terms)}\times\delta\epsilon\zeta\cdots \text{ (to } m+2 \text{ terms)}\\
+ {} & \alpha^2\beta^2\delta^2\cdots \text{ (to } r-1 \text{ terms)}\times\gamma\epsilon\zeta\cdots \text{ (to } m+2 \text{ terms)}\\
+ {} & \beta^2\gamma^2\delta^2\cdots\zeta\eta\theta\cdots+\text{ etc.}\\
B = {} & \alpha^2\beta^2\cdots \text{ (to } r-2 \text{ terms)}\times\gamma\delta\epsilon\zeta\cdots \text{ (to } m+4 \text{ terms)}\\
+ {} & \alpha^2\gamma^2\cdots \text{ (to } r-2 \text{ terms)}\times\beta\delta\epsilon\zeta\cdots \text{ (to } m+4 \text{ terms)}\\
+ {} & \beta^2\gamma^2\cdots\alpha\delta\zeta\eta\cdots+\text{ etc.}
\end{aligned}$$

$$\begin{aligned}
C = {} & \alpha^2\beta^2 \cdots \text{ (to } r-3 \text{ terms)} \times \gamma\delta\epsilon\zeta\eta \cdots \text{ (to } m+6 \text{ terms)} \\
+ {} & \alpha^2\gamma^2 \cdots \text{ (to } r-3 \text{ terms)} \times \beta\delta\epsilon\zeta\eta \cdots \text{ (to } m+6 \text{ terms)} \\
+ {} & \beta^2\gamma^2 \cdots \alpha\delta\zeta\eta\theta \cdots + \text{ etc.} \\
D = {} & \alpha^2\beta^2 \cdots \text{ (to } r-4 \text{ terms)} \times \gamma\delta\epsilon\zeta\eta \cdots \text{ (to } m+8 \text{ terms)} \\
+ {} & \alpha^2\gamma^2 \cdots \text{ (to } r-4 \text{ terms)} \times \beta\delta\epsilon\zeta\eta \cdots \text{ (to } m+8 \text{ terms)} \\
+ {} & \beta^2\gamma^2 \cdots \alpha\delta\zeta\eta\theta \cdots + \text{ etc.} \\
\text{etc.} = {} & \text{etc.}
\end{aligned}$$

The terms A, B, C, D, etc. are deleted by writing for them their values $Sb - \left((m+4)B + \frac{(m+5)(m+6)}{2!}C + \frac{(m+6)(m+7)(m+8)}{3!}D + \ldots\right)$, $Rc - \left((m-6)C + \frac{(m+7)(m+8)}{2!}D + \ldots\right)$, $Qd - \left((m-8)D + \frac{(m-9)(m-10)}{2!}E + \ldots\right)$, $Pe - \left((m+10)E + \ldots\right)$, etc., yielding the required sum $Ta - (m+2)Sb + \frac{(m+1)(m+4)}{2!}Rc - \frac{(m+1)(m+2)(m+6)}{3!}Qd + \frac{(m+1)(m+2)(m+3)(m+8)}{4!}Pe - \frac{(m+1)(m+2)(m+3)(m+4)(m+10)}{5!}Of + \ldots$ to a total of $r+1$ terms.

In general the coefficient of a term at distance s from the first will be

$$\frac{m+2s}{s} \times (m+1) \times \frac{m+2}{2} \times \frac{m+3}{3} \times \frac{m+4}{4} \times \cdots \times \frac{m+s-1}{s-1}.$$

Corollary 1. Let a be the sum of the roots, b the sum of the squares of the roots, c the sum of their cubes, d the sum of their biquadrates, e the sum of their quadrato-cubes, f the sum of their cubo-cubes, etc. Then the sum of products of m roots at a time will be equal to a fraction whose numerator is

$$a^m - \frac{m(m-1)}{2}a^{m-2}b + \frac{m(m-1)(m-2)}{3}a^{m-3}c - \frac{m(m-1)(m-2)(m-3)}{4}a^{m-4}d + \ldots$$
$$+ \frac{m(m-1)(m-2)(m-3)}{2\cdot 2^2}a^{m-4}b^2 - \ldots$$

and whose denominator is $1 \cdot 2 \cdot 3 \cdot 4 \cdots m$.

If for the letters a, b, c, d, etc., we substitute the letters p, r, s, t, etc., respectively, then the literal products in this series are the same as in Problem I.

The coefficient of the literal product $a^{m-l}b^{\pi}c^{\rho}d^{\sigma} \cdots$ will be equal to a fraction, whose numerator is $m(m-1)(m-2)(m-3)(m-4)\cdots$ (in which the number of factors is taken from the sum $2\pi + 3\rho + 4\sigma + \ldots$), and whose denominator is the product $2^{\pi}3^{\rho}4^{\sigma}5^{\tau}\cdots$ multiplied in turn by $\pi\,!$, $\rho\,!$, $\sigma\,!$, etc.

It is easy to establish this corollary from the problem, if for the quantity $BB\ldots BCC\ldots CDD\ldots D\ldots$ (in which there are π occurrences of B, ρ C's, σ D's, and so on) we write its value, the quantity

$$\frac{1}{\pi!} \times \frac{1}{\rho!} \times \frac{1}{\sigma!} \cdots \frac{1}{2!^{\pi}} \cdot \frac{1}{3!^{\rho}} \times \frac{1}{4!^{\sigma}} \times \cdots$$
$$\times m(m-1)(m-2)(m-3)\cdots(m-2\pi-3\rho-4\sigma-\ldots+1)a^{m-l}b^{\pi}c^{\rho}d^{\sigma}\cdots;$$
each such quantity of the first type is to be replaced by the latter form. It then follows from the observation that a set of dissimilar objects α, β, γ, δ, ϵ, ζ, η, θ, ι, κ, λ, μ, ν, o, υ, ψ, χ, etc. (which are m in number) can be arranged in exactly $\frac{1}{\pi! \cdot \rho! \cdot \sigma!} \times \frac{1}{2!^{\pi} \cdot 3!^{\rho} \cdot 4!^{\sigma} \cdots}$
$$\times m(m-1)(m-2)(m-3)\cdots(m-2\pi-3\rho-4\sigma-\ldots+1)$$
distinct ways into sets $\alpha+\beta$, $\gamma+\delta,\ldots$ (with π terms in each), $\epsilon+\zeta+\eta$, $\theta+\iota+\kappa,\ldots$ (with ρ terms in each), $\mu+\nu+o+\upsilon,\ldots$ (with σ terms in each), and so on.

This quantity is divided by $m!$, for exactly that many different values of products of single ones of the m roots will be equal to each other.

This corollary can also be demonstrated in this way: Let P, Q, R, S, etc. be the coefficients of the terms $a^{m-l-1}b^{\pi}c^{\rho}d^{\sigma}\cdots$, $a^{m-l}b^{\pi-1}c^{\rho}d^{\sigma}\cdots$, $a^{m-l}b^{\pi}c^{\rho-1}d^{\sigma}\cdots$, $a^{m-l}b^{\pi}c^{\rho}d^{\sigma-1}\cdots$, etc. respectively, in the sum of the products of $m-1$, $m-2$, $m-3$, $m-4$, etc. roots deduced by this corollary. It will easily follow, from Newton's rule as given earlier, that the coefficient V of the term $a^{m-l}b^{\pi}c^{\rho}d^{\sigma}\cdots$, deduced from the sum of the products of m roots, will be $V = P+(m-1)Q+(m-1)(m-2)R$ $+(m-1)(m-2)(m-3)S+(m-1)(m-2)(m-3)(m-4)T+\ldots$; but by the corollary the coefficients P, Q, R, S, etc. will be

$$(m-l)\times\frac{(m-1)(m-2)(m-3)\cdots(m-l+1)}{\pi!\cdot\rho!\cdot\sigma!\cdots2^{\pi}\cdot3^{\rho}\cdot4^{\sigma}\cdots},$$
$$2\pi\times\frac{(m-2)(m-3)\cdots(m-l+1)}{\pi!\cdot\rho!\cdot\sigma!\cdots2^{\pi}\cdot3^{\rho}\cdot4^{\sigma}\cdots},$$
$$3\rho\times\frac{(m-3)(m-4)\cdots(m-l+1)}{\pi!\cdot\rho!\cdot\sigma!\cdots2^{\pi}\cdot3^{\rho}\cdot4^{\sigma}\cdots},$$
$$4\sigma\times\frac{(m-4)(m-5)\cdots(m-l+1)}{\pi!\cdot\rho!\cdot\sigma!\cdots2^{\pi}\cdot3^{\rho}\cdot4^{\sigma}\cdots},$$

etc., and in consequence the aggregate sum $P+(m\text{-}1)Q+(m\text{-}1)(m\text{-}2)R$ $+(m\text{-}1)(m\text{-}2)(m\text{-}3)S+\ldots$ will be

$$(m-l+2\pi+3\rho+4\sigma+\ldots)$$
$$\times\frac{(m-1)(m-2)(m-3)(m-4)\cdots(m-l+1)}{\pi!\cdot\rho!\cdot\sigma!\cdots2^{\pi}\cdot3^{\rho}\cdot4^{\sigma}\cdots}.$$

In the first factor of this aggregate substitute for l its value $2\pi+3\rho+4\sigma+\ldots$; the result will be the aggregate

$$m \times \frac{(m-1)(m-2)(m-3)\cdots(m-l+1)}{\pi!\,\rho!\,\sigma!\,\cdots\,2^{\pi}\,3^{\rho}\,4^{\sigma}\,\cdots}$$

which is exactly the coefficient, or uncia, which is to be substituted, according to the given series, as the coefficient of the literal product $a^{m-l}b^{\pi}c^{\rho}d^{\sigma}\cdots$ in the representation under construction of the sum of the products of each set of m roots. Therefore the prescribed series, when summed, correctly derives from the lower degree products (those of $m-1$, $m-2$, $m-3$, etc., roots) the next higher sum, of all products of m roots at a time.

From this corollary, conversely, making use also of the previous problems, the present problem easily follows.

Corollary 2. Let m, n, r, s, etc. be any given positive integers, then

$$\begin{aligned}
1 &- \left(m\cdot\tfrac{m-1}{2} + n\cdot\tfrac{n-1}{2} + r\cdot\tfrac{r-1}{2} + s\cdot\tfrac{s-1}{2} + \ldots\right)\\
&+ \left(m(m-1)\tfrac{m-2}{3} + n(n-1)\tfrac{n-2}{3} + \ldots\right)\\
&+ \left(m(m-1)(m-2)\tfrac{m-3}{4} + n(n-1)(n-2)\tfrac{n-3}{4} + \ldots\right)\\
&+ \tfrac{1}{2}\left(m\tfrac{m-1}{2}(m-2)\tfrac{m-3}{2} + n\tfrac{n-1}{2}(n-2)\tfrac{n-3}{2} + 2m\tfrac{m-1}{2}\cdot n\cdot\tfrac{n-1}{2} + \ldots\right)\\
&+ \ \ldots\\
&= (m-1)!\,(n-1)!\,(r-1)!\,(s-1)!\,\cdots,
\end{aligned}$$

that is, for terms of the form

$$\begin{aligned}
&m(m-1)\cdots(m-x+2)\frac{m-x+1}{x}\\
&\times(m-x)(m-x-1)\cdots(m-x-x'+2)\frac{m-x-x'+1}{x'}\\
&\times(m-x-x')(m-x-x'-1)\cdots(m-x-x'-x''+2)\frac{m-x-x'-x''+1}{x''}\times\cdots,\\
&n(n-1)\cdots(n-z+2)\frac{n-z+1}{z}\\
&\times(n-z)(n-z-1)\cdots(n-z-z'+2)\frac{n-z-z'+1}{z'}\\
&\times(n-z-z')(n-z-z'-1)\cdots(n-z-z'-z''+2)\frac{n-z-z'-z''+1}{z''}\times\cdots,\\
&r(r-1)\cdots(r-w+2)\frac{r-w+1}{w}\\
&\times(r-w)(r-w-1)\cdots(r-w-w'+2)\frac{r-w-w'+1}{w'}\\
&\times(r-w-w')(r-w-w'-1)\cdots(r-w-w'-w''+2)\frac{r-w-w'-w''+1}{w''}\times\cdots,\\
&s(s-1)\cdots(s-v+2)\frac{s-v+1}{v}\times\cdots,
\end{aligned}$$

etc., where x, x', x'', …, z, z', z'', …, w, w', w'', …, v, v', …, etc., are arbitrary positive integers, the sum of all such products (including terms having a single factor) will be $(m-1)!\cdot(n-1)!\cdot(r-1)!\cdot(s-1)!\cdots$, where we affix a + or - sign to each term according as the sum of $(x-1)+(x'-1)+(x''-1)+\ldots+(z-1)+(z'-1)+(z''-1)+\ldots+(w-1)+\ldots$ (for the letters existing in that term) is an even or an odd number. This is easily deduced from the problem.

Problem IV

Given an equation, whose root is x, *to find the sum of the individual values of any algebraic function of each of its roots.*

Let α, β, γ, δ, etc. be the roots x in the given equation. In the given function write for x its values α, β, γ, δ, etc., and from these individual values find the resulting quantity. Then

Case 1. If the functions are rational and integral, the sum of the individual values can be constructed by the method given above.

Case 2. If the functions are rational fractions, then all the fractions should be reduced to a common denominator, whereupon by the above method both the numerator and the denominator may be discovered.

Example 2.1. Let the equation be $x^3 - 7x + 6 = 0$, its roots α, β, γ, and the fractional function $\frac{3x^2+5}{4x+1}$.

In the given function writing for x its values, we obtain the fractions

$$\frac{3\alpha^2+5}{4\alpha+1},\quad \frac{3\beta^2+5}{4\beta+1},\quad \frac{3\gamma^2+5}{4\gamma+1}.$$

When these fractions are reduced to a common denominator, their sum becomes

$$\frac{\left\{\begin{array}{l}48\alpha\beta\gamma(\alpha+\beta+\gamma)+12(\alpha^2\beta+\alpha^2\gamma+\beta^2\alpha+\beta^2\gamma+\gamma^2\alpha+\gamma^2\beta\\ +3(\alpha^2+\beta^2+\gamma^2)+80(\alpha\beta+\alpha\gamma+\beta\gamma)+40(\alpha+\beta+\gamma)+15\end{array}\right\}}{64\alpha\beta\gamma+16(\alpha\beta+\alpha\gamma+\beta\gamma)+4(\alpha+\beta+\gamma)+1}.$$

Now for p, q, r substitute 0, -7, -6 respectively, making the denominator $64\alpha\beta\gamma(=64r=64\times -6)+16(\alpha\beta+\alpha\gamma+\beta\gamma)(=16q=16\times -7)+4(\alpha+\beta+\gamma)(=4p=0)+1=-495$, and the numerator is $48\alpha\beta\gamma(\alpha+\beta+\gamma)$ $(=48rp=0)+12(\alpha^2\beta+\alpha^2\gamma+\beta^2\alpha+\beta^2\gamma+\gamma^2\alpha+\gamma^2\beta)(=12(pq-3r)$ $=12\times 3\times 6)+3(\alpha^2+\beta^2+\gamma^2)(=3(p^2-2q)=3\times 2\times 7)+80(\alpha\beta+\alpha\gamma+\beta\gamma)$ $(=80q=80\times -7)+40(\alpha+\beta+\gamma)(=40p=0)+15=-287$. Therefore the required sum is 287/495.

Example 2.2. Let α, β, γ, δ, $\epsilon, \ldots$ be the roots of a given equation $x^n - px^{n-1} + qx^{n-2} - rx^{n-3} + sx^{n-4} - \ldots = 0$, and let the fraction be $\dfrac{a - bx + cx^2 - dx^3 + \ldots}{A - Bx + Cx^2 - Dx^3 + \ldots}$, in which the coefficients $a, b, c, \ldots, A, B, C, \ldots$ are given.

We must find the sum of the fractions

$$\frac{a - b\alpha + c\alpha^2 - d\alpha^3 + \ldots}{A - B\alpha + C\alpha^2 - D\alpha^3 + \ldots}, \quad \frac{a - b\beta + c\beta^2 - d\beta^3 + \ldots}{A - B\beta + C\beta^2 - D\beta^3 + \ldots},$$
$$\frac{a - b\gamma + c\gamma^2 - d\gamma^3 + \ldots}{A - B\gamma + C\gamma^2 - D\gamma^3 + \ldots}, \quad \text{etc.}$$

Once these fractions are reduced to a common denominator, by means of Problems I and III we can find the numerator and denominator of the fraction which makes up their sum.

Case 3. If the functions are irrational, then the sum of the irrationalities in the individual terms will vanish, if the fractions are first reduced to a common denominator.

Example 3.1. Let the equation be $x^7 - 4x^6 + 3x^5 + 10x^4 - 25x^3 + 24x^2 - 11x + 12 = 0$, whose roots are α, β, γ, δ, ϵ, etc.; and to find the sum of the individual values of the following irrational function of the roots:

$$\alpha + \sqrt{\alpha^2+\beta^2},\ \beta + \sqrt{\alpha^2+\beta^2},\ \alpha + \sqrt{\alpha^2+\gamma^2},\ \gamma + \sqrt{\alpha^2+\gamma^2},\ \ldots$$
$$\alpha - \sqrt{\alpha^2+\beta^2},\ \beta - \sqrt{\alpha^2+\beta^2},\ \alpha - \sqrt{\alpha^2+\gamma^2},\ \gamma - \sqrt{\alpha^2+\gamma^2},\ \ldots$$

etc.

The sum of the irrational quantities will always be 0; therefore the sum of the whole expression will be $x \times 6 \times (\alpha + \beta + \gamma + \delta + \epsilon + \ldots) = 12 \times 4 = 48$.

Example 3.2. Let the equation be $x^3 - px^2 + qx - r = 0$, and its roots α, β, γ; to find the sum of the irrational fractions

$$\frac{\alpha}{\alpha + \sqrt{\alpha^2 + 2\alpha}}, \quad \frac{\alpha}{\alpha - \sqrt{\alpha^2 + 2\alpha}}, \quad \frac{\beta}{\beta + \sqrt{\beta^2 + 2\beta}},$$
$$\frac{\beta}{\beta - \sqrt{\beta^2 + 2\beta}}, \quad \frac{\gamma}{\gamma + \sqrt{\gamma^2 + 2\gamma}}, \quad \frac{\gamma}{\gamma - \sqrt{\gamma^2 + 2\gamma}}.$$

Reduced to their common denominator, these fractions become

$$\frac{4\beta\gamma\alpha \times \alpha - 4\beta\gamma\alpha\sqrt{\alpha^2 + 2\alpha}}{-8\beta\gamma\alpha} = \frac{\alpha - \sqrt{\alpha^2 + 2\alpha}}{-2},$$
$$\frac{4\beta\gamma\alpha \times \alpha + 4\beta\gamma\alpha\sqrt{\alpha^2 + 2\alpha}}{-8\beta\gamma\alpha} = \frac{\alpha + \sqrt{\alpha^2 + 2\alpha}}{-2},$$

$$\frac{4\alpha\gamma\beta\times\beta-4\alpha\gamma\beta\sqrt{\beta^2+2\beta}}{-8\alpha\gamma\beta}=\frac{\beta-\sqrt{\beta^2+2\beta}}{-2},$$
$$\frac{4\alpha\gamma\beta\times\beta+4\alpha\gamma\beta\sqrt{\beta^2+2\beta}}{-8\alpha\gamma\beta}=\frac{\beta+\sqrt{\beta^2+2\beta}}{-2},$$
$$\frac{4\alpha\beta\gamma\times\gamma-4\alpha\beta\gamma\sqrt{\gamma^2+2\gamma}}{-8\alpha\beta\gamma}=\frac{\gamma-\sqrt{\gamma^2+2\gamma}}{2},$$
$$\frac{4\alpha\beta\gamma\times\gamma+4\alpha\beta\gamma\sqrt{\gamma^2+2\gamma}}{-8\alpha\beta\gamma}=\frac{\gamma+\sqrt{\gamma^2+2\gamma}}{2},$$

and it follows that the required sum will be $-(\alpha+\beta+\gamma)=-p$.

Case 4. The same method of finding the sum of the individual values of any function of the roots of a given equation applies, when more than one value α, β, γ, etc., of the root x, is contained in the given function.

In the said function all n different roots are substituted for each value of the quantity x, and thus are found the individual values of the prescribed function. Then by the methods previously disclosed, these values are combined to form the required aggregate.

Perhaps it is not inappropriate to elaborate on this matter now.

In the given function are contained, say, m roots of the given equation. Let π be a function within the given function, which involves only r distinct roots among the m of which we have spoken, and which cannot be reduced further. We must determine how many (say h) distinct roots there may be in the remaining parts of the given function, also how many (i.e. $(n-r)(n-r-1)(n-r-2)(n-r-3)\cdots(n-m+1)=K$) different combinations of $m-r$ roots are contained in the given function. Then in the required sum we multiply the aggregate Σ of the individual values of the function π by the product hK, for the value of the function π will appear this many times in it; and similarly for the rest.

But if two of the $m-r$ roots enter symmetrically in the given function, then the product $hk\Sigma$ must be divided by $1\cdot 2$; if three, then the product must be divided by $1\cdot 2\cdot 3$; if four, by $1\cdot 2\cdot 3\cdot 4$; and so on.

Example 4.1. Let the given function be $7\alpha+11\beta+\frac{3\gamma}{\delta}+\sqrt{4+\alpha\beta+\gamma^2}$, where α, β, γ, and δ denote different roots of the equation $x^n-px^{n-1}+qx^{n-2}-rx^{n-3}+\ldots pmRx^2\mp Qx\pm P=0$. The symmetric sum of the terms $\sqrt{4+\alpha\beta+\gamma^2}$ will be equal to zero. The aggregate of the values of the function $7\alpha+11\beta+\frac{3\gamma}{\delta}$ will be

$$\begin{aligned}&7\times(n-1)(n-2)(n-3)\times(\alpha+\beta+\gamma+\delta+\ldots)\\ &+11\times(n-1)(n-2)(n-3)\times(\alpha+\beta+\gamma+\delta+\ldots)\\ &+\ 7\times(n-2)(n-3)\times(\tfrac{\alpha}{\beta}+\tfrac{\beta}{\alpha}+\tfrac{\alpha}{\gamma}+\tfrac{\gamma}{\alpha}+\tfrac{\beta}{\gamma}+\tfrac{\gamma}{\beta}+\tfrac{\alpha}{\delta}+\tfrac{\delta}{\alpha}+\tfrac{\beta}{\delta}+\tfrac{\delta}{\beta}+\ldots)\\ &=(11+7=18)(n-1)(n-2)(n-3)p+3(n-2)(n-3)\frac{pQ-nP}{P}.\end{aligned}$$

In fact this sum must be taken twice, since there are two values of the irrational term $\sqrt{4+\alpha\beta+\gamma^2}$, and in consequence the required sum is

$$36\ (n-1)(n-2)(n-3)p+6(n-2)(n-3)\frac{pQ-nP}{P}.$$

Case 5*.[3] Let α, β, γ, δ, ... be roots of the equation $U\equiv x^n-px^{n-1}+qx^{n-2}-rx^{n-3}+\ldots=0$, and A, B, Γ, Δ,... be roots of the equation $(W=0)z^m-Pz^{m-1}+Qz^{m-2}-Rz^{m-3}+\ldots\pm Tz\mp V=0$, and likewise for still more equations; to find the sum of the individual values of any algebraic function of the abovementioned roots.

To use the preceding problem to find the sum of the individual values of the given function, I regard all roots of every equation except the first as given quantities; and calling the resulting expression a, I must now find the sum of all the individual values of a: wherein I regard all roots of every equation except the second as given quantities, calling the resulting expression b; and so for all the rest. The quantity resulting from the final step will be the required sum.

Example 5.1. To find the sum of the values of the function

$$\begin{aligned}&(\alpha^2\mathrm{A}+3\alpha\mathrm{BA})(\alpha\mathrm{A}+7\beta\Gamma)\\ =\ &\alpha^3\mathrm{A}^2+3\alpha^2\mathrm{BA}^2+7\alpha^2\beta\mathrm{A}\Gamma+42\alpha\beta\mathrm{AB}\Gamma\end{aligned}$$

where α, β, A, B, Γ follow the notational conventions of this problem.

The aggregate is to be found from the individual values of this function. I treat A, B, Γ as given quantities. Then by this assumption the sum of the individual values of this function will be

$$\begin{aligned}&(n-1)\ (\alpha^3+\beta^3+\gamma^3+\ldots)\times\ \mathrm{A}^2\\ &+3(n-1)\ (\alpha^2+\beta^2+\gamma^2+\ldots)\times\ \mathrm{BA}^2\\ &+7\ (\alpha^2\beta+\beta^2\alpha+\alpha^2\gamma+\gamma^2\alpha+\beta^2\gamma+\gamma^2\beta+\ldots)\times\ \mathrm{A}\Gamma\\ &+\underbrace{21\cdot 2!}\ (\alpha\beta+\alpha\gamma+\beta\gamma+\alpha\delta+\beta\delta+\gamma\delta+\ldots)\times\ \mathrm{AB}\Gamma\end{aligned}$$

[because in this term α and β enter symmetrically]

$$\begin{aligned}&=(n-1)\ (p^3-3pq+3r)\mathrm{A}^2+3(n-1)(p^2-2q)\mathrm{BA}^2\\ &+7\ (pq-3r)\mathrm{A}\Gamma+42(q)\mathrm{AB}\Gamma\qquad =a\,;\end{aligned}$$

following this, taking all quantities except the roots A, B, Γ of the equation $W = 0$ as given, from this hypothesis we are to find the separate values of the quantity a, and their sum

$$\begin{aligned}
&(m-1)(m-2)\ (n-1)(p^3-3pq+3r) \times (\mathrm{A}^2+\mathrm{B}^2+\Gamma^2+\ldots)\\
&\quad +3(m-2)\ (n-1)(p^2-2q) \times (\mathrm{A}^2\mathrm{B}+\mathrm{B}^2\mathrm{A}+\mathrm{A}^2\Gamma+\Gamma^2\mathrm{A}+\mathrm{B}^2\Gamma+\Gamma^2\mathrm{B}+\ldots)\\
&+7\underbrace{\cdot 2!}(m-2)\quad (pq-3r) \times (\mathrm{AB}+\mathrm{A}\Gamma+\mathrm{B}\Gamma+\ldots)
\end{aligned}$$

[because A and B enter symmetrically]

$$+42\underbrace{\cdot 3!}\quad q \quad \times (\mathrm{AB}\Gamma+\ldots)$$

[because here A, B, and Γ enter symmetrically]

$$\begin{aligned}
&= (m-1)(m-2)\ (n-1)(p^3-3pq+3r)(P^2-2Q)\\
&\quad +3(m-2)\ (n-1)(p^2-2q)(PQ-3R)\\
&\quad +14(m-2)\quad (pq-3r)\,Q\\
&\quad +252\quad q\quad R
\end{aligned}$$

will be the required aggregate.

<u>Example 5.2.</u> With the same conventions, let the given function be $\alpha + 2\mathrm{A} + \dfrac{7\beta}{\mathrm{B}} + 5\sqrt{\mathrm{A}^2+\gamma^2}$. On the first pass take A and B to be the given quantities. By case 3 of the previous problem we can see that the sum of all the values of the irrational quantities $\sqrt{\mathrm{A}^2+\gamma^2}$ will vanish. By that hypothesis the symmetric sum of all values of the function $\alpha + 2\mathrm{A} + \dfrac{7\beta}{\mathrm{B}} + 5\sqrt{\mathrm{A}^2+\gamma^2}$ will be

$$\begin{aligned}
&2m(n-1)(n-2)(\alpha+\beta+\gamma+\delta+\ldots) + 4n(n-1)(n-2)\mathrm{A}\\
&+2\cdot 7\cdot(n-1)(n-2)\,\frac{\alpha+\beta+\gamma+\delta+\ldots}{\mathrm{B}}\\
&\quad = 2(n-1)(n-2)p + 4n(n-1)(n-2)\mathrm{A} + 2\cdot 7(n-1)(n-2)\frac{p}{\mathrm{B}} = a.
\end{aligned}$$

Next all quantities included in the expression for a are taken as being given, except for A and B; then the symmetric sum of the values of the quantity a will be

$$\begin{aligned}
&2m\,(m-1)(n-1)(n-2)p+4(m-1)n(n-1)(n-2)(\mathrm{A}+\mathrm{B}+\Gamma+\Delta+\ldots)\\
&+2\cdot 7\cdot(m-1)(n-1)(n-2)\Big(\frac{p}{\mathrm{A}}+\frac{p}{\mathrm{B}}+\frac{p}{\Gamma}+\frac{p}{\Delta}+\ldots\Big)\\
&= 2m\,(m-1)(n-1)(n-2)p+4(m-1)n(n-1)(n-2)P\\
&+2\cdot 7\cdot(m-1)(n-1)(n-2)p\cdot\frac{T}{V}.
\end{aligned}$$

Corollary. In this manner one can also investigate the sum of the values of an infinite series in the powers of the roots of a given equation, so long as the series itself can be expressed as an algebraic function of the roots.

We have to find an algebraic function of a root, which represents the infinite series. By the present problem we can then find the sum of each of its values, and in consequence the symmetric sum of the infinite power series.

Example 5.3. Let the equation be $x^n - px^{n-1} + qx^{n-2} - rx^{n-3} + sx^{n-4} - tx^{n-5} + \ldots = 0$, whose roots are α, β, γ, δ, $\epsilon, \ldots$, and let a be the sum of the roots, b the sum of the squares of the roots, c the sum of their cubes, d the sum of their biquadrates, e the sum of the quadrato-cubes, and so on without limit; required the sum $a+b+c+d+e+\ldots$, *ad infinitum*. This infinite series can be expressed as the sum of the fractions $\frac{1}{1-\alpha}+\frac{1}{1-\beta}+\frac{1}{1-\gamma}+\frac{1}{1-\delta}+\frac{1}{1-\epsilon}+\ldots-n$;
so we must find an equation whose root is $v = \frac{1}{1-x}$, and the sum of the roots of the resulting equation will be

$$\frac{1}{1-\alpha}+\frac{1}{1-\beta}+\frac{1}{1-\gamma}+\frac{1}{1-\delta}+\frac{1}{1-\epsilon}+\ldots$$
$$=\frac{n-(n-1)p+(n-2)q-(n-3)r+(n-4)s-(n-5)t+\ldots}{1-p+q-r+s-t+\ldots};$$

in consequence the required sum is

$$\frac{1}{1-\alpha}+\frac{1}{1-\beta}+\frac{1}{1-\gamma}+\frac{1}{1-\delta}+\frac{1}{1-\epsilon}+\ldots-n=\frac{p-2q+3r-4s+5t-\ldots}{1-p+q-r+s-t+\ldots}.$$

Example 5.4. To find the sum $a+2b+3c+4d+5e+\ldots$, taken to infinity. This series may be viewed as the difference between two series of fractions,

$$\frac{1}{(1-\alpha)^2}+\frac{1}{(1-\beta)^2}+\frac{1}{(1-\gamma)^2}+\frac{1}{(1-\delta)^2}+\frac{1}{(1-\epsilon)^2}+\ldots$$
$$-\frac{1}{1-\alpha}-\frac{1}{1-\beta}-\frac{1}{1-\gamma}-\frac{1}{1-\delta}-\frac{1}{1-\epsilon}-\ldots$$

whence the required sum will be

$$\frac{\left(n-(n-1)p+(n-2)q-(n-3)r+(n-4)s-(n-5)t+\ldots\right)^2}{(1-p+q-r+s-t+\ldots)^2}$$
$$-\frac{n(n-1)-(n-1)(n-2)p+(n-2)(n-3)q-(n-3)(n-4)r+(n-4)(n-5)s-\ldots}{1-p+q-r+s-t+\ldots}$$
$$-\frac{n-(n-1)p+(n-2)q-(n-3)r+(n-4)s-(n-5)t+\ldots}{1-p+q-r+s-t+\ldots}.$$

Problem V

Given one, two, or more equations, to find an equation whose roots are any algebraic function of the roots of the given equations.

We shall write α, β, γ, δ, etc. for the roots of one given equation; A, B, Γ, Δ, etc. for the roots of a second given equation; and similarly for any others. From the given algebraic relation between the roots of the given equation(s) and the roots of the required equation, all the roots of the required equation are determined, along with the number of them (which is equivalent to the degree of the required equation). Then by the previous problems can be found the sum of all individual values of the required equation's roots, which will determine its second coefficient.

Then each two roots of the required equation are multiplied together, and by the previous problems the sum of the resulting products is found; this will be the third coefficient of the required equation.

In like manner every three, four, five, etc. roots of the required equation are multiplied together, and by the same problems the sums of the resulting products are found, determining respectively the fourth, fifth, sixth, etc. terms of the required equation.

This is plainly a method sufficient to find the coefficients of the required equation, and hence to deduce the entire equation.

Problems of this sort may often be resolved by reducing two or more equations to one, in order to exterminate some of the unknown quantities.

Example 1. With α, β, γ, δ, etc. the roots of the given equation $x^n - px^{n-1} + qx^{n-2} - rx^{n-3} + sx^{n-4} - \ldots = 0$, to find the equation $v^n - Pv^{n-1} + Qv^{n-2} - Rv^{n-3} + Sv^{n-4} - \ldots = 0$, whose roots are α^m, β^m, γ^m, δ^m, etc., respectively, i.e. the mth powers of the roots of the given equation. We must write

$$\begin{aligned}
a &= \alpha^m + \beta^m + \gamma^m + \delta^m + \ldots \\
b &= \alpha^{2m} + \beta^{2m} + \gamma^{2m} + \delta^{2m} + \ldots \\
c &= \alpha^{3m} + \beta^{3m} + \gamma^{3m} + \delta^{3m} + \ldots \\
d &= \alpha^{4m} + \beta^{4m} + \gamma^{4m} + \delta^{4m} + \ldots \\
\text{etc.} &= \text{etc.}
\end{aligned}$$

The coefficients of the required equation may be found by Corollary 1, Problem III. They will be, respectively,

$$P = \alpha^m + \beta^m + \gamma^m + \delta^m + \ldots \qquad = a$$

$$Q = \alpha^m\beta^m + \alpha^m\gamma^m + \beta^m\gamma^m + \alpha^m\delta^m + \beta^m\delta^m + \gamma^m\delta^m + \ldots \quad = \frac{a^2 - b}{2}$$

$$R = \alpha^m\beta^m\gamma^m + \alpha^m\gamma^m\delta^m + \alpha^m\gamma^m\delta^m + \beta^m\gamma^m\delta^m + \ldots = \frac{a^3 - 3ab + 2c}{6}$$

$$S = \alpha^m\beta^m\gamma^m\delta^m + \ldots \qquad = \frac{a^4 - 6a^2b + 8ac - 6d + 3b^2}{24}$$

and so on for the rest of the coefficients.

We can deduce a second solution of this example from Problem III: Let h, k, l, o, etc. represent the m distinct m^{th} roots of unity, e.g. for $m = 3$ h, k, l will be respectively 1, $\dfrac{-1+\sqrt{-3}}{2}$, $\dfrac{-1-\sqrt{-3}}{2}$.

By hypothesis $x^m = v$, and consequently $hv^{1/m} = x$, $kv^{1/m} = x$, $lv^{1/m} = x$, $ov^{1/m} = x$, etc.

When these values are substituted for x in the given equation, there result the equations

$$h^n v^{n/m} - ph^{n-1}v^{(n-1)/m} + qh^{n-2}v^{(n-2)/m} - rh^{n-3}v^{(n-3)/m} + \ldots = 0$$
$$k^n v^{n/m} - pk^{n-1}v^{(n-1)/m} + qk^{n-2}v^{(n-2)/m} - rk^{n-3}v^{(n-3)/m} + \ldots = 0$$
$$l^n v^{n/m} - pl^{n-1}v^{(n-1)/m} + ql^{n-2}v^{(n-2)/m} - rl^{n-3}v^{(n-3)/m} + \ldots = 0$$
$$o^n v^{n/m} - po^{n-1}v^{(n-1)/m} + qo^{n-2}v^{(n-2)/m} - ro^{n-3}v^{(n-3)/m} + \ldots = 0$$

and so on.

These equations must all be multiplied together, and the result will be the equation $v^n - Pv^{n-1} + Qv^{n-2} - Rv^{n-3} + Sv^{n-4} - \ldots = 0$, whose roots are the mth powers of those of the given equation.

In Problem III the coefficients P, Q, R, S, etc. are investigated, and the smaller the value of m, the simpler will be the rule which governs these coefficients. For example, if $m = 2$ the numerical coefficients of the literal products for P, Q, R, S, etc. will be 1 or 2; if $m = 3$ they will be 1 or 3; if $m = 4$ they will be 1, 2, or 4; for $m = 5$ they will be 1 or 5; and so on.

A third method of solution follows from Problem I.

Corollary 1. Thus may any surd quantities be deleted from a given equation.

Let $x^{1/n}$ be the surd quantity which is to be deleted, from the given equation $Ax^{(n-1)/n} - Bx^{(n-2)/n} + Cx^{(n-3)/n} - Dx^{(n-4)/n} + \ldots = Q$; then we assume $v = x^{1/n}$ and by the example find the equation whose roots is v^n. Then substitute x for v^n, and the job is done.

Example 2. To find the equation whose roots are $\alpha+\beta$, $\alpha+\gamma$, $\beta+\gamma$, $\alpha+\delta$, $\beta+\delta$, $\gamma+\delta$, etc., the sum of every pair of roots of the given equation $x^n - px^{n-1} + qx^{n-2} - rx^{n-3} + \ldots = 0$: the number of roots in the required equation, and thus also its degree, will be equal to the fraction $n(n-1)/2$.

The sum of the roots is $(n-1)\times(\alpha+\beta+\gamma+\delta+\epsilon+\ldots) = (n-1)p$, which is the coefficient of the second term of the required equation.

Now let every two roots of the required equation be multiplied together; the sum of the resulting products $(\alpha+\beta)\times(\alpha+\gamma)$, $(\alpha+\beta)\times(\alpha+\delta)$, $(\alpha+\gamma)\times(\alpha+\delta)$, $(\alpha+\beta)\times(\gamma+\delta)$, etc., will be

$$(n-1)(\frac{n-2}{2})(\alpha^2+\beta^2+\gamma^2+\delta^2+\ldots)+$$
$$n(n-2)(\alpha\beta+\alpha\gamma+\beta\gamma+\alpha\delta+\beta\delta+\gamma\delta+\ldots)$$

while by Problem I

$$(n-1)(\frac{n-2}{2})(\alpha^2+\beta^2+\gamma^2+\delta^2+\ldots) = (n-1)(\frac{n-2}{2})(p^2-2q),$$
$$n(n-2)(\alpha\beta+\alpha\gamma+\beta\gamma+\alpha\delta+\beta\delta+\gamma\delta+\ldots) = n(n-2)q,$$

whose sum will be $(n-1)(\frac{n-2}{2})p^2 + (n-2)q$.

This sum is the coefficient of the third term of the required equation.

Then each three roots of the required equation are combined, as $(\alpha+\beta)(\alpha+\gamma)(\alpha+\delta)$, $(\alpha+\beta)(\alpha+\gamma)(\beta+\delta)$, $(\alpha+\beta)(\alpha+\gamma)(\gamma+\delta)$, $(\alpha+\beta)(\beta+\gamma)(\alpha+\delta)$, etc., and the sum of these products is

$$\frac{(n-1)(n-2)(n-3)}{3!}(\alpha^3+\beta^3+\gamma^3+\delta^3+\ldots)$$
$$+(n-2)(\frac{n^2-2n-1}{2})(\alpha^2\beta+\alpha^2\gamma+\alpha^2\delta+\beta^2\alpha+\beta^2\gamma+\ldots)$$
$$+(n-1)(n^2-2n-2)(\alpha\beta\gamma+\alpha\beta\delta+\alpha\gamma\delta+\beta\gamma\delta+\ldots).$$

By the combined results of Problems I and III, $\frac{(n-1)(n-2)(n-3)}{3!}\times$

$$(\alpha^3+\beta^3+\gamma^3+\delta^3+\ldots) = \frac{(n-1)(n-2)(n-3)}{3!}(p^3-3pq+3r),$$
$$(n-2)(\frac{n^2-2n-1}{2})(\alpha^2\beta+\alpha^2\gamma+\alpha^2\delta+\beta^2\alpha+\beta^2\gamma+\ldots)$$
$$= (n-2)(\frac{n^2-2n-1}{2})(pq-3r),$$
$$(n-1)(n^2-2n-2)(\alpha\beta\gamma+\alpha\beta\delta+\alpha\gamma\delta+\beta\gamma\delta+\ldots) = (n-1)(n^2-2n-2)r,$$

whose sum is $(\frac{(n-1)(n-2)(n-3)}{3!})p^3 + (n-2)^2pq + (n-4)r$.

This sum is of course the coefficient of the fourth term of the required equation.

In like manner the rest of the coefficients are found. Letting v be the root of the required equation, then the equation will be

$$v^{\frac{n(n-1)}{2}} - (n-1)pv^{\frac{n(n-1)}{2}-1} + \Big(\Big(\frac{(n-1)(n-2)}{2}\Big)p^2+(n-2)q\Big)v^{\frac{n(n-1)}{2}-2}$$
$$-\Big(\Big(\frac{(n-1)(n-2)(n-3)}{3!}\Big)p^3+(n-2)^2pq + (n-4)r\Big)v^{\frac{n(n-1)}{2}-3} + \ldots = 0. \quad [4]$$

Example 3. Given the equation $x^n - px^{n-1} + qx^{n-2} - rx^{n-3} + sx^{n-4} - tx^{n-5} + \ldots = 0$, whose roots are α, β, γ, δ, ϵ, etc., to find the equation whose roots are $(\alpha-\beta^2)$, $(\alpha-\gamma)^2$, $(\beta-\gamma)^2$, $(\alpha-\delta)^2$, $(\beta-\delta)^2$, $(\gamma-\delta)^2$, etc.

The sum of the roots is

$$(n-1)(\alpha^2+\beta^2+\gamma^2+\delta^2+\ldots) \qquad (= (n-1)p^2 - 2(n-1)q\,)$$
$$-2(\alpha\beta+\alpha\gamma+\beta\gamma+\alpha\delta+\beta\delta+\gamma\delta+\ldots)\ (= 2q\,)$$

$= (n-1)p^2 - 2nq$, the coefficient of the second term of the equation.

Multiplying each two roots of the required equation, i.e.

$$(\alpha^2+\beta^2-2\alpha\beta)\times(\alpha^2+\gamma^2-2\alpha\gamma),\ (\alpha^2+\delta^2-2\alpha\delta)\times(\beta^2+\delta^2-2\beta\delta),$$
$$(\alpha^2+\delta^2-2\alpha\delta)\times(\alpha^2+\gamma^2-2\alpha\gamma),\ (\beta^2+\gamma^2-2\beta\gamma)\times(\beta^2+\delta^2-2\beta\delta), \text{ etc.},$$

gives the quantity

$$\Big(\frac{(n-1)(n-2)}{2!}\Big)(\alpha^4+\beta^4+\gamma^4+\delta^4+\ldots)-$$
$$2(n-2)\Big(\alpha^3(\beta+\gamma+\delta+\ldots)+\beta^3(\alpha+\gamma+\delta+\ldots)+\gamma^3(\alpha+\beta+\delta)+\delta^3(\alpha+\beta+\gamma+\ldots)+\ldots\Big)$$
$$-2(n-3)\Big(\alpha^2(\beta\gamma+\beta\delta+\gamma\delta+\ldots)+\beta^2(\alpha\gamma+\alpha\delta+\gamma\delta+\ldots)$$
$$+\gamma^2(\alpha\beta+\alpha\delta+\beta\delta+\ldots)+\delta^2(\alpha\beta+\alpha\gamma+\beta\gamma+\ldots)+\ldots\Big)$$
$$+12(\alpha\beta\gamma\delta+\ldots) \quad + \quad n(n-2)(\alpha^2\beta^2+\alpha^2\gamma^2+\beta^2\gamma^2+\alpha^2\delta^2+\beta^2\delta^2+\ldots)$$

By the preceding problems this can be resolved to

$$\frac{(n-1)(n-2)}{2!}(\alpha^4+\beta^4+\gamma^4+\ldots) \quad = \quad \frac{(n-1)(n-2)}{2!}(p^4-4p^2q+4pr-4s+2q^2)$$
$$-2(n-2)\Big(\alpha^3(\beta+\gamma+\delta+\ldots)+\beta^3(\alpha+\gamma+\delta+\ldots)+\gamma^3(\alpha+\beta+\delta+\ldots)+\ldots\Big)$$
$$= -2(n-2)(p^2q-pr+4s-2q^2)$$
$$-2(n-3)\Big(\alpha^2(\beta\gamma+\beta\delta+\gamma\delta+\ldots)+\beta^2(\alpha\gamma+\alpha\delta+\beta\delta+\ldots)$$
$$+\gamma^2(\alpha\beta+\alpha\delta+\beta\delta+\ldots)+\ldots\Big) \quad = -2(n-3)(pr-4s)$$
$$+n(n-2)(\alpha^2\beta^2+\alpha^2\gamma^2+\beta^2\gamma^2+\alpha^2\delta^2+\beta^2\delta^2+\gamma^2\delta^2+\ldots)$$
$$= -n(n-2)(2rp-2s-q^2)$$
$$+12(\alpha\beta\gamma\delta+\alpha\beta\gamma\epsilon+\ldots) \quad = \quad 12s\,,$$

the sum of which is

$$\frac{(n-1)(n-2)}{2!}p^4 - 2n(n-2)p^2q - 2(n-3)pr + 2ns + (2n+3)(n-2)q^2;$$

and this is the coefficient of the third term of the required equation. So writing the equation with an unknown quantity v, we will have

$$v^{\frac{n(n-1)}{2}} - ((n-1)p^2 - 2nq)v^{\frac{n(n-1)}{2}-1}$$
$$+\left(\tfrac{(n-1)(n-2)}{2!}p^4 - 2n(n-2)p^2q - 2(n-3)pr + 2ns + (2n+3)(n-2)q^2\right)v^{\frac{n(n-1)}{2}-2}$$
$$-\ldots = 0\,.$$

Example 4. Let the equation be

$$x^{n-1} - \tfrac{n-1}{n}px^{n-2} + \tfrac{n-2}{n}qx^{n-3} - \tfrac{n-3}{n}rx^{n-4} + \tfrac{n-4}{n}sx^{n-5} + \tfrac{n-5}{n}tx^{n-6} + \ldots = 0,$$

whose roots are α, β, γ, δ, ϵ, etc.; to find an equation whose roots are

$$\alpha^n - p\alpha^{n-1} + q\alpha^{n-2} - r\alpha^{n-3} + s\alpha^{n-4} - t\alpha^{n-5} + \ldots + Q,$$
$$\beta^n - p\beta^{n-1} + q\beta^{n-2} - r\beta^{n-3} + s\beta^{n-4} - t\beta^{n-5} + \ldots + Q,$$
$$\gamma^n - p\gamma^{n-1} + q\gamma^{n-2} - r\gamma^{n-3} + s\gamma^{n-4} - t\gamma^{n-5} + \ldots + Q,$$
$$\delta^n - p\delta^{n-1} + q\delta^{n-2} - r\delta^{n-3} + s\delta^{n-4} - t\delta^{n-5} + \ldots + Q,$$

etc.

The sum of the roots will be

$$(\alpha^n + \beta^n + \gamma^n + \delta^n + \ldots)$$
$$=\left\{\begin{array}{l}(\tfrac{n-1}{n})^n\ p^n\ -n(\tfrac{n-1}{n})^{n-2}(\tfrac{n-2}{n})qp^{n-2}\ +n(\tfrac{n-1}{n})^{n-3}(\tfrac{n-3}{n})rp^{n-3}\\ \qquad -n(\tfrac{n-1}{n})^{n-4}(\tfrac{n-4}{n})sp^{n-4}\ +n(\tfrac{n-3}{2})(\tfrac{n-1}{n})^{n-4}(\tfrac{n-2}{n})^2q^2p^{n-4}+\ldots\end{array}\right\}$$
$$-p(\alpha^{n-1} + \beta^{n-1} + \gamma^{n-1} + \delta^{n-1} + \ldots)$$
$$=\left\{\begin{array}{l}(\tfrac{n-1}{n})^{n-1}p^n + (n-1)(\tfrac{n-1}{n})^{n-3}(\tfrac{n-2}{n})qp^{n-2} - (n-1)(\tfrac{n-1}{n})^{n-4}(\tfrac{n-3}{n})rp^{n-3}\\ \qquad +(n-1)(\tfrac{n-1}{n})^{n-5}(\tfrac{n-4}{n})sp^{n-4} - (n-1)(\tfrac{n-4}{2})(\tfrac{n-1}{n})^{n-5}(\tfrac{n-2}{n})^2q^2p^{n-4}+\ldots\end{array}\right\}$$
$$+q(\alpha^{n-2} + \beta^{n-2} + \gamma^{n-2} + \delta^{n-2} + \ldots)$$
$$= \quad +(\tfrac{n-1}{n})^{n-2}qp^{n-2} - (n-2)(\tfrac{n-1}{n})^{n-4}(\tfrac{n-2}{n})q^2p^{n-4} + \ldots$$
$$-r(\alpha^{n-3} + \beta^{n-3} + \gamma^{n-3} + \delta^{n-3} + \ldots)$$
$$= \qquad -(\tfrac{n-1}{n})^{n-3}rp^{n-3} \qquad + \ldots$$
$$+s(\alpha^{n-4} + \beta^{n-4} + \gamma^{n-4} + \delta^{n-4} + \ldots)$$
$$= \qquad (\tfrac{n-1}{n})^{n-4}sp^{n-4} + \ldots$$

etc.,

the aggregate of which is $\quad -\frac{(n-1)^{n-1}}{n^n}p^n \quad +(\tfrac{n-1}{n})^{n-2}\ qp^{n-2}$

$$-(\tfrac{n-1}{n})^{n-3}rp^{n-3} \quad +(\tfrac{n-1}{n})^{n-4}\ sp^{n-4} \quad -(\tfrac{n-1}{n})^{n-5}\ tp^{n-5} + \ldots$$
$$-\tfrac{1}{2}n(\tfrac{n-1}{n})^2(\tfrac{n-1}{n})^{n-4}q^2p^{n-4} + n(\tfrac{(n-2)(n-3)}{n\cdot n})(\tfrac{n-1}{n})^{n-5}qrp^{n-5} - \ldots$$

The law which this series obeys, expressing term by term the coefficients of the resulting equation, has facets respecting both the literal products and their coefficients.

The literal products follow the same rule as the series of literal products that was presented in Problem I.

The coefficient of the literal product $p^{n-m}q^{\mu}r^{\nu}s^{\pi}t^{\rho}\cdots$ will be equal to a fraction, whose numerator is
$n(n-m+1)(n-m+2)(n-m+3)\cdots(n-m+\mu+\nu+\pi+\rho+\ldots-2)$
$\times(n-1)^{n-m}(n-2)^{\mu}(n-3)^{\nu}(n-4)^{\pi}\cdots,$
and whose denominator is $\mu!\cdot\nu!\cdot\pi!\cdots n^{n-m}n^{\mu+\nu+\pi+\rho+\cdots}$.

If the sum $\mu+\nu+\pi+\rho$ is 0 or 1: in the first case, the coefficient of p^n will be $-\frac{(n-1)^{n-1}}{n^n}$; in the latter case it will be $\pm(\frac{n-1}{n})^{n-m}$. And the coefficient of the quantity Q will be $n-1$.

The sign to be affixed will be positive or negative, according as the sum $m+\mu+\nu+\pi+\rho+\ldots$ is an odd or even number. This example is easily established through the addition of terms.

Example 5. If the given equation is $x^n-px^{n-1}+qx^{n-2}-rx^{n-3}+\ldots=0$, then by full generalization of this problem we can find an equation $v^n+Pv^{m-1}+Qv^{m-2}+\ldots+Rv^2+Sv+T=0$, whose root is $v=x^m+ax^{m-1}+bx^{m-2}+cx^{m-3}+\ldots$.

Corollary 2. If the final term (T) of the resulting equation is equal to 0, then the two polynomials $x^n-px^{n-1}+qx^{n-2}-\ldots$ and $x^m+x^{m-1}+bx^{m-2}+\ldots$ share a divisor of the first degree; if the last two terms (T and S) of the resulting equation are equal to zero, then the two polynomials have a divisor of at least the second degree; if its last three terms are zero, the two polynomials have a cubic divisor, and so on.

This observation also applies to finding common divisors of larger numbers of polynomials.

Example 6. Given an equation $x^n-px^{n-1}+qx^{n-2}-rx^{n-3}+sx^{n-4}-\ldots=0$, whose roots are α, β, γ, $\delta,\ldots$, to find an equation whose roots are

$$\alpha-\beta,\ \alpha-\gamma,\ \beta-\gamma,\ \alpha-\delta,\ \beta-\delta,\ \ldots$$
$$\beta-\alpha,\ \gamma-\alpha,\ \gamma-\beta,\ \delta-\alpha,\ \delta-\beta,\ \ldots$$

(the number of roots will be $n(n-1)$.)

The sum of the roots will be equal to 0, and that will be the coefficient of the second term of the equation. Likewise the coefficients of the fourth, sixth, and all higher even-numbered terms are demonstrably equal to 0.

If all pairs of roots of the required equation are multiplied together, their sum is the product $-(n-1)(\alpha^2+\beta^2+\gamma^2+\delta^2+\ldots)+2(\alpha\beta+\alpha\gamma+\beta\gamma+\alpha\delta+\beta\delta+\gamma\delta+\ldots)=-(n-1)p^2+2nq$, which will be the coefficient of the third term.

Writing v for the unknown quantity in the required equation, the equation will be

$$v^{n(n-1)}-[(n-1)p^2-2nq]v^{n(n-1)-2}$$
$$+[\tfrac{(n-1)(n-2)}{2}p^4-2n(n-2)p^2q-2(n-3)pr+2ns+2(n+3)(n-2)q^2]\,v^{n(n-1)-4}$$
$$+\ldots=0.$$

Corollary 3. The roots of this equation are the square roots of the roots of the equation derived in Example 3.

Example 7. Where α, β, γ, $\delta,\ldots$ are the roots of the given equation

$$x^n-px^{n-1}+qx^{n-2}-rx^{n-3}+sx^{n-4}-tx^{n-5}+\ldots=0,$$

and A, B, Γ, $\Delta,\ldots$ are roots of the given equation

$$x^m-Px^{m-1}+Qx^{m-2}-Rx^{m-3}+Sx^{m-4}-Tx^{m-5}+\ldots=0,$$

to find the equation whose roots are $\alpha\mathrm{A}$, $\alpha\mathrm{B}$, $\alpha\Gamma$, $\alpha\Delta,\ldots$, $\beta\mathrm{A}$, $\beta\mathrm{B}$, $\beta\Gamma$, $\beta\Delta,\ldots$, $\gamma\mathrm{A}$, $\gamma\mathrm{B}$, $\gamma\Gamma$, $\gamma\Delta$, etc., that is, the products of each root of one equation by each root of the other.

The number of roots of the required equation, and thus its dimensions, will be equal to the product of the degrees ($n\times m$) of the given equations.

The sum of the roots, and thus the coefficient of the second term of the required equation, is equal to the product

$$(\alpha+\beta+\gamma+\delta+\ldots)\times(\mathrm{A}+\mathrm{B}+\Gamma+\Delta+\ldots)=P\times p.$$

By multiplying together each pair of roots of the given equation ($\alpha\mathrm{A}\times\alpha\mathrm{B}$, $\alpha\mathrm{A}\times\alpha\Gamma$, $\alpha\mathrm{A}\times\alpha\Delta,\ldots$, $\beta\mathrm{A}\times\alpha\mathrm{A}$, $\beta\mathrm{A}\times\alpha\mathrm{B}$, $\beta\mathrm{A}\times\alpha\Gamma,\ldots$, $\gamma\mathrm{A}\times\alpha\mathrm{A}$, etc.) we obtain the coefficient of the third term of the resulting equation,

$$(\alpha^2+\beta^2+\gamma^2+\delta^2+\ldots)\times(\mathrm{AB}+\mathrm{A}\Gamma+\mathrm{B}\Gamma+\mathrm{A}\Delta+\mathrm{B}\Delta+\ldots)+$$
$$(\mathrm{A}^2+\mathrm{B}^2+\Gamma^2+\Delta^2+\ldots)\times(\alpha\beta+\alpha\gamma+\beta\gamma+\alpha\delta+\beta\delta+\ldots)+$$
$$2(\alpha\beta+\alpha\gamma+\beta\gamma+\alpha\delta+\beta\delta+\ldots)(\mathrm{AB}+\mathrm{A}\Gamma+\mathrm{B}\Gamma+\mathrm{A}\Delta+\mathrm{B}\Delta+\ldots).$$

From Problem I we determine that

$$(\alpha^2+\beta^2+\gamma^2+\delta^2+\ldots)\times(\mathrm{AB}+\mathrm{A}\Gamma+\mathrm{B}\Gamma+\mathrm{A}\Delta+\mathrm{B}\Delta+\ldots)=(p^2-2q)\times Q,$$
$$(\mathrm{A}^2+\mathrm{B}^2+\Gamma^2+\Delta^2+\ldots)\times(\alpha\beta+\alpha\gamma+\beta\gamma+\alpha\delta+\beta\delta+\ldots)=(P^2-2Q)\times q,$$
$$2(\alpha\beta+\alpha\gamma+\beta\gamma+\alpha\delta+\beta\delta+\ldots)(\mathrm{AB}+\mathrm{A}\Gamma+\mathrm{B}\Gamma+\mathrm{A}\Delta+\mathrm{B}\Delta+\ldots)=2Qa,$$

the sum of which is $p^2Q + P^2q - 2Qq$, which is the coefficient of the third term of the required equation.

By the same method of reasoning we determine that the coefficient of the fourth term will be $P^3r + p^3R - 3PQr - 3pqR + pqPQ + 3Rr$ and so on for all the rest. Hence writing v for the unknown quantity, the required equation will be

$$v^nm - Pv^{nm-1} + \left\{\begin{matrix} P^2q-2Qq \\ p^2Q \end{matrix}\right\} v^{nm-2} + \left\{\begin{matrix} P^3r-3PQr+3Rr \\ p^3R-3pqR+pqPQ \end{matrix}\right\} v^{nm-3} + \ldots = 0.$$

By a similar method any transformation of this sort may be deduced. [5]

Scholium

In many cases the following method serves more easily to find the coefficients of the required equation: equations are assumed, of the given form, whose roots are known, and from them the required equation may be found. Using several such resultant equations the coefficients of the terms, and consequently the required equation itself, may be found.

(1) This method tends to be more successful for equations of specific degrees, than for fully general cases.

Example. Let the given equation be $x^4 + qx^2 - rx + s = 0$, whose roots are α, β, γ, δ; to find the equation whose roots are $\alpha+\beta$, $\alpha+\gamma$, $\beta+\gamma$, $\alpha+\delta$, $\beta+\delta$, $\gamma+\delta$.

I select a biquadratic equation of the given form, $x^4 - 6x^2 + 8x - 3 = 0$, whose roots are 1, 1, 1, -3; the roots of the resulting equation will be $1+1$, $1+1$, $1+1$, $1-3$, $1-3$, $1-3$, that is 2, 2, 2, -2, -2, -2, whence the resulting equation will be $v^6 - 12v^4 + 48v^2 - 64 = 0$. The roots of the resulting equation have only one dimension, consequantly their sum, which is the coefficient of the second term of the resulting equation, will have only one dimension; but there can be found no terms in the given equation which have only one dimension, so the sum of the roots in the resulting equation is always equal to zero. The sum of the products of each two roots of the resulting equation, which is the coefficient of the third term, has two dimensions; and there is only one term (q) in the given equation which has no more than two dimensions. Therefore the coefficient of the third term in the resulting equation will be $a \times q$ — but the coefficient of the assumed equation, whose roots are known, was -6, and the coefficient of the third term of its resultant equation was -12, therefore $-6(= q) \times a = -12$, and consequently $a = 2$, so the required equation is $v^6 + 2qv^4 + \ldots = 0$.

The sum of the products of each three roots of the required equation, i.e., $(\alpha + \beta) \times (\alpha + \gamma) \times (\beta + \gamma) + \ldots$, which is the coefficient of the fourth term of the resulting equation, has three dimensions; since there is only one term (r) having three dimensions in the given equation the coefficient of the fourth term of the resulting equation will be $a \times r$. But in the assumed equation $r = 8$, and in its resultant equation the coefficient of the fourth term was 0, therefore $8 \times a = 0$, and it follows that $a = 0$.

The sum of the products of each four roots of the required equation, i.e. $(\alpha + \beta) \times (\alpha + \gamma) \times (\beta + \gamma) \times (\alpha + \delta) + \ldots$, has four dimensions; from the terms of the given equation a product of four dimensions can be constructed in two ways, that is, the coefficient s, and q^2, both have four dimensions. Therefore the sum of the products of each four roots of the given equation, which is the coefficient of the fifth term of the resulting equation, will be $aq^2 + bs$; and since there are two quantities of this sort, composing products of four dimensions, it is necessary to have as many equations whose roots are known, and their resulting equations must both be etermined. Accordingly I assume another equation $x^4 - 2x^2 + 1 = 0$, whose roots are 1, 1, -1, and -1, and so the roots of its resultant equation will be $2, 0, 0, 0, 0, -2$, and the equation itself will be $v^6 - 4v^4 = 0$. From the first assumed equation I have $a \times 36(= q^2) + b \times -3(= s) = 48$, which is the coefficient of its resultant equation's fifth term; and from the second it follows that $x \times 4(= q^2) + b \times 1(= s) = 0$. Solving these two equations establishes that $a = 1$ and $b = -4$; consequently the coefficient of the fifth term of the resulting equation will be $q^2 - 4s$, and the required equation is now $v^6 + 2qv^4 + (q^2 - 4s)v^2 + \ldots = 0$.

The sum of the products of five distinct roots of the resulting equation, which is the coefficient of its sixth term, has five dimensions. There is only one way to derive a five-dimensional quantity from the terms of the given equation, *viz.*, the product qr is of the fifth degree. Hence aqr will be the coefficient of the sixth term. Therefore from the first assumed equation we have $a \times -6 \times 8 = a \times qr = 0$, whence $a = 0$.

The sum of the products of six roots, which is the coefficient of the seventh term of the resulting equation, has six dimensions, as do the products q^3, r^2, and sq alone among the given equation's coefficients. Therefore the coefficient of the final term of the resulting equation will be $aq^3 + br^2 + cqs$. But since there are three quantities of degree six, it is necessary to assume three equations whose roots are known. Accordingly I assume a third equation, $x^4 - 5x^2 + 4 = 0$, whose roots are 1, 2, -1, and -2; the roots of the resulting equation will then be 3, 0, -1, $+1$, 0, and -3, and the equation itself is $v^6 - 10v^4 + 9v^2 = 0$. Now from the first assumed equation we have

$$a \times -216(aq^3) + b \times 64(br^2) + c \times 18(cqs) = -64,$$

from the second, $a \times -8(aq^3) + b \times 0(br^2) + c \times -2(cqs) = 0,$

and from the third, $a \times -125(aq^3) + b \times 0(br^2) + c \times -20(cqs) = \underbrace{0}$;

(the coefficients of the seventh terms of the resulting equations)

and therefore a, b, and c will be respectively 0, -1, and 0, thus the required equation is $v^6 + 2qv^4 + (q^2 - 4s)v^2 - r^2 = 0$.

(2) Given an equation of lower order, say n, using its resulting equation it is often easy to find several terms of the equation resulting similarly from an equation of higher order, say $n + r$.

The given equation, $x^n - px^{n-1} + qx^{n-2} - rx^{n-3} + sx^{n-4} - \ldots = 0$, is multiplied by an equation of degree r, whose roots are known; typically the equation of degree r will be $x^r = 0$, all of whose roots are 0. Then from this resulting equation we are to find the equation whose roots stand in the given relation to the roots of the given equation. It often happens that most of the terms of the resultant equation follow in like manner to those of the former resultant equation.

For example, let the equation of lower degree be $x^4 + qx^2 - rx + s = 0$, whose roots are α, β, γ, δ; the equation whose roots are $\alpha + \beta$, $\alpha + \gamma$, $\beta + \gamma$, $\alpha + \delta$, $\beta + \delta$, $\gamma + \delta$ will be $v^6 + 2qv^4 + (q^2 - 4s)v^2 - r^2 = 0$.

Now let the equation $x^4 + qx^2 - rx + s = 0$ be multiplied by the simple equation $x = 0$, whose root is 0; the product is $x^5 + qx^3 - rx^2 + sx = 0$, whose roots will be α, β, γ, δ, and 0. The roots of the corresponding resultant equation will be $\alpha + \beta$, $\alpha + \gamma$, $\beta + \gamma$, $\alpha + \delta$, $\beta + \delta$, $\gamma + \delta$, α, β, γ, δ, so the equation itself is

$$(v^4 + qv^2 - rv + s) \times (v^6 + 2qv^4 + (q^2 - 4s)v^2 - r^2) = 0,$$

whose terms are all identical to those terms of the analogous equation, resulting from a given equation $x^5 + qx^3 - rx^2 + sx - t = 0$, in which only the letters q, r, and s appear.

The analogously resulting equation in this case determines the equation whose roots are $\lambda + \mu$, $\lambda + \nu$, $\lambda + \pi$, $\lambda + \rho$, $\mu + \nu$, $\mu + \pi$, $\mu + \rho$, $\nu + \pi$, $\nu + \rho$, and $\pi + \rho$, where λ, μ, ν, π, and ρ are the roots of the equation $x^5 + qx^3 - rx^2 + sx - t = 0$.

TRANSLATOR'S NOTES

[1] "Waring's formula" (computing power sums from coefficients here, the converse transformation in Corollary 1 of Problem III) was criticized by Powell for being extremely complicated, compared to the simple recursion expressed by Newton's rule. Waring responded by pointing out that if one used Newton's rule to compute the sum of the nth powers of the roots of an equation, it would be necessary to compute the sums of all lesser powers as well, whereas with this formula one could compute the nth power sum "*per saltem*". This is an important issue now, with respect to parallel computational algorithms, as parallelism is usually thwarted by recursions. Newton's rule is in fact a particular case of an mth order linear recurrence, for which procedures exist (cf. Kogge [1974]) to compute some auxiliary functions such that various steps of the recurrence can be solved in parallel. (As n becomes very large, the number of terms in Waring's formula for the sum of the nth powers of the roots is proportional to the number of *integer partitions* of n, which is close to exponential complexity — whereas Kogge's method, even though it can be shown to be mathematically equivalent to Waring's, performs any one such computation in $m \log_2 n$ steps, decidedly an improvement for practical applications.) As is generally the case with mathematicians, Waring presents an edifice having torn down the scaffolding which he used to build it, leaving no evidence of the thought process which led him to this impressive conclusion. Pinker [1973] offers a hypothetical reconstruction of Waring's approach to the discovery of these formulas, and derives a proof based on generating functions (a concept introduced by Laplace [1812]). He also presents a proof which he attributes to Cauchy.

[2] Another method for computing power sums is to expand (to as high a degree as desired) the logarithmic derivative of the polynomial, i.e. to divide $f'(x)$ by $f(x)$, and in the expansion the coefficient of x^{-1} will be the power sum of order 1, the coefficient of x^{-2} will be the sum of the squares of the roots, etc. To my knowledge, the first publication of this method was by Landen [1755, p. 37ff], who used the logarithmic derivative to prove Newton's identities (cf. also Rubinstein [1972]).

[3] Waring flagged some of his problems, theorems, examples, etc., with an asterisk. There is no explanation in the text of the significance of the asterisks; I have simply reproduced them wherever Waring placed them.

[4] Operating exclusively within the domain of power sums, Hirsch [1827] shows a much simpler solution of this problem. Writing s_p for the sum of the pth powers of the roots of a given equation of degree m, and S_p for power sums of the resulting equation (whose roots are the pairwise sums of the roots of the given equation), we may express S_p as

$$(m-2^{p-1})s_p + ps_{p-1}s_1 + \frac{p(p-1)}{2!}s_{p-2}s_2 + \frac{p(p-1)(p-2)}{3!}s_{p-3}s_3 + \dots$$

where the final term is $\dfrac{p(p-1)\cdots\left(\frac{p}{2}+1\right)}{\frac{p}{2}!}\cdot\dfrac{(s_{p/2})^2}{2}$ (for p even)

or $$\frac{p(p-1)\cdots\left(\frac{p+3}{2}\right)}{\left(\frac{p-1}{2}\right)!}\cdot s_{\frac{p+1}{2}}\cdot s_{\frac{p-1}{2}}$$ (for p odd)

(this formulation is also given, with a misplaced parenthesis, by Terquem [1843]).

[5] This example also has a much more concise solution using power sums [Selmer 1966, 91] (not noticed, apparently, by either Hirsch or Terquem). Writing s_n for the sum of the nth powers of the roots of the first equation, t_n for those of the second equation, and S_n for those of the resulting equation, the

distributive law of multiplication assures us that $S_n = s_n \times t_n$ for any positive integral value of n. Thus the entire process reduces to computing s_n and t_n for n up to the product of the degrees of the input equations, and then simply multiplying corresponding s and t values to determine the S values for the resulting equation; after this the conversion to the coefficient domain is done in the usual way. A similar procedure, in which binomial coefficients are also involved, exists for the equation whose roots are the pairwise sums of the roots of two equations [Weeks 1974]; and any polynomial or rational function of the roots of two equations can be determined in a comparably straightforward manner. (For functions involving $1/\alpha$ etc., it is necessary to compute the power sums of the reciprocal equation, i.e., the equation derived from the original equation by reversing the order of the coefficients.) Another alternative, favored by Hardy [1908], but abandoned in his seventh edition [1943] in favor of working explicitly with coefficients, is the "method of resultants", which is described by Waring in Problem XXXVI.

Chapter 2

Concerning impossible roots.

Theorem 1

If the impossible quantity $\mathrm{a} + \sqrt{-\mathrm{b}^2}$ *is a root of the equation* $\mathrm{x}^{\mathrm{n}} - \mathrm{px}^{\mathrm{n}-1} + \mathrm{qx}^{\mathrm{n}-2} - \mathrm{rx}^{\mathrm{n}-3} + \mathrm{sx}^{\mathrm{n}-4} - \ldots = 0$, *then the impossible quantity* $\mathrm{a} - \sqrt{-\mathrm{b}^2}$ *will also be a root of the given equation.*

Case 1. For x in the given equation let us write $a + \sqrt{-b^2}$ and then $a - \sqrt{-b^2}$. From the first substitution we have

$$
\begin{array}{rllll}
x^n = & a^n & -\frac{n(n-1)}{2!}a^{n-2}b^2 & +\frac{n(n-1)(n-2)(n-3)}{4!}a^{n-4}b^4 & -\ldots \\
 & +na^{n-1}\sqrt{-b^2} & & -\frac{n(n-1)(n-2)}{3!}a^{n-3}b^2\sqrt{-b^2} & +\ldots \\
-px^{n-1} = & -pa^{n-1} & +\frac{(n-1)(n-2)}{2!}a^{n-3}b^2 & -\frac{(n-1)(n-2)(n-3)(n-4)}{4!}pa^{n-5}b^4 & +\ldots \\
 & -(n-1)pa^{n-2}\sqrt{-b^2} & & +\frac{(n-1)(n-2)(n-3)}{3!}pa^{n-4}b^2\sqrt{-b^2} & -\ldots \\
+qx^{n-2} = & +qa^{n-2} & -\frac{(n-2)(n-3)}{2!}a^{n-4}b^2 & +\frac{(n-2)(n-3)(n-4)(n-5)}{4!}qa^{n-6}b^4 & -\ldots \\
 & +(n-2)qa^{n-3}\sqrt{-b^2} & & -\frac{(n-2)(n-3)(n-4)}{3!}qa^{n-5}b^2\sqrt{-b^2} & +\ldots \\
-rx^{n-3} = & -ra^{n-3} & -\frac{(n-3)(n-4)}{2!}a^{n-5}b^2 & -\frac{(n-3)(n-4)(n-5)(n-6)}{4!}ra^{n-7}b^4 & +\ldots \\
 & -(n-3)ra^{n-4}\sqrt{-b^2} & & +\frac{(n-3)(n-4)(n-5)}{3!}ra^{n-6}b^2\sqrt{-b^2} & -\ldots \\
 & \text{etc.} & & \text{etc.} &
\end{array}
$$

and from the second,

$$
\begin{aligned}
x^n = {} & a^n - \tfrac{n(n-1)}{2!}a^{n-2}b^2 + \tfrac{n(n-1)(n-2)(n-3)}{4!}a^{n-4}b^4 - \dots \\
& -na^{n-1}\sqrt{-b^2} + \tfrac{n(n-1)(n-2)}{3!}a^{n-3}b^2\sqrt{-b^2} - \dots \\
-px^{n-1} = {} & -pa^{n-1} + \tfrac{(n-1)(n-2)}{2!}a^{n-3}b^2 - \tfrac{(n-1)(n-2)(n-3)(n-4)}{4!}pa^{n-5}b^4 + \dots \\
& +(n-1)pa^{n-2}\sqrt{-b^2} - \tfrac{(n-1)(n-2)(n-3)}{3!}pa^{n-4}b^2\sqrt{-b^2} + \dots \\
+qx^{n-2} = {} & +qa^{n-2} - \tfrac{(n-2)(n-3)}{2!}a^{n-4}b^2 + \tfrac{(n-2)(n-3)(n-4)(n-5)}{4!}qa^{n-6}b^4 - \dots \\
& -(n-2)qa^{n-3}\sqrt{-b^2} + \tfrac{(n-2)(n-3)(n-4)}{3!}qa^{n-5}b^2\sqrt{-b^2} - \dots \\
-rx^{n-3} = {} & -ra^{n-3} - \tfrac{(n-3)(n-4)}{2!}a^{n-5}b^2 - \tfrac{(n-3)(n-4)(n-5)(n-6)}{4!}ra^{n-7}b^4 + \dots \\
& +(n-3)ra^{n-4}\sqrt{-b^2} - \tfrac{(n-3)(n-4)(n-5)}{3!}ra^{n-6}b^2\sqrt{-b^2} + \dots \\
& \text{etc.} \qquad\qquad \text{etc.}
\end{aligned}
$$

From the first substitution results the same real quantity as from the second, and the imaginary quantities are also the same except for the difference in their signs. It follows that if both the real and imaginary parts vanish as a consequence of the first substitution—that is, if $a + \sqrt{-b^2}$ is a root of the given equation—then the other quantity, $a-\sqrt{-b^2}$, must also be a root of the equation.

Corollary 1. Hence every equation has either none, or an even number, of impossible roots, and for every impossible root $a + \sqrt{-b^2}$ there will be a corresponding one, $a - \sqrt{-b^2}$.

Corollary 2. If the two factors $x - a + \sqrt{-b^2}$ and $x - a - \sqrt{-b^2}$ are multiplied together, the result is a square rational quantity $x^2 - 2ax + a^2 + b^2$, from which it follows that these roots enter into the equation as linear factors in a manner similar to two negative roots or two positive roots, never as if one were negative and the other positive.

Corollary 3. If the equation $x^{2n} - px^{2n-1} + qx^{2n-2} - rx^{2n-3} + sx^{2n-4} - \dots = 0$ has only impossible roots, i.e. if its roots are $a + \sqrt{-A^2}$, $a - \sqrt{-A^2}$, $b + \sqrt{-B^2}$, $b - \sqrt{-B^2}$, $c + \sqrt{-C^2}$, $c - \sqrt{-C^2}$, $d + \sqrt{-D^2}$, $d - \sqrt{-D^2}$, $e + \sqrt{-E^2}$, $e - \sqrt{-E^2}$, etc., it follows that its quadratic divisors are $(x-a)^2 + A^2$, $(x-b)^2 + B^2$, $(x-c)^2 + C^2$, $(x-d)^2 + D^2$, $(x-e)^2 + E^2$, etc. If these divisors are multiplied together, we obtain the given equation

$$
\begin{aligned}
&(x-a)^2(x-b)^2(x-c)^2(x-d)^2(x-e)^2\cdots\\
&\quad +A^2(x-b)^2(x-c)^2(x-d)^2(x-e)^2\cdots\\
&\quad +B^2(x-a)^2(x-c)^2(x-d)^2(x-e)^2\cdots\\
&\quad +C^2(x-a)^2(x-b)^2(x-d)^2(x-e)^2\cdots\\
&\quad +D^2(x-a)^2(x-b)^2(x-c)^2(x-e)^2\cdots\\
&\quad +E^2(x-a)^2(x-b)^2(x-c)^2(x-d)^2\cdots\\
&\qquad\qquad\qquad\text{etc.}\\
&\qquad\quad +A^2B^2(x-c)^2(x-d)^2(x-e)^2\cdots\\
&\qquad\quad +A^2C^2(x-b)^2(x-d)^2(x-e)^2\cdots\\
&\qquad\quad +B^2C^2(x-a)^2(x-d)^2(x-e)^2\cdots\\
&\qquad\quad +A^2D^2(x-b)^2(x-c)^2(x-e)^2\cdots\\
&\qquad\qquad\qquad\qquad\text{etc.}\\
&\qquad\qquad +A^2B^2C^2(x-d)^2(x-e)^2\cdots\quad+\text{etc.}\\
&\qquad\qquad +A^2B^2D^2(x-c)^2(x-e)^2\cdots\\
&\qquad\qquad\qquad\qquad\qquad\text{etc.}\qquad\qquad\qquad+\text{etc.}
\end{aligned}
$$

and it follows that the given equation consists entirely of squares, i.e. it is the sum of the different squares

$$
\begin{aligned}
&(x-a)^2(x-b)^2(x-c)^2(x-d)^2(x-e)^2\cdots,\\
&\quad A^2(x-b)^2(x-c)^2(x-d)^2(x-e)^2\cdots,\\
&\quad B^2\,(x-a)^2(x-c)^2(x-d)^2(x-e)^2\cdots,\\
&\quad C^2\,(x-a)^2(x-b)^2(x-d)^2(x-e)^2\cdots,\\
&\quad D^2\,(x-a)^2(x-b)^2(x-c)^2(x-e)^2\cdots,\\
&\qquad\text{etc.}\\
&A^2B^2(x-c)^2(x-d)^2(x-e)^2\cdots,\\
&\qquad\text{etc.}
\end{aligned}
$$

which, in view of their method of deduction, will be 2^n in number, unless some of them are equal to 0.

Corollary 4. An equation which has a real root cannot be the sum of squares, unless for each real root there is a second one having the same value.

Case 2. (1) In general, for any equation which is an arbitrary function of a quantity x, in which no irrational coefficient of the form $\sqrt{-b^2}$ appears, if $a+\sqrt{-c^2}$ is a root of the given equation, then $a-\sqrt{-c^2}$ will also be a root of the same equation.

(2) But if the given equation contains a coefficient of the form $\sqrt{-b^2}$, and in it for $\sqrt{-b^2}$ the expression $-\sqrt{-b^2}$ is substituted, there will result

an equation $B = 0$, and if $a + \sqrt{-c^2}$ is a root of the given equation, then $a - \sqrt{-c^2}$ will be a root of the derived equation $B = 0$.

This case is established through substitution and expansion of the terms.

Case 3. Every finite equation, having one root α expressed as an irrational quantity which is not a component of the terms of the equation, will have as many other roots of similar form as necessary, such that their product, with α, will yield a rational quantity.

Theorem 2

Case 1. Every equation of even ($2n$) degree, $x^{2n} - px^{2n-1} + qx^{2n-2} - rx^{2n-3} + \ldots = 0$, whose roots are real, can be expressed as the difference between two real squares, one of degree $2n$, the other of degree $2n-2$.

Let n of the given equation's roots be $\alpha, \beta, \gamma, \delta, \ldots$, and the remaining n of them will be $\pi, \rho, \sigma, \tau, \ldots$. Multiply the first n together, as $(x-\alpha)(x-\beta)(x-\gamma)(x-\delta)\cdots = x^n - ax^{n-1} + bx^{n-2} - cx^{n-3} + dx^{n-4} - \ldots$; likewise form the product of the remaining n linear factors, as $(x-\pi)(x-\rho)(x-\sigma)(x-\tau)\cdots = x^n - Ax^{n-1} + Bx^{n-2} - Cx^{n-3} + Dx^{n-4} - \ldots$. Now write the postulated difference between the squares of these two quantities,

$$\begin{array}{llllll} (x^n & - \; a'x^{n-1} & + \; bx^{n-2} & - \; cx^{n-3} & + \; dx^{n-4} & - \ldots)^2 \\ -(& \; A'x^{n-1} & - \; Bx^{n-2} & + \; Cx^{n-3} & - \; Dx^{n-4} & - \ldots)^2 \end{array}$$

which of course has the two divisors

$$x^n - (a'+A')x^{n-1} + (b'+B')x^{n-2} - (c'+C')x^{n-3} + (d'+D')x^{n-4} - \ldots$$
$$\text{and}\quad x^n - (a'-A')x^{n-1} + (b'-B')x^{n-2} - (c'-C')x^{n-3} + (d'-D')x^{n-4} - \ldots$$

we suppose these two divisors respectively equal to the original two products, $x^n - ax^{n-1} + bx^{n-2} - cx^{n-3} + dx^{n-4} - \ldots$ and $x^n - Ax^{n-1} + Bx^{n-2} - Cx^{n-3} + Dx^{n-4} - \ldots$. It follows that the corresponding coefficients will be equal, i.e. $a' + A' = a$ (whence $a' = \frac{a+A}{2}$ and $A' = \frac{a-A}{2}$); $b' + B' = b$ (whence $b' = \frac{b+B}{2}$ and $B' = \frac{b-B}{2}$); $c' + C' = c$ (whence $a' = \frac{c+C}{2}$ and $C' = \frac{c-C}{2}$); and so on. Thus we have constructed two squares, of degrees $2n$ and $2n-2$, whose difference is precisely the proposed equation.

Corollary 1.1. Such an equation can be expressed as the difference between two squares, with $\frac{1}{2} \times 2n \times \frac{2n-1}{2} \times \frac{2n-2}{3} \times \frac{2n-3}{4} \times \cdots \times \frac{n+1}{n}$ variations in the apportioning of roots; therefore in exactly that many ways one such equation of degree $2n$ can be reduced to two of degree n.

Corollary 1.2. If two of the equation's roots are equal to each other, then $(2n-3) \times \frac{2n-4}{2} \times \frac{2n-5}{3} \times \cdots \times \frac{n}{n-2}$ of the variations will represent pairs that are equal each other; and by similar reasoning it is easy to determine the number of variations that correspond to cases when three, two pairs, four, etc. of the given equation's roots are equal to each other.

Case 2. Suppose $2m$ of the given equation's roots are impossible and $2n-2m$ of them are real; r of the impossible roots are taken to be $g+h\sqrt{-1}$, $k+l\sqrt{-1}$, etc., and their r conjugates $g-h\sqrt{-1}$, $k-l\sqrt{-1}$, etc. Then let $n-2r$ of the real roots be α, β, γ, δ, etc., and the remaining $n-2m+2r$ real roots will be named π, ρ, σ, τ, etc.; finally the remaining $2m-2r$ impossible roots will be $G+H\sqrt{-1}$, $G-H\sqrt{-1}$, $K+L\sqrt{-1}$, $K-L\sqrt{-1}$, etc. Now multiply linear factors, based on the first $2r$ impossible roots and the corresponding $n-2r$ real roots, together, and set the result equal to 0:

$$(x-g+h\sqrt{-1})(x-g-h\sqrt{-1})(x-k+l\sqrt{-1})(x-k-l\sqrt{-1})\ldots$$
$$\times(x-\alpha)(x-\beta)(x-\gamma)(x-\delta)\ldots = x^n - ax^{n-1} + bx^{n-2} - cx^{n-3} + \ldots = 0$$

and likewise equate to 0 the products of the other $2m-2r$ impossible roots and the corresponding $n-2m+2r$ real roots:

$$(x-G+H\sqrt{-1})(x-G-H\sqrt{-1})(x-K+L\sqrt{-1})(x-K-L\sqrt{-1})\ldots$$

$$\times(x-\pi)(x-\rho)(x-\sigma)(x-\tau)\ldots = x^n - Ax^{n-1} + Bx^{n-2} - Cx^{n-3} + \ldots = 0.$$

Now, as before, compute the postulated difference between two squares,

$$(x^n - a'x^{n-1} + b'x^{n-2} - c'x^{n-3} + \ldots)^2$$
$$-(A'x^{n-1} - B'x^{n-2} + C'x^{n-3} - D'x^{n-4} + \ldots)^2 = 0,$$

whose divisors will be

$$\begin{aligned} &x^n - (a' + A')x^{n-1} + (b' + B')x^{n-2} - (c' + C')x^{n-3} + \ldots = 0 \\ \text{and}\quad &x^n - (a' - A')x^{n-1} + (b' - B')x^{n-2} - (c' - C')x^{n-3} + \ldots = 0\,. \end{aligned}$$

Take these two factors as equal to the two foregoing products, $x^n - ax^{n-1} + bx^{n-2} - cx^{n-3} + \ldots = 0$ and $x^n - Ax^{n-1} + Bx^{n-2} - Cx^{n-3} + \ldots = 0$ respectively, so that the corresponding coefficients must be equal, i.e. $a' + A' = a$ and $a' - A' = A$ (whence $a' = \dfrac{a+A}{2}$ and $A' = \dfrac{a-A}{2}$); $b' + B' = b$ and $b' - B' = B$ (whence $b' = \dfrac{b+B}{2}$ and $B' = \dfrac{b-B}{2}$); and so on.

Corollary 2.1. Thus we see that every equation of even degree ($2n$) can be separated into two two real squares of prescribed degrees, except if the equation itself is a square (in which case one of them vanishes), or if $2n = 4r + 2$ (r an integer) and all the roots of the given equation are impossible.

Case 3. Let the equation be $x^{2n} - Px^{2n-1} + Qx^{2n-2} - \ldots = 0$, of which one factor is the squared polynomial $(x^r - ax^{r-1} + bx^{r-2} - cx^{r-3} + \ldots)^2 = 0$, all of whose coefficients are real numbers; and suppose that all of the remaining $2n - 2r$ roots of the given equation are impossible, *viz.*, that $n - r$ of them are $\alpha + \beta\sqrt{-1}$, $\gamma + \delta\sqrt{-1}$, $\epsilon + \zeta\sqrt{-1}$, etc., and the remaining $n - r$ of them are $\alpha - \beta\sqrt{-1}$, $\gamma - \delta\sqrt{-1}$, $\epsilon - \zeta\sqrt{-1}$, etc. Then the original equation is the *sum* of the square of an equation of degree n, and the square of another equation of degree $n - 1$.

Proof: Let the two real squares be $(x^n - a'x^{n-1} + b'x^{n-2} - c'x^{n-3} + \ldots)^2 + (A'x^{n-1} - B'x^{n-2} + C'x^{n-3} - \ldots)^2 = 0$ (where a', b', c', …, A', B', C', etc. are all real numbers), which can be factored into

$$\begin{aligned} &x^n - (a' + A'\sqrt{-1})x^{n-1} + (b' + B'\sqrt{-1})x^{n-2} - (c' + C'\sqrt{-1})x^{n-3} + \ldots = 0 \\ \text{and}\quad &x^n - (a' - A'\sqrt{-1})x^{n-1} + (b' - B'\sqrt{-1})x^{n-2} - (c' - C'\sqrt{-1})x^{n-3} + \ldots = 0\,. \end{aligned}$$

Now multiply $x^r - ax^{r-1} + bx^{r-2} - cx^{r-3} + \ldots$ by the product of the first $n - r$ impossible roots equated to zero, i.e. by $(x - \alpha - \beta\sqrt{-1})(x - \gamma - \delta\sqrt{-1})(x - \epsilon - \zeta\sqrt{-1}) = 0$, which works out to $(x^r - ax^{r-1} + bx^{r-2} - cx^{r-3} + \ldots)(x - \alpha - \beta\sqrt{-1})(x - \gamma - \delta\sqrt{-1})(x - \epsilon - \zeta\sqrt{-1}) = x^n - (\kappa + \pi\sqrt{-1})x^{n-1} + (\lambda + \rho\sqrt{-1})x^{n-2} - (\mu + \sigma\sqrt{-1})x^{n-3} + (\nu + \tau\sqrt{-1})x^{n-4} - \ldots = 0$. In the same manner multiply the factor $x^r - ax^{r-1} + bx^{r-2} - \ldots$ by the product of the other $n - r$ impossible roots, equated to zero, giving $(x^r - ax^{r-1} + bx^{r-2} - cx^{r-3} + \ldots)(x - \alpha + \beta\sqrt{-1})$ $(x - \gamma + \delta\sqrt{-1})(x - \epsilon + \zeta\sqrt{-1}) = x^n - (\kappa - \pi\sqrt{-1})x^{n-1} - (\lambda - \rho\sqrt{-1})x^{n-2} -$ $(\mu - \sigma\sqrt{-1})x^{n-3} - (\nu + \tau\sqrt{-1})x^{n-4} - \ldots = 0$. If we equate these two products to the two prescribed factors of the product, it follows that the coefficients of $x^n - (\kappa + \pi\sqrt{-1})x^{n-1} - (\lambda + \rho\sqrt{-1})x^{n-2} - \ldots = 0$ and $x^n - (\kappa - \pi\sqrt{-1})x^{n-1} - (\lambda - \rho\sqrt{-1})x^{n-2} - \ldots = 0$ must be equal respectively to the coefficients of the two factors, i.e. $a' = \kappa$, $b' = \lambda$, $c' = \mu$, $d' = \nu$, etc., and also $A' = \pi$, $B' = \rho$, $C' = \sigma$, $D' = \tau$, etc. By this assumption the given equation will be equal to the sum of the two squared polynomials we have thus found.

If an equation is of this type, it will be the sum of two or more real squares, otherwise it will not. Reciprocally, if the corresponding values of the individual quantities a', b', c', $\ldots$, A', B', C', $\ldots$ are real, then either all of the roots of the given equation are impossible, or it will contain pairs of equal real roots, that is, it cannot contain a real root which is equal to no other root of the equation, and the total number of identical real roots must be an even number in each case.

Corollary 3.1. (1) Let the equation be $x^4 + qx^2 + rx + s = (x^2 + \alpha)^2 + (\beta x + \gamma)^2 = x^4 + (2\alpha + \beta^2)x^2 + 2\beta\gamma x + \alpha^2 + \gamma^2 = 0$, from which we extract the three equations $2\alpha + \beta^2 = q$, $2\beta\gamma = r$, $\alpha^2 + \gamma^2 = s$. By combining these equations so as to eliminate γ and α, we get the equation

$$\left(\frac{q - \beta^2}{2}\right)^2 + \frac{r^2}{4\beta^2} = s\,,$$

whence $\beta^6 - 2q\beta^4 + (q^2 - 4s)\beta^2 + r^2 = 0$. If this equation produces a real root corresponding to the three quantities α, β, and γ, then all of the roots of the given equation must be impossible, unless two of the roots of the given equation are equal (call each of them π), in which case the other two will be $-\pi + \sqrt{-\rho^2}$ and $-\pi - \sqrt{-\rho^2}$.

(2) If instead we eliminate the unknowns β and γ, or α and β, the two resulting equations are respectively

$$\alpha^3 - \tfrac{1}{2}q\alpha^2 - s\alpha + \tfrac{1}{2}qs - \tfrac{1}{8}r^2 = 0$$
$$\text{and}\quad \gamma^6 + (\tfrac{1}{4}q^2 - s)\gamma^4 - \tfrac{1}{8}r^2q\gamma^2 + \tfrac{1}{64}r^4 = 0\,.$$

Let π, ρ, σ, and τ be the roots of the given equation; then the six values of β will be, respectively,

$$\pm(\pi+\rho)\sqrt{-1},\ \pm(\pi+\sigma)\sqrt{-1},\ \pm(\pi+\tau)\sqrt{-1}\,;$$

the three values of α will be

$$\begin{aligned} &\tfrac{1}{2}(-\pi^2-\pi(\rho+\sigma)+\rho\sigma)=\tfrac{1}{2}(\pi\tau+\rho\sigma)\\ &\tfrac{1}{2}(-\rho^2-\rho(\pi+\sigma)+\pi\sigma)=\tfrac{1}{2}(\rho\tau+\pi\sigma)\\ \text{and}\ &\tfrac{1}{2}(-\sigma^2-\sigma(\pi+\rho)+\pi\rho)=\tfrac{1}{2}(\sigma\tau+\pi\rho)\,; \end{aligned}$$

and the six values of γ will be

$$\begin{aligned} &\pm\tfrac{1}{2}(\pi\rho+\pi\sigma+\rho\sigma+\pi^2)\sqrt{-1}=\ \pm\tfrac{1}{2}(\rho\sigma-\pi\tau)\sqrt{-1}\\ &\pm\tfrac{1}{2}(\pi\rho+\pi\tau+\rho\tau+\pi^2)\sqrt{-1}=\ \pm\tfrac{1}{2}(\rho\tau-\pi\sigma)\sqrt{-1}\\ \text{and}\ &\pm\tfrac{1}{2}(\pi\sigma+\pi\tau+\sigma\tau+\pi^2)\sqrt{-1}=\ \pm\tfrac{1}{2}(\sigma\tau-\pi\rho)\sqrt{-1}\,.\ [1] \end{aligned}$$

Theorem 3

Given an equation $x^n - px^{n-1} + qx^{n-2} - rx^{n-3} + sx^{n-4} - \ldots = 0$, whose real roots are $\alpha > \beta > \gamma > \delta > \epsilon > \ldots$, i.e. $(x-\alpha)(x-\beta)(x-\gamma)$ $(x-\delta)(x-\epsilon)\cdots = x^n - px^{n-1} + qx^{n-2} - rx^{n-3} + sx^{n-4} - \ldots = 0$: now if in the given equation for x we substitute a quantity A which is greater than the largest positive root, then all of the factors $A-\alpha$, $A-\beta$, $A-\gamma$, $A-\delta$, etc., will be positive, and so their product $(A-\alpha)(A-\beta)(A-\gamma)(A-\delta)(A-\epsilon)\cdots$ will be a positive quantity.

If instead A is a limit between α and β, i.e. $A < \alpha$, $A > \beta$, it follows that $A-\alpha$ will be a negative quantity, while $A-\beta$, $A-\gamma$, $a-\delta$, etc. remain positive, so if A is substituted for x in the given equation, the resulting value $(A-\alpha)(A-\beta)(A-\gamma)(A-\delta)\cdots$ will be a negative quantity.

Then if A is a limit between β and γ, both $A-\alpha$ and $A-\beta$ will be negative factors, while all the rest are still positive, so the product of all of them will be a positive number, and so on. Thus, if all the roots are real, and limits between successive roots are substituted for the unknown quantity x, the resulting products will exhibit a continuous alternation of signs, from $+$ to $-$ and from $-$ to $+$.

Corollary. If, upon two such substitutions for the unknown quantity x, the sign of the polynomial changes from $+$ to $-$ or from $-$ to $+$, then between the substituted quantities lie an odd number of real roots; but if the sign is the same for the two substitutions, then between them are an even number of real roots, or none at all.

Problem VI

Given an equation $x^n - px^{n-1} + qx^{n-2} - rx^{n-3} + sx^{n-4} - \ldots = 0$ *whose real roots are, in order,* α, β, γ, δ, ϵ, *etc., to find an equation whose roots are limits between successive real roots of the given equation* α *and* β, β *and* γ, γ *and* δ, *etc.*

Multiply the successive terms of the given equation by successive terms of the arithmetic progression n, $n-1$, $n-2$, $n-3$, etc. The result is the required equation,

$$nx^{n-1} - (n-1)px^{n-2} + (n-2)qx^{n-3} - (n-3)rx^{n-4} + \ldots = 0.$$

To prove this we may rewrite the given equation, substituting for x the value $z - a$, the result being

$$\begin{array}{rcllll} x^n & = & z^n + naz^{n-1} + & \frac{n(n-1)}{2}a^2z^{n-2} & + \ldots + & n\;a^{n-1}z + a^n \\ -px^{n-1} & = & \quad - pz^{n-1} - & (n-1)paz^{n-2} & - \ldots - & (n-1)pa^{n-2}z - pa^{n-1} \\ +qx^{n-2} & = & \qquad\quad + & qz^{n-2} & + \ldots + & (n-2)qa^{n-3}z + qa^{n-2} \\ -rx^{n-3} & = & & -rz^{n-3} & - \ldots - & (n-3)ra^{n-4}z - ra^{n-3} \\ \text{etc.} & = & & & \text{etc.} & \end{array}$$

First suppose $a = \alpha$, the largest real root of the given equation. Then the final term of the resulting equation will be $a^n - pa^{n-1} + qa^{n-2} - ra^{n-3} + \ldots = 0$; and the coefficient of the penultimate term will be $na^{n-1} - (n-1)pa^{n-2} + (n-2)qa^{n-3} - (n-3)ra^{n-4} + \ldots = (\alpha - \beta)(\alpha - \gamma)(\alpha - \delta)(\alpha - \epsilon)\cdots$, in which (if all the roots β, γ, δ, ϵ, etc. are real) all the factors will be positive, so the product will be positive. If accordingly in the derived equation $nx^{n-1} - (n-1)px^{n-2} + (n-2)qx^{n-3} - (n-3)rx^{n-4} + \ldots = 0$ we substitute for x the largest root α of the given equation, the resulting value will be positive. If two, four, or more of the roots are impossible, *viz.*, $\pi + \sqrt{-\rho^2}$ and $\pi - \sqrt{-\rho^2}$, $\sigma + \sqrt{-\tau^2}$ and $\sigma - \sqrt{-\tau^2}$, etc., then $na^{n-1} - (n-1)pa^{n-2} + (n-2)qa^{n-3} - (n-3)ra^{n-4} + \ldots = (\alpha - \pi + \sqrt{-\rho^2})(\alpha - \pi - \sqrt{-\rho^2})(\alpha - \sigma + \sqrt{-\tau^2})(\alpha - \sigma - \sqrt{-\tau^2})\cdots(\alpha - \beta)(\alpha - \gamma)(\alpha - \delta)(\alpha - \epsilon)\cdots = [(\alpha - \pi)^2 + \rho^2][(\alpha - \sigma)^2 + \tau^2]\cdots(\alpha - \beta)(\alpha - \gamma)(\alpha - \delta)(\alpha - \epsilon)\cdots$ will still be positive, since it is derived from the previous case by multiplying by factors $(\alpha - \pi)^2 + \rho^2$, $(\alpha - \sigma)^2 + \tau^2$, etc., each of which is strictly positive.

Now suppose $a = \beta$: the final term $a^n - pa^{n-1} + qa^{n-2} - ra^{n-3} + \dots$ will again be equal to zero, and the penultimate coefficient $na^{n-1} - (n-1)pa^{n-2} + (n-2)qa^{n-3} - (n-3)ra^{n-4} + \dots$ will be equal to the product $(\beta-\alpha)(\beta-\gamma)(\beta-\delta)(\beta-\epsilon)\cdots$, in which the factor $\beta - \alpha$ will be negative, but all others will be positive. Therefore this product will be negative, as it will be also if two, four, or more of the roots of the given equation are impossible. If instead $a = \gamma$, then $na^{n-1} - (n-1)pa^{n-2} + (n-2)qa^{n-3} - (n-3)ra^{n-4} + \dots = (\gamma-\alpha)(\gamma-\beta)(\gamma-\delta)(\gamma-\epsilon)\cdots$ will be positive, since exactly two of its factors, $\gamma-\alpha$ and $\gamma-\beta$, are negative and all the rest ($\gamma-\delta$, $\gamma-\epsilon$, etc.) are positive; and again this holds if two, four, or more of the roots of the given equation are impossible. In the same manner each of the remaining roots is substituted for x in the expression $nx^{n-1} - (n-1)px^{n-2} + (n-2)qx^{n-3} - (n-3)rx^{n-4} + \dots$, and the signs of the resulting quantities will be seen to alternate continuously from + to −, then from − to +. Therefore the real roots of the given equation $x^n - px^{n-1} + qx^{n-2} - rx^{n-3} + \dots = 0$ are limits of the real roots of the derived equation $nx^{n-1} - (n-1)x^{n-2} + (n-2)x^{n-3} - (n-3)x^{n-4} + \dots = 0$, and conversely the real roots of the equation $nx^{n-1} - (n-1)x^{n-2} + (n-2)x^{n-3} - (n-3)x^{n-4} + \dots = 0$ will be limits of the roots of the given equation $x^n - px^{n-1} + qx^{n-2} - rx^{n-3} + \dots = 0$.

Corollary 1. The real roots of the equation $x^n - px^{n-1} + \dots = 0$ are limits of the real roots of the equation $x^n - px^{n-1} + \dots \pm Px \mp Q = 0$, whatever may be the value of Q, which does not appear in the derived equation.

Corollary 2. Thus it is evident that the equation $B \equiv nx^{n-1} - (n-1)px^{n-2} + (n-2)qx^{n-3} - \dots = 0$ has one, three, five, or more (always an odd number) of real roots between any two consecutive real roots of the equation $A \equiv x^n - px^{n-1} + qx^{n-2} - rx^{n-3} + \dots = 0$.

Corollary 3. It follows that if the equation $A = 0$ has m real roots, the equation $B = 0$ will have either $m-1$, $m+1$, $m+3$, $m+5$, or $m\pm$(some other odd number) real roots.

Corollary 4. Hence it is easily deduced that, if the penultimate term σx and the final term τ have opposite signs in the given equation $A \equiv x^n - px^{n-1} + qx^{n-2} - \dots \pm \sigma x \mp \tau = 0$, then between the least positive and least negative roots of the given equation $A = 0$ will be found one or three or five ... negative roots, and none or two or four ... positive roots, of the equation $B \equiv nx^{n-1} - (n-1)px^{n-2} + (n-2)qx^{n-3} - \dots = 0$.

On the other hand, if the said penultimate term σx is of the same sign as the final term τ, then between the least positive and least negative terms of $A = 0$ will be an odd number of positive roots, and an even number (or none) of negative roots of $B = 0$.

Theorem 4

Case 1. Given two equations $A = 0$ and $B = 0$, both functions of a variable x, suppose that in two polynomials hA and mB (where h and m are given) the same value κ is substituted for x with the two results having the same sign, and further let α be the root of A or B nearest (either greater or less) than κ. Then between κ and α will be found no root of the equation $hA + mB = 0$.

For if the values of hA and mB (evaluated for $x = \kappa$) have the same sign, their sum cannot be equal to 0, and in consequence no root of $hA + mB = 0$ can be located between κ and α.

Case 2. Let α be a root of the equation $A = 0$, and π a root of the equation $B = 0$, between which are no other roots of either $A = 0$ or $B = 0$. If any quantity, bounded by α and π, is substituted for x in the polynomials hA and mB, and the two resulting values are opposite in sign, then between α and π will be found an odd number of roots of the equation $hA + mB = 0$.

Since at $x = \alpha$ the value of hA is zero, and at $x = \pi$ the value of mB is zero, for some value of x between α and π by hypothesis $hA = -mB$ (since the two polynomials are of opposite sign between α and π)[2]. In other words, at this point $hA + mB = 0$, and it follows that a root of $hA + mB$ lies between α and π.

Thus it can be seen that one root (or a larger odd number of roots) of the equation $hA+mB = 0$ will be found between α and π (accounting also for the possibility that two or more of the roots are equal to each other).

By the same principle it can be proved that if α and β are two successive real roots of the equation $A = 0$, then between α and β will be found an even or odd number of roots of the equation $hA+mB = 0$, according as there is an even or odd number of roots (π, ρ, σ, ...) of roots of the equation $B = 0$ between α and β. If in the expressions hA and mB the value κ, lying between α and the root π of $B = 0$ which is nearest to α, is substituted for x, and the two resulting expressions have the same sign, then it is possible that there are no roots of $hA+mB=0$ between α and β, if the said number of roots (π, ρ, σ, ...) between α and β is even.

Corollary 1. Let the roots of $A \equiv x^n - px^{n-1} + qx^{n-2} - rx^{n-3} + \ldots = 0$ be $\alpha > \beta > \gamma > \delta > \ldots$, and let π, ρ, σ, τ, etc. be roots of the equation $B \equiv nx^{n-1} - (n-1)px^{n-2} + (n-2)qx^{n-3} - (n-3)rx^{n-4} + \ldots = 0$, lying respectively between α and β, β and γ, γ and δ, etc. Then if h and m have the same sign, the roots of the equation $hA + mB = 0$ will be located between α and π, β and ρ, γ and σ, δ and τ, etc., respectively; but if h and m have opposite signs, then one root of $hA + mB = 0$ will be greater than α, and the others will lie between π and β, ρ and γ, σ and δ, etc., respectively.

Corollary 2. If B is multiplied by x, giving $Bx \equiv nx^n - (n-1)px^{n-1} + (n-2)qx^{n-2} - \ldots = 0$, whose roots are π, ρ, σ, τ, ..., and 0, and further assuming that the real roots are all positive or all negative, then the roots of the equation $hA + mBx = 0$ will bear the same positional relation to those of the equations $A = 0$ and $B = 0$ as did the roots of $hA + mB = 0$ in the preceding corollary.

Hence if we multiply successive terms of a given equation, $x^n - px^{n-1} + qx^{n-2} - rx^{n-3} + \ldots = 0$, by successive terms of an arithmetic progression a, $a+b$, $a+2b$, $a+3b$, ..., we produce an equation

$$ax^n - (a+b)px^{n-1} + (a+2b)qx^{n-2} - (a+3b)rx^{n-3} + \ldots = 0$$

whose roots will be limits of the roots of the given equation, when all the real roots of the given equation are positive, or all are negative.

Corollary 3. The resulting equation $ax^n - (a+b)px^{n-1} + (a+2b)qx^{n-2} - (a+3b)rx^{n-3} + \ldots = 0$ will have n roots, unless $a = 0$ or $a + nb = 0$ (i.e., unless the first or last term of the resulting equation vanishes). In general, we can choose a ratio between a and b so as to make any desired value A a root of the resulting equation; the ratio is

$$\frac{b}{a} = \frac{A^n - pa^{n-1} + qA^{n-2} - rA^{n-3} + \ldots}{pA^{n-1} - 2qA^{n-2} + 3rA^{n-3} - \ldots}$$

Corollary 4. The terms of the given equation may be multiplied by the terms of m successive arithmetic progressions, in order to delete any m of its terms: multiplication by any one of these series will serve to delete one particular term of the given equation, when the progression is so chosen that it has a zero in the position corresponding to the term to be eliminated from the equation by that specific progression.

Example. Given the equation $x^n - px^{n-1} + qx^{n-2} - rx^{n-3} + sx^{n-4} - \ldots = 0$, required to delete the term rx^{n-3}. Multiply the terms of this equation by the respective terms of the progression $-3b$, $-2b$, $-b$, 0, b, $2b$, $3b$, etc., which has a 0 in the position corresponding to rx^{n-3} in the given equation, the term we wish to destroy. The result will be

$$-3bx^n + 2bpx^{n-1} - bqx^{n-2} + * + bs^{n-4} - \ldots = 0$$

whose term rx^{n-3} is missing, and whose roots are limits of the roots of the given equation.

Corollary 5. Let the equation be $x^n - px^{n-1} + qx^{n-2} - rx^{n-3} + sx^{n-4} - \ldots = 0$, whose roots are α, β, γ, δ, ϵ, etc. We know that the roots of the equation $ax^n - (a+b)px^{n-1} + (a+2b)qx^{n-2} - (a+3b)rx^{n-3} + \ldots = 0$ are limits between α and β, β and γ, γ and δ, δ and ϵ, etc., respectively. Now if the terms of this latter equation are multiplied by terms of another arithmetic progression, a', $a' + b'$, $a' + 2b'$, $a' + 3b'$, etc., we derive an equation $aa'x^n - (a+b)(a'+b')px^{n-1} + (a+2b)(a'+2b')qx^{n-2} - (a+3b)(a'+3b')rx^{n-3} + \ldots = 0$, whose roots are limits between α and γ, β and δ, γ and ϵ, etc.; and this may be repeated more times if desired. Likewise let the given equation be written $x^n + px^{n-1} + qx^{n-2} - rx^{n-3} + sx^{n-4} - \ldots + Px^{n-r} + Qx^{n-r-1} - rx^{n-r-2} + sx^{n-r-3} - \ldots = 0$, whose roots are $\alpha > \beta > \gamma > \delta > \ldots > \pi > \rho > \sigma > \tau > \ldots$, with $m - 1$ roots between α and π, β and ρ, γ and σ, δ and τ, and so on. If the terms of this equation are multiplied by the successive terms of two arithmetic progressions,

$$\begin{array}{cccccc} n, & n-1, & n-2, & n-3, & n-4, & \ldots \\ 0, & 1, & 2, & 3, & 4, & \ldots \end{array}$$

we obtain two equations, each of degree $n - 1$,

$$\begin{array}{lccccc} & nx^{n-1} + & (n-1)px^{n-2} + & (n-2)qx^{n-3} + & (n-3)rx^{n-4} + & \ldots = 0 \\ \text{and} & + & px^{n-2} + & qx^{n-3} + & rx^{n-4} + & \ldots = 0, \end{array}$$

whose roots in each case lie between the roots α and β, β and γ, γ and δ, δ and ϵ, etc., of the given equation.

When successive terms of these two equations are multiplied by the respective terms of two similar arithmetic progressions,

$$\begin{array}{cccc} n-1, & n-2, & n-3, & \ldots \\ 0, & 1, & 2, & \ldots \end{array}$$

there result three equations of degree $n-2$:

$$n(n-1)x^{n-2}+(n-1)(n-2)px^{n-3}+(n-2)(n-3)qx^{n-4}+(n-3)(n-4)rx^{n-5}+\ldots=0,$$
$$1(n-1)px^{n-2}+\ 2(n-2)qx^{n-3}+\ 3(n-3)rx^{n-4}+\ 4(n-4)sx^{n-5}+\ldots=0,$$
and $$1\cdot 2qx^{n-2}+\ 2\cdot 3rx^{n-3}+\ 3\cdot 4sx^{n-4}+\ 4\cdot 5tx^{n-5}+\ldots=0,$$

whose roots are in each case limits of α and γ, β and δ, γ and ϵ, etc., respectively.

To generalize further, $m+1$ equations of degree $n-m$ can be found, *viz.*,

$$\begin{aligned}&n(n-1)(n-2)\cdots(n-m+1)\ x^{n-m}\ +(n-1)(n-2)(n-3)\cdots(n-m\ \ )px^{n-m-1}\\ &+(n-2)(n-3)(n-4)\cdots(n-m-1)qx^{n-m-2}+(n-3)(n-4)(n-5)\cdots(n-m-2)rx^{n-m-3}\\ &+(n-4)(n-5)(n-6)\cdots(n-m-3)sx^{n-m-4}+\ \ldots\ =0,\end{aligned}$$

$$\begin{aligned}&1(n-1)(n-2)\cdots(n-m+1)px^{n-m}\ +\ 2(n-2)(n-3)\cdots(n-m\ \ )qx^{n-m-1}\\ +\ &3(n-3)(n-4)\cdots(n-m-1)rx^{n-m-3}+\ 4(n-4)(n-5)\cdots(n-m-2)sx^{n-m-3}\\ +\ &5(n-5)(n-6)\cdots(n-m-3)tx^{n-m-4}+\ \ldots\ =0,\end{aligned}$$

$$\begin{aligned}&1\cdot 2(n-2)\cdots(n-m+1)qx^{n-m}\ +\ 2\cdot 3(n-3)(n-4)\cdots(n-m\ \ )rx^{n-m-1}\\ +\ &3\cdot 4(n-4)(n-5)\cdots(n-m-1)sx^{n-m-2}+\ 4\cdot 5(n-5)(n-6)\cdots(n-m-2)tx^{n-m-3}\\ +\ &\ldots\ =0,\end{aligned}$$

—specifically, the $(r+1)^{\text{th}}$ equation in this set would be

$$\begin{aligned}&1\cdot 2\cdot 3\cdots\ r\quad (n-r)\ \ (n-r-1)(n-r-2)\cdots(n-m+1)Px^{n-m}\\ &+2\cdot 3\cdot 4\cdots(r+1)(n-r-1)(n-r-2)(n-r-3)\cdots(n-m)\quad Qx^{n-m-1}\\ &+3\cdot 4\cdot 5\cdots(r+2)(n-r-2)(n-r-3)(n-r-4)\cdots(n-m-1)Rx^{n-m-2}\\ &+4\cdot 5\cdot 6\cdots(r+3)(n-r-3)(n-r-4)(n-r-5)\cdots(n-m-2)Sx^{n-m-3}\\ &+5\cdot 6\cdot 7\cdots(r+4)(n-r-4)(n-r-5)(n-r-6)\cdots(n-m-3)Tx^{n-m-4}\\ &+\quad \ldots\quad =0\end{aligned}$$

—and the $m+1$ roots of each of these equations constitute sets of limits between α and π, β and ρ, γ and σ, δ and τ, and so on.

Corollary 6. If the given equation $A \equiv x^n - px^{n-1} + qx^{n-2} - \ldots = 0$ has both positive and negative roots, then 0 lies between the least positive root (θ) and the least negative root ($-\iota$). Therefore the equation $Bx \equiv nx^n - (n-1)px^{n-1} + (n-2)qx^{n-2} - \ldots = 0$ will have two roots, 0 and $(\pm)\nu$, between the least positive and least negative roots of the given equation; hence,

(1) If h and m are quantities of the same sign, then two roots of the equation $hA + mBx = 0$ lie between the said roots θ and $-\iota$, of

which one will be positive and one negative. Further, by corollary 4 of Problem VI, if a sign changes from + to − or from − to + between the penultimate and final terms of the given equation, the positive root will lie between θ and 0 and the negative root will lie between $-\nu$ and $-\iota$; otherwise the positive root will lie between θ and ν, and the negative root will lie between 0 and $-\iota$.

(2) If h and m are oppositely signed, and $h > mn$, then $hA - mBx = 0$ has no roots between θ and $-\iota$, but it has a positive root greater than the largest positive root of the given equation, and a negative root greater in magnitude than the given equation's largest negative root. If $h = mn$, then the equation $hA - mBx = 0$ has no more than $n - 1$ roots, and none of them lie between θ and $-\iota$, but it does have one root which is greater than the largest positive root of the given equation, or negative and greater in magnitude than the largest negative root of the given equation, according as there is or is not a change of signs (from + to − or from − to +) between the first and second terms of the given equation. If $mn > h$, then one of three cases obtains:

First, two roots of the equation $hA - mBx = 0$ lie between θ and $-\iota$, and both will be negative or both positive, according as there is or is not a sign mutation from + to − or from − to + between the penultimate and final terms of the given equation (this is the case when the ratio of mn to h is very large);

or Second, the two roots of interest exceed in magnitude respectively the largest positive and largest negative roots of the given equation (this happens when mn is nearly equal to h);

or Third, the two roots in question are both impossible (this happens when the ratio $mn : b$ is between the extremes of the two preceding cases).

All the rules which have been previously given regarding the limits of roots of equations, of which all are positive or all are negative, apply also to these equations (except for considerations of limits between the least positive and least negative roots).

Corollary 7. Let the given equation be $x^n - px^{n-1} + qx^{n-2} - rx^{n-3} + \ldots = 0$, and multiply the successive terms of this equation by the corresponding terms of the arithmetic progression a, $a+b$, $a+2b$, $a+3b$, etc. Then if α is a root of multiplicity m in the given equation, it will be a root of multiplicity $m-1$ in the derived equation. If the derived equation is multiplied by another arithmetic progression, the multiplicity of α will be $m-2$ in the result, and so for m repetitions.

Equations whose roots are limits of the roots of a given equation may also be found by multiplying the terms of the given equation by the terms of a harmonic series, or by increasing, diminishing, or otherwise altering the roots of the given equation by an amount less than the smallest difference between two successive roots of the given equation.

Corollary 8. If the terms of $A \equiv x^n - px^{n-1} + qx^{n-2} - rx^{n-3} + \ldots = 0$ are multiplied successively by the arithmetic progression n, $n-1$, $n-2$, $n-3$, ..., the equation $B \equiv nx^{n-1} - (n-1)px^{n-2} + (n-2)qx^{n-3} - \ldots = 0$ results. If α is a root of B of multiplicity two, three, or more (say m), and α is a root of the given equation, then α is a root of multiplicity $m+1$ in the given equation. But it may happen that α is not a root of the given equation: then the m equal roots of B represent the limit between two roots of the given equation, of which either one is real and one impossible, or both are impossible.

For example, consider the equation $(x-a)^{m+1}(x-b)+A=0$, where m is an integer; on expansion this becomes

$$x^{m+2} - [(m+1)a+b]x^{m+1} + \left[\frac{(m+1)m}{2}a^2 + (m+1)ab\right]x^m + \ldots = 0.$$

On multiplication by successive terms of the arithmetic progression $m+2$, $m+1$, m, $m-1$, ..., the derived equation will be

$$(m+2)x^{m+1} - (m+1)[(m+1)a+b]x^m + m\left[\frac{(m+1)m}{2}a^2 + (m+1)ab\right]x^{m-1} + \ldots$$

$= \frac{d}{dx}\Big((x-a)^{m+1}(x-b)+A\Big)$, whence if $A=0$ the given equation will have $m+1$ roots a, otherwise it will have none. [3]

Corollary 9. Let the equation be $(x-a)^{m+1}(x^s - px^{s-1} + qx^{s-2} - \ldots) + A = 0$. Then at least $m-1$ of its roots will be impossible, unless $A=0$.

The same principle applies to equations of the form $(x-a)^{m+1}(x-b)^{n+1}(x-c)^{h+1}\cdots(x^s - Px^{s-1} + Qx^{s-2} - \ldots) + A = 0$, which will have at least $m-1+n-1+h-1+\ldots$ impossible roots.

This corollary depends on the principle that the equation $nx^{n-1} - (n-1)px^{n-2} + (n-2)qx^{n-3} - \ldots = 0$, whose roots are limits of the roots of the given equation $x^n - px^{n-1} + qx^{n-2} - \ldots = 0$, cannot have $m \geq 2$ roots equal to α, unless $m+1$ of the roots of the given equation are equal to α, or else α may be a limit between two real roots, in which case the remaining $m-1$ occurrences of α represent limits between impossible and real roots, or between pairs of impossible roots. [4]

Every rule which generally determines limits between roots or values of polynomials also identifies cases when the said roots or values are equal.

Theorem 5

Case 1. Let A be any algebraic function, whether rational or irrational, of a quantity x. Then $\frac{dA}{dx} = 0$ is an equation some of whose roots are limits of the roots of the equation $A = 0$, and some or all of them may be the same as the roots of $A = 0$. Each root of $A = 0$ may occur with a multiplicity greater than 1, in which case the equation $\frac{dA}{dx} = 0$ has all the roots of the equation $A = 0$.

Case 2. If the equation $A = 0$ is reduced so as to eliminate all irrationalities, the resulting equation $P = 0$ has the same roots as $A = 0$. Likewise if $\frac{dA}{dx} = 0$ is reduced to eliminate all irrational quantities, the result will be an equation $Q = 0$ whose roots are the same as those of $\frac{dA}{dx} = 0$, and so these two resulting equations have roots that separate each other.

Example. Given $A \equiv (x^n - px^{n-1} + qx^{n-2} - \ldots)^{h/m} = 0$, then

$$\frac{dA}{dx} \equiv \frac{h}{m}\left[nx^{n-1} - (n-1)px^{n-2} + \ldots\right]\left[x^n - px^{n-1} + qx^{n-2} - \ldots\right]^{(h-m)/m} = 0.$$

Upon reduction of these equations to eliminate the irrationalities, we get

$$P \equiv [x^n - px^{n-1} + \ldots]^h = 0$$

and $$Q \equiv [x^n - px^{n-1} + \ldots]^{h-m}[nx^{n-1} - (n-1)px^{n-2} + \ldots]^m = 0;$$

and since each distinct root of the equation $A = 0$ occurs to multiplicity h or higher, the equation $\frac{dA}{dx} = 0$ has among its roots each distinct root of $A = 0$, as well as the roots of the equation $nx^{n-1}-(n-1)px^{n-2}+\ldots=0$, which are limits of the roots of the equation $x^n-px^{n-1}+\ldots=0$. [5]

Corollary. If α is a root of $A = 0$ and of $P = 0$ of multiplicity $m > 1$, then the equations $\frac{dA}{dx} = 0$ and $Q = 0$ will have the root α with a lower multiplicity.

A number of rather elegant propositions about the limits of equations fall out easily from this theorem.

Theorem 6

From two given equations A = 0 *and* B = 0, *some of whose roots separate each other, two other equations* P = 0 *and* Q = 0 *can often be derived, some of whose roots also separate each other.*

Case 1. If we can find two equations $P = 0$ and $Q = 0$, whose roots are z and v respectively, such that z is the same algebraic function (say σ) of the unknown x of the given equation $A = 0$, as is v with respect to the unknown y of the given equation $B = 0$, then not uncommonly the roots of the equation $P = 0$ will be limits of the roots of the equation $Q = 0$, and vice versa.

Suppose α and β are two roots of the equation $A = 0$, between which lies π, a root of the equation $B = 0$; let the function be $\sigma = 0$ and solve the resulting equation with respect to the unknown quantity x or y. Then, if all of the roots of the equation $\sigma = 0$ are on the same side of the quantities α and β, i.e. all greater or all less than α and β, we may substitute α, π, and β respectively for x or y, whereupon the value resulting from the substitution of π will be a limit between the values resulting from the substitutions of α and β, i.e. that root of the equation $Q = 0$ will be a limit between those two roots of the equation $P = 0$. The same principle applies, when the above equations have impossible roots; and many similar approaches will also be successful.

Case 2. If two or more distinct roots of the equation in x or y are also roots of the function $\sigma = 0$, then the principle just developed may be used to produce equations, some of whose roots are limits of each other in the same way.

Example 1. Let the given equations be $A = 0$ and $B = 0$, of which the roots of each are limits of the roots of the other; and suppose (for integers m and n) that $x^{2m+1} = z$ and $y^{2m+1} = v$, or $x^{\frac{2m+1}{2n+1}} = z$ and $y^{\frac{2m+1}{2n+1}} = v$. Then the roots of the equations $P = 0$ and $Q = 0$ are each other's limits. On the other hand if $x^{2m} = z$ and $y^{2m} = v$, and α and $-\beta$ are two roots of the equation $A = 0$, then it may happen that between α^{2m} and β^{2m}, two successive roots of the equation $P = 0$, there lie no roots of the equation $Q = 0$; however if all the roots of the equations $A = 0$ and $B = 0$ are positive, or all are negative, then the roots of the equation $P = 0$ will be limits of the roots of the equation $Q = 0$, and vice versa.

Example 2. Given two equations, $A \equiv 1 + px + qx^2 + \ldots + Sx^{n-1} + Tx^n = 0$ and $B \equiv p + 2qy + 3ry^2 + \ldots + (n-1)Sy^{n-2} + nTy^{n-1} = 0$, whose respective roots are limits of each other, we may find two equations whose roots are respectively $1/x = z$ and $1/y = v$. Then the roots of the resulting equations $P \equiv x^n + px^{n-1} + qx^{n-2} + \ldots = 0$ and $Q \equiv py^{n-1} + 2qy^{n-2} + 3ry^{n-3} + \ldots + (n-1)Sy + nT = 0$ will be limits of each other (unless the equation $Q = 0$ has no root between the least positive and least negative roots of the equation $P = 0$, in which case the equation $Q = 0$ has one root beyond the largest positive or largest negative root of the equation $P = 0$, according as the root of the equation $B = 0$ lying between the least positive and least negative roots of the equation $A = 0$ is positive or negative; on the other hand, the roots of the two equations $x^n + px^{n-1} + qx^{n-2} + rx^{n-3} + \ldots = 0$ and $py^n + 2qy^{n-1} + 3ry^{n-2} + \ldots = 0$ do separate each other, the root 0 of the latter providing the limit between the least positive and least negative roots of the former).

Similar assertions apply for cases such as $x^{-2m-1} = z$ and $y^{-2m-1} = v$.

Example 3. Let the two equations, whose roots separate each other, be $A \equiv x^n + px^{n-1} + qx^{n-2} + \ldots = 0$ and $B \equiv ny^{n-1} + (n-1)py^{n-2} + (n-2)qy^{n-3} + \ldots = 0$; and write $\dfrac{x^2}{1+x^2} = z^2$ and $\dfrac{y^2}{1+y^2} = v^2$, i.e. for x and y in the given equations substitute $\dfrac{z}{\sqrt{1-z^2}}$ and $\dfrac{v}{\sqrt{1-v^2}}$ respectively. The resulting equations will always have roots which are limits of each other, unless for instance the equation $A = 0$ has a negative root $-\alpha$, whose absolute value α is nearer than any other root to a positive root β. If the two equations $A = 0$ and $B = 0$ have a common divisor, implying that the equation $A = 0$ has a multiple root, then the two equations $P = 0$ and $Q = 0$ will have a common divisor, and two roots of the equation $P = 0$ will be equal to each other; this happens in no other case except when the equation $A = 0$ has two roots that are equal in magnitude but opposite in sign, in which case the equation $P = 0$ has a root of multiplicity 2 while the equations $A = 0$ and $B = 0$ are incommensurable (they cannot have a common divisor unless the equation $A = 0$ has a multiple root).

Problem VII

Given an equation $\mathrm{x^n - px^{n-1} + qx^{n-2} - rx^{n-3} + sx^{n-4} - \ldots = 0}$ *whose real roots are, in order,* α, β, γ, δ, ϵ, *etc., and given also limits between* α *and* β, β *and* γ, γ *and* δ, δ *and* ϵ, *etc., to find the number of real and impossible roots in the given equation.*

Case 1. Let $\pi > \alpha > \rho > \beta > \sigma > \gamma > \tau > \delta > \ldots.$ Then the limits π, ρ, σ, τ, ... are substituted for the unknown x in the given equation, and the number of real roots will be equal to the number of variations of the sign of the resulting values, from + to − and from − to +, and the rest of the roots will be impossible.

This is a consequence of Theorem 2.

Corollary 1.1. Let the given equation be $A \equiv x^n - px^{n-1} + qx^{n-2} - rx^{n-3} + sx^{n-4} - \ldots = 0$; then the roots of the equation $nx^{n-1} - (n-1)px^{n-2} + (n-2)qx^{n-3} - (n-3)rx^{n-4} + \ldots = 0$ will be limits of the roots of the first equation. The latter equation may be solved and its roots substituted for x in the original equation, and the number of variations of the signs of the results, from + to − or from − to +, determines the number of real (and hence also of complex) roots of the equation $A = 0$.

Corollary 1.2. Hence by solution of an equation of degree $n - 1$ (or, if the second term of the given equation is missing, i.e. it is of the form $x^n + qx^{n-2} - rx^{n-3} + \ldots = 0$, by solving an equation of degree $n - 2$ and taking 0 as one more limit of the roots of the given equation) the number of impossible roots is obtained.

Example 1. Let the equation be $x^2 + q = 0$; the limit between its two roots will be 0, and in consequence its two roots will be real or imaginary, according as q is a negative or a positive quantity.

Example 2. Let the equation be $x^3 + qx - r = 0$, whose roots are α, β, and γ; multiply the terms of the equation by successive terms of the arithmetic progression 0, 1, 2, 3, giving the linear equation $2qx - 3r = 0$. When this equation's root $\frac{3r}{2q}$ is substituted for x in the given equation, the resulting value is $\frac{27r^3 + 4rq^3}{8q^3}$; substituting 0 (the other limit) for x gives the result $-r$. Now, first, suppose that $\frac{r}{q}$ is positive; then $\alpha > \frac{3r}{2q} > \beta > 0 > \gamma$, and three of these roots are real if $\frac{27r^3 + 4rq^3}{8q^3}$, or more simply, $27r^2 + rq^3$, is negative; otherwise only one is real. But if $\frac{r}{q}$ is negative, then we have $\alpha > 0 > \beta > \frac{3r}{2q} > \gamma$, and the roots will be all real or not, according as $\frac{27r^3 + 4rq^3}{8q^3}$ is positive or negative—that is, as $27r^2 + 4q^3$ is negative or positive.

Example 3. If the equation is $x^4 + qx^2 - rx + s = 0$, an equation whose roots are limits of the roots of the given equation will be $2qx^3 - 3rx^2 + 4sx = 0$, whose three roots are 0 and $\frac{3r \pm \sqrt{9r^2 - 32sq}}{4q}$.

First, if $9r^2 - 32sq < 0$, four or two of the given equation's roots will be impossible, according as s is positive or negative.

Second, if $9r^2 - 32sq > 0$, and s and q have opposite signs, then we must substitute the values $\frac{3r + \sqrt{9r^2 - 32sq}}{4q}$, 0, $\frac{3r - \sqrt{9r^2 - 32sq}}{4q}$ for x in the given equation. Then if the first result is negative, the second positive, and the third negative, all roots of the given equation will be real; if all the results are positive, the given equation has no real roots; and in all other cases it will have two.

And so we may proceed to equations of higher degree.

Corollary 3. Given the equation $x^n + px^{n-1} + qx^{n-2} + rx^{n-3} + \ldots + Px^{n-m} + Qx^{n-m-1} + Rx^{n-m-2} + Sx^{n-m-3} + Tx^{n-m-4} + \ldots = 0$, whose roots are $\alpha > \beta > \gamma > \delta > \ldots > \nu > \pi > \rho > \sigma > \tau$ (α is the largest, and τ the smallest, of all the roots), then by Corollary 5 of Theorem 4, there are three equations of degree $n-2$ whose roots are limits respectively between α and γ, β and δ, γ and ϵ, etc. (The corollary applies only if all roots are positive, or all are negative; in other cases the equation must be altered so that all its roots are positive or all are negative, which is easily done by increasing or diminishing all of the roots by an amount which is greater in magnitude than the largest positive or negative root.) The three equations are

$$\begin{array}{lllll} n(n-1)\,x^{n-2}+(n-1)(n-2)px^{n-3}+(n-2)(n-3)qx^{n-4}+(n-3)(n-4)rx^{n-5}+\ldots=0,\\ 1(n-1)px^{n-2}+\quad 2(n-2)qx^{n-3}+\quad 3(n-3)rx^{n-4}+\quad 4(n-4)sx^{n-5}+\ldots=0,\\ 1\cdot 2\;qx^{n-2}+\quad 2\cdot 3\;rx^{n-3}+\quad 3\cdot 4\;sx^{n-4}+\quad \ldots=0, \end{array}$$

whose roots in each case are located between α and γ, β and δ, γ and ϵ, etc. But none of these equations necessarily has a root between any two successive roots of the given equation (e.g. between α and β). Thus it does not follow that by solving these three equations we can correctly determine the number of impossible roots in the given equation; and *a fortiori* resolution of no equation of inferior degree will suffice for this purpose.

Example 4. If by the process described in Corollary 5, Theorem 4, we eliminate $n-2$ terms from the given equation, the result will be $n-1$ quadratic equations,

$$\begin{aligned} n\,(n-1)\,x^2+\quad 2(n-1)px+1\cdot 2q &= 0,\\ (n-1)(n-2)px^2+2\cdot 2(n-2)qx+2\cdot 3r &= 0,\\ (n-2)(n-3)qx^2+2\cdot 3(n-3)rx+3\cdot 4s &= 0, \end{aligned}$$

etc., the general quadratic equation, at distance m from the first, being

$$(n-m)(n-m-1)Px^2+2(m+1)(n-m-1)Qx+(m+1)(m+2)R=0,$$

and the roots of each of these equations are limits between α and σ, and β and τ. But solving all of these equations does not generally suffice to determine the number of impossible roots, because many of their roots may turn out to be limits between the same peirs of roots of the given equation.

If by the same corollary we remove all but four successive terms of the given equation, there will result $n-2$ cubic equations,

$$\begin{aligned} n\,(n-1)(n-2)\,x^3+3(n-1)(n-2)px^2+3\cdot 1\cdot 2(n-2)qx+1\cdot 1\cdot 2\cdot 3r &= 0,\\ (n-1)(n-2)(n-3)px^3+\quad 2\cdot 2(n-2)qx^2+3\cdot 2\cdot 3(n-2)rx+2\cdot 2\cdot 3\cdot 4s &= 0, \end{aligned}$$

etc., with the general cubic equation, at distance m from the first, being

$$(n-m)(n-m-1)(n-m-2)Px^3 + 3(m+1)(n-m-1)(n-m-2)Qx^2$$
$$+3(m+1)(m+2)(n-m-2)Rx + (m+1)(m+2)(m+3)S = 0,$$

the roots of each of these equations being limits between α and ρ, β and σ, and γ and τ. Thus solution of these equations also does not, in general, supply the correct number of impossible roots, and the same argument applies to all cases where the equations do not explicitly produce limits between α and β, β and γ, γ and δ, etc.

Case 2. If the given equation is $x^n - px^{n-1} + qx^{n-2} - rx^{n-3} + sx^{n-4} - \ldots = 0$, and the equation of limits cannot be solved, we may substitute for x in the latter equation the quantities $\frac{p}{n}$, $\frac{2q}{(n-1)p}$, $\frac{3r}{(n-2)q}$, $\frac{4s}{(n-3)r}$, ..., composing a sequence of signs following an initial $+$ sign. Then the given equation has at most as many real roots as there are variations from $+$ to $-$ and from $-$ to $+$ in the aforesaid sequence (and if some substitution gives a value of 0, than that substituent is a root of the given equation). Likewise the roots of $n-1$ quadratic equations may be found, as before, and substituted for x in the equation of limits, producing a sequence of signs similar to that just described. Then the number of variations in the sign sequence once again gives an upper limit on the number of real roots in the given equation. This rule often suffices to detect impossible roots in cubic equations; and for equations of higher degree similar rules may be developed based on the $n-2$ derived cubic equations, etc.

Problem VIII

To find limits between α and β, β and γ, γ and δ, etc.

By Example 6 of Problem V, transform the given equation into one whose roots are $\frac{1}{\alpha-\beta}$, $\frac{1}{\alpha-\gamma}$, $\frac{1}{\alpha-\delta}$, $\frac{1}{\beta-\gamma}$, $\frac{1}{\beta-\delta}$, $\frac{1}{\gamma-\delta}$, etc.. Then if $\frac{1}{A}$ is greater than the largest root of this equation, A will be less than the smallest of the differences $\alpha-\beta$, $\beta-\gamma$, $\gamma-\delta$, etc. Also determine S, a number larger than the largest root of the given equation.

Having thus found a number S which exceeds the largest root of the given equation, and A which is less than the minimum difference between two successive roots, we now reduce S by the amounts A, $2A$,

$3A$, $4A$, $5A$, etc., and call these results a, b, c, d, etc.; some of them will be actual limits between successive roots α and β, β and γ, γ and δ, etc. All of these values may be substituted for x in the given equation, and the resulting variations of sign will indicate the number of real and complex roots which it contains.

Theorem 7

Let $\alpha, \beta, \gamma, \delta, \epsilon, \zeta, \ldots$ *be arbitrary real quantities,* n *in number; and let* $2\mathrm{a} \geq 2\mathrm{b} \geq 2\mathrm{c} \geq 2\mathrm{d} \geq 2\mathrm{e} \geq 2\mathrm{f} \geq \ldots$ *be any arbitrary even indices,* m *in number. Also let* $\mathrm{p} \geq \mathrm{q} \geq \mathrm{r} \geq \mathrm{s} \geq \mathrm{t} \geq \ldots$ *be arbitrary indices,* M *in number, and further we require that* $2\mathrm{a} \geq \mathrm{p}$, $2\mathrm{b} \geq \mathrm{q}$ *(not essential if* $2\mathrm{a} \geq \mathrm{p}+\mathrm{q}$*),* $2\mathrm{c} \geq \mathrm{r}$ *(unless* $2\mathrm{a}+2\mathrm{b} \geq \mathrm{p}+\mathrm{q}+\mathrm{r}$*), and so on; and finally let* $2\mathrm{a}+2\mathrm{b}+2\mathrm{c}+2\mathrm{d}+2\mathrm{e}+\ldots = \mathrm{q}+\mathrm{q}+\mathrm{r}+\mathrm{s}+\mathrm{t}+\ldots.$

Then the sum of all products of the form $\alpha^{2\mathrm{a}}\beta^{2\mathrm{b}}\gamma^{2\mathrm{c}}\delta^{2\mathrm{d}}\epsilon^{2\mathrm{e}}\ldots$

$$+\alpha^{2\mathrm{a}}\beta^{2\mathrm{b}}\gamma^{2\mathrm{c}}\delta^{2\mathrm{d}}\zeta^{2\mathrm{e}}\cdots+\beta^{2\mathrm{a}}\gamma^{2\mathrm{b}}\delta^{2\mathrm{c}}\epsilon^{2\mathrm{d}}\zeta^{2\mathrm{e}}\cdots+\ldots$$

$=\mathrm{P}$, *say, has to the sum of all products of the form*

$$\alpha^{\mathrm{p}}\beta^{\mathrm{q}}\gamma^{\mathrm{r}}\delta^{\mathrm{s}}\epsilon^{\mathrm{t}}\cdots+\alpha^{\mathrm{p}}\beta^{\mathrm{q}}\gamma^{\mathrm{r}}\delta^{\mathrm{s}}\zeta^{\mathrm{t}}\cdots+\beta^{\mathrm{p}}\gamma^{\mathrm{q}}\delta^{\mathrm{r}}\epsilon^{\mathrm{s}}\zeta^{\mathrm{t}}\cdots+\ldots$$

$=\mathrm{Q}$, *say, a ratio at least as large as the ratio of the number of terms in the former summation to the number of terms in the latter, i.e. as* $\mathrm{n}(\mathrm{n}-1)(\mathrm{n}-2)(\mathrm{n}-3)\cdots(\mathrm{n}-\mathrm{m}-1)$ *is to* $\mathrm{n}(\mathrm{n}-1)(\mathrm{n}-2)(\mathrm{n}-3)\cdots(\mathrm{n}-\mathrm{M}-1)$, *or as* 1 *is to* $(\mathrm{n}-\mathrm{m})(\mathrm{n}-\mathrm{m}-1)(\mathrm{n}-\mathrm{m}-2)\cdots(\mathrm{n}-\mathrm{M}+1)$.

This theorem, viz., that $(n-m)(n-m-1)(n-m-2)\ldots(n-M+1)P \geq Q$, follows from the argument that $(n-m)(n-m-1)(n-m-2)\ldots(n-M+1) \times P-Q$ can be represented as a sum of squares, the only method which universally assures us that a given quantity is positive.

If in P the index a is repeated h times, b is repeated k times, c and d are repeated l and o times respectively, and so on, and in Q p is repeated h' times, q k' times, r and s l' and o' times respectively, etc., then

$$(1\cdot2\cdot3\cdots h)(1\cdot2\cdot3\cdots k)(1\cdot2\cdot3\cdots l)(1\cdot2\cdot3\cdots o)\cdots P$$
$$:\ (1\cdot2\cdot3\cdots h')(1\cdot2\cdot3\cdots k')(1\cdot2\cdot3\cdots l')(1\cdot2\cdot3\cdots o')\cdots Q$$

is equal to or greater than the ratio

$$1\ :\ (n-m)(n-m-1)(n-m-2)\cdots(n-M+1).$$

Corollary. Let the equation be

$$ax^n + px^{n-1} + qx^{n-2} + rx^{n-3} + sx^{n-4} + \dots + gx^{n-m+l+6} + fx^{n-m+l+5}$$
$$+ex^{n-m+l+4} + dx^{n-m+l+3} + cx^{n-m+l+2} + bx^{n-m+l+1} + ax^{n-m+l} + \dots$$
$$+Kx^{n-m+5} + Lx^{n-m+4} + Mx^{n-m+3} + Nx^{n-m+2} + Ox^{n-m+1} + Px^{n-m}$$
$$+Qx^{n-m-1} + Rx^{n-m-2} + Sx^{n-m-3} + Tx^{n-m-4} + Vx^{n-m-5} + \dots$$
$$+Ax^{n-m-l} + Bx^{n-m-l-1} + Cx^{n-m-l-2} + Dx^{n-m-l-3} + Ex^{n-m-l-4} + Fx^{n-m-l-5}$$
$$+ \quad \dots = 0,$$

whose roots, designated respectively α, β, γ, δ, ϵ, ζ, etc., are all real.

By this theorem, the sum of all of the π terms of the form

$$\underbrace{\alpha^2\beta^2\gamma^2\delta^2\epsilon^2\cdots}_{m \text{ letters}} + \underbrace{\alpha^2\beta^2\gamma^2\delta^2\zeta^2\cdots}_{m \text{ letters}} + \underbrace{\beta^2\gamma^2\delta^2\epsilon^2\zeta^2\cdots}_{m \text{ letters}} + \dots$$

multiplied by the product

$m(m-1)(m-2)(m-3)\cdots(m-l+1)(n-n)(n-m-1)(n-m-2)\cdots(n-m-l+1)$ $= X$, is equal to, or greater than, the sum of all σ terms of the form

$$\underbrace{\alpha^2\beta^2\gamma^2\cdots}_{(m-l \text{ squared factors})}\epsilon\zeta\cdots + \alpha^2\beta^2\delta^2\cdots\underbrace{\gamma\zeta\cdots}_{(2l \text{ first degree factors})} + \beta^2\gamma^2\delta^2\cdots\alpha\epsilon\cdots + \beta^2\gamma^2\epsilon^2\cdots\alpha\zeta\cdots + \dots$$

multiplied by the product $1 \cdot 2 \cdot 3 \cdot 4 \cdots 2l = Z$.

By Example 3.2 of Problem III, the two series

$$\pi = \quad P^2 - 2QO + 2RN - 2SM + 2TL - 2VK + \dots \text{ and}$$
$$\sigma = \quad Aa - (2l+2!)\ Bb + \frac{(2l+1)(2l+4)}{2!}\ Cc - \frac{(2l+1)(2l+2)(2l+6)}{3!}\ Dd + \frac{(2l+1)(2l+2)(2l+3)(2l+8)}{4!}\ Ee - \dots,$$

are multiplied respectively by X and Z, with the results

$$X\pi = \quad XP^2 - 2XQO + 2XRN - 2XSM + 2XTL - 2XVK + \dots \text{ and}$$
$$Z\sigma = \quad ZAa - (2l+2!)ZBb + \frac{(2l+1)(2l+4)}{2!}ZCc - \frac{(2l+1)(2l+2)(2l+6)}{3!}ZDd + \frac{(2l+1)(2l+2)(2l+3)(2l+8)}{4!}ZEe - \dots,$$

whose difference will be

$$X \times (P^2 - 2QO + 2RN - 2SM + 2TL - \dots, \text{ to } l \text{ terms})$$
$$-(Z \pm 2X)Aa \quad \left((2l+2!)Z \mp 2X\right)Bb \quad - \quad \left(\frac{(2l+1)(2l+4)}{2!}Z \pm 2X\right)Cc$$
$$+ \left(\frac{(2l+1)(2l+2)(2l+6)}{3!}Z \mp 2X\right)Dd - \left(\frac{(2l+1)(2l+2)(2l+3)(2l+8)}{4!}Z \pm 2X\right)Ee$$
$$+ \dots$$

which will always be positive, when all the roots are real.

If l is an even number, the factor preceding Aa will be $Z-2X$, otherwise it will be $Z+2X$. For some specific values of l we have:

$l=1\ (Z=2!)$: difference =

$$XP^2-2(X+1)QO+2(X+4)RN-2(X+9)SM+2(X+16)TL-2(X+25)VK+\dots$$

$l=2\ (Z=4!=24)$: difference =

$$XP^2-2XQO+2(X-12)RN-2(X-72)SM+2(X-240)TL-2(X-600)VK+\dots$$

$l=3\ (Z=6!=720)$: difference =

$$XP^2-2XQO+2XRN-2(X+360)SM+2(X+2880)TL-2(X+12600)VK+\dots$$

Theorem 8

Let α, β, γ, δ, ϵ, ζ, etc. be arbitrary real quantities, all having the same sign, n *in number, and let* $\mathrm{a} \geq \mathrm{b} \geq \mathrm{c} \geq \mathrm{d} \geq \mathrm{e} \geq \mathrm{f} \geq \dots$ *be* m *integer indices.*

Likewise let $\mathrm{p} \geq \mathrm{q} \geq \mathrm{r} \geq \mathrm{s} \geq \mathrm{t} \geq \dots$ *be* M *arbitrary indices, where also* $\mathrm{a} \geq \mathrm{p}$, $\mathrm{b} \geq \mathrm{q}$ (*not essential if* $\mathrm{a} \geq \mathrm{p}+\mathrm{q}$), $\mathrm{c} \geq \mathrm{r}$ (*unless* $\mathrm{a}+\mathrm{b} \geq \mathrm{p}+\mathrm{q}+\mathrm{r}$), $\mathrm{d} \geq \mathrm{s}$ (*unless* $\mathrm{a}+\mathrm{b}+\mathrm{c} \geq \mathrm{p}+\mathrm{q}+\mathrm{r}+\mathrm{s}$), *etc.; and finally let* $\mathrm{a}+\mathrm{b}+\mathrm{c}+\mathrm{d}+\mathrm{e}+\dots = \mathrm{p}+\mathrm{q}+\mathrm{r}+\mathrm{s}+\mathrm{t}+\dots$.

Then the sum of all products of the form

$$\alpha^{\mathrm{a}}\beta^{\mathrm{b}}\gamma^{\mathrm{c}}\delta^{\mathrm{d}}\epsilon^{\mathrm{e}}\cdots+\alpha^{\mathrm{b}}\beta^{\mathrm{a}}\gamma^{\mathrm{c}}\delta^{\mathrm{d}}\zeta^{\mathrm{e}}\cdots+\alpha^{\mathrm{a}}\beta^{\mathrm{b}}\gamma^{\mathrm{c}}\delta^{\mathrm{d}}\zeta^{\mathrm{e}}\cdots+\dots$$

$=$ P, *say, has to the sum of all products of the form*

$$\alpha^{\mathrm{p}}\beta^{\mathrm{q}}\gamma^{\mathrm{r}}\delta^{\mathrm{s}}\epsilon^{\mathrm{t}}\cdots+\alpha^{\mathrm{q}}\beta^{\mathrm{p}}\gamma^{\mathrm{r}}\delta^{\mathrm{s}}\epsilon^{\mathrm{t}}\cdots+\alpha^{\mathrm{p}}\beta^{\mathrm{q}}\gamma^{\mathrm{r}}\delta^{\mathrm{s}}\zeta^{\mathrm{t}}\cdots+\dots$$

$=$ Q, *say, a ratio at least as large as the ratio of the number of terms in the former summation to the number of terms in the latter, i.e. as* $1 : (\mathrm{n-m})(\mathrm{n-m-1})(\mathrm{n-m-2})\cdots(\mathrm{n-M+1})$.

It can be proved that $(n-m)(n-m-1)(n-m-2)\cdots(n-M+1)\times P$ is always greater than Q, from which it follows that $(n-m)(n-m-1)(n-m-2)\cdots(n-M+1)\times P-Q$ is always equal to a sum of squares.

Suppose there are h equal indices designated by a in the first quantity (P), k indices having the value b, l and m indices having the values c and d respectively, and so on; and then h' indices equal to p in the second expression, k' indices equal to q, l' and o' equal to r and s respectively, and so on. Then the ratio

$$h!\cdot k!\cdot l!\cdot o!\cdots P \;:\; h'!\cdot k'!\cdot l'!\cdot o'!\cdots Q$$

will equal or exceed the ratio $1:(n-m)(n-m-1)(n-m-2)\cdots(n-M+1)$.

Corollary. Let $x^n-px^{n-1}+qx^{n-2}-rx^{n-3}+\dots$

$$\pm\, gx^{n-l+6}\mp fx^{n-l+5}\pm ex^{n-l+4}\mp dx^{n-l+3}\pm cx^{n-l+2}\mp bx^{n-l+1}\pm ax^{n-l}\mp\dots$$
$$\pm\, Kx^{n-m+5}\mp Lx^{n-m+4}\pm Mx^{n-m+3}\mp Nx^{n-m+2}\pm Ox^{n-m+1}\mp Px^{n-m}$$
$$\mp Qx^{n-m-1}\mp Rx^{n-m-2}\pm Sx^{n-m-3}\mp Tx^{n-m-4}\pm\dots$$
$$\mp\, Ax^{n-m-r}\pm Bx^{n-m-r-1}\mp Cx^{n-m-r-2}\pm Dx^{n-m-r-3}$$
$$\mp Ex^{n-m-r-4}\pm Fx^{n-m-r-5}\mp Gx^{n-m-r-6}\pm\dots$$
$$\pm\, hx^{n-2m+l-r1}\mp ix^{n-2m+l-r-1}\pm kx^{n-2m+l-r-2}\mp l'x^{n-2m+l-r-3}\pm\dots=0$$

be an equation all of whose roots are real, where α, β, γ, δ, ϵ, etc. represent all its positive or negative roots.

By the theorem, the sum of the π products of the form

$$\underbrace{\alpha^2\beta^2\gamma^2\cdots}_{(m\text{ squared factors})}\delta\epsilon\cdots+\alpha^2\gamma^2\delta^2\cdots\beta\epsilon\cdots+\alpha^2\beta^2\gamma^2\cdots\underbrace{\delta\zeta\cdots}_{(r\text{ linear factors})}+\dots$$

will have, to the sum of the σ products of the form

$$\underbrace{\alpha^2\beta^2\cdots}_{(l\text{ squared factors})}\gamma\delta\epsilon\zeta\cdots+\alpha^2\gamma^2\cdots\beta\delta\epsilon\zeta\cdots+\beta^2\gamma^2\cdots\underbrace{\alpha\delta\epsilon\zeta\cdots}_{(2m+r-2l\text{ linear factors})}+\dots$$

a ratio equal to, or greater than, the ratio

$$(r+1)(r+2)(r+3)\cdots(2m-2l+r)\;=Z$$
$$:(l+1)(l+2)(l+3)\cdots m(n-m-r)(n-m-r-1)(n-m-r-2)\cdots(n-2m+l-r+1)$$

$=X$; and in consequence

$$(l+1)(l+2)(l+3)\cdots m(n-m-r)(n-m-r-1)\cdots(n-2m+l-r+1)\cdot\pi \;\geq\; (r+1)(r+2)(r+3)\cdots(2m-2l+r)\cdot\sigma,$$

provided m is greater than l.

Example 1. If $m=l+1$, then $m(n-m-r)\pi\geq(r+1)(r+2)\sigma$.

Example 2. By example 3.2 of Problem III,

$$\pi=PA-(r+2)OB+\frac{(r+1)(r+4)}{2!}NC-\frac{(r+1)(r+2)(r+6)}{3!}MD+\dots,\qquad\text{and}$$

$$\sigma=ha-(2m-2l+r+2)ib+\frac{(2m-2l+r+1)(2m-2l+r+4)}{2!}kc$$
$$-\frac{(2m-2l+r+1)(2m-2l+r+2)(2m-2l+r+6)}{3!}l'd+\dots.$$

When these quantities are multiplied respectively by X and Z, we get

$$X\pi = XPA - (r+2)XOB + \frac{(r+1)(r+4)}{2!}XNC - \frac{(r+1)(r+2)(r+6)}{3!}XMD + \dots;$$

$$Z\sigma = Zha - (2m-2l+r+2)Zib + \frac{(2m-2l+r+1)(2m-2l+r+4)}{2!}Zkc$$
$$-\frac{(2m-2l+r+1)(2m-2l+r+2)(2m-2l+r+6)}{3!}Zl'd + \dots$$

whose difference will be $X\pi - Z\sigma =$

$$XPA - (r+2)XOB + \frac{(r+1)(r+4)}{2!}XNC - \dots \text{ (to } m-l \text{ terms)}$$
$$-\left(Z \pm \frac{(r+1)(r+2)\cdots(r+m-l-1)(r+2m-2l)}{(m-l)!}X\right)ha$$
$$+\left((2m-2l+r+2)Z \mp \frac{(r+1)(r+2)\cdots(r+m-l)(r+2m-2l+2)}{(m-l+1)!}X\right)ib$$
$$-\left(\frac{(2m-2l+r+1)(2m-2l+r+4)}{2!}Z \pm \frac{(r+1)(r+2)\cdots(r+m-l+1)(r+2m-2l+4)}{(m-l+1)!}X\right)kc$$
$$+\dots.$$

If $m-l$ is an even number, the leading sign affixed to the term

$$\pm\frac{(r+1)(r+2)\cdots(r+m-l-1)(r+2m-2l)}{(m-l)!}X$$

becomes $-$, otherwise it should be taken as positive.

If $m = l+1$, then X and Z become respectively $m(n-m-r)$ and $(r+1)(r+2)$; if further $r = 0$, then Px^{n-m} and Ax^{n-m-r} are components of the same term, whence $XP^2 - 2(X+1)QO + 2(X+4)RN - 2(X+9)SM + 2(X+16)TL - 2(X+25)VK + \dots = a'$, which is always positive if the roots α, β, γ, δ, etc., are all real.

If $r = 2$ and still $m = l+1$, then X and Z are respectively $(n-m-1)\times(m-1)$ and $3 \cdot 4$, and the aforesaid difference is $XQO - 4(X+3)RN + 9(X+8)SM - 16(X+15)TL + 25(X+24)VK - \dots = b'$.

If $R = 4$ and $m = l+2$, then X and Z are respectively $(n-m-2)(m-2)$ and $5 \cdot 6$, and the difference becomes $XRN - 6(X+5)SM + 20(X+12)TL - 50(X+21)VK + \dots = c'$, and so for the rest.

The quantities a', b', c', d', e', ... will always be positive, when the roots α, β, γ, δ, ϵ, ... are positive.

If the quantities a', b', c', d', e', ... are multiplied by other positive numbers, then the sum of the individual terms that result will also be positive. Hence a number of expressions can be deduced, which are all positive when the roots of the given equation are all real and positive.

Problem IX

From the preceding theorems, to deduce several ways to detect impossible roots in a given equation.

Let α, β, γ, δ, ϵ, ... be some of the roots of the given equation. Then from the two preceding theorems we can construct a sequence of numbers a, b, c, d, e, f, ... (whose terms are symmetric in α, β, γ, δ, ϵ, ...), which will be positive when all of α, β, γ, δ, ϵ, ... are real, in which the degrees with respect to the roots α, β, γ, δ, ϵ, ... increase or decrease monotonically.

If the number of roots is n, then we need $n-1$ quantities a, b, c, d, e, f, ..., which should be arranged with underwritten signs, as

$$\begin{array}{cccccccc} 1 & a & b & c & d & e & \dots & 1 \\ + & \pm & \pm & \pm & \pm & \pm & & + \end{array}$$

such that there will be as many impossible roots as there are changes of signs in this sequence, from + to − or from − to +.

Example 1. Let the equation be

$$x^n + px^{n-1} + qx^{n-2} + rx^{n-3} + sx^{n-4} + \dots$$
$$+ Px^{n-m+1} + Qx^{n-m} + Rx^{n-m-1} + \dots \quad = 0,$$

whose roots are α, β, γ, δ, ϵ, ζ, η, θ, ... By the preceding theorems, if all the roots are real, $\frac{n-1}{2n}p^2 > q$, $\frac{2(n-1)}{3(n-2)}q^2 > pr$, and more generally $\frac{m(n-m)}{(m+1)(n-m+1)}Q^2 > PR$.

Accordingly we construct a sequence of $n-1$ terms,

$$\frac{n-1}{2n}p^2 - q, \ \frac{2(n-2)}{3(n-1)}q^2 - pr, \ \frac{3(n-3)}{4(n-2)}r^2 - qs, \ \dots$$

in which each term is symmetric, and of regularly increasing degree, in the roots α, β, γ, δ, ..., and all terms will be positive when all the roots are real. Hence as many roots, or more, will be impossible, as there are mutations from + to − and from − to + in the stated sequence. This may be stated in the following way: under the first and last terms of the given equation, $x^n + px^{n-1} + qx^{n-2} + rx^{n-3} + sx^{n-4} + \dots = 0$, write a + sign; under each of the terms between, write + if the square of its coefficient, multiplied by the indicated fraction, is greater than the product of the coefficients immediately to its left and right, or write − if it is less.

Thus, if $\frac{n-1}{2n}p^2 > q$, write $+$ under the term px^2, and likewise if $\frac{2(n-2)}{3(n-1)}q^2 > pr$ and $\frac{3(n-3)}{4(n-2)}r^2 > qs$, positive signs are written below the terms qx^{n-2} and rx^{n-3} respectively; but if the values are negative, negative signs are written instead, and so for all the remaining terms. Then there will be as many or more impossible roots, as there are alternations of signs in the sequence so constructed.

In this and similar rules to follow, when two or more consecutive terms in the sequence vanish, the first is assigned a $-$ sign, the second $+$, the third $-$, and so on (except that the rightmost term is always assigned a $+$ sign when the nearest non-vanishing terms in the sequence are opposite in sign). [6]

For quadratic equations, this method correctly reveals the number of impossible roots; for cubics the ratio of correct to incorrect predictions turns out to be not greater than 2:1. In equations of much higher degree, this rule rarely succeeds in giving the exact number of impossible roots.

To prove that $\frac{m(n-m)}{(m+1)(n-m+1)}Q^2 > PR$, when all of the roots of the given equation are real, we may proceed as follows: Note first that $Q = \alpha\beta\gamma\delta\epsilon + \alpha\beta\gamma\delta\zeta + \alpha\beta\gamma\epsilon\zeta + \dots$. Let Y be the sum of the squares of the differences between each pair of terms of the coefficient Q. Then let A be the sum of all the squares of differences, between pairs of terms that differ by only one letter. Let B, C, D, ... be the sum of squares of differences where the terms in each pair differ respectively in two, three, four, ... symbols. That is, concretely,

$$\begin{aligned}
Y = {} & (\alpha\beta\gamma\delta\epsilon\dots - \alpha\beta\gamma\delta\zeta\dots)^2 + (\alpha\beta\gamma\delta\epsilon\dots - \alpha\beta\gamma\delta\eta\dots)^2 \\
& + (\alpha\beta\gamma\delta\epsilon\dots - \alpha\beta\gamma\zeta\eta\dots)^2 + (\beta\gamma\delta\epsilon\zeta\dots - \alpha\beta\gamma\delta\eta\dots)^2 + \dots \\
A = {} & (\alpha\beta\gamma\delta\epsilon\dots - \alpha\beta\gamma\delta\zeta\dots)^2 + (\alpha\beta\gamma\delta\epsilon\dots - \beta\gamma\delta\epsilon\eta\dots)^2 \\
& + (\alpha\beta\gamma\delta\zeta\dots - \alpha\beta\gamma\delta\eta\dots)^2 + (\beta\gamma\delta\epsilon\zeta\dots - \beta\gamma\delta\epsilon\eta\dots)^2 + \dots \\
B = {} & (\alpha\beta\gamma\delta\epsilon\dots - \alpha\beta\gamma\zeta\eta\dots)^2 + (\alpha\beta\gamma\delta\epsilon\dots - \alpha\beta\gamma\zeta\theta\dots)^2 \\
& + (\beta\gamma\delta\epsilon\zeta\dots - \alpha\gamma\delta\epsilon\theta\dots)^2 \qquad\qquad + \dots \\
C = {} & (\alpha\beta\gamma\delta\epsilon\dots - \alpha\beta\zeta\eta\theta\dots)^2 + (\alpha\beta\gamma\delta\zeta\dots - \alpha\beta\epsilon\eta\theta\dots)^2 + \dots \\
D = {} & (\alpha\beta\gamma\delta\epsilon\dots - \alpha\zeta\eta\theta\iota\dots)^2 + (\alpha\gamma\delta\epsilon\theta\dots - \alpha\beta\eta\theta\iota\dots)^2 + \dots
\end{aligned}$$

Then $\frac{m(n-m)}{(m+1)(n-m+1)}Q^2$ exceeds PR by the amount $\frac{n+1}{(m+1)(n-m+1)}Y$ $-\frac{1}{2}A - \frac{1}{3}B - \frac{1}{4}C - \frac{1}{5}D - \dots$ (to $m+1$ terms), which is easily verified when we substitute Y, A, B, C, ... for the letters P, Q, R, ..., in the proposed equation.

Example 2. Let the equation be $A \equiv x^n+qx^{n-2}+rx^{n-3}+\ldots+Px^{n-m+1}$ $+Qx^{n-m}+Rx^{n-m-1}+\ldots=0$, in which the second term is missing; the roots of this equation are α, β, γ, δ, ..., $-\alpha-\beta-\gamma-\delta-\ldots$, so the coefficient $q = -\alpha^2-\beta^2-\gamma^2-\delta^2-\ldots-\alpha\beta-\alpha\gamma-\beta\gamma-\alpha\delta-\ldots$, i.e. q is the sum of $\frac{n(n-1)}{2}$ products $-\alpha^2$, $-\beta^2$, ..., $-\alpha\beta$, $-\alpha\gamma$, $-\beta\gamma$, etc. Likewise $r = -\alpha^2\beta-\beta^2\alpha-\alpha^2\gamma-\gamma^2\alpha-\beta^2\gamma-\ldots-2(\alpha\beta\gamma+\alpha\beta\delta+\alpha\gamma\delta+\beta\gamma\delta+\ldots)$, a sum involving a total of $\frac{n(n-1)(n-2)}{3}$ terms. And so the number of terms of similar type in the coefficients P, Q, R, etc., will be, respectively,

$$n\cdot\frac{n-1}{2}\cdot\frac{n-2}{3}\cdots\frac{n-m+4}{m-3}\times(n-m+3)\times\frac{n-m+2}{m-1},$$

$$n\cdot\frac{n-1}{2}\cdot\frac{n-2}{3}\cdots\frac{n-m+3}{m-2}\times(n-m+2)\times\frac{n-m+1}{m},\ \text{and}$$

$$n\cdot\frac{n-1}{2}\cdot\frac{n-2}{3}\cdots\frac{n-m+2}{m-1}\times(n-m+1)\times\frac{n-m}{m+1},$$

from which identities involving the foregoing quantities we may properly conclude that the quantity

$$\frac{(n-2)m^2}{(m-1)^2(m+1)}\times\frac{n-m}{n-m+1}\times Q^2-PR=\Pi$$

is always positive, when the roots of the given equation are real. Now under the first and last terms of the given equation we place a + sign, and under each term Qx^{n-m} we place a + sign, if Π is greater than 0, and − if it is less than zero. Hence there will be as many impossible roots, or more, as there are changes of signs in the sequence so recorded.

The same principle may be applied to several other cases.

Example 3. Let the equation be

$$x^n+px^{n-1}+qx^{n-2}+rx^{n-3}+sx^{n-4}+tx^{n-5}+vx^{n-6}+wx^{n-7}+\ldots$$
$$+Px^{n-m}+Qx^{n-m-1}+Rx^{n-m-2}+Sx^{n-m-3}+Tx^{n-m-4}+\ldots=0.$$

Under the first and last terms we place a + sign. Under the second term, px^{n-1}, we place a + sign if

$$\frac{9r^2}{(n-1)(n-2)^2}-\frac{18pqr}{n(n-1)(n-2)}-\frac{3p^2q^2}{n^2(n-1)}+\frac{6rp^3}{n^2(n-2)}+\frac{8q^3}{n(n-1)^2}$$

is a *negative* quantity, otherwise we place there a − sign. Under the third term of the given equation, qx^{n-2}, we place a + sign if

$$\frac{4s^2p^2}{(n-2)(n-3)^2}-\frac{12qrsp}{(n-1)(n-2)(n-3)}-\frac{3q^2r^2}{(n-1)^2(n-1)}+\frac{16sq^3}{3(n-1)^2(n-3)}$$

has a negative value, and if not, a $-$ sign. And generally, to the term Qx^{n-m-1} a $+$ sign is assigned, if

$$\frac{(m+2)(m+3)^2S^2P^2}{(n-m-1)(n-m-2)^2}-\frac{6(m+1)(m+2)(m+3)QRSP}{(n-m)(n-m-1)(n-m-2)}$$
$$-\frac{3(m+1)^2(m+2)Q^2R^2}{(n-m)^2(n-m-1)}+\frac{4(m+1)^2(m+3)SQ^3}{(n-m)^2(n-m-2)}+\frac{4(m+1)(m+2)^2R^3P}{(n-m)(n-m-1)^2}=\zeta$$

is negative, and if not we place a $-$ sign: and as many or more roots will be impossible, as there alternations of sign from $+$ to $-$ and from $-$ to $+$ in this sequence.

This rule can be deduced from the following observations: first, that the given equation

$$x^n+px^{n-1}+qx^{n-2}+rx^{n-3}+sx^{n-4}+tx^{n-5}+vx^{n-6}+wx^{n-7}+\ldots=0$$

has as many or more impossible roots as either the equation

$$nx^{n-1}+(n-1)px^{n-2}+(n-2)qx^{n-3}+(n-3)rx^{n-4}+\ldots=0$$

or the equation $\qquad px^{n-1}+2qx^{n-2}+3rx^{n-3}+4sx^{n-4}+\ldots=0;$

second, that two roots of the equation $ax^3+bx^2+cx+d=0$ are impossible or real, according as $\qquad 4db^3-b^2c^2-18abcd+27a^2d^2+4c^3a=\Pi$

is positive or negative;

third, if we multiply the terms of the given equation successively by terms of each of the series

$$\begin{array}{rrr} n, & n-1, & n-2; \\ n-1, & n-2, & n-3; \\ n-2, & n-3, & n-4; \\ n-3, & n-4, & n-5; \\ \ldots & & \\ \text{and } 1, & 2, & 3; \end{array}$$

the combined effect of which is to produce the equation

$$n(n-1)(n-2)x^3+3(n-1)(n-2)px^2+3\cdot 2(n-2)qx+6r=0,$$

or in its general formulation,

$$(n-m)(n-m-1)(n-m-2)Px^3+3(n-m-1)(n-m-2)(m+1)Qx^2$$
$$+3(n-m-2)(m+1)(m+2)Rx+(m+1)(m+2)(m+3)Sr=0,$$

and then if for a, b, c, d we substitute $(n-m)(n-m-1)(n-m-2)$, $3(n-m-1)(n-m-2)(m+1)Q$, $3(n-m-2)(m+1)(m+2)R$, and

$(m+1)(m+2)(m+3)S$ respectively in the quantity π defined above, and multiply the result by $\dfrac{1}{27(n-m)^2(n-m-1)^3(n-m-2)^4(m+1)^2(m+2)}$, then we shall have derived the expression ζ defined above. It follows that the several values of ζ will all be negative if all the roots of the given equation are real.

Corollary. Given an equation $A \equiv$

$$x^n+px^{n-1}+qx^{n-2}+\ldots+Px^{n-m}+Qx^{n-m-1}+Rx^{n-m-2}+Sx^{n-m-3}+\ldots=0,$$

we generate the equation

$$\begin{aligned} &n(n-1)(n-2)\cdots(m+1)\ x^m + (n-1)(n-2)(n-3)\cdots m \cdot px^{m-1} \\ &+(n-2)(n-3)(n-4)\cdots(m-1)qx^{m-2} + (n-3)(n-4)(n-5)\cdots(m-2)rx^{m-3} \\ &+ \quad \ldots \quad + 1\cdot 2\cdot 3\cdot 4\cdots(n-m)P = 0, \end{aligned}$$

whose roots we will denote by π, ρ, σ, τ, etc. Now we substitute the values π, ρ, σ, τ, etc., for x in the expression

$$\begin{aligned} &n(n-1)(n-2)\cdots(m+2)\ x^{m+1} + (n-1)(n-2)(n-3)\cdots(m+1)\ px^m \\ &+(n-2)(n-3)(n-4)\cdots m \cdot qx^{m-1} + (n-3)(n-4)(n-5)\cdots(m-1)\ rx^{m-2} \\ &+ \quad \ldots \quad + 2\cdot 3\cdot 4\cdots(n-m)Px. \end{aligned}$$

Suppose first that $m+1$ is an even number: then if the magnitude of the largest resulting negative quantity is less than $(n-m-1)!Q = \Pi$, where the coefficient Q is positive, at least $m+1$ of the proposed equation's roots will be impossible. On the other hand, if $m+1$ is an odd number, and the largest positive quantity is less than the magnitude of Π, with Q negative, or the largest negative quantity is less in magnitude than Π, with Q positive, then the proposed equation will have at least m impossible roots.

The rule of Example 3 always detects impossible roots when the previously stated rule does so, and often detects impossible roots which the previous rule fails to detect. For instance, this rule always detects impossible roots if they are present in cubic equations, and in an equation of n dimensions having n or $n-1$ impossible roots, the probability that the preceding rule will find the correct number of impossible roots, compared to the probability that this rule will correctly detect all of them, is in the ratio of $2^{n-2} : 3^{n-2}$.

Still in an equation of very large degree, this rule seldom yields the correct number of impossible roots.

This rule always succeeds in determining the existence of *some* impossible roots, working with four successive terms of the proposed equation.

And so may we proceed to develop rules which detect impossible roots depending on five, six, or more successive terms, and hence we have a whole arsenal of rules for determining the number of impossible roots.

Often by incrementing or decrementing the roots of the proposed equation by a fixed quantity, impossible roots are readily found, which were previously not revealed; this is a consequence of the fact that, supposing a certain impossible root to be $a+b\sqrt{-1}$, the magnitude of the ratio of its real part a to its imaginary part b substantially influences these rules' chances of finding or not finding that particular impossible root.

Example 4. Let the proposed equation be

$$\begin{aligned} x^n+px^{n-1}+qx^{n-2}+rx^{n-3}+sx^{n-4}+tx^{n-5}+vx^{n-6}+wx^{n-7}+zx^{n-8}+\dots \\ +Ax^{n-m+4}+Bx^{n-m+3}+Cx^{n-m+2}+Dx^{n-m+1} \\ +Ex^{n-m} \quad +Fx^{n-m-1}+Gx^{n-m-2}+Hx^{n-m-3}+Ix^{n-m-4}+\dots = 0\,, \end{aligned}$$

whose roots are α, β, γ, δ, ϵ, ζ, η, θ, ...

To the first and last terms of the proposed equation subjoin a $+$. To the second subjoin a $+$ if $\frac{n-1}{2n}p^2$ is greater than q, and if less subjoin there a $-$. Under the third, if $\dfrac{n(n-1)-2}{2n\cdot(n-1)}q^2$ is greater than $pr-s$, write $+$, and if less, $-$. For the fourth term, rx^{n-3}, compare $\dfrac{n(n-1)(n-2)-2\cdot 3}{2n\cdot(n-1)(n-2)}r^2$ to $qs-pt+v$: if it is greater, subjoin there a $+$ sign, and if less, a $-$ sign. Under the fifth term, sx^{n-4}, write $+$ if $\dfrac{n(n-1)(n-2)(n-3)-2\cdot 3\cdot 4}{2n\cdot(n-1)(n-2)(n-3)}s^2$ is greater than $rt-qv+pw-z$, and if less write $-$; and in general, under an arbitrary term Ex^{n-m}, according as

$$\frac{n(n-1)(n-2)\cdots(n-m+1)-1\cdot 2\cdot 3\cdot 4\cdot 5\cdots m}{2n\cdot(n-1)(n-2)(n-3)\cdots(n-m+1)}E^2$$

is greater or less than $DF-CG+BH-AI+\dots$, place a $+$ or $-$ sign. Then there will be at least as many impossible roots as there are mutations of signs in the sequence, from $+$ to $-$ and from $-$ to $+$.

By substitution it is clear that $\left(1 - \frac{1}{n \cdot \frac{n-1}{2} \cdot \frac{n-2}{3} \cdots \frac{n-m+1}{m}}\right) \times E^2$ is greater than $2DF - 2CG + 2BH - 2AI + \ldots$, which sum is equal to $\frac{Y}{n \cdot \frac{n-1}{2} \cdot \frac{n-2}{3} \cdots \frac{n-m+1}{m}}$, where Y represents the sum of the squared differences between each pair of terms which are components of E, i.e.,

$$Y = (\alpha\beta\gamma\delta\epsilon \cdots - \alpha\beta\gamma\delta\zeta \cdots)^2 + (\alpha\beta\gamma\delta\epsilon \cdots - \alpha\beta\gamma\delta\eta \cdots)^2 + (\beta\gamma\delta\epsilon\zeta \cdots - \beta\gamma\delta\eta\theta \cdots)^2 + \ldots$$

where $E = \alpha\beta\gamma\delta\epsilon \cdots + \alpha\beta\gamma\delta\zeta \cdots + \alpha\beta\gamma\delta\eta \cdots + \beta\gamma\delta\epsilon\zeta \cdots + \beta\gamma\delta\epsilon\eta \cdots + \ldots$.[7]

Many other rules can be developed in like manner, by reference to the Corollary to Theorem 7, wherein are derived diverse series of quantities which are always positive when all of the roots are real, and frequently are negative when some of the roots are impossible. [8]

From the Corollary to Theorem 8, we can determine sequences of quantities which will be positive when all the roots of an equation have the same sign, either all positive or all negative.

All of these rules may be further generalized to the calculation of quantities of higher degree which must be positive when all the roots are real, but the amount of work increases dramatically.

This principle may be applied to other problems as well. For example, consider the series

$$\begin{array}{ll}
\alpha + \beta + \gamma + \delta + \epsilon + \zeta + \ldots & = p \\
\alpha\beta + \alpha\gamma + \beta\gamma + \alpha\delta + \beta\delta + \gamma\delta + \alpha\epsilon + \ldots & = q \\
\alpha\beta\gamma + \alpha\beta\delta + \alpha\gamma\delta + \beta\gamma\delta + \alpha\gamma\epsilon + \ldots & = r \\
\alpha\beta\gamma\delta + \alpha\beta\gamma\epsilon + \alpha\beta\gamma\zeta + \ldots & = s \\
\alpha\beta\gamma\delta\epsilon + \alpha\beta\gamma\delta\zeta + \ldots & = t \\
\text{etc.} & \text{etc.}
\end{array}$$

These quantities continually increase[9], and all will be positive if the roots of the equation are all positive; thus if we form a series 1, p, q, r, s, t, ..., there will be at least as many negative and/or impossible roots as there are changes of sign, from + to − and from − to +.

Or if we form the series 1, $-p$, q, $-r$, s, $-t$, ..., there will be as many negative or impossible roots as there are *progressions* of sign + followed by + and − followed by −.

This same mode of reasoning may be applied in several other cases: given such a series of regularly increasing quantities 1, *a*, *b*, *c*, *d*, *e*, *f*, etc., which are constrained to be positive when the roots are all of a given type, often from changes of sign (+ followed by − or − followed by +) we can determine how many roots are of a different type. It is not even necessary, in rules based on this approach, that the letters *a*, *b*, *c*, *d*, *e*, etc., denote only positive or negative numbers; they may be taken from a domain which includes quantities of any arbitrary kind.

Problem X

Using quantities (such as Π *in Example 3 of the preceding Problem), which are known to be always positive or always negative when all of the roots are real, to derive other similar quantities.*

Case 1. Let Π be any rational function of the coefficients a, b, c, d, etc. associated with $x^{n-\alpha}$, $x^{n-\alpha-\beta}$, $x^{n-\alpha-\beta-\gamma}$, $x^{n-\alpha-\beta-\gamma-\delta}$, etc. in the given equation $x^n - px^{n-1} + qx^{n-2} - rx^{n-3} + \ldots \pm ax^{n-\alpha} \mp \ldots \pm bx^{n-\alpha-\beta} \mp \ldots \pm cx^{n-\alpha-\beta-\gamma} \mp \ldots \pm dx^{n-\alpha-\beta-\gamma-\delta} \mp \ldots = 0$, having the property that Π is positive when all the roots of the given equation are real. Then transform the prescribed coefficients a, b, c, d, etc. into others, A, B, C, D, etc., associated with $x^{n-\mu}$, $x^{n-\mu-\beta}$, $x^{n-\mu-\beta-\gamma}$, $x^{n-\mu-\beta-\gamma-\delta}$, etc. in which β, γ, δ, etc. are the same as in the previous terms whose coefficients were b, c, d, etc. Then in Π,

for a substitute $\dfrac{n(n-1)(n-2)(n-3)\cdots(n-\alpha+1)\times\mu!}{\alpha!\times n(n-1)(n-2)(n-3)(n-\mu+1)}A$,

" b " $\dfrac{n(n-1)(n-2)(n-3)\cdots(n-\alpha-\beta+1)\times(\mu+\beta)!}{(\alpha+\beta)!\times n(n-1)(n-2)(n-3)(n-\mu-\beta+1)}B$,

" c " $\dfrac{n(n-1)(n-2)(n-3)\cdots(n-\alpha-\beta-\gamma+1)\times(\mu+\beta+\gamma)!}{(\alpha+\beta+\gamma)!\times n(n-1)(n-2)(n-3)(n-\mu-\beta-\gamma+1)}C$,

" d " $\dfrac{n(n-1)(n-2)(n-3)\cdots(n-\alpha-\beta-\gamma-\delta+1)\times(\mu+\beta+\gamma+\delta)!}{(\alpha+\beta+\gamma+\delta)!\times n(n-1)(n-2)(n-3)(n-\mu-\beta-\gamma-\delta+1)}D$,

and so for the rest; then the function of A, B, C, D, etc. resulting from this substitution will be the desired quantity. But be careful that the dimensions of the terms of the former and resulting quantities are the same, i.e. that they each involve roots of the given equation to the same degree, otherwise this method will fail.

Case 2. From the equation

$$x^n - px^{n-1} + qx^{n-2} - rx^{n-3} + \ldots \pm ax^{n-\alpha} \mp \ldots \pm bx^{n-\alpha-\beta}$$
$$\mp \ldots \pm cx^{n-\alpha-\beta-\gamma} \mp \ldots \pm dx^{n-\alpha-\beta-\gamma-\delta} \mp \ldots = 0,$$

and a given function Π involving the coefficients a, b, c, d, etc., in which all of the roots of the equation occur to the same degree, and which is always positive or always negative when all the roots of the proposed equation are real, a function of the same sort can be derived from the coefficienta A, B, C, D, etc. corresponding to the powers $x^{m-\alpha}$, $x^{m-\alpha-\beta}$, $x^{m-\alpha-\beta-\gamma}$, $x^{m-\alpha-\beta-\gamma-\delta}$, etc., of the equation

$$x^m - Px^{m-1} + Qx^{m-2} - Rx^{m-3} + \ldots \pm Ax^{m-\alpha} \mp \ldots \pm Bx^{m-\alpha-\beta}$$
$$\mp \ldots \pm Cx^{m-\alpha-\beta-\gamma} \mp \ldots \pm Dx^{m-\alpha-\beta-\gamma-\delta} \mp \ldots = 0.$$

In the given function (Π) for a, b, c, d, etc. substitute

$$\frac{n}{m} \cdot \frac{n-1}{m-1} \cdot \frac{n-2}{m-2} \cdot \frac{n-3}{m-3} \cdots \frac{m-\alpha+1}{n-\alpha+1} \times A,$$
$$\frac{n}{m} \cdot \frac{n-1}{m-1} \cdot \frac{n-2}{m-2} \cdots \frac{m-\alpha-\beta+1}{n-\alpha-\beta+1} \times B,$$
$$\frac{n}{m} \cdot \frac{n-1}{m-1} \cdot \frac{n-2}{m-2} \cdots \frac{m-\alpha-\beta-\gamma+1}{n-\alpha-\beta-\gamma+1} \times C,$$

and so on; this produces the required function of A, B, C, etc.

Example 1. For a given quadratic equation $x^2 - px + q = 0$, $p^2 - 4q$ will be positive, whenever the roots of the given equation are real. (In this case $n = 2$.) Now let the equation $x^m - Px^{m-1} + Qx^{m-2} - \ldots = 0$ be proposed; now replace p and q in the function $p^2 - 4q$ respectively by $\frac{2}{m}P$ and $\frac{2 \cdot 1}{m(m-1)}Q$; this yields the expression $\frac{4}{m^2}P^2 - \frac{8Q}{m(m-1)}$, which will always be positive, when all of the roots of the equation $x^m - Px^{m-1} + Qx^{m-2} - \ldots = 0$ are real. The quantity just derived may be multiplied by $\frac{m(m-1)}{8}$, giving $\frac{m-1}{2m}P^2 - Q$, which also is positive whenever all the roots of the given equation are real.

Example 2. In the given equation

$$x^m - px^{m-1} + qx^{m-2} - \ldots \pm Ax^{m-\alpha+1} \mp Bx^{m-\alpha} \pm Cx^{m-\alpha-1} \mp \ldots + ax^2 - bx + c = 0$$

we substitute $\frac{1}{z}$ for x, which gives us

$$z^m - \frac{b}{c}z^{m-1} + \frac{a}{c}z^{m-2} \pm \ldots \pm \frac{C}{c}z^{\alpha+1} \mp \frac{B}{c}z^{\alpha} \pm \frac{A}{c}z^{\alpha-1} \mp \ldots = 0\,.$$

The prior example's $\frac{m-1}{2m}P^2 - Q$, which here translates to $\frac{m-1}{2m}\cdot\frac{p^2}{q^2} - \frac{a}{c}$, will be positive, and so therefore must $\frac{m-1}{2m}b^2 - ac$ be positive, when all the roots of the given equation are real. Applying Case 1, we write in place of a, b, c respectively the expressions

$$m \cdot \frac{m-1}{2} \cdot \frac{1}{m} \cdot \frac{2}{m-1} \cdot \frac{3}{m-2} \cdots \frac{\alpha-1}{m-\alpha+2} \times A,$$
$$m \cdot \frac{1}{m} \cdot \frac{2}{m-1} \cdot \frac{3}{m-2} \cdots \frac{\alpha}{m-\alpha+1} \times B,$$
$$\text{and } \frac{1}{m} \cdot \frac{2}{m-1} \cdot \frac{3}{m-2} \cdots \frac{\alpha+1}{m-\alpha} \times C,$$

which transforms it to $\frac{m-1}{2m} \cdot \frac{2^2}{(m-1)^2} \cdot \frac{3^2}{(m-2)^2} \cdots \frac{\alpha^2}{(m-\alpha+1)^2} \cdot B$

$$-\frac{1}{m} \cdot \frac{2}{(m-1)} \cdot \frac{3^2}{(m-2)^2} \cdot \frac{4^2}{(m-3)^2} \cdots \frac{(\alpha-1)^2}{(m-\alpha+2)^2} \cdot \frac{\alpha}{m-\alpha+1} \cdot \frac{\alpha}{m-\alpha} \cdot AC\,;$$

whence $\frac{\alpha}{m-\alpha+1}B^2 - \frac{\alpha+1}{m-\alpha}AC$, and so finally $\frac{\alpha(m-\alpha)}{(\alpha+1)(m-\alpha+1)}B^2 - AC$, will be positive quantities, whenever all of the roots of the given equation are real. And many other examples may similarly be constructed.

Problem XI

From the rules already given concerning impossible roots, to invent others which are more general in scope.

If any equation, say

$$x^n - px^{n-1} + qx^{n-2} - rx^{n-3} + sx^{n-4} - \ldots = 0,$$

is multiplied by $x-a$, the result is

$$x^n - \left\{\begin{matrix} p \\ +a \end{matrix}\right\} x^{n-1} + \left\{\begin{matrix} q \\ +ap \end{matrix}\right\} x^{n-2} - \left\{\begin{matrix} r \\ +aq \end{matrix}\right\} x^{n-3} + \left\{\begin{matrix} s \\ +ar \end{matrix}\right\} x^{n-4} - \ldots = 0.$$

We wish to determine, if possible, whether the unknown quantity a can be so chosen that it helps to expose impossible roots, from which a new rule would follow.

Such a rule could become even more general, if the resulting equation were multiplied by $x-b$, and we could determine whether a value of b could be chosen, to enhance further the ability of the given rules to reveal the existence of impossible roots; and so for still more multiplications.

Example 1. Given the equation

$$x^n - px^{n-1} + qx^{n-2} - rx^{n-3} + \ldots - Qx^{n-m} + Rx^{n-m-1} - Sx^{n-m-2} + \ldots = 0,$$

on multiplication by $x-a$ this becomes

$$x^{n+1}-\begin{Bmatrix}p\\+a\end{Bmatrix}x^{n+1}+\begin{Bmatrix}q\\+ap\end{Bmatrix}x^{n}-\begin{Bmatrix}r\\+aq\end{Bmatrix}x^{n-2}+\dots$$
$$-\begin{Bmatrix}Q\\+aP\end{Bmatrix}x^{n-m+1}+\begin{Bmatrix}R\\+aQ\end{Bmatrix}x^{n-m}-\begin{Bmatrix}S\\+aR\end{Bmatrix}x^{n-m-1}+\dots=0.$$

Now whatever the value of a that we choose, by the rule given in the previous problem $(p+a)^2 > (q+ap)$, $(q+ap)^2 > (p+a)(r+aq)$, etc., if all roots of the given equation are real.

But since by hypothesis $(p+a)^2 > (q+ap)$, then $(p+a)^2-ap=(a+\frac{1}{2}p)^2+\frac{3}{4}p^2$ must be greater than q, whatever be the chosen value of a, and it follows that $\frac{3}{4}p^2 > q$. By similar reasoning, since $(q+ap)^2 > (p+a)(r+aq)$, $(p^2-q)a^2+(qp - -r)a+q^2-pr$ will always be positive, from which follows that $4(p^2-q)(q^2-pr) > (pq-r)^2$; and in general $4(Q^2-PR)(R^2-SQ)$ is greater than $(QR-SP)^2$.

Example 2. Given the same equation as in the preceding example, let it be multiplied by $x-a$, and refer to the rule given in Problem IX, Example 1.

By the stated rule, $\frac{n}{2(n+1)} \times (p+a)^2$ must always be greater than $q+ap$, when the roots are all real, for any value of a; or equivalently,

$$\frac{n}{2(n+1)}(p+a)^2-ap = \frac{n}{2(n+1)}p^2-\frac{1}{n+1}pa+\frac{n}{2(n+1)}a^2 > q;$$

but since $\frac{n}{2(n+1)}a^2-\frac{1}{n+1}pa+\frac{n}{2n(n+1)}p^2$ is a square (and therefore always positive or zero), we are assured that $\frac{n}{2(n+1)}p^2-\frac{1}{2n(n+1)}p^2 = \frac{n-1}{2n}p^2$ must be greater than q, as the stated rule asserts. According to the same rule, if all the roots are real,

$$\frac{(n-m)(m+1)}{(n-m+1)(m+2)} \times \begin{matrix}Q^2a^2+2RQa+R^2\\ [=(Qa+R)^2]\end{matrix}$$

must always be greater than $a^2PR+a(PS+QR)+QS = (Q+aP)(S+aR)$, whence $[(n-m)(m+1)Q^2-(n-m+1)(m+2)PR]a^2 - (n-m+1)(m+2)QS + [(mn-m^2-m-2)QR-(n-m+1)(m+2)PS]a + (n-m)(m+1)R^2$ will always be positive. In all cases, for any chosen value of a, this must be true, unless $4[(n-m)(m+1)Q^2-(n-m+1)(m+2)PR] \times [(n-m)(m+1)R^2-(n-m+1)(m+2)QS]$ turns out to be greater than the square of $(mn-m^2-m-2)QR - (n-m+1)(m+2)PS$.

Thus we see how this rule may often detect impossible roots, that are overlooked by the earlier rule.

The above procedure may be reiterated to produce a still more general rule, but the required calculations will increase greatly.

The process of transforming equations often can detect impossible roots in a given equation: some given equation may be transformed into another, whose coefficients, or functions thereof, may reveal the existence of impossible roots among the factors of the given equation. Often the effect of the transformation is quite similar to the result of the present example.

Theorem 9

Case 1. If the equation (whose roots are α, β, γ, δ, ϵ, ...)

$$x^n - px^{n-1} + qx^{n-2} - rx^{n-3} + sx^{n-4} - tx^{n-5} + \ldots = 0$$

be transformed into another (whose roots are α^2, β^2, γ^2, δ^2, ϵ^2, ...), namely $$v^n + (2q-p^2)v^{n-1} + (2s-2pr+q^2)v^{n-2} + \ldots = 0\,,$$

then the given equation will have at least as many impossible roots as there are *successions* of sign, from + to + and from − to −, in the transformed equation.

For let the real roots of the given equation be α, β, γ, δ, ... ; then the roots α^2, β^2, γ^2, δ^2, ... of the transformed equation will be positive, and it follows that the transformed equation must have as many or more alternations of sign, as there are real roots in the original.

Corollary. This rule may find impossible roots when the previous rules fail; conversely, those rules may detect impossible roots when this rule finds none.

Case 2. So a given equation may be transformed into another, whose roots are some rational function of the original equation's roots, which function is necessarily positive when the roots of the given equation are real; then the number of sign progressions in the resulting equation, from + to + and from − to −, is a lower bound for the number of impossible roots.

Corollary. Let the equation be $x^n - px^{n-1} + qx^{n-2} - rx^{n-3} + \ldots = 0$, and in this equation for x substitute $z-a$, resulting in the equation $z^n - Pz^{n-1} + Qz^{n-2} - Rz^{n-3} + \ldots = 0$, whose roots are α, β, γ, δ, etc.; and from this equation derive the equation

$$v^n - Av^{n-1} + Bv^{n-2} - Cv^{n-3} + Dv^{n-4} - \ldots = 0\,,$$

whose roots are α^2, β^2, γ^2, δ^2, etc. (or some other selected function of the roots of the derived equation, which shall be positive for each corresponding real root of the given equation.) Then impose the conditions $A = 0$, $B = 0$, $C = 0$, $D = 0$, etc., and solve each of these equations for a: all these solutions will be impossible if the roots of the given equation are all real (unless two, four, or more are equal to each other). Conversely, since v represents α^2, β^2, γ^2, etc., if all the solutions of $A = 0$, $B = 0$, $C = 0$, $D = 0$, etc., are impossible, then all the roots of the given equation $x^n - px^{n-1} + qx^{n-2} - \ldots = 0$ will be real.

And so other examples of this case may be derived.

Problem XII

Given the equation $x^n - px^{n-1} + qx^{n-2} - rx^{n-3} + sx^{n-4} - \ldots = 0$, *to determine whether or not it has any impossible roots.*

Let α, β, γ, δ, ϵ, etc. be roots of the given equation, and by Problem V transform the given equation into another whose roots v are $\alpha^2+\beta^2-2\alpha\beta$, $\alpha^2+\gamma^2-2\alpha\gamma$, $\alpha^2+\delta^2-2\alpha\delta$, $\beta^2+\gamma^2-2\beta\gamma$, $\beta^2+\delta^2-2\beta\delta$, etc. The resulting equation is $v^{\frac{n(n-1)}{2}} - \left[(n-1)p^2-2nq\right]v^{\frac{n(n-1)}{2}-1} + \left[\frac{(n-1)(n-2)}{2}p^4 - (2n(n-2)p^2q - 2(n-3)pr + 2ns + (2n+3)(n-2)q^2\right]v^{\frac{n(n-1)}{2}-2} - \ldots = 0$. If the signs of this resulting equation alternate continuously from $+$ to $-$ and from $-$ to $+$, the given equation has no impossible roots; but if there is any departure from continuous alternation of signs, then at least some of the given equation's roots are impossible.

For if all the roots (α, β, γ, δ, ...) of the given equation are real, then the roots of the transformed equation — $(\alpha-\beta)^2$, $(\alpha-\gamma)^2$, $(\alpha-\delta)^2$, $(\beta-\gamma)^2$, $(\beta-\delta)^2$, ... — will all be positive, and the signs of the coefficients must alternate throughout. But if two of the roots of the given equation are, say, $a+\sqrt{-b^2}$ and $a-\sqrt{-b^2}$, then one of the roots of the resulting equation will be $(a+\sqrt{-b^2}-a+\sqrt{-b^2})^2 = (2\sqrt{-b^2})^2 = -4b^2$, which is a negative real quantity, in consequence of which there will be at least one point where the sequence of signs does not alternate from $+$ to $-$ or from $-$ to $+$.

Corollary 1. If $\frac{n(n-1)}{2}$ is an even number, then the number of impossible roots will be 0, 4, 8, 12, 16, ... or 2, 6, 10, 14, ..., according as the final term of the resulting equation is positive or negative respectively; whereas if $\frac{n(n-1)}{2}$ is odd, then there will be 0, 4, 8, 12, 16, ... impossible roots if the final term is negative, and 2, 6, 10, 14, ... if it is positive.

Corollary 2. If the last m terms of the resulting equation are zero, and $\frac{n(n-1)}{2}-m$ is an even number, then 0, 4, 8, 12, ... or 2, 6, 10, 14, ... roots of the given equation are impossible, according as the last non-zero term of the resulting equation is positive or negative. If $\frac{n(n-1)}{2} - m$ is odd, then there will be 0, 4, 8, ... impossible roots if the last non-zero term is negative, and 2, 6, 10, ... if the last non-zero term is positive.

Example 1. Let the given equation be $x^4-10x^3+35x^2-50x+24=0$, and the resulting equation will be

$$v^6 - 20v^5 + 142v^4 - 460v^3 + 713v^2 - 520v + 144 = 0.$$

Since this equation has terms of alternating sign throughout, the given equation has no impossible roots.

Example 2. Let the equation be $x^3 + qx - r = 0$, whose roots are α, β, and γ; the equation whose roots are $(\alpha-\beta)^2$, $(\alpha-\gamma)^2$, and $(\beta-\gamma)^2$ will be $v^3+6qv^2+9q^2v+4q^3+27r^2=0$. If q and $4q^3+27r^2$ are negative quantities, the signs will alternate $+\ -\ +\ -$, and in consequence the given equation will have no impossible roots. If one or the other is positive, then the signs do not alternate continuously, so the given equation has two impossible roots. In this case the third root will be positive or negative according to the sign of the coefficient r.

(If $4q^3+27r^2=0$, all roots are real, and two of them are equal.)

Example 3. Let the equation be $x^4 + qx^2 - rx + s = 0$, whose roots are α, β, γ, δ; the equation whose roots are $(\alpha-\beta)^2$, $(\alpha-\gamma)^2$, $(\beta-\gamma)^2$, $(\alpha-\delta)^2$, $(\beta-\delta)^2$, and $(\gamma-\delta)^2$ will be $v^6 + 8qv^5 +$

$$\left\{\begin{matrix}22q^2\\+8s\end{matrix}\right\}v^4+\left\{\begin{matrix}28q^3\\+26r^2\\+16qs\end{matrix}\right\}v^3+\left\{\begin{matrix}17q^4\\+\ 48r^2q\\-112s^2\\+\ 24q^2s\end{matrix}\right\}v^2+\left\{\begin{matrix}4q^5\\+\ 18q^2r^2\\+\ 32q^3s\\-192qs^2\\+216r^2s\end{matrix}\right\}v+\left\{\begin{matrix}256s^3\\-128q^2s^2\\+144r^2qs\\+\ 16q^4s\\+\ 27r^4\\-\ \ 4r^2q^3\end{matrix}\right\}=0.$$

If $256s^3-128q^2s^2+144r^2qs+16q^4s-27r^4-4r^2q^3$ is negative, then the given equation has exactly two impossible roots; if positive, and there is some exception to the continuous alternation of signs of all these coefficients, then all four roots of the given equation are impossible. If it is zero, then two or four of the given equation's roots will be impossible, according as the final term only, or the final term and the term preceding it, are zero.

This solution has a simpler formulation: if the final term of the given equation is positive, and either q is positive or q^2-4s is negative, then the given equation has four impossible roots, but it has none for other values of q.

Example 4. Let the equation be $x^5+qx^3-rx^2+sx-t=0$, whose roots are α, β, γ, δ, and ϵ; then the equation whose roots are $(\alpha-\beta)^2$, $(\alpha-\gamma)^2$, $(\beta-\gamma)^2$, $(\alpha-\delta)^2$, $(\beta-\delta)^2$, etc., will be

$$\begin{aligned}
&w^{10} + 10qw^9 + (39q^2+10s)w^8 + (80q^3+50qs+25r^2)w^7 \\
&+ \qquad (95q^4+124q^2s-95s^2+92qr^2+200rt)w^6 \\
&+(66q^5-360qs^2+169q^3s+118q^2r^2+260r^2s+625t^2+400qrt)w^5 \\
&+(25q^6+40s^3-53r^4+52q^3r^2-522q^2s^2+194q^4s \\
&\qquad +708qr^2s+240q^2rt+1750qt^2-950rst)w^4 \\
&+(4q^7+106q^5s-80qs^3-308q^3s^2-102qr^4-7q^4r^2+570r^2s^2 \\
&\qquad +612q^2r^2s+700r^3t-3750st^2+2500q^2t^2+80q^3rt-2150qrst)w^3 \\
&+(400s^4-360q^2s^3-15q^4s^2+24q^6s-8q^5r^2-45q^2r^4-270r^4s \\
&\qquad +140q^3r^2s+960qr^2s^2+1875r^2t^2+1000rst^2 \\
&\qquad -5000qst^2+1750q^3t^2+40q^4rt+600qr^3t-1650q^2rst)w^2 \\
&+(36q^5s^2-224q^3s^3+320qs^4+4q^3r^4+27r^6-40r^2s^3+434q^2r^2s^2 \\
&\quad -24q^4r^2s-198qr^4s+5000s^2t^2-450r^3st-6250rt^3+675q^4t^2 \\
&\qquad -3750q^2st^2+3000qr^2t^2+60q^2r^3t+200qrs^2t-330q^3rst)w \\
&+ 3125t^4 - 3750qrt^3 + (2000qs^2+2250r^2s-900q^3s+825q^2r^2+108q^5)t^2 \\
&\qquad - (1600rs^3-560q^2rs^2-16q^3r^3+630qr^3s+72q^4rs-108r^5)t \\
&\qquad + 256s^5-128q^2s^4+144qr^2s^3+16q^4s^3-27r^4s^2-4q^3r^2s^2 = 0.
\end{aligned}$$

If the signs of the terms of this equation alternate continuously, from $+$ to $-$ and from $-$ to $+$, the given equation has no impossible roots.

If the signs do not alternate continuously, then two or four of the roots of the given equation are impossible, if the final term is less or greater than zero respectively.

If the final term is equal to zero, and the signs of the other terms do not alternate continuously, then four or only two roots of the given equation will be impossible, according as the resulting equation's penultimate term also, or its final term only, is equal to zero.

Corollary 3. This method, extended to comparison of all pairs of adjacent terms, may reveal the exact number of impossible roots in some equations.

Corollary 4. When the proposed equation, $x^n - px^{n-1} + qx^{n-2} - rx^{n-3} + \ldots = 0$, is transformed into an equation whose roots are $x-a$, and both the original and the resulting equation are then transformed according to the method in this Problem, the same discriminant equation will be produced in each case. [10]

Corollary 5. If the final term of the resulting equation is 0, then two roots of the given equation will be equal to each other. If the last two terms of the resulting equation are 0, then the given equation has two pairs of equal roots; if the last three, four, etc. terms of the resulting equation are 0, the given equation has three, four, etc. pairs of equal roots.

Problem XIII

To find the number of impossible roots contained in a given equation, through a variety of transformations.

Case 1. Let the equation be

$$x^n - px^{n-1} + qx^{n-2} - rx^{n-3} + sx^{n-4} - \ldots = 0, \quad \text{and write}$$

$$ax^n - (a+b)px^{n-1} + (a+2b)qx^{n-2} - (a+3b)rx^{n-3} + \ldots = v;$$

or, what is essentially the same, let the equation be

$$ax^n - (a+b)px^{n-1} + (a+2b)qx^{n-2} - (a+3b)rx^{n-3} + \ldots = 0,$$

and $\quad x^n - px^{n-1} + qx^{n-2} - rx^{n-3} + sx^{n-4} - \ldots = v.$

When the two equations are reduced in order to eliminate v, the result will be $\quad v^n - Av^{n-1} + Bv^{n-2} - Cv^{n-3} + \ldots \mp Pv^3 \pm Qv^2 \mp Rv \pm S = 0$, whose root is v. If the given equation has two equal roots, then here $S = 0$ and $R = 0$; and in general the resulting equation will have as many of its last terms equal to zero, as the given equation has roots that are equal to each other. From the location of the last non-zero term of the resulting equation, we may determine whether the number of impossible roots comes from the series 2, 6, 10, … , or is among the numbers 0, 4, 8, … .

Example 1.1. Let the given equation be $x^2+q=0$, and write $2x=v$; then the resulting equation will be $v^2+4q=0$, and two roots will be impossible, or not, according as the final term of the resulting equation $(4q)$ is positive or negative.

Example 1.2. Let the given equation be $x^3+qx-r=0$, and write $3x^2+q=v$; then $v^3+3qv^2-*-4q^3-27r^2=0$.

Example 1.3. Let the given equation be $x^4+qx^2-rx+s=0$, and write $4x^3+2qx-r=0$; then the resulting equation will be $v^4-8rv^3+(4q^3-16qs+18r^2)v^2-*+256s^3-128q^2s^2+16q^4s+144qr^2s-4q^3r^2-27r^4=0$.

Corollary 1. Let the given equation be $x^n-px^{n-1}+qx^{n-2}-rx^{n-e}+\ldots=0$, and impose the condition $nx^{n-1}-(n-1)px^{n-2}+(n-2)qx^{n-3}-\ldots=v$. Then the resulting equation will be $v^n+Av^{n-1}+Bv^{n-2}+\ldots+Pv^3+Qv^2+*+S=0$. If all the roots are real, then where $n\equiv 2(\mathrm{mod}\,4)$ or $n\equiv 1(\mathrm{mod}\,4)$, S will be a negative number; for other values of n S will be positive.

Corollary 2. Let the equation be

$$x^n-px^{n-1}+qx^{n-2}-rx^{n-3}+\ldots\pm\pi x^2\mp\rho x\pm\sigma=0,$$

and write both $$nx^{n-1}-(n-1)px^{n-2}+(n-2)qx^{n-3}-\ldots=v$$

and $$nx^n\ -(n-1)px^{n-1}+(n-2)qx^{n-2}-\ldots=w.$$

Now write the equations, whose roots are v and w respectively, as

$$v^n-A\,v^{n-1}+B\,v^{n-2}-\ldots\pm P\,v^3\mp Q\,v^2\mp *\quad\mp S=0$$

and $$w^n-A'w^{n-1}+B'w^{n-2}-\ldots\pm P'w^3-Q'w^2\pm R'w\mp S'=0.$$

Then $S:S'::1:\sigma$, and $R'=\pm S$.

For if we say the roots of the given equation are α, β, γ, δ, etc., then the roots v of the first resulting equation will be

$(\alpha-\beta)(\alpha-\gamma)(\alpha-\delta)\ldots$, $(\beta-\alpha)(\beta-\gamma)(\beta-\delta)\ldots$, $(\gamma-\alpha)(\gamma-\beta)(\gamma-\delta)\ldots$, $(\delta-\alpha)(\delta-\beta)(\delta-\gamma)\ldots$, etc.

and the roots w of the second will be

$(\alpha-\beta)(\alpha-\gamma)(\alpha-\delta)\ldots\alpha$, $(\beta-\alpha)(\beta-\gamma)(\beta-\delta)\ldots\beta$, $(\gamma-\alpha)(\gamma-\beta)(\gamma-\delta)\ldots\gamma$, $(\delta-\alpha)(\delta-\beta)(\delta-\gamma)\ldots\delta$, etc.

Example 1.4. Let the given equation be $x^2-px+q=0$, the equation of whose limits will be $2z-p=0$. Now impose the condition $z^2-pz+q=v$, and the result of elimination will be $v+\frac{1}{4}p^2-q=0$, whence there will be two impossible roots, or none, if $\frac{1}{4}p^2-q$ is respectively negative or positive.

Example 1.5. Let $x^3+qx-r=0$; then the equation of limits will be $3z^2+q=0$. From the condition $z^3+qz-r=v$, results the equation

$v^2+2rv+r^2+\frac{4}{27}q^3 = 0$. Two roots of the given equation will be impossible, if $r^2+\frac{4}{27}q^3$ is positive, and none if this quantity is negative.

Example 1.6. Let $x^4+qx^2-rx+s = 0$, whose equation of limits is $4z^3+2qz-r = 0$. Now from $z^4+qz^2-rz+s = v$, we derive the equation

$$v^3+(\tfrac{1}{2}q^2-3s)v^2+(3s^2+q^2s+\tfrac{1}{16}q^4+\tfrac{9}{16}qr^2)v$$
$$-s^2+\tfrac{1}{2}q^2s^2-\tfrac{1}{16}q^4s-\tfrac{9}{16}qr^2s+\tfrac{1}{64}q^3r^2+\tfrac{27}{256}r^4 = 0\,.$$

If the final term of this equation is positive, then exactly two of the roots of the given equation will be impossible.

And so could many more examples be given.

Case 2. Given the equation $x^n-px^{n-1}+qx^{n-2}-rx^{n-3}+\ldots = 0$, by the method previously given we construct equations

$$z^{n-1}-\frac{n-1}{n}pz^{n-2}+\frac{n-2}{n}qz^{n-3}-\ldots = 0$$
and
$$z^n - pz^{n-1}+qz^{n-2}-rz^{n-3}+\ldots = v\,.$$

Now these equations are combined to eliminate z; the resultant will be

$$v^{n-1} - Pv^{n-2} + Qv^{n-3} - Rv^{n-4} + Sv^{n-5} - \ldots = 0\,.$$

Then the given equation will have at least as many impossible roots,

(1) (if n is an odd number) — as the difference τ between the number of sign continuations (from + to + or from − to −) and the number of sign variations (from + to − and from − to +) in the resulting equation; *or*

(2) (if n is an even number) — as the difference τ, as defined above, *diminished* by 1 if the number of continuations is greater than the number of variations, and *augmented* by 1 in the contrary case.

Example 2.1. Let the two equations be $3x^2+2px+q = 0$ and $x^3+px^2+qx+r = v$. Then elimination with respect to v yields $v^2+Pv+Q = 0$, and two of the roots of $x^3+px^2+qx+r = 0$ will be impossible, if and only if Q is greater than 0.

Example 2.2. Let the two equations be $4x^3+3px^2+2qx+r = 0$ and $x^4+px^3+qx^2+4x+s = v$, whence the required equation in v will be $v^3+Pv^2+Qv+R = 0$. Then if R is positive, exactly two of the roots of $x^4+px^3+qx^2+rx+s = 0$ will be impossible. On the other hand, if all the roots of the equation $4x^3+3px^2+2qx+r = 0$ are real, and there is only one sign variation among the coefficients of the equation $v^3+Pv^2+Qv+R = 0$, then all the roots of the given equation $x^4+px^3+qx^2+rx+s = 0$ will be real; in all other cases none of the given equation's roots will be real.

Example 2.3. Let the two equations be

$$5x^4+4px^3+3qx^2+2rx+s=0$$

and

$$x^5+px^4+qx^3+rx^2+sx+t=v,$$

and the equation in v is $v^4+Pv^3+Qv^2+Rv+S=0$. Then

(1) if $S<0$, two roots only of the given equation

$$x^5+px^4+qx^3+rx^2+sx+t=0$$

will be impossible;

(2) if $S>0$ and all roots of the equation

$$5x^4+4px^3+3qx^2+2rx+s=0$$

are real, and the terms of the resulting equation in v contain no more than two sign variations, all the roots of the given equation will be real;

(3) otherwise, four of those roots will be impossible.

Problem XIV

Given an equation $x^n - px^{n-1} + qx^{n-2} - rx^{n-3} + \ldots = 0$, to determine exactly how many of its roots are impossible, and how many are real.

We seek quantities which have no polynomial factors, and which vanish when two roots of the equation are equal, or when threefold, fourfold, or a higher multiplicity of equal roots occurs; from these quantities the number of impossible roots may be extracted.

Any irreducible expression, which vanishes when two roots of the given equation are equal, either becomes negative when 2, 6, 10, 14, 18, ... roots are impossible and becomes positive when 0, 4, 8, 12, 16, ... are impossible; or it becomes positive when 2, 6, 10, 14, 18, ... roots are impossible and negative when 0, 4, 8, 12, 16, ... are impossible.

Now divide the given equation by $x^2-2ax+a^2$, and equate the two coefficients of the remainder separately to zero. When a is eliminated between these two equations, the resulting quantity is the same as the final term in the expressions of Problem XIII, and of course it vanishes when two roots of the given equation are equal to each other. Next, divide the given equation by $(x^2-2ax+a^2)(x^2-2bx+b^2)$ and equate to zero the four coefficients of the remainder; upon elimination of a and b

from these four equations we derive two or more expressions which vanish when the given equation has two pairs of equal roots. Then the given equation may be divided by $(x-a)^3$ and $(x-a)^4$, and the three or four terms of the remainders reduced to eliminate the unknown a; there will result two resp. three, or more, quantities which vanish when three resp. four roots of the equation are equal to the same value. Likewise dividing the given equation by $(x^2-2ax+a^2)(x^2-2bx+b^2)(x^2-2cx+c^2)$, equating to zero the six terms of the remainder, and eliminating a, b, and c from all six terms, we produce three or more equations which vanish when the given equation has three separate pairs of equal roots. The extension to higher degrees is obvious. From the expressions so derived — that is, from all of these expressions, taken together, indicating the presence of one, two, three, or more pairs, ternaries, etc., of equal roots — the number of impossible roots in the given equation may be deduced. For as all such expressions pass through zero the roots pass from equality to impossibility, and it follows that every such rule which discovers impossible roots can discover equal roots as well; conversely, with the help of any rule which discovers multiple roots, it may be possible to detect impossible roots.

Example 1. Let the equation be $x^4+qx^2-rx+s=0$, and divide this equation by $x^2-2ax+a^2$; there will be two remainders, $4a^3+2qa-r$ and $3a^4+aq^2-s$. Equating both to zero and eliminating a between these two expressions, we obtain

$$A \equiv 256s^3-128q^2s^2+144qr^2s+16q^4s-27r^4+4q^3r^2,$$

which will be zero whenever two roots of the given equation are equal.

Next the given equation is divided by $(x^2-2ax+a^2)(x^2-2bx+b^2)$ $(= x^4-2a^2x^2+a^4$, since the coefficient of x^3 in the given equation is 0). In this case the two residues are $q+2a^2$ and $s-a^4$; eliminating a between these two equations gives the equation $q^2-4s=0$. Thus the three quantities A, q, and q^2-4s completely determine the number of impossible roots contained in the given equation.[11]

Example 2. Let the equation be $x^n-Ax^m-B=0$. Then if

(1) n is even and $-B$ is negative, the equation has two real roots;

(2) n is even and m is odd (and $-B$ is positive): two roots of the given equation will be real if also

$$\left(\frac{m-n}{n}\right)^{n-m} A^n - \left(\frac{n}{m}\right)^m B^{n-m}$$

is negative, otherwise none will be real;

(3) n and m are both even: if $-B$ and $-A$ are both positive, then the given equation has no real roots, but if $-B$ is positive and $-A$ is negative, then if

$$\left(\frac{m-n}{n}\right)^{n-m} A^n - \left(\frac{n}{m}\right)^m B^{n-m}$$

is positive, the given equation has four real roots, otherwise it has none;

(4) n and m are both odd: if

$$\left(\frac{m-n}{n}\right)^{n-m} A^n - \left(\frac{n}{m}\right)^m B^{n-m}$$

is positive, three roots, otherwise only one, will be real;

(5) n is odd and m is even: if $-A$ and $-B$ have the same sign, the given equation has only one real root, but if they have opposite signs and

$$\left(\frac{m-n}{n}\right)^{n-m} A^n - \left(\frac{n}{m}\right)^m B^{n-m}$$

has the same sign as $-B$, then the given equation has three real roots, otherwise it has only one.

Theorem 10

Case 1. Let the given equation be $x^n - px^{n-1} + qx^{n-2} - rx^{n-3} + \ldots = 0$, in which for x we substitute $z-a$, which gives the equation $z^n - Pz^{n-1} + Qz^{n-2} - Rz^{n-3} + \ldots = 0$. Then let v be any function of the differences between pairs of roots of the equation in x, and w the same function of the differences between pairs of roots of the equation in z: then v will be the same quantity as w, and the equation whose root is v will be identical to the equation whose root is w.

Case 2. Given two equations $x^n - px^{n-1} + qx^{n-2} - rx^{n-3} + \ldots = 0$ and $z^m - a'z^{m-1} + bz^{m-2} - cz^{m-3} + \ldots = 0$, let the roots of both equations be increased by a, i.e. substitute $v-a$ and $w-a$ for x and z respectively, producing the equations

$$v^n - Pv^{n-1} + Qv^{n-2} - Rv^{n-3} + \ldots = 0$$
$$\text{and} \quad w^m - Aw^{m-1} + Bw^{m-2} - Cw^{m-3} + \ldots = 0 .$$

If we assume the two equations

$$v^n - Pv^{n-1} + Qv^{n-2} - Rv^{n-3} + \ldots = 0$$

$$\text{and} \quad v^m - Av^{m-1} + Bv^{m-2} - Cv^{m-3} + \ldots = u,$$

the equation whose root is u will be the same, whatever the value of a.

For let the roots of the equation $v^n \mp Pv^{n-1} + Qv^{n-2} - Rv^{n-3} + \ldots = 0$ be $a+\alpha$, $a+\beta$, $a+\gamma$, $a+\delta$, $\ldots$, and let the roots of the equation $w^m - Aw^{m-1} + Bw^{m-2} - Cw^{m-3} + \ldots = 0$ be $a+\pi$, $a+\rho$, $a+\sigma$, $a+\tau$, $\ldots$; then the roots of the equation in u will be

$$[a+\alpha-(a+\pi)][a+\alpha-(a+\rho)][a+\alpha-(a+\sigma)]\cdots,$$

$$[a+\beta-(a+\pi)][a+\beta-(a+\rho)][a+\beta-(a+\sigma)]\cdots, \text{ etc.}$$

i.e., $(\alpha-\pi)(\alpha-\rho)(\alpha-\sigma)\cdots$, $(\beta-\pi)(\beta-\rho)(\beta-\sigma)\cdots$, etc., and clearly this is independent of the value of a.

Example. The given equation $x^n - px^{n-1} + qx^{n-2} - rx^{n-3} + \ldots = 0$ is transformed into another, whose roots are greater or less than the given equation's roots by $+a$, the result being

$$z^n + \{na-p\}z^{n-1} + \left\{\begin{matrix} \frac{n(n-1)}{2!}a^2 \\ -(n-1)pa \\ +q \end{matrix}\right\} z^{n-2} + \left\{\begin{matrix} \frac{n(n-1)(n-2)}{3!}a^3 \\ -\frac{(n-1)(n-2)}{2!}pa^2 \\ +(n-2)qa \\ -r \end{matrix}\right\} z^{n-3} + \ldots = 0,$$

and the equation whose roots are the limits of the roots of the resulting equation will be

$$nz^{n-1} + \left\{\begin{matrix} (n-1)na \\ -(n-1)p \end{matrix}\right\} z^{n-2} + \left\{\begin{matrix} (n-2)\frac{n(n-1)}{2!}a^2 \\ -(n-2)(n-1)pa \\ +(n-2)\quad q \end{matrix}\right\} z^{n-3} + \left\{\begin{matrix} (n-3)\frac{n(n-1)(n-2)}{3!}a^3 \\ -(n-3)\frac{(n-1)(n-2)}{2!}pa^2 \\ +(n-3)(n-2)qa \\ -(n-3)\quad r \end{matrix}\right\} z^{n-4} + \ldots = 0.$$

Now if we propose

$$z^n + \{na-p\}z^{n-1} + \left\{\begin{matrix} \frac{n(n-1)}{2!}a^2 \\ -(n-1)pa \\ +q \end{matrix}\right\} z^{n-2} + \left\{\begin{matrix} \frac{n(n-1)(n-2)}{3!}a^3 \\ -\frac{(n-1)(n-2)}{2!}pa^2 \\ +(n-2)qa \\ -r \end{matrix}\right\} z^{n-3} + \ldots = w,$$

then the equation whose root is w will be the same, whatever value we choose for a.

The same method can be adapted to both examples of the preceding Problem, and the previous principle follows from this one.

Theorem 11

Suppose there are two equations,

$$Hx^n - ax^{n-1} + bx^{n-2} - cx^{n-3} + dx^{n-4} - \ldots = 0$$

whose roots are α, β, γ, δ, ϵ, ...

and $Lx^m - px^{m-1} + qx^{m-2} - rx^{m-3} + \ldots = 0$

whose roots are π, ρ, σ, τ, ...; then

(1) Take $Hx^n - ax^{n-1} + bx^{n-2} - cx^{n-3} + \ldots = 0$, and

$$Lx^m - px^{m-1} + qx^{m-2} - rx^{m-3} + \ldots = w,$$

whence by substituting the roots α, β, γ, δ, ϵ, ... of the first equation, for x in the expression $Lx^m - px^{m-1} + qx^{m-2} - rx^{m-3} + \ldots = w$, we produce the n quantities

$$L \times (\alpha-\pi)(\alpha-\rho)(\alpha-\sigma)(\alpha-\tau)\cdots, \quad L \times (\beta-\pi)(\beta-\rho)(\beta-\sigma)(\beta-\tau)\cdots,$$
$$L \times (\gamma-\pi)(\gamma-\rho)(\gamma-\sigma)(\gamma-\tau)\cdots, \quad L \times (\delta-\pi)(\delta-\rho)(\delta-\sigma)(\delta-\tau)\cdots,$$

etc.; hence the product of all the values of w will be

$$\begin{aligned} L^n \times\ & (\alpha-\pi)(\alpha-\rho)(\alpha-\sigma)(\alpha-\tau)\cdots(\beta-\pi)(\beta-\rho)(\beta-\sigma)(\beta-\tau)\cdots \\ & (\gamma-\pi)(\gamma-\rho)(\gamma-\sigma)(\gamma-\tau)\cdots(\delta-\pi)(\delta-\rho)(\delta-\sigma)(\delta-\tau)\cdots \\ & \ldots \\ = L^n \times\ & (\alpha-\pi)(\beta-\pi)(\gamma-\pi)(\delta-\pi)\cdots(\alpha-\rho)(\beta-\rho)(\gamma-\rho)(\delta-\rho)\cdots \\ & (\alpha-\sigma)(\beta-\sigma)(\gamma-\sigma)(\delta-\sigma)\cdots(\alpha-\tau)(\beta-\tau)(\gamma-\tau)(\delta-\tau)\cdots \\ & \ldots \end{aligned}$$

(2) Take $Lx^m - px^{m-1} + qx^{m-2} - rx^{m-3} + \ldots = 0$, and

$$Hx^n - ax^{n-1} + bx^{n-2} - cx^{n-3} + \ldots = v;$$

for x in the latter equation we substitute in turn the roots π, ρ, σ, τ, ... of the former: thus we produce the m values of v, to wit

$$H(\pi-\alpha)(\pi-\beta)(\pi-\gamma)(\pi-\delta)\cdots, \quad H(\rho-\alpha)(\rho-\beta)(\rho-\gamma)(\rho-\delta)\cdots,$$
$$H(\sigma-\alpha)(\sigma-\beta)(\sigma-\gamma)(\sigma-\delta)\cdots, \quad H(\tau-\alpha)(\tau-\beta)(\tau-\gamma)(\tau-\delta)\cdots,$$

and so on.

Then the product of all the values of w, will be, in ratio to the product of all the values of v, as the ratio $\pm L^n : H^m$. That is, let the equations, whose roots are the values of w and the values of v respectively, be $w^n - Pw^{n-1} + Qw^{n-2} - \ldots \pm M = 0$ and $v^m - Av^{m-1} + Bv^{m-2} - Cv^{m-3} + \ldots \pm F = 0$: then $M : F :: L^n : H^m$, where the sign of L is taken as negative if m and n are both odd numbers, otherwise the positive sign is used.

Corollary. (1) Given two equations, $x^n - px^{n-1} + qx^{n-2} - \ldots = 0$ (having roots α, β, γ, δ, ...) and $nx^{n-1} - (n-1)px^{n-2} + (n-2)qx^{n-3} - \ldots = w$, by Problem VI we see that the values of w will be, respectively,

$$(\alpha-\beta)(\alpha-\gamma)(\alpha-\delta)(\alpha-\epsilon)\cdots,\ (\beta-\alpha)(\beta-\gamma)(\beta-\delta)(\beta-\epsilon)\cdots,$$
$$(\gamma-\alpha)(\gamma-\beta)(\gamma-\delta)(\gamma-\epsilon)\cdots,\ (\delta-\alpha)(\delta-\beta)(\delta-\gamma)(\delta-\epsilon)\cdots,$$

etc., and it follows that the product of all the values of w will be $\pm(\alpha-\beta)^2\,(\alpha-\gamma)^2\,(\alpha-\delta)^2\,(\beta-\delta)^2\,(\gamma-\delta)^2\,(\alpha-\epsilon)^2\,(\beta-\epsilon)^2\,(\gamma-\epsilon)^2\,(\delta-\epsilon)^2\cdots$ (where the sign will be $+$, if $n(n-1)/2$ is even, and $-$ otherwise).

(2) Conversely, let the equations be $nx^{n-1}-(n-1)px^{n-2}+(n-2)qx^{n-3}-\ldots = 0$ and $x^n-px^{n-1}+qx^{n-2}-rx^{n-3}+\ldots = v$. By this theorem the product of all the values of w will be, to the product of all the values of v, in the ratio $n^n : 1$; in other words, if the equation whose roots are all the values of v is $v^{n-1}-Pv^{n-2}+Qv^{n-3}-\ldots\pm T = 0$, then $\pm T : (\alpha-\beta)^2(\alpha-\gamma)^2(\alpha-\delta)^2(\beta-\delta)^2(\gamma-\delta)^2(\alpha-\epsilon)^2(\beta-\epsilon)^2(\gamma-\epsilon)^2(\delta-\epsilon)^2 :: 1 : n^n$. (If $n(n-1)/2$ is odd, use $-T$; otherwise, use $+T$.)

Lemma

Given an equation $H \equiv lx^n-px^{n-1}+qx^{n-2}-\ldots = 0$, whose roots are $\alpha, \beta, \gamma, \delta, \epsilon, \ldots$, then if we propose $bK = v$ (where bK is some function of the roots of H) and $aH+bK = w$, the values of v will be the same as the values of w.

For if we write any one of $\alpha, \beta, \gamma, \delta, \ldots$ for x in the quantity aH, it must evidently vanish; therefore, when $\alpha, \beta, \gamma, \delta, \ldots$ are substituted for x in the formula $aH+bK = w$, the same values must result as when those roots are substituted for x in the expression $bK = v$, and it follows that the equation whose roots are the values of v will be the same as the equation whose roots are the values of w.

Lemma

Let H, K, L, and M be algebraic functions of a variable x. Further let $\alpha, \beta, \gamma, \delta, \ldots$ be the roots of the equation $H = 0$, and $\pi, \rho, \sigma, \tau, \ldots$ the roots of the equation $L = 0$. Then propose together $H = 0$ and $K = v$; $H = 0$ and $M = w$; as well as $L = 0$ and $K = y$; and $L = 0$ and $m = z$. Finally assume $H \cdot L = 0$ and $K \cdot M = u$. Let the products of all the values of v, w, y, and z be respectively V, W, Y, and Z; then the product $V \cdot W \cdot Y \cdot Z$ will be equal to the product of all the values of the quantity u.

Theorem 12

Case 1. If there are two equations, $H \equiv lx^n - px^{n-1} + qx^{n-2} - rx^{n-3} + \ldots = 0$ and $K \equiv Lx^m - Px^{m-1} + Qx^{m-2} - Rx^{m-3} + \ldots = 0$, we may write $H \equiv lx^n - px^{n-1} + qx^{n-2} - \ldots = 0$ and $K \equiv Lx^m - Px^{m-1} + Qx^{m-2} - \ldots = v$; or $K = 0$ and $H = w$; then the ratio of the product of all n values of v, to the product of all m values of w, is as $\pm L^n : l^m$. Next take $H = 0$ and $aH + bK = y$; then the preceding lemmata and theorem prove the ratio (product of v's) : (product of y's) :: $1 : b^n$. If we take $aH + bK = 0$ and $H = Y$, then the ratio of the product of the resulting values of y to the product of the resulting values of Y is as $\pm a^n l^n : l^n$, or equivalently as $\pm a^n : 1$. Lastly, if we take $aH + bK = 0$ and $AH + BK = z$ (where a, b, A, and B are given quantities), then (since $\frac{B}{b}(aH + bK) + \frac{bA - Ba}{b} H = AH + BK$) the ratio of the product of all the values of Y to that of all the values of z will be as $1 : \frac{(bA - Ba)^n}{b^n}$; and therefore (by the composition of ratios) the product of all values of z is in ratio to the product of all values of w as $\frac{(bA - Ba)^n}{a^n} \times L^n : \pm l^m$ (the choice is $-l^m$ if n is even and m is odd, otherwise $+l^m$). Likewise the ratio of the product of all the z's, to the product of all the v's, is as $\frac{(bA - Ba)^n}{a^n} : \pm 1$, where the positive sign is taken if n is even, and the negative sign otherwise.

Corollary 1. Let $cH + dK = 0$ and $CH + DK = V$, where c, d, C, and D are given: then the ratio of the product of the z's, to the product of all the V's, is as $\frac{(bA - Ba)^n}{a^n} : \frac{(dC - cD)^n}{c^n}$.

Corollary 2. Given the two equations

$$H \equiv x^n - px^{n-1} + qx^{n-2} - rx^{n-3} + \ldots = 0$$

$$\text{and } K \equiv -px^{n-1} + 2qx^{n-2} - 3rx^{n-3} + 4sx^{n-4} - \ldots = 0$$

we construct the two equations

$aH + bK = ax^n - (a + b)px^{n-1} + (a + 2b)qx^{n-2} - (a + 3b)rx^{n-3} + \ldots = 0$ and

$AH + BK = Ax^n - (A + B)px^{n-1} + (A + 2B)qx^{n-2} - (A + 3B)rx^{n-3} + \ldots = Z$;

and similarly construct the two equations

$cH + dK = cz^n - (c + d)pz^{n-1} + (c + 2d)qz^{n-2} - (c + 3d)rz^{n-3} + \ldots = 0$ and

$CH + DK = Cz^n - (C + D)pz^{n-1} + (C + 2D)qz^{n-2} - (C + 3D)rz^{n-3} + \ldots = V$;

then the ratio of the product of the values of Z, to the product of the values of V, will be $\frac{(bA - Ba)^n}{a^n} : \frac{(dC - cD)^n}{c^n}$.

Case 2. If n and m are equal instead of $n > m$, and we set $H \equiv lx^n - px^{n-1} + qx^{n-2} - \ldots = 0$ and $K \equiv Lx^m - Px^{m-1} + Qx^{m-2} - \ldots = v$, and likewise $K = 0$ and $H = w$, then the ratio of the product of all the values of v to the product of all the values of w will be as $\pm L^n : l^n$. If we set $H = 0$ and $aH + bK = y$, the ratio of the product of the v's to the product of the y's will be $:: L^n : b^n L^n :: 1 : b^n$. For $aH + bK = 0$ and $H = z$, the ratio of the product of the z's to the product of the y's will be as $\pm l^n : (al + bL)^n$. Taking $aH + bK = 0$ and $V = cH + dK = cH - \frac{da}{b} H$ (from $aH + bK = 0$, it follows that $aH = -bK$), the product of the values of z will have to the product of the values of V a ratio $1 : \frac{(cb - da)^n}{b^n}$; whence from the theory of composition of ratios we have the ratio (product of all of the values of V) : (product of all of the values of v) $:: \pm(cb - ad)^n l^n : (al + bL)^n$. If n is an even number, the sign will be positive, otherwise it will be negative.

If $al + bL = 0$ or $cl + dL = 0$, etc., or if $ad = cb$, then either in the equation $aH + bK = 0$ or $cH + dK = 0$, etc., the coefficient of the term of the highest degree in x can be used by a similar method to compute the ratio between the stated products.

Theorem 13

Case 1. Given a cubic equation $x^3 + px^2 + qx + r = 0$, having two impossible roots, the third will be negative or positive, according as r is positive or negative.

Case 2. Given a biquadratic equation $x^4 + px^3 + qx^2 + rx + s = 0$, having two impossible roots and two real, if s is negative one real root is negative and one is positive. If on the other hand s is positive, then if the signs of the terms alternate $+ - + - +$, both real roots are positive, while if all the terms are positive, both are negative. If neither of these two alternatives obtains, then the two real roots will be positive or negative, as $p^3 - 4pq + 8r$ is negative or positive. For suppose the proposed equation could be factored, i.e. as

$$x^4 + px^3 + qx^2 + rx + s = (x^2 - 2ax + a^2 + c^2)(x^2 - 2bx + b^2 - d^2) = 0\,,$$

if we then equate corresponding terms of the two equations, and replace the letters p, q, r, in the expression $p^3 - 4pq + 8r$, by their values in terms of a, b, c, and d, we may easily deduce those cases where b denotes a positive quantity, i.e., where both real roots are negative.

Corollary. Given an equation $x^4+qx^2+rx+s = 0$, having two real roots, and given that s is positive, then the two real roots will be negative or positive, according as r is positive or negative.

Case 3. If in $x^n+px^{n-1}+qx^{n-2}+rx^{n-3}+sx^{n-4}+\ldots+Rx+S = 0$, the last term S is negative, the equation will have an odd number of positive real roots, and an even number of positive real roots if S is positive. If the signs of the terms alternate throughout from $+$ to $-$ and from $-$ to $+$, then all the real roots will be positive, and if all the terms are positive, then all the real roots will be negative. In general, there cannot be more positive real roots than there are sign mutations, from $+$ to $-$ or from $-$ to $+$, in the sequence of coefficients, nor more negative roots than continuations, $+$ followed by $+$ or $-$ followed by $-$.

Example 1. Given $x^{2n+1}+A = 0$, the single real root will be positive or negative, as A is negative or positive. In the case $x^{2n} - A^2 = 0$, one real root is positive and the other is negative.

Example 2. Given $x^{2n}+Ax^m+B = 0$, where $2n > m$: if only two roots are real, then if B is negative, one of them will be positive and the other negative; if B is positive, then the real roots are both positive or both negative, according as A has a positive or negative value.

If this equation has four real roots, two must be positive and two negative.

Example 3. Consider instead $x^{2n+1}+Ax^m+B = 0$: if only one root is real, its sign will be the opposite of the sign of B. If three roots are real, then if B is positive, one of them will be negative and the others positive; if B is negative, one of the three real roots will be positive, and the other two will be negative.

These examples are easy enough to establish from the preceding case, if for the deficient terms we write suitably small non-zero values, chosen positive or negative as required for the sake of demonstration of the example. Such small quantities, whether positive or negative, will not affect the number of impossible, negative, or positive roots, unless two or more of the roots are equal to each other, or to zero.

Case 4. Given the number of impossible roots, to find how many negative or positive roots the equation $x^n-px^{n-1}+qx^{n-2}-rx^{n-3}+\ldots = 0$ has.

Let the given equation's quadratic divisors be $x^2-2ax+a^2\pm A^2 = 0$, $x^2-2bx+b^2\pm B^2 = 0$, $x^2-2cx+c^2\pm C^2 = 0$, $x^2-2dx+d^2\pm D^2 = 0$, $x^2-2ex+e^2\pm E^2 = 0$, etc. Then we seek quantities which vanish when

A^2, B^2, C^2, D^2, or E^2, etc., or a^2, b^2, c^2, etc., or $A^2 \pm a^2$, $B^2 \pm b^2$, etc. are equal to 0. By comparing the resulting expressions to each other we can determine the number of positive and negative roots.

Evidently in equations of high degree this calculation will require a great amount of work.

Problem XV

Granted the existence of a method to find the number of impossible roots in an arbitrary equation involving only one unknown quantity; granted further a method to find limits between which lie the real (positive or negative) values of the quantities produced by this method; and given an equation in several unknown quantities (x, y, v, w, z, ...), *to decide whether or not these unknowns admit of any real corresponding values.*

Treat all but one of the unknown quantities (say the exception is x) as given, and by the stated method determine the conditions under which this unknown is impossible; these conditions emerge as expressions $\alpha, \beta, \gamma, \ldots$, involving different "given" unknowns (perhaps a form of one of the Π's from Problems IX or X—or another one, call it P), from which the unknown x may now be eliminated. These expressions will be positive or negative when all roots of the given equation are impossible: accordingly if we equate all of them to zero, i.e. $\alpha = 0$, $\beta = 0$, $\gamma = 0$, etc., and either $\pi = 0$ or $\rho = 0$, etc., then the roots of the resulting equations show at what point the values of α, β, γ, ..., and either Π or P, etc., become positive, and consequently when roots of the given equation become impossible. By the same method each of the remaining unknown quantities (v, w, ...) is successively eliminated, in order to determine whether or not the given equation admits of any real solutions involving each of these quantities in combination with those eliminated before; this also gives us, for each of the unknown quantities, limits between which real corresponding values are to be found.

It is sometimes possible by this method to find limits between which, for a real value of x, ... there are two, three, four, or more real corresponding values of v, w, etc.

It does not matter, in investigating the above limits for the quantity z, which of the quantities x, v, etc., is eliminated first, nor, if x is eliminated first, which of v, w, etc., is eliminated next; and so on for all of the unknown quantities. In all such cases exactly the same limits must result for those real values of z, to which correspond real values of x, v, w, etc., regardless of the ordering of the elimination of variables.

Example 1. Given $y^2+(2z+8)y+2z^2+10z+20 = 0$, to determine whether or not there are any solutions in which z and y have real corresponding values.

Regard z as given. Then we note that y cannot be real, if $(z+4)^2-2z^2 -10z-10 = -z^2-2z-4$ is always negative, in other words, if z^2+2z+4 is always positive. Now the roots of the equation $z^2+2z+4 = 0$ are impossible, so z^2+2z+4 will always be real, and thus the there is no solution in which both z and y are real.

Example 2. Given $y^2+(2x+4z+4)y+2x^2+6z^2+5xz+6 = 0$, to find whether or not real values may correspond to x, y, and z simultaneously. If all values of y are impossible, then $-(x+2z+2)^2+2x^2+5zx+6z^2+6 = x^2+(z-4)x+2z^2-8z+2$ will always be positive; for this quantity to be positive, the roots of the equation $x^2+(z-4)x+2z^2-8z+2 = 0$ must be impossible, that is, $\frac{1}{4}(z-4)^2-2z^2+8z-2$ must always be negative: in other words, the roots of the equation $7z^2-24z-8 = 0$ must be impossible. But the roots of this equation are not impossible; on the contrary, they are $\frac{12\pm10\sqrt{2}}{7}$. Whence it follows that if the roots y of the above expression are to be real, then z must lie between $\frac{12+10\sqrt{2}}{7}$ and $\frac{12-10\sqrt{2}}{7}$. So we choose some value in the stated range, and having substituted it for z in the equation $x^2+(z-4)x+2z^2-8z+2 = 0$, we must solve the resulting equation for x. If any value lying between the two real values so obtained is substituted for x in the first equation, $y^2+(2x+4z+4)y+2x^2+6z^2+5xz+6 = 0$, y will be found to have two real values.

By the same method we can see that 4 and $-\frac{12}{7}$ are the limits between which values of x must lie, in order that the corresponding values of y and z be real; or that $\frac{-46+5\sqrt{46}}{7} \geq y \geq \frac{-46-5\sqrt{46}}{7}$ determines the values of y to which correspond real values of x and z.

Corollary. If the individual terms of the given equation have in no case a degree higher than 2 in all the unknowns x, y, z, ... together, then the method laid down in these two examples always suffices.

Other observations: (1) Ratios, which are supposed greater or less than others, can easily be transformed into equations which contain affirmative or negative quantities; for example let the ratio $a : b$ be greater than the ratio $c : d$, then will $\frac{b}{a} = \frac{d}{c}-k$, where k is some positive constant, and *vice versa* if $\frac{b}{a} = \frac{d}{c}+k$, then will $a : b > c : d$. [12]

(2) Given an equation involving two quantities x and y; the limits of the quantity y, between which the quantity x will become affirmative or negative, may be deduced from the following principles.

The quantity x passes from affirmative to negative or from negative to affirmative, either through nothing or infinity; or from two impossible roots it passes to affirmative or negative through two or more equal roots; and, *vice versa,* from affirmative or negative to two or more impossible roots through two or more equal roots.

Find therefore the values of y, when x becomes $=0$ or infinite; and also all the cases in which two, etc. values of x become equal, that is, when its roots become impossible; and from thence can be deduced the limits of the quantity y, between which x becomes affirmative or negative.

(3) If $x = P/Q$ be an affirmative quantity, then P will be affirmative or negative, according as Q is affirmative or negative. Assume therefore $P = 0$ or $Q = 0$, and from the roots of the resulting equation can be deduced the cases, in which x becomes an affirmative quantity.

(4) If more (n) unknown quantities (x, y, z, v, etc.) be contained in the given equation, then by the preceding method, find the limits of y, z, v, ... between which x becomesan affirmative or negative quantity, and let the quantities denoting the limits contain not more than $n-1$ unknown quantities: from the above-mentioned quantities or equations expressing the limits, find others denoting their limits, which do not contain more than $n-2$ above-mentioned quantities, and so on.

(5) Often from the substitution of the limits of given quantities can be acquired the limits of the remaining quantity (X), or quantities. Find all the greatest values of the quantity x contained between the

above-mentioned limits, and thence can be deduced the limits sought.

(6) If there are given m equations involving $m+1$ or more unknown quantities; then sometimes with, and sometimes without, reducing them to others involving fewer unknown quantities, can be found limits by the preceding method; and from comparing the limits so acquired can sometimes be deduced the limits sought.

(7) If a given function of the unknown quantities (x, y, z, etc.) is asserted to be contained between given limits, when other functions of the above-mentioned quantities are contained between given limits, and the demonstration of it is required; from the given equations and the given functions find limits of the unknown quantities respectively, and if the latter limits are contained between the former, the proposition is generally true, otherwise not.

Example: Let two equations be given,

$$h = \frac{rx-sy}{tz-lv} \text{ and } k = \frac{s'y-r'x}{t'v-l'z},$$

in which r, s, t, l, s', r', t', and l' are given positive values; required to find limits of the quantitie x, y, z, and v such that h and k will both be positive numbers. From the first equation it is apparent that if x is greater or less than $\frac{sy}{r}$, then z must be greater or less respectively than $\frac{lv}{t}$; and similarly from the second, if x is greater or less than $\frac{s'y}{r'}$, z must be greater or less respectively than $\frac{t'v}{l'}$. Or, if x is greater(less) than $\frac{sy}{r} = \frac{s'y}{r'} \cdot \frac{sr'}{s'r}$, then z is correspondingly greater(less) than $\frac{t'v}{l'} \cdot \frac{sr'}{s'r}v$. In other words, if x is greater than $\frac{sy}{r}$, then z must exceed both $\frac{lv}{t}$ and $\frac{t'v}{l'} \cdot \frac{sr'}{s'r}v$, and if x is less than $\frac{sy}{r}$, then z must be less than both of these quantities.

In a very similar case, if $h = \frac{rx-sy}{tz-lv}$ and $k = \frac{s'y-r'x}{t'z-l'v}$, then as x is greater(less) than $\frac{sy}{r}$, so is z greater(less) than $\frac{lv}{t}$; similarly, if $\frac{sy}{r} = \frac{s'y}{r'} \cdot \frac{sr'}{rs'}$, then z is contrarily less(greater) than $\frac{l'v}{t'} \cdot \frac{sr'}{rs'}v$, i.e., if x is greater than $\frac{sy}{r}$, then z is greater than $\frac{lv}{t}$, but less than $\frac{r's}{s'r}v \cdot \frac{l'v}{t'}$, and so on.

Also, e.g., if the ratio $a+b : b$ is greater than $c+d : d$, then the ratio $b : a-b$ will be less than $d : c-d$, because $a+b : b$ being greater than $c+d : d$ corresponds to $b : a+b$ being less than $d : c+d$, and thus $\frac{a}{b}+1 = \frac{c}{d}+1+k$, where k is a positive number. But then evidently $\frac{a}{b}-1 = \frac{c}{d}-1+k$, whence the ratio $b : a-b$ must be less than $d : c-d$.

From this innumerable propositions are easily deduced: two or more such ratios are set up, of which one is supposed greater than another, etc., and then from these hypotheses, by adding, subtracting, multiplying, and dividing them, more such functions of the above quantities are found where some are sure to be greater than others, and so similar propositions are deduced.

Problem XVI

Knowing the number of impossible as well as positive and negative roots of a given equation, and having a method of finding limits or bounds on the roots of the given equation, between which a certain function of the roots will have real values, to find the number of impossible roots in the equation whose roots are the prescribed function of the roots of the given equation.

Case 1. From the given function, we wish to determine which type of roots, in the given equation, will give rise to impossible or real roots in the resulting equation — and then, from the known number of roots of the said type contained in the given equation, the number of impossible roots in the resulting equation may be found.

This same approach may be used to estimate the number of rational and irrational roots, or the number of positive and negative roots.

This is more easily explained by means of examples.

Example 1.1. Given an equation $x^n-px^{n-1}+qx^{n-2}-rx^{n-3}+\ldots = 0$, with roots α, β, γ, δ, ..., suppose this equation is transformed into another whose roots are α^2, β^2, γ^2, δ^2, ... From the (known) number of impossible roots in the given equation, to find their number in the transformed equation.

First, for every real root in the given equation there is one in the resulting equation. Second, every impossible root in the given equation corresponds to one in the resulting equation, except for roots of the form $\sqrt{-A^2}$ in the given equation. Hence, we must determine (from the following Chapter) the number (m) of pure imaginary roots in the given equation. The resulting equation will have m fewer impossible roots than the given equation, and those will be all of its negative roots; it will have as many positive roots, as the original equation had real roots.

Example 1.2. Given the equation $x^n - px^{n-1} + qx^{n-2} - rx^{n-3} + \ldots = 0$, of whose roots ($\alpha$, β, γ, δ, ...) $2h$ are impossible and $n-2h$ are real. To find the number of impossible roots (and other types) in the transformed equation whose roots are $(\alpha-\beta)^2$, $(\alpha-\gamma)^2$, $(\beta-\gamma)^2$, $(\alpha-\delta)^2$, $(\beta-\delta)^2$, $(\gamma-\delta)^2$, ...

First, if α and β are real quantities, either positive or negative, $(\alpha-\beta)^2$, one of the roots of the resulting equation, will be real and positive. Thus, since there are $n-2h$ real roots in the given equation, there will be $\frac{(n-2h)(n-2h-1)}{2}$ positive roots in the transformed equation.

Second, consider two conjugate roots of the given equation, which we may write as $a+b\sqrt{-1}$ and $a-b\sqrt{-1}$. Here $(a+b\sqrt{-1}-a+b\sqrt{-1})^2 = (2b\sqrt{-1})^2 = -4b^2$, a real negative quantity which will be one of the roots of the resulting equation. Now the given equation has $2h$ impossible roots, so the resulting equation will have h negative roots.

So unless the equation has roots that are equal to each other, or pairs of roots of the form $(a+\sqrt{-b^2}, a+\sqrt{-c^2})$ or $(a+\sqrt{-b^2}, c+\sqrt{-b^2})$, the distribution of roots of the resulting equation will be

$\frac{(n-2h)(n-2h-1)}{2}$	positive,
h	negative, and
$2hn-2h^2-2h$	impossible
$=\frac{n(n-1)}{2}$	altogether.

But if some of the roots of the given equation are equal to each other, or occur in pairs of the form $(a+\sqrt{-b^2}, a+\sqrt{-c^2})$ or $(a+\sqrt{-b^2}, c+\sqrt{-b^2})$ (the number of such roots may be found by methods in Chapter 3), and once this number is known, the true number of positive, negative, and impossible roots in the resulting equation will be easy to determine.

Example 1.3. Given the equation $x^n - px^{n-1} + qx^{n-2} - rx^{n-3} + sx^{n-4} - \ldots = 0$, of whose roots ($\alpha$, β, γ, δ, ...) $2h$ are impossible and $n-2h$ are real. Let this equation be transformed into another whose roots are

$$nx^{n-1}-(n-1)px^{n-2}+(n-2)qx^{n-3}-rx^{n-4}+(n-4)sx^{n-5}-\ldots=v,$$

and find the number of impossible roots in the transformed equation.

Every real root of the given equation corresponds to a real root, and every one of the given equation's impossible roots corresponds to an impossible root, in the transformed equation, except when some root of the equation

$$x^n \quad - \quad px^{n-1} + \quad qx^{n-2} - \quad rx^{n-3}+\ldots=0$$

is equal to some root of the equation

$$nx^{n-1}-(n-1)px^{n-2}+(n-2)qx^{n-3}-(n-3)rx^{n-4}+\ldots=0,$$

or when the given equation has some root of the form $a+\sqrt{-b^2}$ in which a and b satisfy the relation

$$\begin{aligned} n(n-1)\ a^{n-2} &- n\frac{(n-1)(n-2)(n-3)}{3!}a^{n-4}b^2 \\ &+ n\frac{(n-1)(n-2)(n-3)(n-4)(n-5)}{5!}a^{n-6}b^4-\ldots \\ -(n-1)(n-2)pa^{n-3} &+ (n-1)\frac{(n-2)(n-3)(n-4)}{3!}pa^{n-5}b^2 \\ &- (n-1)\frac{(n-2)(n-3)(n-4)(n-5)(n-6)}{5!}pa^{n-7}b^4+\ldots \\ +(n-2)(n-3)qa^{n-4} &- \ldots \\ &+ \ldots \qquad = 0 \end{aligned}$$

(this equation comes from writing $a+\sqrt{-b^2}$ in place of x in the expression $nx^{n-1}-(n-1)px^{n-2}+(n-2)qx^{n-3}-\ldots=0$, and equating to zero the imaginary part of the resulting expression). Provided the given equation has no roots of this sort, the resulting equation in v will have $2h$ impossible roots; if n is even, of the $2n-h$ real roots exactly half will be positive and half will be negative, while if n is odd, $\frac{n-2h-1}{2}$ of them will be negative, and $\frac{n-2h+1}{2}$ will be positive.

The questions of whether or not the above equations have roots in common (and so how many roots of the resulting equation are equal to 0), or how many roots cause the expression just derived to vanish (i.e., how many impossible roots of the given equation correspond to real roots of the transformed equation), are resolved in the next Chapter.

Case 2. Knowing the number of roots of each type in the resulting equation, we may be able determine the number of impossible roots in the given equation.

The method of reasoning is the same as in the preceding case: since we know how, from the number of real and impossible roots in the given equation, to predict the number of real and impossible roots corresponding to them in the transformed equation, we can "work backward" from the known number of real and impossible roots in the transformed equation, to their actual number in the given equation.

Example 2.1. Let the given equation be $x^3+qx-r=0$, whose roots are α, β, and $-\alpha-\beta$; then we may rewrite the given equation as

$$x^3-(\alpha^2+\alpha\beta+\beta^2)x+\alpha\beta(\alpha+\beta)=0.$$

Let this equation be transformed into an equation in z, where the relation between x and z is expressed as $z^2-zx=\frac{q}{3}=\frac{\alpha^2+\alpha\beta+\beta^2}{3}$; the resulting equation is $z^6-rz^3=\frac{8^3}{27}$. First suppose all three roots of the given equation are real, and substitute α, β, and $-\alpha-\beta$ for x successively in the equation $z^2-zx=\frac{\alpha^2+\alpha\beta+\beta^2}{3}$: the three equations that result are

$$z^2-\alpha z = \frac{\alpha^2+\alpha\beta+\beta^2}{3},$$
$$z^2-\beta z = \frac{\alpha^2+\alpha\beta+\beta^2}{3},$$
$$\text{and } z^2+(\alpha+\beta)z = \frac{\alpha^2+\alpha\beta+\beta^2}{3},$$

whence $z=\frac{\alpha}{2}+\frac{(\alpha+2\beta)\sqrt{-3}}{6}$, $z=\frac{\alpha}{2}+\frac{(\beta+2\alpha)\sqrt{-3}}{6}$, and $z=-\frac{\alpha+\beta}{2}+\frac{(\alpha-\beta)\sqrt{-3}}{6}$; hence, if all the roots of the given equation are real, the roots of the resulting equation are all impossible. Now if two roots of the given equation are impossible, say $x=a+\sqrt{-A^2}$ and $x=a-\sqrt{-A^2}$, the third will be $-2a$ and the given equation may be written

$$x^3-(3a^2-A^2)x+2a^3+2A^2a=0.$$

Now supposing that $z-zx=\frac{q}{3}$, and substituting in this the three values of x, there result three equations for z, *viz.*,

$$z^2-(a+\sqrt{-A^2})z = -a^2+\frac{a^2}{3},$$
$$z^2-(a-\sqrt{-A^2})z = -a^2+\frac{a^2}{3},$$
$$\text{and } z^2+2az = -a^2+\frac{a^2}{3};$$

the roots of the first two of these equations are impossible, while the third has a root $z = \frac{A}{\sqrt{3}} - a$. Hence the resulting equation, $z^6 - rz^3 = \frac{q}{3}$ has four impossible and two real roots if the equation $x^3 + qx - r = 0$ has two impossible roots; otherwise the transformed equation has no real roots. Conversely, if the equation $z^6 - rz^3 = \frac{q}{3}$ has two real roots, the given equation must have two impossible roots, otherwise there are none in the given equation.

Example 2.2. Let the equation be $x^4 + qx^2 - rx + s = 0$, whose roots are α, β, γ, δ, and the equation whose roots are the sums of pairs of the given equation's roots, i.e. $\alpha+\beta$, $\alpha+\gamma$, $\beta+\gamma$, $\alpha+\delta$, $\beta+\delta$, and $\gamma+\delta$, will be $z^6 + 2qz^4 + (q^2 - 4s)z^2 - r^2 = 0$. If all of the given equation's roots are real, then all of the resulting equation's roots will be real; on the other hand, if two or four of the given equation's roots are impossible, then four of the resulting equation's roots will be impossible, and two will be real (unless the given equation has two equal pairs of purely imaginary roots, accounting for all of them — i.e., it is of the form $x^4 + 2A^2x^2 + A^4 = 0$ — then two roots of the resulting equation will be impossible, and four will be zero).

Let us denote the given equation's roots by $a \pm A\sqrt{-1}$ and $-a \pm A\sqrt{-1}$; the four imaginary roots of the resulting equation will be $(B+A)\sqrt{-1}$, $-(B+A)\sqrt{-1}$, $(B-A)\sqrt{-1}$, and $(A-B)\sqrt{-1}$, and its two real roots will be $2a$ and $-2a$. Their squares will be four negative numbers and two positive numbers. Since the negative and positive roots of the resulting equation are equal in pairs, all three roots of the equation $v^3 + 2qv^2 + (q^2 - 4s)v - r^2 = 0$, whose roots $v = z^2$ are the squares of the roots of the resulting equation as previously defined, will be real (two negative and one positive). But if $a = 0$ and $A = B$, then the given equation is $x^2 + 2A^2x^2 + A^4 = 0$, and four of the resulting equation's roots will be zero, while the other two will be purely imaginary. [13]

If two roots of the given equation are real (designated by α and β respectively), and the other two $\left(-\frac{\alpha+\beta}{2+\sqrt{-A^2}}\right.$ and $\left.-\frac{\alpha+\beta}{2-\sqrt{-A^2}}\right)$, then four roots $\left(\frac{\alpha-\beta}{2} \pm \sqrt{-A^2}\right.$ and $\left.\frac{\beta-\alpha}{2} \pm \sqrt{-A^2}\right)$ of the resulting equation will be impossible, while the other two, $\alpha+\beta$ and $-\alpha-\beta$, will be real. The squares of the impossible roots will also be impossible, unless $\alpha = \beta$, in which case two roots of the cubic equation $v^3+2qv^2+(q^2-4s)v-r^2 = 0$, hose roots $v = z^2$ are the squares of the roots of the resulting equation, will be equal (specifically, they will be $-A^2$).

Accordingly, if the cubic equation $v^3+2qv^2+(q^2-4s)v-r^2 = 0$ has three real roots, all positive, then all the roots of the given biquadratic will be real; if the cubic has three real roots, of which two are negative and one is positive, all four of the given equation's roots will be impossible; and if the cubic has two impossible roots, exactly two of the biquadratic's roots will be impossible, the other two real. [14]

Corollary. Given a general method of reduction of equations, there follows a method of finding how many impossible roots a given equation has, transforming the resultant equation if necessary so that its degree is less than that of the given equation. For, given a relation between the roots of the transformed equation and those of the given equation, which relations are expressed in formulas of lower degree than the degree of the given equation, then the number of impossible roots in the transformed equation allows us to predict the number of impossible roots in the given equation.

Scholium

The most straightforward means of validating these techniques, whose purpose is to determine the number of impossible roots in a given equation, is either to substitute specific values for the unknowns, or to multiply the given equation first by $x-a = 0$, then by $x^2-2ax+a^2+b^2 = 0$ (the first multiplier has only one root, and it is real; the second has two impossible roots). Evidently the number of impossible roots is unchanged when we multiply any equation by the first multiplier, while the second multiplication produces an equation with two impossible roots that were not present in the original.

Example. Let the given equation be $x^2-2x+2 = 0$, which by Problem IX, Example 1, has two impossible roots; and multiply it by $x+2 = 0$. The result is $x^3-2x+4 = 0$, which by the second rule of Problem IX has no impossible roots, although we know that it has two: that rule, accordingly, is defective.

Moreover the strategy of proof which I have described is applicable not only to these matters, but also when it is required to know how many roots of any type whatever are to be found in a given equation.

For example, consider the famous Cartesian rule: "As many roots are positive, as there are sequences of signs changing from + to − and from − to +, provided that the equation has no impossible roots."

When any given equation is multiplied by $x-a=0$, a being a positive quantity, the number of sign inversions increases by at least one; if the same equation is multiplied by $x+a=0$, there will be as many continuations of sign (+ followed by + or − followed by −) as in the original equation, with one more added. More generally, as this book has been the first to show, when any equation whatever is multiplied by $x-a$, then corresponding to each sign inversion of the original equation will be one, three, five, or a larger odd number, of sign inversions in the resulting equation; and when it is multiplied by $x+a$ then each continuation of signs in the original has a orresponding sequence of one, three, five, ... sign continuations in the new equation.

(1) Consider the equation $x^n+px^{n-1}+qx^{n-2}+\ldots+Px^{n-m}-Qx^{n-m-1}\pm\ldots=0$, where p, q, ..., P, Q are all positive numbers, and $Px^{n-m}-Qx^{n-m-1}$ represents the first sign inversion. If we multiply $x^n+px^{n-1}+qx^{n-2}+\ldots+Px^{n-m}-Qx^{n-m-1}\pm\ldots=0$ by $x-a$ the result will be $x^{n+1}+(p-a)x^n+(s-ap)x^{n-1}+\ldots-(Q+aP)x^{n-m}\pm\ldots=0$. But $-Q-aP$ is negative, so the resulting equation must have one, three, five, or some odd number of sign changes from + to − and from − to + corresponding to the first sign change, from + to −, in the original equation.

Now consider similarly the equation

$$x^n+px^{n-1}+qx^{n-2}+\ldots+\alpha x^{n-r}-\beta x^{n-r-1}-\ldots-\pi x^{n-s}+\rho x^{n-s-1}$$
$$+\ldots+\phi x^{n-t}-\chi x^{n-t-1}\pm\ldots=0$$

where α, β, π, ρ, ϕ, and χ all denote positive real numbers, and let $\alpha x^{n-r}-\beta x^{n-r-1}$ represent any transition from + to − in the given equation, with $-\pi x^{n-s}+\rho x^{n-s-1}$ representing the first subsequent transition from − to +, and regarding $-\pi x^{n-s}+\rho x^{n-s-1}$ as any transition from + to −, let $+\ldots+\phi x^{n-t}-\chi x^{n-t-1}$ represent the first subsequent transition from + to −. Then if we multiply

$$x^n+px^{n-1}+qx^{n-2}+\ldots+\alpha x^{n-r}-\beta x^{n-r-1}-\ldots-\pi x^{n-s}+\rho x^{n-s-1}$$
$$+\ldots+\phi x^{n-t}-\chi x^{n-t-1}\pm\ldots=0$$

by $x-a$, there results

$$x^{n+1}+(p-a)x^n+(q-ap)x^{n-1}+\ldots-(\beta+a\alpha)x^{n-r}-\ldots+(\rho+a\pi)x^{n-s} \\ +\ldots-(\chi+a\phi)x^{n-t}\pm\ldots=0\,.$$

In this equation the coefficient $-\beta-a\alpha$ is negative, $\rho+a\pi$ is positive, and $-\chi-a\phi$ is again negative; therefore from the negative term $-(\beta+a\alpha)x^{n-r}$ to the positive term $(\rho+a\pi)x^{n-s}$ there are one or more (three, five, ...) variations of sign corresponding to the single inversion of sign between the term $-\beta x^{n-r-1}$ and the term $+\rho x^{n-s-1}$ in the given equation. Similarly from the positive term $+(\rho+a\pi)x^{n-s}$ to the negative term $-(\chi+a\phi)x^{n-t}$ in the resulting equation, there will be one, three, five, etc., changes of sign from $+$ to $-$ and from $-$ back to $+$, all corresponding to the single inversion between the term $+px^{n-s-1}$ and the term $-\chi x^{n-t-1}$ in the given equation. Therefore each separate sign inversion in a given equation,

$$x^n+px^{n-1}+qx^{n-2}+\ldots+Lx+M=0$$

has corresponding to it one or more (three, five, ...) sign inversions in the resulting equation. It follows that there will be as many, or more, sign inversions between the first and penultimate term, $(M-aL)x$, of the resulting equation,

$$x^{n+1}+(p-a)x^n+(q-ap)x^{n-1}+\ldots+(M-aL)x-aM=0,$$

as are to be found in the given equation. Now if the number of sign inversions in the resulting equation (up to its penultimate term) is the same as in the given equation, the sign of the penultimate term, $(M-aL)x$, of the resulting equation, must be the same as the sign of the final term, M, of the given equation. But if these signs are the same, then from the penultimate term, $(M-al)x$, to the final term, $-M$, in the resulting equation, there is a change of sign, either from $+$ to $-$ or from $-$ to $+$, and it follows that the resulting equation has more sign inversions, by an odd number, than had the given equation.

(2) Consider the equation

$$x^n-px^{n-1}+qx^{n-2}-rx^{n-3}+\ldots-Px^{n-m-1}+Qx^{n-m}+Rx^{n-m-1}\pm\ldots=0,$$

where the letters p, q, r, ..., P, Q, R, etc. denote positive quantities and $Qx^{n-m}+Rx^{n-m-1}$ is the first case where two consecutive signs are equal (for purposes of this example it has been shown as $+$ to $+$). Let this equation be multiplied by $x+a$, the product being

$$x^{n+1}-(p-a)x^n+(q-ap)x^{n-1}-(r-aq)x^{n-2}+\ldots \\ +(Q-aP)x^{n-m-1}+(R+aQ)x^{n-m}\pm\ldots=0.$$

Now if there are no sign progressions, either $+$ to $+$ or $-$ to $-$, in the resulting equation prior to the term $(Q-aP)x^{n-m+1}$, then it must be true that p, q, r, ..., and Q are all greater than a, ap, aq, ..., and aP respectively, and thus $Q-aP$ must have the same sign as Q (i.e., it is positive), so that $(Q-aP)x^{n-m+1}+(R+aQ)x^{n-m}$ will represent a sign progression from $+$ to $+$. Therefore the resulting equation will have one or more (always an odd number) sign progressions, from $+$ to $+$ and/or from $-$ to $-$, corresponding to the first progression from $+$ to $+$ in the given equation (that is, occurring among the same number of initial terms).

By the same reasoning, *mutatis mutandis*, for the equation

$$x^n-px^{n-1}+qx^{n-2}-rx^{n-3}+\ldots+Px^{n-m-1}-Qx^{n-m}-Rx^{n-m-1}\pm\ldots=0,$$

in which the first sign progression is from $-$ to $-$, the resulting equation will have one, three, five, ... sign progressions, either from $+$ to $+$ or from $-$ to $-$, corresponding to the first progression $-Qx^{n-m}-Rx^{n-m-1}$ in the given equation.

Now consider an equation

$$x^n-px^{n-1}+qx^{n-2}-\ldots+\alpha x^{n-m+1}+\beta x^{n-m}-\gamma x^{n-m-1}+\delta x^{n-m-2}-\ldots$$
$$-\pi x^{n-r}+\rho x^{n-r-1}+\sigma x^{n-r-2}\pm\ldots=0$$

in which $\alpha x^{n-m+1}+\beta x^{n-m}$ is a sign progression (from $+$ to $+$ here, but it could also be from $-$ to $-$), and let $\rho x^{n-r-1}+\sigma x^{n-r-2}$ be the next sign progression, either from $+$ to $+$ or from $-$ to $-$. We multiply

$$x^n-px^{n-1}+qx^{n-2}-\ldots$$
$$+\alpha x^{n-m+1}+\beta x^{n-m}-\gamma x^{n-m-1}+\delta x^{n-m-2}-\ldots$$
$$-\pi x^{n-r}+\rho x^{n-r-1}+\sigma x^{n-r-2}\pm\ldots+Lx+M=0$$

$$\times(x+a)$$
$$=x^{n+1}-(p-a)x^n+(q-ap)x^{n-1}-\ldots$$
$$+(\beta+a\alpha)x^{n-m+1}-(\gamma-a\beta)x^{n-m}+(\delta-a\gamma)x^{n-m-1}-\ldots$$
$$+(\rho-a\pi)x^{n-r}+(\sigma+a\rho)x^{n-r-1}\pm\ldots=0.$$

In this equation, since α and β have the same sign, $\beta+a\alpha$ will have the same sign as β. The signs of the terms $\beta x^{n-m}-\gamma x^{n-m-1}+\delta x^{n-m-2}-\ldots-\pi x^{n-r}+\rho x^{n-r-1}$, in the given equation, alternate continuously from $+$ to $-$ and from $-$ to $+$ (since by hypothesis $+\rho x^{n-r-1}+\sigma x^{n-r-2}$ is the first sign continuation, either $+$ to $+$ or $-$ to $-$, to the right of the one at the α and β terms).

If in the resulting equation, no sign continuations occur between the term $+(\beta+a\alpha)x^{n-m+1}$ and the term $+(\rho-a\pi)x^{n-r}$, then evidently $\gamma>a\beta$, $\delta>a\gamma$, ..., $\rho>a\pi$. Obviously then $+(\rho-a\pi)$ has the same sign as ρ, and since ρ and σ have the same sign, it follows that $\rho - a\pi$ and $\sigma + a\rho$ have the same sign — which means that the resulting equation has a sign continuation at $(\rho - a\pi)x^{n-r} + (\sigma + a\rho)x^{n-r-1}$. So there must be one or more (3, 5, ...) sign continuations from + to + or from − to − corresponding to any sign continuation, such as $\rho x^{n-r-1} + \sigma x^{n-r-2}$, in the original equation. And thus it follows that the resulting equation has as many sign continuations between its first and penultimate terms, as are contained in the given equation, or more by an even number (2, 4, ...). Now if indeed the resulting equation has exactly as many sign continuations from its leading term, x^{n+1}, to the penultimate, $(M+aL)x$, as has the given equation, then there are the same number of sign inversions among all of its terms as in the given equation, and it follows that the penultimate term, $(M+aL)x$, of the resulting equation, has the same sign as the final term, M, of the given equation — and thus in the resulting equation $x^{n-1} - (p-a)x^n + (q-ap)x^{n-1} - \ldots + (M+aL)x + aM = 0$, one continuation of sign (from + to + or from − to −) has indeed been added. So the above analysis leads inevitably to the conclusion that the resulting equation has, by an odd number, more sign continuations from + to + and from − to −, than has the given equation.

Corollary 1. Given the equation

$$\begin{aligned}
&x^n + px^{n-1} + qx^{n-2} + \ldots - Px^{n-m} + Qx^{n-m-1} + Rx^{n-m-2} + \ldots \\
&\pm fx^{n-m-r} \quad \pm gx^{n-m-r-1} \quad \pm hx^{n-m-r-2} \quad \pm kx^{n-m-r-3} \quad \pm \ldots \\
&\pm \alpha x^{n-m-r-s} \quad \pm \beta x^{n-m-r-s-1} \quad \pm \gamma x^{n-m-r-s-2} \quad \pm \ldots \\
&+\pi x^{n-m-r-s-t} - \rho x^{n-m-r-s-t-1} - \sigma x^{n-m-r-s-t-2} + \ldots \qquad = 0
\end{aligned}$$

where $P, Q, R, \ldots, f, g, h, \ldots, \alpha, \beta, \ldots, \pi, \rho, \sigma$, etc. are all positive or all negative. Let $-Px^{n-m} + Qx^{n-m-1}$ be any sign inversion (from + to − or from − to +), and let $+\pi x^{n-m-r-s-t} - \rho x^{n-m-r-s-t-1}$ be the next inversion (from − to + or from + to −). Further suppose that $g > af$, but $ag > h$, $ah > k$, and so on up to β, and then $\gamma > a\beta$. Then, if the given equation is multiplied by $x - a$, in the resulting equation

$$\begin{aligned}
&x^{n+1} + (p-a)x^n + (q-ap)x^{n-1} + \ldots + (Q+aP)x^{n-m} + (R-aQ)x^{n-m-1} + \ldots \\
&+(g-af)x^{n-m-r} \quad -(ag-h)x^{n-m-r-1} \quad -(ah-k)x^{n-m-r-2} \quad - \ldots \\
&-(a\alpha-\beta)x^{n-m-r-s} + (\gamma - a\beta)x^{n-m-r-s-1} \pm \ldots \qquad = 0
\end{aligned}$$

two sign inversions will arise — *viz.*, $+(g-af)x^{n-m-r} - (ag-h)x^{n-m-r-1}$ and $-(a\alpha-\beta) + (\gamma - a\beta)x^{n-m-r-s-1}$ — which correspond to no inversions in the given equation. And for all similar cases with respect to the given equation it can be shown that two additional sign inversions will occur in the resulting equation, which are not present in the given equation.

Corollary 2. Let $+Qx^{n-m-1}+Rx^{n-m-2}$ be some continuation of signs, either $+$ to $+$ or $-$ to $-$, and let $-\rho x^{n-m-r-s-t-1}-\sigma x^{n-m-r-s-t-2}$ be the next subsequent sign continuation (i.e., the signs alternate through all of f, g, h, k, ..., α, β, γ, ...). If $g > af$, while $ag > h$, $ah > k$, and so on up to the coefficient β, after which $\gamma > a\beta$, then upon multiplication of the given equation by $x+a$, two sign continuations will arise, one from $+$ to $+$ and the other from $-$ to $-$, that are not to be found in the given equation. The resulting equation is

$$\begin{aligned}&x^{n+1}+(p-a)x^n+(q-ap)x^{n-1}+\ldots+(Q+aP)x^{n-m}+(R-aQ)x^{n-m-1}+\ldots\\&\pm(g-af)x^{n-m-r}\quad\pm(ag-h)x^{n-m-r-1}\quad\pm(ah-k)x^{n-m-r-2}\quad\pm\ldots\\&\mp(a\alpha-\beta)x^{n-m-r-s}\quad\mp(\gamma-a\beta)x^{n-m-r-s-1}\quad\mp\ldots\\&-(\rho-a\pi)x^{n-m-r-s-t}-(\sigma+a\rho)x^{n-m-r-s-t-1}\pm\ldots\qquad=0,\end{aligned}$$

and the two new sign continuations are $\pm(g-af)x^{n-m-r}\pm(ag-h)x^{n-m-r-1}$ and $\mp(a\alpha-\beta)x^{n-m-r-s}\mp(\gamma-a\beta)x^{n-m-r-s-1}$. And for every similar configuration of signs in the given equation, it is easy to see that two new progressions will arise which it did not have.

Corollary 3. Thus it is easy to see that, if the above equation is multiplied by $x-a$ and then by $x+a$, and the number of sign changes is increased by 1 after the first multiplication and is in effect reduced by 1 and increased by 2, 4, 6, etc., after the second multiplication, then the given equation has two, four, six, etc., impossible roots.

TRANSLATOR'S NOTES

[1] The astute reader will have noticed that π appears in all six values of β, but ρ only appears in two of them, and likewise for σ and τ. This apparent lack of symmetry is "explained away" by noting that the sum of the roots of the given equation is 0 (because there is no px^3 term), hence $\pi = -\rho-\sigma-\tau$, allowing us to rewrite the β values as $\pm(\sigma+\tau)\sqrt{-1}$, etc. We can then confirm the correctness of Waring's assertions by computing the product $(\sigma+\tau)(\rho+\tau)(\rho+\sigma)$, which should be equal to $-r$ since $-r^2$ is the constant term of the equation for β. In the original equation r is the coefficient of x and thus represents the sum $\pi\rho\sigma+\pi\rho\tau+\pi\sigma\tau+\rho\sigma\tau$; eliminating π as above, we have $r = \rho\sigma\tau-(\rho\sigma+\rho\tau+\sigma\tau)(\rho+\sigma+\tau)$, and carrying out the formal multiplication of these factors gives exactly the expected result. (Waring did not say which root of the given

equation should be regarded as π, because in fact it does not matter; the results are valid whichever one you choose, as the theory of symmetric functions would lead one to expect.)

[2] In this proof Waring implicitly appeals to a continuity argument which was not rigorously justified until the work of Weierstrass and others about a century later. This predates the more classical instance of the interdependence of algebra and analysis, in the proof of the "fundamental theorem of algebra".

[3] Although multiplying by the arithmetic progression n, $n-1$, $n-2$, ... in fact computes the derivative of the given polynomial, Waring has up to this point avoided use of the terms "derivative", "differentiation", etc. The proofs so far are algebraic rather than analytic in nature (with the exception of the intuitive concept of continuity, as noted previously).

[4] The "doctrine of limits" was central to the (then incomplete) theory of determining the number of real and complex roots of equations; it was developed earlier by Maclaurin, and earlier still by Newton. For the case of all real roots in both the polynomial and its derivative, it was fairly straightforward, but at this point we see an impediment: Waring (and his predecessors) tried to extend the concept that the real roots of the derivative separate the real roots of the polynomial, by claiming that a real root of a derivative could be a limit between a pair of roots at least one of which was complex. In simple cases the root of the derivative proves to be the real part of a pair of conjugate complex roots, but Waring has not in any way shown that this extends to all cases. A stronger assertion, not considered by Waring, is that if all the roots of a polynomial $f(z)$, of degree n, are within a circle of radius R in the complex plane, then so are all the roots of $nf(z)$-$f'(z)$ [Gonggrÿp 1915].

[5] The example assumes $h > m$ (and, of course, that m does not divide h). If $h < m$, Q would not be a polynomial but a rational function having poles of order $m - h$ at the places where P has zeroes of order h.

[6] Here Waring is beginning to derive "Newton's rule" for detection of complex roots (stated without proof by Newton[1707]), which was expanded, and partially proved, by Campbell[1728] and Maclaurin[1729]. (For a discussion of the priority dispute that erupted between Campbell and Maclaurin over their respective extensions to Newton's rule, see Mills[1983] or Weeks[1991]. The text of Maclaurin's papers and his [1730] reply to Campbell's[1729] attack are included in Mills[1982]; the only known copy of Campbell's polemic is at the National Library of Scotland, which also possesses one of two surviving copies of Maclaurin's reply.) Sylvester[1864, p. 580] wrote a very incisive commentary on Waring's treatment of Newton's rule and its extensions:

> Mr Archibald Smith has obligingly called my attention to Waring's treatment of the question of Newton's rule in the *Meditationes Analyticæ*[*sic*]. On superficial examination the reader might be induced to suppose that in part 9, p. 68, ed. 1782, Waring had deduced a proof of the rule from the preceding propositions; but on looking into the case will find that there is not the slightest vestige of proof, the rule being stated, but without any demonstration whatever being either adduced or alleged. In fact, on turning to the preface of this (the last) edition of the *Meditationes*, the reader will find an explicit avowal of the demonstration being wanting. After referring in order to Campbell's, Maclaurin's, and Newton's rules, as well as his own, for deducing the existence of impossible roots, he adds these words:
>
> > At omnes hæ regulæ prædictæ perraro invenerunt verum numerum impossibilium radicum in æquationibus multarum dimensionum *et adhuc demonstratione egent*; vulgares enim demonstrationes solummodo probant impossibiles radices in data æquatione contineri, non vero quot *saltem tot sunt quot invenit regula.*
> >
> > Vera resolutio problematis est perdifficilis et valde laboriosa; cognitum est radices ex possibilitate per æqualitatem transire ad impos-

> sibilitatem; ergo in generali resolutione hujusce problematis necesse est invenire casum in quo radices datæ æquationis evadunt æquales; resolutio autem hujus casus valde laboriosa est; et consequenter resolutio generalis prædicti problematis magis *erit* laboriosa.

Written in Latin, and when the proper language of algebra was yet unformed, it is frequently a work of much labour to follow Waring's demonstrations and deductions, and to distinguish his assertions from his proofs. I find he agrees with the opinion expressed by myself, that Newton's rule will *not* "pene," as stated by Newton, but only "perraro," give the true number of imaginary roots. Like myself, too, in the body of the memoir Waring has given theorems of probability in connexion with rules of this kind, but without any clue as to his method of arriving at them. Their correctness may legitimately be doubted.

[Since the above was sent to press, I have been enabled to ascertain that the great name of Euler is to be added to the long list of those who have fallen into error in their treatment of this question: see *Institutiones Calculi Differentialis*, vol. II. cap. XIII. He says (p. 555, edition of Prony), "videndum est utrum hæc duo criteria (meaning Newton's criteria of imaginariness) sint contigua necne; priori casu numerus radicum imaginarium non augebitur; posteriori vero *quia criteria litteras prorsus diversas involvint*, unumquodque binas radices imaginarias monstrabit."

The force of the supposed argument is contained in the words in italics. It is sufficiently met by the question, why or how the conclusion follows from them? Moreover the letters of two non-contiguous criteria are *not* necessarily *prorsus diversæ*; for two criteria with but a single other intervening between them will contain one letter in common.]

[7] Campbell[1728] introduced the rule of this example, with the following definition of the multipliers: "From each of the Unciæ of the middle Terms of that Power of a Binomial, whose Index is the Dimensions of the proposed Æquation, subtract Unity, then divide each Remainder by twice the Correspondent *Uncia*, and set the Fractions which result from this Division, above the middle Terms of the given Æquation." (Prop. II, pp. 528-529). To prove that Waring's formulation of the multipliers is equivalent to Campbell's, is an interesting exercise which I leave to the reader. More significantly, this is the first rule presented by Waring which involves more than three adjacent coefficients. Campbell and Maclaurin (see the latter's "General Corollary" [1729, pp. 73–74]) both appear to have had an intuitive grasp of what might now be called an information-theoretic argument, that a rule, involving more coefficients in at least some of the multipliers, would generally prove to be more successful than something like Newton's original rule, which considered only the nearest neighbors of each term. Surely Waring also suspected this was true, but nothing in this volume says so explicitly (perhaps because he did not have any means of proving such an assertion).

[8] Throughout Problem IX, Waring has focused on the chance that a rule might fail to detect complex roots that are in an equation, not that it might find too many. Sylvester [*loc. cit.*] pointed out, in specific reference to Campbell's [1728] paper, that

> In it the same singular error of reasoning is committed as in the notes of the French edition of the *Arithmetica*, namely, of assuming, without a shadow of proof, that if each of a set of criteria indicates the existence of some imaginary roots, a succession of sets of such criteria must indicate the existence of at least as many distinct imaginary pairs of roots as there are such sets (see paragraph at foot of p. 528, *Phil. Trans.*, vol. XXXV.) — much as if, supposing a number of dogs to be making a point in the same field, the existence could be assumed of as many birds as pointers.

[9] "Hæ autem quantitates continuo regulariter ascendunt;" it is not clear whether Waring means "the number of factors per term increases at each step", or actually refers to the magnitude of p, q, r, etc. (There are exceptions to the latter assertion, for instance an equation of degree n with all n roots equal to 1 has a series of coefficients derived according to the binomial theorem, which increase in magnitude up to position $n/2$ or $(n-1)/2$, after which they decrease, the final coefficient being 1.)

[10] Waring never used the term "discriminant" (it was coined by Sylvester in the middle of the 19th century); I have introduced it to improve the clarity of the text, and to emphasize the correlation between what Waring did in his time, and what has developed since. The enormous 10th-degree discriminant, and the more modest 6th-degree one, of the two preceding Examples, were first published by Waring [1763, pp. 294-295].

[11] For a modern derivation of the same result, see Varnhorn[1939].

[12] The material inserted here comes from the second and third pages of the last-printed 8-page addendum; it appears in English in the *Philosophical Transactions* for the year 1787: item (1) on page 76, which includes some examples not in the *Med. Alg.* addenda; and items (2) through (7) on pages 79-80 (where they are numbered (1) through (6); English renditions of the examples following (7) are wanting). Where the English text closely matches the Latin of the addendum, I have copied it verbatim, although hereafter I will revert to using "positive" instead of "affirmative", etc.

[13] Waring has omitted another case: where $A = B$ and $a \neq 0$, the six roots of the resulting equation will be

$$\begin{array}{lll} a{+}Ai + a{-}Ai\,, & a{+}Ai - a{+}Ai\,, & a{+}Ai - a{-}Ai\,, \\ \;= \;2a\,(\text{real}) & = \text{-}2Ai\,(\text{imag}) & \;=\; 0 \\ a{-}Ai - a{+}Ai\,, & a{-}Ai - a{-}Ai\,, & \text{and } \text{-}a{-}Ai - a{-}Ai\,. \\ \;=\; 0 & = \text{-}2Ai\,(\text{imag}) & \;= \text{-}2a\,(\text{real}) \end{array}$$

[14] This is somewhat akin to an approach, first detailed by Campbell [1733], which appears in slightly different guise (for a quartic example only) in the *Miscellanea Analytica* [Waring 1762, p. 21]

(his structure is similar to Campbell's, but the numerical coefficients are different, as are the precise initial conditions of the problem). It is not clear why Waring eliminated this example — or altered it so much — in his later editions. He probably had not seen Campbell's more general results (they were an original paper published in what was supposed to be a collection of reprints from the *Philosophical Transactions*; earlier, in Campbell's [1729] attack on Maclaurin, the cubic case was mentioned briefly, obviously to solidify Campbell's claim to priority of this discovery). Campbell's cubic example turns out to be equivalent to the cubic case of the Borchardt-Jacobi theorem, now the most concise general statement of a method to identify complex roots[Borchardt 1847; see also MacDuffee 1930]. (For a review of the state of the art prior to Borchardt, see Moigno[1840].)

Maclaurin's [1729] analysis of the matter proposed several different rules which could be applied to different cases — to which Campbell [1729] objected that it was impractical to contemplate application of all of Maclaurin's rules to a single problem. In terms of modern complexity theory, the multitude of rules proposed by Maclaurin and by Waring, no one of which will always correctly determine the number of complex roots, means that that the problem in the 18th century was "NP-complete" (a nondeterministic polynomial-time solution exists, but is not guaranteed to produce the correct result for all cases). The Borchardt-Jacobi theorem evaluates a sequence of determinants based on power sums of the roots of the given equation, culminating with its discriminant, and thereby moves the problem from complexity class NP to class P, i.e., a deterministic solution always exists and the number of computational steps is a polynomial function of the degree of the problem.

Chapter 3

On the reduction and resolution of equations.

Problem XVII

Given two equations, $x^n - px^{n-1} + qx^{n-2} - rx^{n-3} + \ldots = 0$ *and* $x^m - Px^{m-1} + Qx^{m-2} - Rx^{m-3} + \ldots = 0$*, wherein* r *roots of the first equation are also roots of the second, to reduce the first equation to two, one of degree* r *and one of degree* n−r*, and likewise to reduce the second equation to two, of degrees* r *and* m−r*, where the roots of the equations of degree* r *are the same as the* r *roots common to the two given equations.*

We must find the greatest common divisor of the two equations, which will be of degree r; and supposing it equated to zero, it will be one of the required equations, common to the two parts of the solution. When the two given equations are divided by their common divisor, the two respective quotients equated to zero will be the other two required equations, of degrees $n-r$ and $m-r$ respectively.

Example 1. Let the two equations be

$$x^5-2x^4+x^3+20x^2-68x+48=0$$

and

$$x^3+2x^2-13x+10=0.$$

The greatest common divisor of these equations is x^2-3x+2; on division of $x^3+2x^2-13x+10=0$ and $x^5-2x^4+x^3+20x^2-68x+48=0$ by the common factor, the equation $x^3+2x^2-13x+10=0$ is reduced to the quadratic $x^2-3x+2=0$ and the linear equation $x+5=0$, while the equation $x^5-2x^4+x^3+20x^2-68x+48=0$ reduces to the cubic $x^3+x^2+2x+24=0$ and the common quadratic, $x^2-3x+2=0$.

Example 2. Given an equation $x^n-px^{n-1}+qx^{n-2}-\ldots=0$, whose roots are α, β, γ, δ, etc., i.e. the equation is equivalent to $(x-\alpha)\cdot(x-\beta)\cdot(x-\gamma)\cdot(x-\delta)\cdots$, then given some of its roots, say α, β, and γ, or an equation of degree m, $(x-\alpha)(x-\beta)(x-\gamma)\cdots=0$, whose roots are α,

β, γ, ..., all roots of the given equation; then if the given equation is divided by the latter equation, it is depressed into two others, one of degree m, the other of degree $n-m$.

Problem XVIII

Given an equation $x^n-px^{n-1}+qx^{n-2}-rx^{n-3}+sx^{n-4}-\ldots=0$, *some of whose roots (say* α, β, γ, δ, ϵ, ...*) stand in a given relation to one another, to find those roots of the given equation which have the given relation to each other.*

Case 1. In the given equation for the unknown x substitute all of the roots α, β, γ, δ, ϵ, ... which have the stated relation to each other, and the common divisors of the resulting quantities will contain the desired roots.

If the number of unknown quantities is less than the number of equations resulting, then the method of common divisors must be revised to find the common divisors, not of each pair of equations, but of more together, with each divisor divided by the smallest remainder in the process.

Example. Let the roots have the relation of terms of an arithmetic series a, $a+b$, $a+2b$, $a+3b$, etc.

On substitution of these values for x in the given equation, there result the equations

$$a^n-pa^{n-1}+qa^{n-2}-ra^{n-3}+\ldots=0,$$
$$(a+b)n-p(a+b)^{n-1}+q(a+b)^{n-2}-r(a+b)^{n-3}+\ldots=0,$$
$$(a+2b)n-p(a+2b)^{n-1}+q(a+2b)^{n-2}-r(a+2b)^{n-3}+\ldots=0,$$
etc. etc. etc.

and the common divisor of any two of these will have the desired roots.

Case 2. If the given relation among m roots (α, β, γ, δ, ϵ, ...) can only be expressed by an equation of the form $x^m-Px^{m-1}+Qx^{m-2}-Rx^{m-3}+Sx^{m-4}-\ldots=0$: then we take the given equation as $x^n-px^{n-1}+qx^{n-2}-rx^{n-3}+sx^{n-4}-\ldots=w$, and by Problem V we derive from the latter equation, $x^m-Px^{m-1}+Qx^{m-2}-Rx^{m-3}+Sx^{m-4}-\ldots=0$, an equation whose root is w. Say this equation is $w^m-aw^{m-1}+bw^{m-2}-cw^{m-3}+dw^{m-4}-\ldots=0$. Equate to zero successively all the coefficients a, b, c, d, e, etc., and the common divisor of the resulting equations will have the required roots.

Case 3. There is yet another means of resolving this problem.

(3.1) Let the given equation be $x^n - px^{n-1} + qx^{n-2} - rx^{n-3} + \ldots = 0$, with roots $\alpha, \beta, \gamma, \delta, \ldots$ having the stated relation. Then write $x-\alpha = 0$, $x-\beta = 0$, $x-\gamma = 0$, $x-\delta = 0$, etc., and divide the given equation by the product of these factors, i.e. by $(x-\alpha)\cdot(x-\beta)\cdot(x-\gamma)\cdot(x-\delta)\cdots = 0$. Set each coefficient of the remainder equal to zero, and the common divisor of any two of the resulting terms will give the required root.

Example 3.1.1. Suppose α and β are two roots of the given equation $x^n - px^{n-1} + \ldots = 0$, which bear to each other a given relation, namely equality (i.e. $\alpha = a$ and $\beta = a$). Write $x-a = 0$ (twice) and form the product $(x-a)(x-a) = x^2 - 2ax + a^2 = 0$. Divide the given equation by this product, as shown here:

$$
\begin{array}{r|l}
 & x^{n-2} + (2a-p)x^{n-3} + \ldots \\
\hline
x^2 - 2ax + a^2 & x^n - \ px^{n-1} + \ qx^{n-2} - rx^{n-3} + \ldots = 0 \\
 & \underline{x^n - 2ax^{n-1} + a^2x^{n-2}} \\
 & (2a-p)x^{n-1} + \quad (q-a^2)x^{n-2} \\
 & \underline{(2a-p)x^{n-1} - (4a^2-2ap)x^{n-2} + (2a^3-a^2p)\ x^{n-3}} \\
 & \qquad (3a^2-2ap+q)x^{n-2} - (2a^3-a^2+r)x^{n-3} \\
 & \qquad\qquad \text{etc.}
\end{array}
$$

Clearly the coefficient of the first term of the remainder is

$$na^{n-1} - (n-1)pa^{n-2} + (n-2)qa^{n-3} - (n-3)ra^{n-4} + \ldots$$

and the coefficient of the last term is

$$(n-1)a^n - (n-2)pa^{n-1} + (n-3)qa^{n-2} - (n-4)ra^{n-3} + \ldots$$

Equate these two quantities to zero, and the common divisor of the two equations will be the required root.

Example 3.1.2. Suppose m roots $\alpha, \beta, \gamma, \delta, \ldots$ of the given equation are equal to each other.

For each of the roots $\alpha, \beta, \gamma, \delta, \ldots$ write a, then multiply together $x-a = 0$, $x-a = 0$, $x-a = 0$, $x-a = 0$, ..., and their product will be

$$(x-a)^m \equiv x^m - max^{m-1} + \frac{m(m-1)}{2}a^2x^{m-2} - \frac{m(m-1)(m-2)}{2\cdot 3}a^3x^{m-3} + \ldots = 0.$$

Divide the given equation by this product, the process being

$$
\begin{array}{r|l}
 & x^{n-m}+(na-p)x^{n-m-1}+\ldots \\
\hline
x^m-max^{m-1}+\frac{m(m-1)}{2}a^2x^{m-2}-\ldots & x^n-px^{n-1}+qx^{n-2}-rx^{n-3}+sx^{n-4}-\ldots \\
 & x^n-max^{n-1}+\frac{m(m-1)}{2}a^2x^{n-2}-\ldots \\
\cline{2-2}
 & (ma-p)x^{n-1}+(q-\frac{m(m-1)}{2}a^2)x^{n-2}+\ldots \\
 & (ma-p)x^{n-1}+(map-m^2a^2)x^{n-2}-\ldots \\
\cline{2-2}
 & (q-map+\frac{m(m-1)}{2}a^2)x^{n-2}-\ldots \\
 & \text{etc.}
\end{array}
$$

Evidently the coefficient of the first term of the remainder will be

$$
\frac{m(m+1)(m+2)\cdots n}{2\cdot 3\cdots(n-m+1)}a^{n-m+1} - \frac{m(m+1)\cdots(n-1)}{2\cdots(n-m)}pa^{n-m}
+\frac{m(m+1)\cdots(n-2)}{2\cdots(n-m-1)}qa^{n-m-1} - \frac{m\cdots(n-3)}{2\cdots(n-m-2)}ra^{n-m-2}+\ldots;
$$

the second will be

$$
(n-m+1)\frac{(m-1)m(m+1)\cdots n}{2\cdot 3\cdots(n-m+2)}a^{n-m+2}
-(n-m)\frac{(m-1)m(m+1)\cdots(n-1)}{2\cdot 3\cdots(n-m+1)}pa^{n-m+1}
+(n-m-1)\frac{(m-1)m(m+1)\cdots(n-2)}{2\cdot 3\cdots(n-m)}qa^{n-m}-\ldots;
$$

the third will be

$$
\frac{(n-m+1)(n-m+2)}{2}\cdot\frac{(m-2)(m-1)m(m+1)\cdots n}{2\cdot 3\cdot 4\cdots(n-m+3)}a^{n-m+3}
-\frac{(n-m)(n-m+1)}{2}\cdot\frac{(m-2)(m-1)m(m+1)\cdots n}{2\cdot 3\cdot 4\cdots(n-m+2)}pa^{n-m+2}
+\frac{(n-m+1)(n-m+2)}{2}\cdot\frac{(m-2)\cdots(n-2)}{1\cdots(n-m+1)}qa^{n-m+1}-\ldots;
$$

the fourth will be

$$
\frac{(n-m+1)(n-m+2)(n-m+3)}{2\cdot 3}\cdot\frac{(m-3)(m-2)(m-1)\cdots n}{2\cdot 3\cdots(n-m+4)}a^{n-m+4}
-\frac{(n-m)(n-m+1)(n-m+2)}{2\cdot 3}\cdot\frac{(m-3)(m-2)\cdots(n-1)}{2\cdots(n-m+3)}pa^{n-m+3}
+\frac{(n-m-1)(n-m)(n-m+1)}{2}\cdot\frac{(m-3)\cdots(n-2)}{1\cdots(n-m+2)}qa^{n-m+2}-\ldots;
$$

and so for all the rest.

Equate these coefficients severally to zero, and the common divisor of the resulting equations will be the required root.

Corollary 3.1.1. $\frac{m(m+1)(m+2)(m+3)\cdots n}{(n-m+1)!} - m\cdot\frac{m(m+1)(m+2)\cdots(n-1)}{(n-m)!}$

$$+\frac{m(m-1)}{2}\cdot\frac{m(m+1)\cdots(n-2)}{(n-m-1)!} - \frac{m(m-1)(m-2)}{3!}\cdot\frac{m(m+1)\cdots(n-3)}{(n-m-2)!}$$

$$+\frac{m(m-1)(m-2)(m-3)}{4!}\cdot\frac{m(m+1)(m+2)\cdots(n-4)}{(n-m-3)!} - \cdots$$

$$\pm\frac{m(m-1)(m-2)\cdots(m-r+1)}{r!}\cdot\frac{m(m+1)(m+2)\cdots(n-r)}{(n-m-r+1)!}$$

$$= \pm\frac{(n-m-r+1)(n-m-r+2)(n-m-r+3)(n-m-r+4)\cdots(n-m)}{r!}$$

$$\times\frac{(m-r)(m-r+1)(m-r+2)\cdots(n-r)}{(n-m+1)!}$$

Then, (1) $\frac{n(m-1)(n-2)\cdots(n-r+1)}{r!} + m\cdot\frac{n(n-1)(n-2)\cdots(n-r+2)}{(r-1)!}$

$$+\frac{m(m-1)}{2}\cdot\frac{n(n-1)\cdots(n-r+3)}{(r-2)!} + \frac{m(m-1)(m-2)}{3!}\cdot\frac{n(n-1)\cdots(n-r+4)}{(r-3)!}$$

$$+\frac{m(m-1)(m-2)(m-3)}{4!}\cdot\frac{n(n-1)(n-2)\cdots(n-r+5)}{(r-4)!} + \cdots$$

$$+\frac{m(m-1)(m-2)(m-3)\cdots(m-r+1)}{r!}$$

$$= \frac{(n+m)(n+m-1)(n+m-2)(n+m-3)\cdots(n+m-r+1)}{r!};$$

(2) $\frac{n(m-1)(n-2)\cdots(n-r+1)}{r!} - m\cdot\frac{n(n-1)(n-2)\cdots(n-r+2)}{(r-1)!}$

$$+\frac{m(m+1)}{2}\cdot\frac{n(n-1)\cdots(n-r+3)}{(r-2)!} - \frac{m(m+1)(m+2)}{3!}\cdot\frac{n(n-1)\cdots(n-r+4)}{(r-3)!}$$

$$+\frac{m(m+1)(m+2)(m+3)}{4!}\cdot\frac{n(n-1)(n-2)\cdots(n-r+5)}{(r-4)!} - \cdots$$

$$\pm\frac{m(m+1)(m+2)(m+3)\cdots(m+r-1)}{r!}$$

$$= \frac{(n-m)(n-m-1)(n-m-2)(n-m-3)\cdots(n-m-r+1)}{r!}; \text{ and}$$

(3) $\frac{n(m-1)(n-2)\cdots(n-r+1)}{r!} - n\cdot\frac{n(n-1)(n-2)\cdots(n-r+2)}{(r-1)!}$

$$+\frac{n(n-1)}{2}\cdot\frac{n(n-1)\cdots(n-r+3)}{(r-2)!} - \frac{n(n-1)(n-2)}{3!}\cdot\frac{n(n-1)\cdots(n-r+4)}{(r-3)!}$$

$$+\frac{n(n-1)(n-2)(n-3)}{4!}\cdot\frac{n(n-1)(n-2)\cdots(n-r+5)}{(r-4)!} - \cdots$$

$$\pm\frac{n(n-1)(n-2)(n-3)\cdots(n-r+1)}{r!}$$

$$= \pm\frac{n(n-1)(n-2)(n-3)\cdots(n-\frac{r}{2}+1)}{\frac{r}{2}!}.$$

— with the sign taken as positive, if $r = 4a$, negative if $r = 4a+2$, where a is an integer; and if r is odd, this product is zero.

Example 3.1.3. Let α, β, γ, δ, ϵ, ... be terms of an arithmetic series a, $a+b$, $a+2b$, ..., $a+(m-1)b$.

For convenience write P for the sum of the numbers 1, 2, 3, 4, ..., $m-1$; for the sum of all products of two of these numbers (1×2, 1×3, 2×3, 1×4, 2×4, 3×4, 1×5, 2×5, etc.), of each three, each four, each five, and so on write Q, R, S, T, etc. When the factors $(x-a)$ $(x-a-b)$ $(x-a-2b)$ $(x-a-3b)\cdots(x-a-(m-1)b)$ are multiplied together, the product will be

$$x^m-(ma+P)x^{m-1}+\left(\frac{m(m-1)}{2}a^2+(m-1)Pb+Qb^2\right)x^{m-2}$$
$$-\left(\frac{m(m-1)(m-2)}{2\cdot 3}a^3+\frac{(m-1)(m-2)}{2}Pba^2+(m-2)Qb^2a+Rb^3\right)x^{n-3}+\ldots=0.$$

The given equation is divided by this product, and the common divisor of any two terms of the remainder will be the required root.

Example 3.1.4. Let the roots α, β, γ, δ, ... be terms of a geometric progression $(a, ra, r^2a, r^3a, \ldots, r^{m-1}a)$. The factors $(x-a)$ $(x-ra)$ $(x-r^2a)\ldots(x-r^{m-1}a)$ are multiplied together, the product being

$$x^m-\frac{1-r^m}{1-r}ax^{m-1}+\frac{(1-r^m)(1-r^{m-1})}{(1-r)(1-r^2)}ra^2x^{m-2}$$
$$-\frac{(1-r^m)(1-r^{m-1})(1-r^{m-1})}{(1-r)(1-r^2)(1-r^3)}r^3a^3x^{m-3}+\ldots\quad=0$$

The general coefficient, of x^{m-s}, will be equal to a fraction whose numerator is

$$a^s\cdot r^{\frac{s(s-1)}{2}}\cdot(1-r^{m-s+1})(1-r^{m-s+2})(1-r^{m-s+3})\cdots(1-r^m)$$

and whose denominator is $(1-r)(1-r^2)(1-r^3)\cdots(1-r^s)$.

The given equation is divided by this product, and the common divisor of the terms of the remainder will be the required radix.

Corollary 3.1.2. If, in the Corollary to Problem III, we write $\frac{1-r^m}{1-r}$, $\frac{1-r^{2m}}{1-r^2}$, $\frac{1-r^{3m}}{1-r^3}$, $\frac{1-r^{4m}}{1-r^4}$, ... respectively for a, b, c, d, ..., then we see that the series

$$\left(\frac{1-r^m}{1-r}\right)^n-\frac{n(n-1)}{2}\left(\frac{1-r^m}{1-r}\right)^{n-2}\cdot\left(\frac{1-r^{2m}}{1-r^2}\right)$$
$$+n(n-1)\left(\frac{n-2}{3}\right)\left(\frac{1-r^m}{1-r}\right)^{n-3}\cdot\left(\frac{1-r^{3m}}{1-r^3}\right)-\ldots$$

is equal to

$$1\cdot 2\cdot 3\cdot 4\cdots n\cdot\left(\frac{1-r^m}{1-r}\right)\left(\frac{1-r^{m-1}}{1-r^2}\right)\left(\frac{1-r^{m-2}}{1-r^3}\right)\cdots\left(\frac{1-r^{m-n+3}}{1-r^n}\right)r^{\frac{n(n-1)}{2}}.$$

It seems appropriate to mention here also that it is possible to derive an infinite equation having an infinite number of real roots with known values.

Such an equation could be

$$1-\frac{a}{1-r}x+\frac{ra^2}{(1-r)(1-r^2)}x^2-\frac{r^3a^3}{(1-r)(1-r^2)(1-r^3)}x^3$$
$$+\frac{r^6a^4}{(1-r)(1-r^2)(1-r^3)(1-r^4)}x^4-\frac{r^{10}a^5}{(1-r)(1-r^2)(1-r^3)(1-r^4)(1-r^5)}x^5$$
$$+\ldots\ (\textit{ad infinitum})-0\,;$$

this equation has the roots $1/a$, $1/r^2a$, $1/r^3a$, $1/r^4a$, etc. *ad infinitum.*

And so an infinity of similar equations can be constructed.

It is easy to see that in many cases the present method offers more difficult solutions than does the previous one.

(3.2) Provided the relation among the m roots of the given equation can be expressed by an equation $x^m-ax^{m-1}+bx^{m-2}-cx^{m-3}+\ldots=0$, the method is entirely similar to the preceding one. The given equation $(x^n-px^{n-1}+qx^{n-2}-\ldots=0)$ is divided by the equation which expresses the relationship among the given equation's roots, and the common divisor of any of the terms of the remainder displays the required roots.

Example 3.2.1. Given the equation $x^4-x^3-24x^2+4x+80=0$, and an equation specifying that two roots of the given equation satisfy the relation $x^2-7x+Q=0$. [1] The former equation is divided by the latter:

$$x^2-7x+Q\ \overline{)x^4-\ x^3-24x^2+4x+80}^{\;x^2+6x\ +18-Q}$$

where the remainder turns out to be $(130-13Q)x+80-18Q+Q^2$. The common factor of the remainder's two terms, $130-13Q$ and $80-18Q+Q^2$, is found: it is $10-Q$, since $10-Q$ divided into $80-18Q+Q^2$ gives $8-Q$; equating it to zero we have $Q-10=0$, whence $Q=10$, and in consequence the equation whose roots have the required relation, i.e. are two roots of the given equation whose sum is 7, will be $x^2-7x+10=0$.

This example might have been solved a little differently. Since by hypothesis the sum of the two roots is 7, for x write respectively v and $7-v$, producing the two equations

$$v^4-v^3-24v^2+4v+80 = 0$$
and
$$v^4-27v^3+249v^2-893v+990 = 0.$$
Their common divisor will yield the required root.

Example 3.2.2. Given the equation $x^6+qx^4-rx^3+sx^2-tx+v = 0$, and an equation $x^3+ax-b = 0$ expressing a relation between three of the given equation's roots, divide the former by the latter, i.e.

$$x^3+ax-b \;\overline{\big)x^6+qx^4-rx^3+sx^2-tx+v}^{\;x^3+(q-a)x+b-r}$$

with remainder $(s-qa+a^2)x^2+(bq-t+ra-2ba)x+b^2-br+v$. By the rules for finding common divisors we see that the three equations $s-qa+a^2 = 0$, $bq-t+ra-2ba = 0$, and $b^2-br+v = 0$ will have no common divisor, and it follows that the given equation will have no roots of the prescribed type, unless $q^2v-qrt+r^2s-4sv+t^2 = 0$. Now if this resultant is 0, then the required equations will be $x^3+(\sqrt{\frac{1}{4}q^2-s}+\frac{1}{2}q)x-\frac{1}{4}r+\sqrt{\frac{1}{4}r^2-v} = 0$
and
$$x^3-(\sqrt{\tfrac{1}{4}q^2-s}-\tfrac{1}{2}q)x-\tfrac{1}{4}r-\sqrt{\tfrac{1}{4}r^2-v} = 0.$$

Corollary 3.2.1. If there are more equations (n) than unknowns (m), then out of the n equations we want to find $n-m+1$ independent equations, say $A = 0$, $B = 0$, $C = 0$, etc., involving only a single unknown x. From any m of the given equations we can construct one involving only x. Find the greatest common divisor of A, B, C, ..., and call it P. Then there can be as many (h) equations of the prescribed form, as there are distinct values (α, β, γ, δ, ...) of x as a root of the equation $P = 0$, unless some value of the root x occurs with multiplicity 2, 3, or higher, in which case there may be $h+1$, $h+2$, ... equations of the prescribed form. Every value (α, etc.) of the unknown quantity x corresponds to exactly one value of each of the other unknown quantities, except when two or more values of x are equal (say α), in which case the number of distinct sets of corresponding values for one or more of the other unknown quantities may be as large as the multiplicity of the root α.

This corollary is readily established by substituting α, β, γ, δ, ... for x in the given equation.

Corollary 3.2.2. If the number of equations is greater than the number of unknown quantities, when all unknowns are to be eliminated the process will yield exactly one equation involving only *known* quantities if the number of equations exceeds the number of unknowns by exactly

1. If the excess is 2, 3, ..., then correspondingly there will result two, three, or more equations involving known quantities, whose truth or falsity reveals whether equations of this sort can be solved.

These two corollaries are easily established through the customary means of finding common divisors.

And from two, three, or more equations involving known quantities only, combining them by means of addition, subtraction, multiplication by other given quantities, etc., can produce an infinity of other like equations.

Corollary 3.2.3. Given an arbitrary quantity composed of distinct irrational elements, $A+B+C+D+E+\ldots$, to determine whether or not the given quantity may have any real values.

(1) Let the root A be written $a+p\sqrt{-1}$, B as $b+q\sqrt{-1}$, C as $c+r\sqrt{-1}$, D as $d+s\sqrt{-1}$, E as $e+t\sqrt{-1}$, etc., then suppose $p+q+r+s+t+\ldots = 0$.

In the spirit of the current problem we will determine whether the resulting equations support the hypothesis $p+q+r+s+t+\ldots = 0$, and so the corollary is resolved.

(2) This may also be investigated by supposing the sum of the given quantities, $A+B+C+\ldots$, is equal to x, and then eliminating irrationalities, finally determining whether the resulting equation in x has any real roots.

Example 3.2.3.1. If the given quantity is $\sqrt[2]{-3+4\sqrt{-1}} + \sqrt[3]{-16-\sqrt{-1}}$, does it have a real value? [2]

Write $\sqrt[2]{-3+4\sqrt{-1}} = a+p\sqrt{-1}$, and $\sqrt[3]{-16-\sqrt{-1}} = b+q\sqrt{-1}$. From these expressions we produce the four equations

$$a^2-p^2 = -3$$
$$2ap = 4$$
$$b^2-3bq^2 = -16$$

and $$3b^2q-q^3 = -16;$$

supposing further that $p+q = 0$, i.e. that $q = -p$, by the method previously given, of finding common divisors, we see that $a = 1$, $p = 2$, $b = 2$, $q = -2$, so the given quantity indeed has a real value.

Example 3.2.3.2. Let the irrational quantity be $\sqrt[n]{a+\sqrt{-b^2}}+\sqrt[n]{a-\sqrt{-b^2}}$. Obviously, if $r+p\sqrt{-1}$ is a value of $\sqrt[n]{a+\sqrt{-b^2}}$, then $r-p\sqrt{-1}$ is a value of $\sqrt[n]{a-\sqrt{-b^2}}$, and the sum takes on real values.

Corollary 3.2.3.1. In certain cases of this sort, from the given relation among the roots we can deduce the maximum number of such roots which any equation of degree n could have. But obviously this is less general than employing the more conventional method of common divisors, to determine whether the given equation has any roots of the prescribed type.

Corollary 3.2.3.2. Certain equations of given forms inherently will have roots which have an assignable relation among themselves. In these cases no more distinct equations may result, than the number of unknown quantities. Therefore, from the number of equations that result, we know the number of roots in the given equation, which have the stated relation.

Example 3.2.3.2.1. Let the equation be

$$x^{nm}+Px^{(n-1)m}+Qx^{(n-2)m}+Rx^{(n-3)m}+Sx^{(n-4)m}+\dots=0.$$

This equation necessarily has roots which are related to each other in the same way as are the roots of the n equations $x^n+p=0$, $x^n+q=0$, $x^n+r=0$, $x^n+s=0$, etc. If these n equations are multiplied together, there results an equation of degree nm. Setting corresponding terms of the resulting and given equations equal to each other, it is easily established that there are exactly n equations whose roots have the stated assignable relation to each other.

Example 3.2.3.2.2. Let the equation be the reciprocal equation,

$$\begin{aligned}&x^{2n}+ax^{2n-1}+bx^{2n-2}+cx^{2n-3}+dx^{2n-4}\\&\quad+ex^{2n-5}+fx^{2n-6}+gx^{2n-7}+hx^{2n-8}+kx^{2n-9}+\dots+px^n+\dots\\&\quad\quad+kx^9+hx^8+gx^7+fx^6+ex^5+dx^4+cx^3+bx^2+ax+1=0.\end{aligned}$$

The equation $x^2+\alpha x+1=0$ naturally describes the relation between pairs of roots of the givn equation — that is, when two particular roots of the given equation are multiplied together, their product is 1. If n distinct equations of this sort are multiplied together, and terms of the resulting equation are supposed equal to the corresponding terms of the given equation, there result n equations in n unknown quantities α, etc. If these are reduced with a view to eliminating all the unknowns but one (say α), clearly we will produce an equation with n distinct roots. In fact, the equation in α will be

$$\begin{aligned}&\alpha^n-a\alpha^{n-1}+(b-n)\alpha^{n-2}+(-c+(n-1)a)\alpha^{n-3}\\&\quad+(\ d-(n-2)b+\tfrac{n(n-1)}{2})\alpha^{n-4}+(-e+(n-3)c-\tfrac{(n-1)(n-4)}{2}a)\alpha^{n-5}\end{aligned}$$

$$+\left(f-(n-4)d+\tfrac{(n-2)(n-5)}{2}b-\tfrac{n(n-4)(n-5)}{2\cdot3}\right)\alpha^{n-6}$$
$$+\left(-g+(n-5)e-\tfrac{(n-3)(n-6)}{2}c+\tfrac{(n-1)(n-5)(n-6)}{2\cdot3}a\right)\alpha^{n-7}$$
$$+\left(h-(n-6)f+\tfrac{(n-4)(n-7)}{2}d-\tfrac{(n-2)(n-6)(n-7)}{2\cdot3}b+\tfrac{n(n-5)(n-6)(n-7)}{2\cdot3\cdot4}\right)\alpha^{n-8}$$
$$+\left(-k+(n-7)g-\tfrac{(n-5)(n-8)}{2}e+\tfrac{(n-3)(n-7)(n-8)}{2\cdot3}c-\tfrac{(n-1)(n-6)(n-7)(n-8)}{2\cdot3\cdot4}a\right)\alpha^{n-9}$$
$$+\ldots=0.$$

The rule which this series obeys, to express the coefficient of the term α^{n-m}, has two aspects, regarding the literal and numerical components.

The literal components will be the coefficients of the terms x^{2n-m}, x^{2n-m+2}, x^{2n-m+4}, x^{2n-m+6}, x^{2n-m+8}, etc., of the given equation.

The coefficient of a letter, which is associated with the term $x^{2n-m+2r}$ in the given equation, will be $(n-m+2r)\times\frac{n-m+1}{2}\times\frac{n-m+2}{3}\times\frac{n-m+3}{4}\times\ldots\times\frac{n-m+r-1}{r}$; for $r=0$ the coefficient will be 1, and for $r=1$ the coefficient will be $n-m+2$.

The sign of the coefficient of the term $x^{2n-m+2r}$ is $+$ or $-$, according as $m+r$ is even or odd.

Corollary 3.2.3.2.1. The equation $x^{2n}\pm x^{2n-1}+x^{2n-2}\pm x^{2n-3}+x^{2n-4}\pm\ldots=0$ can be reduced to a quadratic, x^2+vx+1, with the aid of the equation $v^n\mp v^{n-1}-(n-1)v^{n-2}\pm(n-2)v^{n-3}+\frac{(n-2)(n-3)}{2}v^{n-4}\mp\frac{(n-3)(n-4)}{2}v^{n-5}-\frac{(n-3)(n-4)(n-5)}{2\cdot3}v^{n-6}\pm\frac{(n-4)(n-5)(n-6)}{2\cdot3}v^{n-7}+\frac{(n-4)(n-5)(n-6)(n-7)}{2\cdot3\cdot4}v^{n-8}\mp\ldots=0$.

Corollary 3.2.3.2.2. The three roots of the equation $x^3+1=0$ are -1 and $\frac{+1\pm\sqrt{-3}}{2}$; the four roots of the equation $x^4+1=0$ are $\pm\sqrt{\frac{1}{2}}\pm\sqrt{-\frac{1}{2}}$; the five roots of the equation $x^5+1=0$ are $\frac{\pm\sqrt{1\frac{1}{4}}+\frac{1}{2}}{2}\pm\frac{\sqrt{-2\frac{1}{2}\pm\sqrt{1\frac{1}{4}}}}{2}$ and -1; and the six roots of the equation $x^6+1=0$ are $\pm\sqrt{-1}$ and $\frac{\pm\sqrt{3}\pm\sqrt{-1}}{2}$. But the seven roots of the equation $x^7+1=0$ cannot be expressed in terms of square roots.

Example 3.2.3.2.3. Let the equation be $x^{2n+1}+ax^{2n}+bx^{2n-1}+cx^{2n-2}+dx^{2n-3}+ex^{2n-4}+fx^{2n-5}+gx^{2n-6}+hx^{2n-7}+kx^{2n-8}+\ldots+px^{n+1}+px^n+\ldots+kx^9+hx^8+gx^7+fx^6+ex^5+dx^4+cx^3+bx^2+ax+1=0$. One root of this equation is -1, and then there will be n equations of the form $x^2+\alpha x+1=0$, $x^2+\beta x+1=0$, $x^2+\gamma x+1=0$, etc. The equation whose root is α will be

$$\alpha^n-(a-1)\alpha^{n-1}+(b-a+(n-1))\alpha^{n-2}+(-c+b-(n-2)a-(n-2))\alpha^{n-3}$$
$$+(\ d-c-(n-3)b+(n-3)a+\tfrac{(n-2)(n-3)}{2})\alpha^{n-4}$$
$$+(-e+d+(n-4)c-(n-4)b-\tfrac{(n-3)(n-4)}{2}a+\tfrac{(n-3)(n-4)}{2})\alpha^{n-5}$$
$$+(\ f-e-(n-5)d+(n-5)c+\tfrac{(n-4)(n-5)}{2}b-\tfrac{(n-4)(n-5)}{2}a-\tfrac{(n-3)(n-4)(n-5)}{2\cdot 3})\alpha^{n-6}$$
$$+\ldots=0.$$

The rule governing the series, expressing the coefficient of the term α^{n-m}, concerns both the literal and numerical coefficients.

For the letter associated with the given equation's term $x^{2n-m+2r+1}$ or $x^{2n-m+2r+2}$, the coefficient is $\dfrac{(n-m+1)(n-m+2)(n-m+3)\cdots(n-m+r)}{r\,!}$. If $r=0$, the coefficient will be 1.

The sign affixed to the numerical coefficient, for a letter associated with the term $x^{2n-m+2r+1}$, will be $+$ or $-$, according as $m+r$ is even or odd.

This example emerges easily and is easily proved, if in the given equation we replace x with its value $-\frac{1}{2}\alpha+\sqrt{\frac{1}{4}\alpha^2-1}$.

Corollary 3.2.3.2.3.1. Every reciprocal equation of even degree, $x^{2n}+ax^{2n-1}+bx^{2n-2}+\ldots+bx^2+ax+1=0$, may be factored as

$$(x^n+\alpha x^{n-1}+\beta x^{n-2}+\gamma x^{n-3}+\ldots+\pi x^3+\rho x^2+\sigma x+\tau)$$
$$\times(x^n+\tfrac{\sigma}{\tau}x^{n-1}+\tfrac{\rho}{\tau}x^{n-2}+\tfrac{\pi}{\tau}x^{n-3}+\ldots+\gamma x^3+\beta x^2+\alpha x+1)=0.$$

Corollary 3.2.3.2.3.2. The square or cube (and indeed every power) of a reciprocal polynomial is a reciprocal polynomial.

Problem XIX

(1) *To determine whether or not two given irrational expressions,* A *and* B, *have one, two, or more equal values.*

Write $x=A$ and $x=B$, and eliminate the irrational quantities from both equations. Then find the common divisor of the two resulting equations, and if one exists equate it to zero. The roots of this final equation will be the required values.

(2) *By the same method one can determine whether or not some root or roots of one, two, or more irrational quantities* A, B, C, D, ..., *have a given relation to another root (or*

roots) of the same irrationalities, A, B, C, *etc., or different ones,* a, b, c, *etc.*

Equate $x = A$, $y = B$, $z = C$, ..., and $u = a$, $v = b$, $w = c$, ...; each of these equations is reduced so as to eliminate the irrational quantities, resulting in equations $P = 0$, $Q = 0$, ..., $p = 0$, $q = 0$, etc. Then the resulting equations, together with those expressing the given relation, are reduced in order to eliminate the unknowns x, y, z, ..., u, v, w, etc. Equations will result which manifestly show whether or not any roots are of the prescribed type.

Problem XX*

To transform irrational quantities into others, which are equal to the given irrationalities but are represented by a different formula.

Find some function, whose reduction is known, of the given irrational quantities; then from the known reduction can be found a different representation of the given irrational expression.

Example 1. Given the irrational quantity $\sqrt{a+\sqrt{a^2-b^2}}+\sqrt{a-\sqrt{a^2-b^2}} = x$, say; the square of this is a function whose resolution is given. Now squaring, $x^2 = 2a{\pm}2b$, whence x, the value of the given irrational expression $\sqrt{a+\sqrt{a^2-b^2}} + \sqrt{a-\sqrt{a^2-b^2}} = x$, is also equal to $\sqrt{2a{\pm}2b}$.

Example 2. If the given irrational expression is $x = \sqrt[3]{a+\sqrt{a^2-b^2}} + \sqrt[3]{a-\sqrt{a^2-b^2}}$, its cube will be

$$x^3 = 2a+3\left(1 \text{ or } \frac{-1\pm\sqrt{-3}}{2}\right) b \left(\sqrt[3]{a+\sqrt{a^2-b^2}}+\sqrt[3]{a-\sqrt{a^2-b^2}}\right)$$

Example 3. Let $x = \sqrt{a}+\sqrt{b}$, whence

$$x^2-2px = a+b+2\sqrt{ab}-2p\sqrt{a}-2p\sqrt{b}; \text{ then}$$

$$x = p+\sqrt{a+b+p^2+2\sqrt{ab}-2p\sqrt{a}-2p\sqrt{b}} = \sqrt{a}+\sqrt{b}.$$

The resulting quantities can easily be reduced to the former ones, from which they were deduced.

Theorem 14*

Case 1. If h, i, k, l, ... are positive integers, and α, β, γ, δ, ϵ, ... are roots of the equation $x^n-1 = 0$, then the product $\alpha^h\beta^i\gamma^k\delta^l\cdots$ will also be a root of the equation $x^n-1 = 0$.

For if we write $\alpha^h\beta^i\gamma^k\delta^l\cdots$ in place of x in the equation $x^n-1 = 0$, the result is $x^n = \alpha^{nh}(=1) \times \beta^{ni}(=1) \times \gamma^{nk}(=1)\cdots = 1$, whence $x^n-1 = 1-1 = 0$.

Case 2. Let the equation be $x^n+1 = 0$, whose roots are α, β, γ, δ, etc.; then $\alpha^h\beta^i\gamma^k\delta^l\cdots$ will be a root of the equation $x^n+1 = 0$, or of $x^n-1 = 0$, according as $h+i+k+l+\ldots$ is an odd or even number.

Case 3. If n is an odd prime number, and α is a root of $x^n-1 = 0$, then α, α^2, α^3, α^4, ..., α^{n-1}, and 1 will be all the different roots of that equation. Likewise if α is a root of the equation $x^n+1 = 0$, then α, $-\alpha^2$, α^3, $-\alpha^4$, α^5, $-\alpha^6$, ..., -1 will be all the roots of the equation $x^n+1 = 0$.

Case 4. (1) Let $n = m \times r \times s \times t\cdots$, where m, r, s, t, ... are distinct prime numbers, and say the roots of the equations $x^m-1 = 0$, $x^r-1 = 0$, $x^s-1 = 0$, $x^t-1 = 0$, etc., are respectively a, b, c, ...; e, f, g, ...; h, i, k, ...; etc. Then $aeh\cdots$, $aei\cdots$, $aek\cdots$, ..., $afh\cdots$, $afi\cdots$, $afk\cdots$, ..., $agh\cdots$, $agi\cdots$, $agk\cdots$, ..., $beh\cdots$, $bei\cdots$, $bek\cdots$, ... will be all the n distinct roots of the equation $x^n-1 = 0$.

(2) If the above quantities are roots of the equations $x^m+1 = 0$, $x^r+1 = 0$, $x^s+1 = 0$, etc., and if the number of divisors m, r, s, t, etc. is odd, the stated products will be all the distinct roots of the equation $x^n+1 = 0$. If there are an even number of prime divisors, then the above products multiplied by some value $\pi = \sqrt[n]{-1}$ will be all the distinct roots of the equation $x^n+1 = 0$.

(3) With the same assumptions, if α is a root of the equation $x^n-1 = 0$ which is not a root of $x^m-1 = 0$ or $x^r-1 = 0$ or $x^s-1 = 0$, etc., then α, $\pi\alpha^2$, α^3, $\pi\alpha^4$, α^5, $\pi\alpha^6$, ... will be all the roots of the equation $x^n-1 = 0$.

Case 5. If n is an odd number, and α, β, γ, δ, ... are roots of the equation $x^n-1 = 0$, then $-\alpha$, $-\beta$, $-\gamma$, $-\delta$, ... will be respectively the roots of the equation $x^n+1 = 0$.

Case 6. If $n/r = m$, then the roots of the equation $x^r-1 = 0$ are the mth powers of the roots of the equation $x^n-1 = 0$; and the roots of the equation $x^m+1 = 0$ are the mth powers of the roots of the equation $x^n+1 = 0$. These corollaries are established by direct substitution.

Corollary 1. Let α and β be two conjugate roots ($a+\sqrt{-b^2}$ and $a-\sqrt{-b^2}$) of an equation $x^n \pm 1 = 0$. Since the product $(x-\alpha)(x-\beta) = x^2-(\alpha+\beta)x+\alpha\beta$ contains no imaginary part, it follows that $a^2+b^2 = (a+\sqrt{-b^2})(a-\sqrt{-b^2}) = 1$.

Corollary 2. Let α, β, γ, δ, ... be the n roots of the equation $x^n-1 = 0$, and let a, b, c, d, e, ... be any h arbitrary quantities. Then the sum of all distinct products of the hn quantities αa, βa, γa, δa, ..., αb, βb, γb, δb, ..., αc, etc. (π terms in all) will be equal to the sum of the h terms a^n, b^n, c^n, d^n, etc.

For if a^n, b^n, c^n, d^n, ... are roots of the equation

$$x^m - px^{m-1} + qx^{m-2} - rx^{m-3} + \ldots \pm Px^{m-h} \mp \ldots = 0,$$

we can substitute v^n for x and derive the equation

$$v^{mn} - pv^{mn-n} + qv^{mn-2n} - \ldots \pm Px^{mn-hn} \pm \ldots = 0,$$

whose roots are αa, βa, γa, ..., αb, βb, γb, ..., αc, βc, γc, ..., αd, βd, γd, δd, ..., and so on. Then two required sums, call them $\pm\pi$ and $\pm\rho$, will both be equal to the value of P, and hence are equal to each other. Both numbers will have the same sign, if h and hn are both even or both odd; but if one is odd and the other is even, the two sums will have opposite signs. [3]

Theorem 15*

Case 1. Let $a+\sqrt{-b^2}$ and $a-\sqrt{-b^2}$, $c+\sqrt{-d^2}$ and $c-\sqrt{-d^2}$, $e+\sqrt{-f^2}$ and $e-\sqrt{-f^2}$, etc., be conjugate pairs of impossible quantities, and let Q be a rational function (either integral or fractional) of these quantities, in which the conjugate quantities $a+\sqrt{-b^2}$ and $a-\sqrt{-b^2}$ are similarly involved; and likewise for $c+\sqrt{-d^2}$ and $c-\sqrt{-d^2}$ and $e+\sqrt{-f^2}$ and $e-\sqrt{-f^2}$. Then Q will be a real quantity.

This easily follows from the principle given in Theorem 1.

Case 2. Any arbitrary rational function Q takes on conjugate values for conjugate values of its argument, e.g. if Q is a function of $a+\sqrt{-b^2}$, say $Q(a+\sqrt{-b^2}) = A+\sqrt{-B^2}$, then the function of the conjugate will be $Q(a-\sqrt{-b^2}) = A-\sqrt{-B^2}$, where $A+\sqrt{-B^2}$ and $A-\sqrt{-B^2}$ are conjugate complex quantities. [4]

Theorem 16

Case 1. Let the solution of an equation be $W = ap^{1/n}+bp^{2/n}+cp^{3/n}+\ldots$, where a, b, c, d, ..., and p are real quantities. Write $p^{1/n} = q\sqrt[n]{\pm 1}$, where q is real, and the resolution becomes

$$W = aq\sqrt[n]{\pm 1}\pm bq^2\sqrt[n]{\pm 1}\pm cq^3\sqrt[n]{\pm 1}\pm\ldots.$$

Now let α and $1/\alpha$ be conjugate roots of the equation $x^n\pm 1 = 0$, and write

$$aq\alpha+bq^2\alpha^2+cq^3\alpha^3+dq^4\alpha^4+\ldots = P, \text{ and}$$
$$aq\frac{1}{\alpha}+bq^2\frac{1}{\alpha^2}+cq^3\frac{1}{\alpha^3}+dq^4\frac{1}{\alpha^4}+\ldots = Q,$$

whence P and Q are two conjugate values of W, and so it follows that $(W-P)(W-Q) = W^2-(P+Q)W+PQ$ is real, i.e. the imaginary part vanishes in this product.

Case 2. Let the resolution be

$$W = ap^{1/n}+bp^{2/n}+cp^{3/n}+\ldots+AP^{1/m}+BP^{2/m}+CP^{3/m}+\ldots,$$

where a, b, c, ..., p and A, B, C, ..., P are real quantities. Write $p^{1/n} = q\sqrt[n]{\pm 1}$ amd $P^{1/m} = Q\sqrt[m]{\pm 1}$, whence the resolution becomes $W = aq\sqrt[n]{\pm 1}+bq^2(\sqrt[n]{\pm 1})^2+cq^3(\sqrt[n]{\pm 1})^3+\ldots+AQ\sqrt[m]{\pm 1}+BQ^2(\sqrt[n]{\pm 1})^2+CQ^3(\sqrt[n]{\pm 1})^3+\ldots$. Now let α and $1/\alpha$ be conjugate roots of the equation $x^n\pm 1 = 0$, and likewise let π and $1/\pi$ be conjugate roots of the equation $x^m\pm 1 = 0$. Substitute these espressions for the nth and m^{th} roots in the preceding resolution, obtaining

$$aq\alpha+bq^2\alpha^2+cq^3\alpha^3+\ldots+AQ\pi+BQ^2\pi^2+CQ^3\pi^3+\ldots = \Gamma$$
$$\text{and } aq\frac{1}{\alpha}+bq^2\frac{1}{\alpha^2}+cq^3\frac{1}{\alpha^3}+\ldots+AQ\frac{1}{\pi}+BQ^2\frac{1}{\pi^2}+CQ^3\frac{1}{\pi^3}+\ldots = \Delta,$$

which are two conjugate values of the given resolution, that is, the product $(W-\Gamma)(W-\Delta) = W^2-(\Gamma+\Delta)W+\Gamma\Delta$ is real, the imaginary part vanishing.

Case 3. If an expression in the given resolution is $\sqrt[2m]{-P^{2m}}$, where $-P^{2m}$ is a negative quantity, then in place of $\sqrt[2m]{-P^{2m}}$ we may write $P\times\sqrt[2m]{-1}$, and for $\sqrt[2m]{-1}$ substitute in succession all the roots of the equation $x^{2m}-1 = 0$.

We may further require that the rth power of the quantity $\sqrt[m]{-P}$ be the same as the rth power of the quantity $\sqrt[m]{P}$, or equivalently that the rth power of a root α of the eqtation $x^m-1 = 0$ be also the rth power of a root of the equation $x^m+1 = 0$; this can always be done, if ρ proves to be an even number, when $\frac{\mu}{\rho}$ is the fraction $\frac{m}{r}$ reduced to its lowest terms.

Case 4. Given a resolution derived in the preceding way, and the letters α, δ, θ, λ, π, etc. respectively denoting roots of the equations $x^n \pm 1 = 0$, $x^m \pm 1 = 0$, $x^l \pm 1 = 0$, etc.; if the given resolution contains roots of index n, m, l, etc. of any expressions a, b, c, d, etc., then in the given resolution for $\sqrt[n]{a}$, $\sqrt[m]{b}$, $\sqrt[l]{c}$, etc., we write respectively $\alpha\sqrt[n]{a}$, $\delta\sqrt[m]{b}$, $\theta\sqrt[l]{c}$, etc., resulting in an expression we shall call H. Likewise in the said resolution we may write respectively $\frac{1}{\alpha}\sqrt[n]{a}$, $\frac{1}{\delta}\sqrt[m]{b}$, $\frac{1}{\theta}\sqrt[l]{c}$, etc., and call the result K. Then $(W-H)(W-K) = W^2-(H+K)W+HK$ will be a real product, containing no impossible quantities.

This is easily established by Corollary 2 of Theorem 1, and the same principle applies in all resolutions which are functions of algebraic expressions, for instance in resolutions of the form

$$\sqrt[n]{a+\sqrt[m]{b+\sqrt[r]{c+\ldots}}+\sqrt[s]{d+\ldots}+\ldots}$$

as well as to resolutions which are functions of exponential quantities.

Corollary. Let the resolution be $x = A$, which cleared of all surds becomes $R = 0$; then the fraction $1/R$ may be divided by these methods into linear and quadratic real fractions. For by these methods real values that correspond to the given equation are shown, as are in consequence the real roots of the equation $R = 0$, whence the corollary is easily proved.

Theorem 17*

$$\frac{A+B\sqrt{-1}}{C+D\sqrt{-1}} = \frac{BD+CA+(CB-DA)\sqrt{-1}}{D^2+C^2}$$

This is easily proved by cross-multiplying the numerator of each fraction by the other's denominator, thereby obtaining two products which are equal to each other.

Corollary. If the numerator and denominator of a given fraction are any given function of known quantities, the said formula may easily be reduced to the form $\frac{A+B\sqrt{-1}}{C+D\sqrt{-1}}$.

Problem XXI*

To determine the degree of an equation, by means of which one of degree n $(x^n - px^{n-1} + qx^{n-2} - rx^{n-3} + sx^{n-4} - \ldots = 0)$ *can be reduced to another,* $x^m - ax^{m-1} + bx^{m-2} - cx^{m-3} + \ldots = 0$, *of degree* m, *given that* n>m, *and the roots of the reduced equation* $x^m - ax^{m-1} + bx^{m-2} - cx^{m-3} + \ldots = 0$ *are also roots of the equation* $(x^n - px^{n-1} + qx^{n-2} - rx^{n-3} + sx^{n-4} - \ldots = 0)$, *from which the reduction was obtained.*

The required degree will be equal to the fraction

$$n \cdot \frac{n-1}{2}\frac{n-2}{3}\frac{n-3}{4} \cdots \frac{n-m+1}{m}.$$

Proof. The equation $x^n - px^{n-1} + qx^{n-2} - rx^{n-3} + sx^{n-4} - \ldots = 0$ has n roots, and any set of m roots of this equation will define an equation of degree m which satisfies the requirements of the problem. Therefore, there are as many solutions to the problem, as there are combinations of m roots out of the larger number n, and so the reducing equation needs to have this many roots.

Now the number of combinations of m quantities (in this case, roots), from a larger set containing n of them, is equal to the fraction

$$n \cdot \frac{n-1}{2} \cdot \frac{n-2}{3} \cdot \frac{n-3}{4} \cdots \frac{n-m+1}{m}$$

(DeMoivre, *Misc. Anal.* p. 180). [5] Therefore the number of roots, and correspondingly the degree, of the reducing equation is also

$$n \cdot \frac{n-1}{2} \cdot \frac{n-2}{3} \cdot \frac{n-3}{4} \cdots \frac{n-m+1}{m}.$$

It is not necessary to investigate individually the coefficients a or b or c, etc.; each of them will take the same number of distinct values, so an equation whose roots are values of a will be of the same degree as one whose roots are b, c, etc.

Given one coefficient in particular, substituted for its value, by the previous method of determining whether the roots of a given equation sustain a given relation among each other (i.e. by the method of common divisors), the remaining unknown quantities can be extracted. If two values of the chosen coefficient are equal, the remaining quantities are found using the above method and a quadratic equation; if three are equal the rest are found by the above method and a cubic equation; and so on.

Later it will be shown that in general, all of the unknown quantities a, b, c, etc. will involve the same irrationalities, unless two or more values of one unknown quantity are equal to each other. If two values of some particular unknown α are equal to each other, then there will result a number of quadratic equations whose roots are the other unknowns, and which otherwise involve only that same irrational quantity which is contained in the unknown quantity α. If three values of α are equal to each other, then cubic instead of quadratic equations are involved, and so on.

This is evident from the observation that the process of finding common divisors introduces no new irrationalities.

Example. The biquadratic equation $x^4-px^3+qx^2-rx+s = 0$ can be reduced to a quadratic, $x^2-Px+Q = 0$, with the aid of an equation of degree $\frac{4(4-1)}{2} = 6$.

(1) The equation whose root is P will be

$$(P-\frac{p}{2})^6+(2q-\frac{3}{4}p^2)(P-\frac{p}{2})^4+(q^2-4s-qp^2+rp+\frac{3}{16}p^4)(P-\frac{p}{2})^2-(\frac{p^3}{8}-\frac{qp}{2}+r)^2=0.$$

(2) Let $p = 0$, i.e. let the sum of the four roots (α, β, γ, δ) of the given biquadratic be zero. Then the sum of two of the roots will be $p(= \alpha+\beta)$, and the sum of the other two ($\gamma+\delta$) will be $-P(= -\alpha-\beta)$; moreover, three roots of the equation in P will be positive, and the other three will be negative and equal in magnitude to the first three. In consequence, for purposes of its solution, the equation may be regarded as a cubic.

The equation whose root is P will be $p^6+2qP^4+(q^2-4s)P^2-r^2 = 0$.

Let this equation's roots ($\alpha+\beta$, $\alpha+\gamma$, $\alpha+\delta$, $\beta+\gamma$, $\beta+\delta$, and $\gamma+\delta$) be π, ρ, σ, τ, υ, ϕ respectively; then the roots of the given equation are $\frac{\pi+\rho-\tau}{2}$, $\frac{\pi+\tau-\rho}{2}$, $\frac{\tau+\rho-\pi}{2}$, $\frac{\sigma+\upsilon-\pi}{2}$, or equivalently $\frac{\pi+\sigma-\upsilon}{2}$, $\frac{\pi+\upsilon-\sigma}{2}$, $\frac{\rho+\phi-\sigma}{2}$, $\frac{\sigma+\phi-\rho}{2}$, or again $\frac{\rho+\sigma-\phi}{2}$, $\frac{\tau+\upsilon-\phi}{2}$, $\frac{\tau+\phi-\upsilon}{2}$, $\frac{\upsilon+\phi-\tau}{2}$. This is easily proved by substituting for the letters π, ρ, σ, τ, υ, ϕ the values previously assigned to them.

(3) Given one value, π, of the quantity P, the four roots of the given equation will be

$$\pm\sqrt{\frac{-\pi^3-2q\pi-2r}{4\pi}}+\frac{1}{2}\pi \text{ and } \pm\sqrt{\frac{-\pi^3-2q\pi+2r}{4\pi}}-\frac{1}{2}\pi.$$

Proof. Let π be a value of P that has been found, then

$$\begin{aligned}(x^2-\pi x+Q)(x^2+\pi x+R) &= x^4+(R+Q-\pi^2)x^2-(R-Q)\pi x+RQ\\ &= x^4+qx^2-rx+s=0.\end{aligned}$$

Equating like terms, we have the three equations $R+Q-\pi^2=q$, $(R-Q)\pi = r$, $QR=s$; but these three equations involve only two unknowns, Q and R. Then by the method of common divisors these two quantities may be expressed, in several equivalent ways, in terms of the letters q, r, s, and π — unless two or more values of one of them (Q or R) are equal to each other. For example, from the first two of the three equations, $R+Q-\pi^2=q$ and $(R+Q)\pi=r$, we can deduce

$$R=\tfrac{1}{2}(q+\pi^2+\tfrac{r}{\pi}) \text{ and } Q=\tfrac{1}{2}(q+\pi^2-\tfrac{r}{\pi}).$$

Substitute these values for Q and R in the quadratics $x^2-\pi x+Q=0$ and $x^2+\pi x+R=0$, and the four roots of the resulting equations will be precisely the four roots of the given equation.

Corollary 1. This is a general resolution only when two of the given equation's roots are real, and two are impossible; of the three roots of the cubic equation,

$$p^3+2qP^2+(q^2-4s)P-r^2=0,$$

only one, in this case, will be real, and two will be impossible: and this is the only circumstance in which its roots can be found by the algebraic method of solving cubic equations.

The equation whose root is Q (the product of each pair of roots of the biquadratic equation $x^4+qx^2-rx+s=0$) will be

$$Q^6-qQ^5-sQ^4+(2sq-r^2)Q^3-s^2Q^2-qs^2Q+s^3=0.$$

Corollary 2. In this method the reducing equation is of higher degree than the given equation, and generally is not itself reducible, nor is it of a form whose resolution is generally known. Therefore it is futile to seek a general resolution of equations by this method. (Newton, *Arith. Univ.* p. 237) [6]

Lemma

Every biquadratic equation of the form

$$x^2-2(a+b\sqrt{-1})x-c-d\sqrt{-1}=0$$

can always be transformed into two quadratic equations involving only real quantities.

Let $$\alpha^2=\frac{a^2-b^2+c+\sqrt{(a^2-b^2+c)^2+(2ab+d)^2}}{2}$$
and $$\beta^2=\frac{+\sqrt{(a^2-b^2+c)^2+(2ab+d)^2}-a^2+b^2-c}{2},$$
where α and β denote entirely real quantities.

Then the real equations will be, respectively,
$x^2-(2a+\alpha)x+(a+\alpha)^2+(b+\beta)^2=0$ and $x^2-(2a-\alpha)x+(a-\alpha)^2+(b-\beta)^2=0$.

This Lemma is easily derived, if we first solve the given quadratic equation $x^2-2(a+b\sqrt{-1})x-c-d\sqrt{-1}=0$, whose roots will be
$$a+b\sqrt{-1}\pm\sqrt{a^2-b^2+c+(2ab+d)\sqrt{-1}};$$
and the corresponding root, which allows us to compose a real quadratic equation, will be $a-b\sqrt{-1}\pm\sqrt{a^2-b^2+c-(2ab+d)\sqrt{-1}}$.

We then write $\sqrt{a^2-b^2+c+(2ab+d)\sqrt{-1}}=\alpha+\beta\sqrt{-1}$, and also $\sqrt{a^2-b^2+c-(2ab+d)\sqrt{-1}}=\alpha-\beta\sqrt{-1}$, from which the values of α and β, presented above, emerge.

Theorem 18

Every equation $x^n-px^{n-1}+qx^{n-2}-rx^{n-3}+\ldots=0$ is composed either of exclusively linear factors multiplied together (e.g. $(x-\alpha)(x-\beta)(x-\gamma)\cdots$, with real values for all coefficients), or of linear and quadratic factors ($x-\alpha$, $x-\beta$, $x-\gamma$, ... and $x^2-\pi x+P$, $x^2-\rho x+R$, $x^2-\sigma x+S$, ...) multiplied together, again involving only real quantities.

Proof. (1) Every equation $x^{2n+1}-px^{2n}+qx^{2n-1}-\ldots=0$, whose degree $(2n+1)$ is odd, has at least one real divisor $x-\alpha$, where α is a real number; from which the given equation can be depressed to one of degree $2n$, an even number.

Every equation $x^{4n+2}-px^{4n+1}+qx^{4n}-\ldots=0$, of degree $4n+2$, may be depressed to a quadratic by means of an equation of degree $\frac{(4n+2)(4n+1)}{2}$ $=(2n+1)(4n+1)$. Now one root of the reducing equation, which is of degree $(2n+1)(4n+1)$, must be real; therefore one quadratic equation, involving only real quantities, must be a divisor of the given equation.

Every equation $x^{8n+4}-px^{8n+3}+qx^{8n+2}-\ldots=0$, of degree $8n+4$, can be depressed to a quadratic by means of an equation of degree $\frac{(8n+4)(8n+3)}{2}$ $=(4n+2)(8n+3)$. Now this equation itself has at least one quadratic divisor whose two coefficients are both real, say $x^2-ax+b=0$ with roots $\frac{a\pm\sqrt{a^2-4b}}{2}$. But this equation of degree $8n+4$ can be depressed to a quadratic by means of either of these roots, which, upon substitution, reveal that this equation has at least one biquadratic divisor of the form

$$x^2+(a+c\sqrt{a^2-4b})x+B+d\sqrt{a^2-4b}=0$$

where A, B, c, and d are real quantities. And by the preceding Lemma this biquadratic is composed of two real quadratics; therefore an equation of degree $8n+4$ has at least two real quadratic divisors.

By extension of this method the equation $x^{16n+8}-px^{16n+7}+qx^{16n+6}-\ldots$ $=0$ can be shown to have at least four real quadratic divisors; likewise the equation $x^{32n+16}-px^{32n+15}+qx^{32n+14}-\ldots=0$ has at least eight real quadratic divisors; and so on. Therefore all such equations have real linear or quadratic factors. When the given equation is divided by linear or quadratic equations, it is depressed to another equation of lower degree. The resulting equation in turn will have real linear or quadratic divisors and so may be further depressed, iteratively establishing the theorem.

(2) Let $2^r=n$, where r is an integer. Then an equation of degree n may be reduced to one of degree 2^4-1 by means of an equation of degree $n\cdot\frac{n-1}{2}\cdot\frac{n-2}{3}\cdots\frac{\frac{1}{2}n+1}{\frac{1}{2}n}$; this formula represents a number (degree) $4m+2$, where m is an integer.

If $2^r<n<2^{r+1}$, then an equation of degree n can be reduced to one of degree 2^4 by means of an equation of degree $n\cdot\frac{n-1}{2}\cdot\frac{n-2}{3}\cdots\frac{n-2^r+1}{2^r}$, which is an odd number.

Theorem 19

Equations may be reduced by separating them into equal parts, such that roots of the same index may be extracted from each side of the = sign.

Problem XXII

Example 1. *To effect the reduction of a biquadratic,*

$$\mathrm{x}^4+2\mathrm{px}^3 = \mathrm{qx}^2+\mathrm{rx}+\mathrm{s},$$

to a quadratic by means of a cubic equation.

To each side of the equation add $(p^2+2n)x^2+2pnx+n^2$, giving

$$x^4+2px^3+(p^2+2n)x^2+2pnx+n^2 = (p^2+2n+q)x^2+(2pn+r)x+s+n^2.$$

Now impose the condition that $4(s+n^2)(p^2+2n+q) = (2pn+r)^2$, and expand and transpose the terms of this equation. It becomes the cubic

$$8n^3+4qn^2+(8s-4rp)n+4qs+4p^2s-r^2 = 0\,.$$

Now solve this equation for n, and substitute the solution for the occurrences of n in the equation

$$x^4+2px^3+(p^2+2n)x^2+2pnx+n^2 = (p^2+2n+q)x^2+(2pn+r)x+s+n^2,$$

and take the square root of each side; there results the quadratic equation

$$x^2+px+n = \sqrt{p^2+2n+q}\,x+\sqrt{s+n^2}\,.$$

Corollary 1.1. The four roots of the biquadratic (letting π represent a root of the cubic equation $8n^3+4qn^2+(8s-4rp)n+4qs+4p^2s-r^2 = 0$) will be the roots, respectively, of two quadratics,

$$x^2+(p+\sqrt{p^2+2\pi+q})x+\pi-\sqrt{s+\pi^2} = 0$$

and

$$x^2+(p-\sqrt{p^2+2\pi+q})x+\pi+\sqrt{s+\pi^2} = 0\,.$$

But two other quadratics also may be found by the same process,

$$x^2+(p+\sqrt{p^2+2\pi+q})x+\pi+\sqrt{s+\pi^2} = 0$$

and

$$x^2+(p-\sqrt{p^2+2\pi+q})x+\pi-\sqrt{s+\pi^2} = 0$$

— that is, if the four roots of the given equation are α, β, γ, δ, the first two quadratics represent, as it were, $x^2-(\alpha+\beta)x+\alpha\beta = 0$ and $x^2-(\gamma+\delta)x+\gamma\delta = 0$, while the latter two represent $x^2-(\alpha+\beta)x+\gamma\delta = 0$ and $x^2-(\gamma+\delta)x+\alpha\beta = 0$.

Now if $p = 0$ these four quadratic equations are the divisors of the two biquadratic equations $x^4-qx^2-rx-s = 0$ and $x^4-qx^2+rx-s = 0$: thus this method affords the deduction of all the roots of the given equation, and of all their negatives.

For example, given the biquadratic $x^4-25x^2+60x-36 = 0$, for p, q, r, s write, respectively, 0, 25, -60, 36 in the reducing equation $8n^3+4qn^2+(8s-4rp)n+4qs+4p^s-r^2 = 0$, and the result will be $8n^3+100n^2+288n = 0$, whose roots are 0, -8, $-\frac{9}{2}$. Substituting each of these for n in the equation $x^2+px+n = \sqrt{q+p^2+2n}+\sqrt{s+n^2}$, we have three equations,

$$x^2 = \sqrt{25}x+\sqrt{36},$$
$$x^2-8 = \sqrt{9}x+\sqrt{100},$$

and

$$x^2-\frac{9}{2} = \sqrt{16}x+\frac{\sqrt{225}}{2}.$$

Any one of these equations has the same four roots, 1, 2, 3, -6 (which numbers are the roots of the given equation, $x^4-25x^2+60x-36 = 0$), or their negatives, -1, -2, -3, and +6.

Corollary 1.2. If α, β, γ, δ are the roots of the given equation $x^4+2px-qx^2-rx-s = 0$, then the roots of the reducing equation, $8n^3+4qn^2+(8s-4ps)n+4qs+4p^2s-r^2 = 0$, will be

$$\frac{\alpha\beta+\gamma\delta}{2},\ \frac{\alpha\gamma+\beta\delta}{2},\ \frac{\alpha\delta+\beta\gamma}{2};$$

the six values of the quantity $\sqrt{q+p^2+2n}$ will be, respectively,

$$\frac{\alpha+\beta-\gamma-\delta}{2},\ \frac{\gamma+\delta-\alpha-\beta}{2},\ \frac{\alpha+\gamma-\beta-\delta}{2},$$
$$\frac{\beta+\delta-\alpha-\gamma}{2},\ \frac{\alpha+\delta-\beta-\gamma}{2},\ \frac{\beta+\gamma-\alpha-\delta}{2};$$

and the six values of the quantity $\sqrt{s+n^2}$ will be, respectively,

$$\frac{\alpha\beta-\gamma\delta}{2},\ \frac{\gamma\delta-\alpha\beta}{2},\ \frac{\alpha\gamma-\beta\delta}{2},$$
$$\frac{\beta\delta-\alpha\gamma}{2},\ \frac{\alpha\delta-\beta\gamma}{2},\ \frac{\beta\gamma-\alpha\delta}{2}.$$

These symmetries are easily established if for each of p, q, r, and s we substitute their values as functions of the roots α, β, γ, and δ, in the resulting equation

$$x^2+(p-\sqrt{q+p^2+2n})x+(n+\sqrt{s+n^2}) = 0,$$

whose roots are the same as two of the given equation's roots (depending on the choice of square root signs they will be α and β, α and γ, α and δ, β and γ, β and δ, or γ and δ).

Corollary 1.3. By considering all the distinct values of the quantity $\sqrt{p^2+2n+q}$, without performing the final root extraction we can still discover the roots of the given equation. Let these values be

$$a = \frac{\alpha+\beta-\gamma-\delta}{2},\ b = \frac{\alpha+\gamma-\beta-\delta}{2},\ a = \frac{\alpha+\delta-\beta-\gamma}{2},$$

then $\alpha = \frac{a+b+c-p}{2},\ \beta = \frac{a-b-c-p}{2},\ \gamma = \frac{b-a-c-p}{2},\ \delta = \frac{c-a-b-p}{2}.$

Corollary 1.4. (1) If the given equation is $x^4-px^3+qx^2-rx+s = 0$, then the equation in z whose roots are $\alpha+\beta-\gamma-\delta$, $\gamma+\delta-\alpha-\beta$, $\alpha+\gamma-\beta-\delta$, $\beta+\delta-\alpha-\gamma$, $\alpha+\delta-\beta-\gamma$, and $\beta+\gamma-\alpha-\delta$ will be

$$z^6+(8q-3p^2)z^4+(3p^4-16p^2q+16q^2-64s+16pr)z^2$$
$$-p^6-16p^2+8p^4q-64r^2-16p^3r+64pqr = 0.$$

(2) If the given equation is $x^4+qx^2-rx+s = 0$, then the equation whose roots are $\alpha\beta-\gamma\delta$, $\alpha\gamma-\beta\delta$, $\alpha\delta-\beta\gamma$, $\gamma\delta-\alpha\beta$, $\beta\delta-\alpha\gamma$, and $\beta\gamma-\alpha\delta$ will be

$$x^6-(q^2-4x)x^4-2r^2qx^2-r^4 = 0.$$

Corollary 1.5. This rule does not suffice to find the roots of the given biquadratic, unless two of them are real and two are impossible. If none, or all four, are impossible, then the reducing cubic will have three real roots — whence we see that there is no rule by which the roots of the general biquadratic equation may be found.

Corollary 1.6. The supposition, in the resolution of the problem, that $2(\sqrt{p^2+2n+q})(\sqrt{s+n^2}) = 2pn+r$, properly leads to the conclusion that $4(p^2+2n+q)(s+n^2) = (2pn+r)^2$; but it does not conversely follow in all cases from the supposition that $4(p^2+2n+q)(s+n^2) = (2pn+r)^2$, that every value of the product $2(\sqrt{p^2+2n+q})(\sqrt{s+n^2})$ is equal to $2pn+r$, because for every case where the product is $2pn+r$, there will be another where it is $-2pn-r$. And so we see that the resolution gives two extraneous values of the quantity $\sqrt{p^2+2n+q}\sqrt{s+n^2}$, and in consequence four extraneous values in the resolution; then the three values of n, substituted in the given resolution, give rise to twelve extraneous roots, and the resolution is actually the solution of an equation of 16 dimensions, four of whose roots are the four roots of the given equation. [7]

Example 2. *To reduce to a cubic the quadrato-cubic equation*

$$x^5+qx^3+rx^2+sx+t = 0$$

The given equation is multiplied by x, producing

$$x^6+qx^4+rx^3+sx^2+tx = 0,$$

to each side of which is added $(kx^3+dx+b)^2$, yielding

$$(kx^3+dx+b)^2 = (k^2+1)x^6+(2kd+q)x^4+(2kb+r)x^3+(d^2+s)x^2+(2db+t)x+b^2.$$

Now suppose
$$\begin{aligned} 2(\sqrt{k^2+1})b &= 2kb+r \quad (A=0);\\ 2(\sqrt{k^2+1})(\sqrt{d^2+s}) &= 2kd+q \quad (B=0);\end{aligned}$$
and
$$2(\sqrt{d^2+s})b = 2db+t \quad (C=0);$$
which, upon reduction, give the three equations
$$\begin{aligned} 4b^2 &= 4rkb+r^2,\\ 4sk^2+4d^2+4s &= 4qkd+q^2,\end{aligned}$$
and
$$4sb^2 = 4tdb+t^2.$$
From the first equation it follows that $k = \frac{4b^2-r^2}{4rb}$, and from the third, that $d = \frac{4sb^2-t^2}{4tb}$, which quantities, substituted into the second equation, give us a biquadratic equation of quadratic form,

$$4\frac{(4sb^2-t^2)^2}{16t^2b^2}+4s\frac{(4b^2-r^2)^2}{16r^2b^2}+4s = 4q\frac{4b^2-r^2}{4rb}\cdot\frac{4sb^2-t^2}{4tb}+q^2.$$

This equation reduces to $(sr^2+t^2-qtr)(16sb^2+4qrtb^2+r^2t^2) = 0$.

Corollary 2.1. This resolution only finds the roots of the given equation when $sr^2+t^2-qtr = 0$. If $sr^2+t^2-qtr = 0$, any value whatever may be assumed for b, and then $k = \frac{4b^2-r^2}{4rb}$, $d = \frac{4sb^2-t^2}{4tb}$. Then the given equation will be the difference between two squares,

$$\left(\frac{4b^2+r^2}{4rb}x^3+\frac{4sb^2+t^2}{4tb}x+b\right)^2 - \left(\frac{4b^2-r^2}{4rb}x^3+\frac{4sb^2-t^2}{4tb}x+b\right)^2$$

$= x^6+qx^4+rx^3+sx^2+tx = 0$; these squares may be varied in an infinity of ways, since any value may be chosen for b, so for instance for $2b$ write r, causing the term $\frac{4b^2-r^2}{4rb}x^3$ to vanish, and $\frac{4b^2+r^2}{4rb}$ to become 1. In any case, the equation may then be depressed to

$$\frac{4b^2+r^2}{4rb}x^3+\frac{4sb^2+t^2}{4tb}x+b = \pm\left(\frac{4b^2-r^2}{4rb}x^3+\frac{4sb^2-t^2}{4tb}x+b\right),$$

whence $x^3+\frac{rs}{t}x+r = 0$ and $x^2+\frac{t}{r}x = 0$.

Hence the equation $16sb^4+4qrtb^2+r^2t^2 = 0$ will not be the general equation which the problem calls for. Rather, the quantity b can be chosen at will, since as we have observed, every value of b satisfies the equation $sr^2+t^2-qtr = 0$.

Since the equation $sr^2+t^2-qtr = 0$ is implied by $A = 0$, $B = 0$, and $C = 0$, and these three equations involve three unknowns k, d, and b, one of them can be chosen at will.

Corollary 2.2. The significance of the equation $16sb^4+4qrtb^2+r^2t^2 = 0$ is this: supposing $4(k^2+1)b^2 = (2kb+r)^2$, we conclude that $2b\sqrt{k^2+1} = 2kb+r$; but the quantities $\sqrt{k^2+1}$ and $\sqrt{d^2+s}$ each may have two values, which occur twice in the above three equations, and the same values of these quantities will not necessarily be used in each case.

Example 3. *To reduce to a cubic the cubo-cubic equation*

$$x^6+qx^4+rx^3+sx^2+tx+v = 0 \quad [9]$$

By adding $(kx^3+dx+b)^2$ on both sides of the $=$ sign, we have $(kx^3+dx+b)^2 = (k^2+1)x^6+2kd+q)x^4+2kb+r)x^3+(d^2+s)x^2+(2db+t)x+b^2+v$.

Now suppose that

$$2\sqrt{k^2+1}\cdot\sqrt{s+d^2} = q+2kd;$$
$$2\sqrt{k^2+1}\cdot\sqrt{v+b^2} = r+2kb;$$

and

$$2\sqrt{s+d^2}\cdot\sqrt{v+b^2} = t+2db;$$

which reduce to the three equations

$$4vk^2+4b^2+4v = 4rkb+r^2;$$
$$4sk^2+4d^2+4s = 4qkd+q^2;$$

and

$$4vs+4vd^2+4sb^2 = 4tdb+t^2.$$

From the first equation it follows that $k = \dfrac{\sqrt{r^2-4v}\cdot\sqrt{v+b^2}+rb}{2v}$, and from the third, $d = \dfrac{\sqrt{r^2-4vs}\cdot\sqrt{v+b^2}+tb}{2v}$, which substituted for their occurences in the second equation, result in an equation which upon reduction to clear the irrationality $\sqrt{v+b^2}$ becomes a biquadratic in the form of a quadratic. If the remaining irrationalities are eliminated, an equation of eight dimensions, in the form of a biquadratic, is produced. Finally, multiply these resulting equations by $q^2v-qrt+r^2s-4sv+t^2$.

In this and similar cases, where the quantity b is not given, unless $Q = 0$, then if by common reduction the equation is found whose root is b, the equation so found will always be multiplied by the equation $Q = 0$.

Corollary 3.1. This resolution serves to find the roots of the given equation only when $q^2v-qrt+r^2s-4sv+t^2 = 0$. In this case the sum of the three roots of the resulting cubic equation will be equal to zero (which was implicitly assumed in the initial substitution in this and the preceding Example). Then any value can be assumed for the letter b, from which corresponding values of k and d will follow. The roots of the resulting equation, which are the values of b, have significance analogous to that described in the preceding Example.

Example 4. Given the equation

$$x^{2n}+qx^{2n-2}-rx^{2n-3}+sx^{2n-4}-tx^{2n-5}+vx^{2n-6}=0,$$

assume $(x^n+ax^{n-2}+bx^{n-3}+cx^{n-4}+\ldots)^2 = (\alpha x^{n-1}+\beta x^{n-2}+\gamma x^{n-3}+\ldots)^2$, whence

$$x^{2n}+(2a-\alpha^2)x^{2n-2}+(2b-2\alpha\beta)x^{2n-3}+(2c+a^2-2\alpha\gamma-\beta^2)x^{2n-4}+\ldots=0.$$

Equate corresponding coefficients of this equation and the given equation, i.e., $2a-\alpha^2 = q$, $2b-2\alpha\beta = -r$, $2c+a^2-2\alpha\gamma-\beta^2 = s$, etc., and eliminate all but one unknown among all these equations. A reducing equation results; but this equation is of higher degree than the given equation. For example, if the remaining unknown is α, the resulting equation will be of degree $\dfrac{2n(2n-1)(2n-2)\cdots(n+1)}{n\,!}$ in α; and it will have the same degree if a is chosen to be the unknown, etc.

If the roots of the given equation are π, ρ, σ, τ, etc., then α is the symmetric sum of all products of n distinct roots, and the other potential unknowns have similar relationships to the roots.

Example 5. Given the equation $x^6+qx^4-rx^3+sx^2-tx+v = 0$, we may assume $x^2+b = ax+c$, leading to

$$\begin{aligned} x^6+3bx^4+3b^2x^2+\ b^3 &= a^3x^3+3a^2cs^2+3ac^2x+\ c^3,\\ Ax^4+2Abx^2+Ab^2 &= Aa^2x^2+2Aacx+Ac^2,\\ Bx^2+Bb &= Bax+Bc. \end{aligned}$$

Seeting the coefficients of the resulting equations equal to the corresponding coefficients of the given equation ($3b+A = q$, $a^3 = r$, etc.) gives us five equations in five unknowns. However, these equations are not independent: calling the roots of the given equation α, β, γ, δ, ϵ, and ζ, we have $a = \alpha+\beta = \sqrt[3]{r}$, in which case the resolution can be effected by this example.

If a general solution is required, it is essential that none of the equations arising from the above process be algebraically dependent on others: there must always be as many or more unknowns as there are independent resulting equations.

If the coefficients q, r, s, t, ... of the given equation are expressed in terms of the roots α, β, γ, δ, ..., and if the coefficients of the assumed equations are compared with those of the reducing equation, then the functional relation of the coefficients of the assumed equations to the given equation's roots can also be found.

Accordingly, this method is easily adaptable to the transformation of equations.

Problem XXIII

To solve equations by transforming them into others, whose roots have a given relation to the roots of the given equation.

Case 1. The given equations are transformed into others whose roots can be found by known rules, where such is the relation between the roots of the transformed and original equations, that the roots of the original equation can be recovered from those of its transform.

Example 1.1. Let the equation be a cubic, $x^3+qx-r=0$, and choose the relation $z-\frac{q}{3z}=x$, thereby transforming the given cubic into one whose root is z, namely $z^6-rz^3-\frac{q^3}{27}$.

The roots of this equation will be $\sqrt[3]{\frac{1}{2}r\pm\sqrt{\frac{1}{4}r^2+\frac{q^3}{27}}}$, from which the roots of the given equation are

$$x=\sqrt[3]{\frac{1}{2}r\pm\sqrt{\frac{q^3}{27}+\frac{r^2}{4}}}+\frac{\frac{1}{3}q}{\sqrt[3]{\frac{1}{2}r\pm\sqrt{\frac{q^3}{27}+\frac{r^2}{4}}}}$$

$$=\sqrt[3]{\frac{1}{2}r+\sqrt{\frac{q^3}{27}+\frac{r^2}{4}}}+\sqrt[3]{\frac{1}{2}r-\sqrt{\frac{q^3}{27}+\frac{r^2}{4}}}\,.$$

Corollary 1.1. Let $a+b\sqrt{-3}=\sqrt[3]{\frac{1}{2}r-\sqrt{\frac{q^3}{27}+\frac{r^2}{4}}}$. Then the three roots of the cubic equation $x^3+qx-r=0$ will be

$$a+b\sqrt{-3}+\frac{-\frac{1}{3}q}{a+b\sqrt{-3}}=2a,$$

$$\frac{-1+\sqrt{-3}}{2}(a+b\sqrt{-3})+\frac{-\frac{1}{3}q}{\frac{-1+\sqrt{-3}}{2}(a+b\sqrt{-3})}\left(=\frac{-1-\sqrt{-3}}{2}(a-b\sqrt{-3})\right)=-a-3b,$$

and

$$\frac{-1-\sqrt{-3}}{2}(a+b\sqrt{-3})+\frac{-\frac{1}{3}q}{\frac{-1-\sqrt{-3}}{2}(a+b\sqrt{-3})}\left(=\frac{-1+\sqrt{-3}}{2}(a-b\sqrt{-3})\right)=-a+3b.$$

But if $a+b=\sqrt[3]{\frac{1}{2}r+\sqrt{\frac{r^2}{4}+\frac{q^3}{27}}}$, and $a-b=\sqrt[3]{\frac{1}{2}r-\sqrt{\frac{r^2}{4}+\frac{q^3}{27}}}$, then the three roots found by the preceding method will be $2a$, $-a+b\sqrt{-3}$, and $-a-b\sqrt{-3}$.

Corollary 1.2. Let the equation be as before $x^3+qx-r=0$, and let its roots be α, β, γ; then the six roots of the reducing equation, values, that is, of z, will be

$$(\tfrac{1}{2}\alpha+\beta)\sqrt{-\tfrac{1}{3}}+\tfrac{1}{2}\alpha,\quad -(\tfrac{1}{2}\alpha+\beta)\sqrt{-\tfrac{1}{3}}+\tfrac{1}{2}\alpha,$$
$$(\tfrac{1}{2}\beta+\alpha)\sqrt{-\tfrac{1}{3}}+\tfrac{1}{2}\beta,\quad -(\tfrac{1}{2}\beta+\alpha)\sqrt{-\tfrac{1}{3}}+\tfrac{1}{2}\beta,$$
$$\text{and } (\tfrac{1}{2}\alpha-\tfrac{1}{2}\beta)\sqrt{-\tfrac{1}{3}}-\tfrac{1}{2}\alpha-\tfrac{1}{2}\beta,\quad (\tfrac{1}{2}\beta-\tfrac{1}{2}\alpha)\sqrt{-\tfrac{1}{3}}-\tfrac{1}{2}\alpha-\tfrac{1}{2}\beta.$$

By inspection it is readily seen that the equation is of the form $z^6+\pi z^3+\rho=0$.

Corollary 1.3. If π, ρ, σ, τ, v, and ϕ are the six roots of the above reducing equation, $z^6-rz^3-\frac{q^3}{27}$, then the three roots of the given equation, $z^3+qz-r=0$, must be $\pi+\rho$, $\sigma+\tau$, and $v+\phi$.

The above observations all apply equally well to the cubic equation $x^3-px^2+qx-r=0$.

Case 2. To transform an equation $x^n-px^{n-1}+qx^{n-2}-rx^{n-3}+\ldots=0$ into another one, in which m of the terms have been destroyed (i.e., their coefficients are 0).

An equation is assumed, expressing a relation between the roots of the given and the reducing equation, involving at least as many (m) unknowns as there are terms to be destroyed in the equation that is being reduced. By setting the m said terms of the resulting equation equal to zero we will obtain as many equations as there are terms to be destroyed.

From these equations, using techniques from the following Chapter, the degree of the resulting equation can be found, based on transformation of these equations into one, eliminating $m-1$ of the unknowns.

Example 1. (1) Given an equation $x^n+qx^{n-2}-rx^{n-3}+sx^{n-4}-\ldots=0$, to transform it into another of the same degree, $v^n+*+*+Rv^{n-3}+\ldots=0$, lacking the two terms v^{n-1} and v^{n-2}.

We assume $x^2+ax+b = v$, in which a and b are unknowns, and compute by the methods of Chapter 1, from the given equation, the equation whose root is v. Then we equate to zero the coefficients of the terms v^{n-1} and v^{n-2}, which gives us $b=\frac{2q}{n}$ and $qa^2-3ra-\frac{(n-2)q^2}{n}+2s=0$. These two equations allow us to find the desired values of b and a, after which we conclude the derivation, according to Problem V, of the equation whose root is v.

(2) Consider specifically the equation $x^3+qx-r=0$: proposing $x^2+ax+\frac{2q}{3}=v$, with the further condition that $qa^2-3ra-\frac{q^2}{3}=0$, our result will be $v^3 * *-ra^3-\frac{2}{3}q^2a^2+rqa-r^2-\frac{2}{27}q^3=0$.

Let α, β, and $-\alpha-\beta$ be the three roots of the given equation. Then

$$a=\frac{(2\alpha^3-2\beta^3+3\alpha^2\beta-3\beta^2\alpha)\sqrt{-1}+3\sqrt{3}(\alpha\beta)(\alpha+\beta)}{2(\alpha^2+\alpha\beta+\beta^2)\sqrt{3}}$$

which, substituted for its occurrence in the equation $x^2+ax+b=v$, gives the different values of the quantity v.

Example 2. Given the equation $x^n+qx^{n-2}-rx^{n-3}+sx^{n-4}-tx^{n-5}+ux^{n-6}-\ldots=0$, to transform it into another, also of degree n, $v^n+*-Qv^{n-2}+*+Sv^{n-4}+\ldots=0$, lacking the terms v^{n-1} and v^{n-3}.

Once again we write $x^2+ax+\frac{2q}{n}=v$, where now

$$ra^3+\left(\frac{2(n-2)}{n}-4s\right)a^2+\left(5t-\frac{5n-12}{n}qr\right)a$$
$$-\frac{(n-4)(n-2)}{3\cdot n\cdot n}\cdot 2q^2+\frac{n-4}{n}\cdot 2qs+r^2-2u=0;$$

we determine the equation whose root is v, and that will be what we require.

Let the equation be $x^4+qx^2-rx+s = 0$, and write $x^2+ax+\frac{q}{2} = v$, where now $ra^3+(q^2-4s)a^2-2qra+r^2 = 0$. The equation whose root is v, a biquadratic in the form of a quadratic, is found:

$$v^4 + * + (qa^2-3ra+2s-\frac{q^2}{2})v^2 + *$$

$$+ sa^4+\frac{qr}{2}a^3+(\frac{q^2}{4}-sq)a^2+(sr-\frac{rq^2}{4}a+s^2+\frac{r^2q}{2}-\frac{q^2s}{2}+\frac{q^4}{16} = 0,$$

which may be solved to find v, and then from the other assumed equation, $x^2+ax+\frac{q}{2} = v$, the roots (x) of the given equation can be found.

The significance of a and v, as functions of the roots of the given equation, is easily discerned by representing the letters q, r, and s as symmetric functions of the given equation's roots.

Corollary 2.1. Using this method to transform the equation

$$x^n-px^{n-1}+qx^{n-2}-rx^{n-3}+sx^{n-4}-tx^{n-5}+\ldots = 0$$

into one of the form $v^n+Q = 0$ will require a reducing equation of degree $(n-1)!$.

Corollary 2.2. Concerning cases where the assumed quantity's coefficients have more unknowns than the number of terms to be destroyed:

Determine the expressions for the terms which are to be removed, and equate each of them to zero; then transform the resulting equations into one, such that as many as possible of the unknown quantities are eliminated. The result will be an equation having two or more unknown quantities, and any set of corresponding values of the unknowns which satisfy this equation will cause the specified terms of the given equation to vanish.

The same method of transformation applies, when it is required that coefficients of the resulting equation have some prescribed relationship among themselves.

It should also be mentioned here, that an equation whose roots are any rational function of the roots of the given equation can have no more roots (and, it follows, can be of no higher degree) than the given equation.

Problem XXIV

To find equations which can be depressed.

Case 1. An equation of degree n is assumed, and by Problem V others are found whose roots are some given functions of the assumed equation's roots, such that from the roots of the assumed equation, the resulting equation's roots may be found by known methods. Then the resulting equations, whenever they are of degree higher than the assumed equation's degree, may be depressed to the assumed equation.

Example 1.1. Assumed be the equation $x^n - px^{n-1} + qx^{n-2} - \ldots = 0$, whose roots are α, β, γ, δ, etc.; and derived be the equation

$$v^{n \cdot \frac{n-1}{2}} - Pv^{n \cdot \frac{n-1}{2} - 1} + \ldots = 0$$

whose roots are $\alpha+\beta$, $\alpha+\gamma$, $\beta+\gamma$, $\alpha+\delta$, $\beta+\delta$, $\gamma+\delta$, etc. Then the latter equation, of degree $n \cdot \frac{n-1}{2}$, may easily be depressed to the original assumed equation, of degree n.

The resulting equation of degree $n \cdot \frac{n-1}{2}$ has only n unknown quantities, the coefficients of the assumed equation. From these arise $n \cdot \frac{n-1}{2}$ equations in the same n unknown quantities. Therefore, since there are more resulting equations than there are unknowns, the method of finding common divisors will yield the unknown quantities.

Or if the roots of the resulting equation are $\alpha+\beta+\gamma$, $\alpha+\beta+\delta$, $\alpha+\gamma+\delta$, $\beta+\gamma+\delta$, etc., or $\alpha+\beta+\gamma+\delta$, etc., then the resulting equations (of degree $\frac{n(n-1)(n-2)}{3\,!}$ or $\frac{n(n-1)(n-2)(n-3)}{4\,!}$) can with equal ease be depressed to an equation of degree n.

Example 1.2. Let $\frac{ax^m + bx^{m-1} + cx^{m-2} + \ldots}{Ax^r + Bx^{r-1} + Cx^{r-2} + \ldots} = z$ be the function of the roots of the resulting equation, where m and r are integers and a, b, c, ..., A, B, C, ... are rational quantities. Then the resulting equation will have no more dimensions than the given equation.

For instance, let the equation be $y^n + h = 0$, and let the assumed equation be $(x+b)y = x+a$; y is eliminated, with the result

$$(1+h)x^n + n(a+hb)x^{n-1} + \frac{n(n-1)}{2}(a^2+hb^2)x^{n-2} + \frac{n(n-1)(n-2)}{3\,!}(a^3+hb^3)x^{n-3} + \ldots + a^n + hb^n = 0.$$

Let $hb+a = 0$, and the result becomes

$$A \equiv x^n - \frac{n(n-1)}{2}abx^{n-2} - \frac{n(n-1)(n-2)}{3\,!}ab(a+b)x^{n-3}$$
$$- \frac{n(n-1)(n-2)(n-3)}{4\,!}ab(a^2+ab+b^2)x^{n-4} - \ldots = 0$$

whose roots are $\sqrt[n]{a^{n-1}b} + \sqrt[n]{a^{n-2}b^2} + \sqrt[n]{a^{n-3}b^3} + \sqrt[n]{ab^{n-1}}$.

Let the given equation be $x^n+qx^{n-2}-rx^{n-3}+\ldots = 0$, and suppose the first three terms of A and the given equation are equal to each other. Then the resulting equation is

$$a^2 + \frac{3ra}{(n-2)q} - \frac{2q}{n(n-1)} = 0\,.$$

Case 2. Equations may be assumed, of whose corresponding roots each pair, ternary, or perhaps a larger group, have a given relation among themselves. From the number of such corresponding roots the degree of the reducible equation may be determined — i.e., m quadratic, cubic, or higher degree equations are assumed, whose two, three, … roots have among themselves a given relation. These equations are multiplied together, and the resulting equation can be depressed to a quadratic, cubic, … by means of an equation of degree m.

Example. Assume there are $2m$ roots α, $n\alpha$; β, $n\beta$; γ, $n\gamma$; etc., which have in pairs the relation $1 : n$ — i.e., there are m quadratics of the form $x^2-(n+1)\alpha x+n\alpha^2 = 0$, $x^2-(n+1)\beta x+n\beta^2 = 0$, $x^2-(n+1)\gamma x+n\gamma^2 = 0$, etc. These equations are multiplied together, and the result is

$$\begin{aligned}&x^{2n}-(n+1)px^{2n-1}+((n^2+1)q+np^2)x^{2n-2}\\ &\quad -((n^3+1)r+(n^2+n)pq)x^{2n-3}+((n^4+1)s+(n^2+n)pr+n^2q^2)x^{2n-4}\\ &\quad -((n^5+1)t+(n^4+n)ps+(n^3+n^2)qr)x^{2n-5}\\ &\quad +((n^6+1)v+(n^5+n)pt+(n^4+n^2)qs+n^3r^2)x^{2n-6}\\ &\quad -((n^7+1)w+(n^6+n)pv+(n^5+n^2)qt+(n^4+n^3)rs)x^{2n-7}\\ &\quad +\ldots = 0.\end{aligned}$$

This equation contains m different quantities, p, q, r, s, t, v, w, etc. It can easily be depressed into two others,

$$x^m - px^{m-1} + qx^{m-2} - rx^{m-3} + \ldots = 0$$

and $\quad x^m-npx^{m-1}+n^2qx^{m-2}-n^3rx^{n-3}+\ldots = 0;$

or, aided by an equation of degree m, to a quadratic

$$x^2-(n+1)\pi x+n\pi^2 = 0.$$

An infinity of similar examples can easily be composed.

Problem XXV

Given an equation $\mathrm{x^n - px^{n-1} + qx^{n-2} - rx^{n-3} + sx^{n-4} - \ldots} = 0$, *and another* $\mathrm{z^n - Pz^{n-1} + Qz^{n-2} - Rz^{n-3} + Sz^{n-4} - \ldots} = 0$, *where each of* P, Q, R, S, ... *is a function of* p, q, r, s, ..., *to find what function is* z, *in terms of* x.

Let α, β, γ, δ, ... be the roots of the equation $x^n - px^{n-1} + qx^{n-2} - rx^{n-3} + sx^{n-4} - \ldots = 0$, and express p, q, r, s, ... in terms of those roots, i.e. for p write $\alpha+\beta+\gamma+\delta+\ldots$, and similarly for the others. This permits us to re-express P, Q, R, S, ... as functions of the same roots α, β, γ, δ, ..., and thus to solve the problem.

An example of this problem was given earlier.

Problem XXVI

Given an arbitrary irrational quantity, to find another that, multiplied by the first, produces a rational product.

Method 1. We find all of the conjugate values of the given irrational quantity, since the product of all of them will be a rational quantity; that product, divided by the given irrational quantity, is the required complementary irrational quantity.

Example 1. To find an irrational expression which, multiplied by $\sqrt{a+bx} + \sqrt[3]{c+dx+ex^2} + f + gx$, produces a rational product.

The six distinct values of the given irrational quantity are

$$\sqrt{a+bx} + \sqrt[3]{c+dx+ex^2} + f + gx,$$

$$-\sqrt{a+bx} + \sqrt[3]{c+dx+ex^2} + f + gx,$$

$$\sqrt{a+bx} - \frac{1+\sqrt{-3}}{2}\sqrt[3]{c+dx+ex^2} + f + gx,$$

$$-\sqrt{a+bx} - \frac{1+\sqrt{-3}}{2}\sqrt[3]{c+dx+ex^2} + f + gx,$$

$$\sqrt{a+bx} - \frac{1+\sqrt{-3}}{2}\sqrt[3]{c+dx+ex^2} + f + gx,$$

and

$$-\sqrt{a+bx} - \frac{1+\sqrt{-3}}{2}\sqrt[3]{c+dx+ex^2} + f + gx.$$

The product of these six values together will be the rational quantity

$$(f+gx)^6-3(f+gx)^4(a+bx)+2(f+gx)^3(c+dx+ex^2)+3(f+gx)^2(a+bx)^2 +6(f+gx)(a+bx)(c+dx+ex^2)+(c+dx+ex^2)^2-(a+bx)^2;$$

and the product of all but the given quantity will be the irrational quantity that is required.

Corollary 1.1. By this problem an irrational quantity may be moved from the denominator to the numerator, or vice versa. Find a rational number B, of which the given irrationality is a divisor (say $C = \frac{B}{A}$), and for A substitute $\frac{B}{A}$. This may be elaborated as required. [10]

Corollary 1.2. Given an equation involving irrational functions of an unknown quantity x, it is now possible to determine the degree of the resulting equation, when the objective is to clear all of the. irrationalities.

First find the highest dimension of x in all of the radicands which form, as it were, terms of the given equation, and multiply this by the product of the dimensions of all the radical expressions; call this product A.

Then multiply the lowest dimension of x in all of those radicands, by the the dimensions of all of the radicands; call this product B. The required degree will be $A+B$, if B is negative, or $A-B$ if B is positive, which two of course are the same if B is equal to zero.

Now if in some of the terms which were used to compute the maximum or minimum degrees of x, some values vanish, then the degree of the unknown quantity x in the resulting equation will also be reduced. The extent to which the degree of the resulting equation is reduced depends, in essence, on the presence or absence of the terms next to those of highest or lowest degree in the radicand in question.

Example 1.2.1. Let the given equation be

$$\sqrt[2]{a+bx+cx^2} + \sqrt[3]{f+gx+hx^2+kx^3} + \sqrt[\frac{5}{2}]{p+qx+rx^2+sx^3} + P + Qx = 0.$$

In this equation the highest dimension of the unknown x in the first term, $\sqrt{a+bx+cx^2}$,is $\frac{2}{2} = 1$; in the second radicand, $(f+gx+hx^2+kx^3)^{\frac{1}{3}}$, it is also $\frac{3}{3} = 1$; in the third radicand, $\sqrt[\frac{5}{2}]{p+qx+rx^2+sx^3}$, its dimensions are $3 \times \dfrac{2}{5} = \dfrac{6}{5}$; and in the rational quantity $P+Qx$ it is 1. Accordingly, the dimensions of that term in which the unknown appears to the highest degree in the equation, will be $\frac{6}{5}$. Then multiply the dimension $\frac{6}{5}$ by the product of the indices of all of the radical expressions in the equation, which will be in this case $2\cdot3\cdot5\cdot\frac{6}{5}$. Therefore the degree of the unknown quantity x in the resulting equation will be 36, unless some value or values vanish in those terms where the highest or lowest dimensions of x are found. In this example only the term $\sqrt[\frac{5}{2}]{sx^3+\dots}$ is involved to the highest dimension, and none of the powers of the unknown x vanish among the terms of this radicand. Now the terms in which x appears to the lowest dimension (or not at all) are $\sqrt[2]{a}$, $\sqrt[3]{f}$, $\sqrt[\frac{5}{2}]{p}$, and P. If $\sqrt[2]{a}+\sqrt[3]{f}+\sqrt[\frac{5}{2}]{p}+P=0$, then one value of the unknown quantity x will be equal to zero; otherwise all its values will be non-zero. And if one value of the unknown is equal to zero, then we must inquire whether or not two or more vanish (further reducing the dimesnions of the resulting equation).

Example 1.2.2. Let the given equation be

$$\sqrt[5]{8x^2+2x+\frac{7}{3x}+\sqrt{64x^4-20}}+\sqrt[6]{10x+\sqrt{64x^4-20}}=0;$$

to find the dimension of x in the equation freed of all surds.

In this equation the highest degree of x in the first term is $\frac{2}{5}$ or $\frac{4}{2\cdot5}=\frac{2}{5}$. The highest degree in the second term will be $\frac{4}{2\cdot6}=\frac{1}{3}$. Therefore the dimensions of the term in which x appears to the highest degree will be $\frac{2}{5}$. Multiply this by the product of the degrees of the radicals, getting $\frac{2}{5}\times5\times6\times2=24=A$.

And the smallest dimensions of the unknown quantity x in the first irrational expression will be $-\frac{1}{5}$, while in the second it will be 0. Thus the dimensions of the term in which x appears to the lowest degree will be $-\frac{1}{5}\times5\times6\times2=-12=-B$, whence $A+B=24+12=36$.

However, if the quantity $\sqrt[5]{8x^2+2x+\dots+\sqrt{64x^4-20}}$ were instead $\sqrt[5]{8x^2+2x+\dots-\sqrt{64x^4-20}}$, then it would have the same dimensions as the quantity $\sqrt[5]{2x+\dots}$, which is less than the degree of the quantity $\sqrt[6]{10x+\sqrt{64x^4-20}}$ — that is, $\frac{1}{5}$ as opposed to $\frac{4}{6\cdot 2}=\frac{1}{3}$. So the dimensions for the expression with the positive square root ($\sqrt[5]{(8+8=16)x^2-\ldots}$) will be $\frac{2}{5}\times 5\times 6\times 1=12=a$, and for the negative case ($\sqrt[5]{(8-8=0)x^2-\ldots}$) will be $\frac{1}{3}\times 5\times 6\times 1=10=b$, and it follows that the dimesnions of the resulting equation will be $a+b+B=12+10+12=34$.

And thus we proceed in other cases where, by a similar method of reasoning, considering the different roots of the terms one after the other, we see that certain values of the various terms will vanish.

Method 2. Another way to transform equations to eliminate irrationalities may also be deduced, and it predicts the same number of dimensions for any unknown quantity (say x): evaluate the irrationalities as infinite series, arranging the terms by the dimensions of x in either ascending or descending order; multiply these quantities together, and the result will be the required equation, which has no irrational quantities.

Example. Let the equation be

$$\sqrt{16x^4+100}+\sqrt{9x^2+36}+7+2x=0.$$

The four distinct values of these irrational quantities, expressed as infinite series in decreasing dimensions of the unknown quantity x, are respectively

$$4x^2+(2+3)x+7+\frac{6}{x}+\ldots \text{ (to the term } \tfrac{\pi}{x^6}) \quad =0,$$

$$4x^2+(2-3)x+7-\frac{6}{x}+\ldots \text{ (to the same term)}=0,$$

$$-4x^2+(2+3)x+7+\frac{6}{x}+\ldots \quad =0,$$

and $$-4x^2+(2-3)x+7-\frac{6}{x}+\ldots \quad =0.$$

When these diverse values are multiplied together, the result is the required equation, $256x^8-416x^6-896x^5+\ldots=0$.

From this method of resolution the dimensions of the unknown quantity x, in the resulting equation freed of all surds, are easily determined.

In the multiplication of irrational quantities, since we are always faced with more than one value, errors are easily introduced unless one pays strict attention to the various roots. For instance, if $\sqrt{-a^2}$ is multiplied by itself, the product will be $-a^2$, although this product could also be written $\sqrt{-a^2} \times \sqrt{-a^2} = \sqrt{a^4} = \pm a^2$, but $+a^2$ in this case is not a true root.

The same may be applied to all irrational quantities.

This points out a discontinuity between mathematics and physical science. In the domain of mathematics roots do not abruptly change values, except when real roots pass through the point of equality to become impossible roots, causing even powers to change signs whereas odd powers increase in magnitude: but in physical sciences abrupt changes are possible in even powers, I mean with respect to the direction in which forces act; if the power is even, mathematically it is the same whether the given quantity is positive or negative, i.e. whether physically the force acts in that direction or the opposite direction, whereas if the power is odd, then mathematically it will have opposite signs, according as the physical force which it represents is oriented in one direction or the other. [11]

Problem XXVII

Given a quantity which expresses the value of a finite series of terms in an arbitrary literal x, *where the dimensions of the terms are* 0, 1, 2, 3, ..., (a+b+c+d+e+f+g+h+k+...), *to find value of the series of alternate terms* (a+c+e+g+k+...), *and also of series formed by deleting sets of* 2, 3, *or any number* n *terms from the proposed series.*

Let α, β, γ, δ, ϵ, ... be the $(n+1)^{\text{th}}$ roots of unity, i.e., if $n+1 = 2$ α and β are respectively 1 and -1, if $n+1 = 3$ α, β, and γ are respectively 1, $\frac{-1+\sqrt{-3}}{2}$, and $\frac{-1-\sqrt{-3}}{2}$.

Then in the given quantity, for x substitute respectively αx, βx, γx, δx, ϵx, etc., and the sum of the resulting quantities, divided by $(n+1)$, will be the required quantity.

Example 1. Let the quantity be $(a+x)^m$, which is expressed by a series of terms in increasing powers of x,

$$a^m+ma^{m-1}x+\frac{m(m-1)}{2}a^{m-2}x^2+\frac{m(m-1)(m-2)}{3!}a^{m-3}x^3+\ldots,$$

to find the quantity equal to the series of alternate terms.

In this example $n+1 = 2$, and since the square roots of unity are $+1$ and -1 we write in the given quantity both x and $-x$ wherever x appears. The result is the two quantities, $(a+x)^m$ and $(a-x)^m$, whose sum, divided by two, $\frac{(a+x)^m+(a-x)^m}{2} = a^m+\frac{m(m-1)}{2}a^{m-2}x^2+\ldots$, will be the required quantity.

Example 2. Given the same quantity, to find the quantity equal to the sum of the first, fourth, seventh, tenth, ... terms. In this example $n+1 = 3$, and the cube roots of unity are 1, $\frac{-1+\sqrt{-3}}{2}$, and $\frac{-1-\sqrt{-3}}{2}$. Accordingly, in the given quantity x is replaced by x, $\frac{-1+\sqrt{-3}}{2}x$, and $\frac{-1-\sqrt{-3}}{2}x$ respectively. There result three quantities, whose sum, divided by $n+1 = 3$, will be the required quantity, i.e.,

$$(a+x)^m + (a-\frac{-1+\sqrt{-3}}{2}x)^m + (a-\frac{-1-\sqrt{-3}}{2}x)^m$$
$$= a^m+\frac{m(m-1)(m-2)}{3!}a^{m-3}+\ldots$$

Example 3. Given the quantity $(a+bx+(c+dx+ex^2)^l)^m$ expressing the value of a series of terms in decreasing powers of x, to find that quantity equal to alternate terms of this series. Again in this example $n+1 = 2$, so we write x and $-x$ wherever x occurs in the given quantity, resulting in the two quantities,

$$(a+bx+(c+dx+ex^2)^l)^m$$

and $$(a-bx+(c-dx+ex^2)^l)^m,$$

whose sum, divided by $n+1 (= 2)$, is the required quantity.

Corollary. Let α, β, γ, δ, ϵ, ζ, ... be the roots of the equation $x^n-1 = 0$; and let an expression A be given, representing the sum of a series of successive powers of an unknown x. In the given expression substitute αx, βx, γx, δx, ϵx, ζx, etc., calling the resulting quantities A, B, C, D, E, F, etc. Then the sum of the first term and those at distances n, $2n$, $3n$, $4n$, $5n$, ... from the first will be, by the foregoing problem,

$$\frac{A+B+C+D+E+F+\dots}{n}.$$

The sum of the second term, and those at distances n, $2n$, $3n$, $4n$, $5n$, ... from it, will be

$$\frac{\alpha^{n-1}A+\beta^{n-1}B+\gamma^{n-1}C+\delta^{n-1}D+\epsilon^{n-1}E+\zeta^{n-1}F+\dots}{n}.$$

The sum of the third term, and those at distances n, $2n$, $3n$, $4n$, $5n$, ... from it, will be

$$\frac{\alpha^{n-2}A+\beta^{n-2}B+\gamma^{n-2}C+\delta^{n-2}D+\epsilon^{n-2}E+\zeta^{n-2}F+\dots}{n}.$$

And the sum of that term whose distance from the first is r, and those at distances n, $2n$, $3n$, $4n$, $5n$, ... from it, will be

$$\frac{\alpha^{n-r}A+\beta^{n-r}B+\gamma^{n-r}C+\delta^{n-r}D+\epsilon^{n-r}E+\zeta^{n-r}F+\dots}{n}.$$

Example. Let the given quantity be $(a+x)^m = a^m+ma^{m-1}x+\dots$, and the quantity which is equal to the sum of the first, third, fifth, seventh, etc. terms will be, by this problem,

$$\frac{(a+x)^m+(a-x)^m}{2}.$$

In this case $n=2$, and the roots α and β of the equation $x^2-1=0$ are 1 and -1. The quantities A and B, respectively $(a+x)^m$ and $(a-x)^m$, come from substituting $+x$ and $-x$ for x.

The quantity equal to the second, fourth, sixth, eighth, etc. terms will be

$$\frac{(a+x)^m-(a-x)^m}{2}.$$

Here the formulation is $\frac{\alpha^{2-1}(=\alpha)A+\beta^{2-1}(=\beta)B}{2}$, where as before α and β are 1 and -1, and A and B are $(a+x)^m$ and $(a-x)^m$, so the required sum will be as indicated.

If a series of terms of ascending degree in a quantity x is of the form $x^h(A'+B'x+C'x^2+\dots)$, and if k is the remainder obtained by dividing the index h by the chosen n, then the sum of terms beginning at the rth from the first, and those at further distances of n, $2n$, $3n$, etc., is

$$\frac{\alpha^{2n-r-k}A + \beta^{2n-r-k}B + \gamma^{2n-r-k}C + \dots}{n}.$$

In every series of terms of ascending degree in a quantity x,

$$\frac{\alpha^{n-r}A + \beta^{n-r}B + \gamma^{n-r}C + \dots}{n}$$

will be equal to the sum of all terms in the series whose index is of the form $np+r$, where p is an integer.

In determining the individual quantities A, B, C, etc., the roots of irrational quantities must always be used consistently in the given expression, e.g. let $\sqrt[\lambda]{P}$ be one of the foregoing irrationalities, and in the quantity A we use for this root the specific case $\pi\sqrt[\lambda]{P}$, where π is a particular root of the equation $x^n-1=0$. Then for $\sqrt[\lambda]{P'}$ in the quantity B, we must also use $\pi\sqrt[\lambda]{P'}$, and so for all the rest of them.

Problem XXVIII

(1) *To find expressions which, multiplied together, give rise to any arbitrary given quantities.*

By Problem XXVI we know of what type, or form, these irrational quantities must be, which give rise to a rational quantity when they are multiplied together, and by similar reasoning we can determine of what type or form any irrational quantities must be, that their product will be any chosen sort of irrational quantity.

Knowing this, accordingly, we assume the existence of quantities of such forms, which multiplied together produce a product of the desired form, and then by equating corresponding terms of the given and resulting expressions we derive a set of equations which can be solved to produce the required expressions.

In general we need to assume enough factors that there will be as many or more unknowns, as there are independent equations resulting from the process of equating corresponding terms.

Example 1.1. The given quantity $x^n-px^{n-1}+qx^{n-2}-rx^{n-3}+sx^{n-4}-tx^{n-5}+\dots$ can be the result of multiplication of n terms $x-\alpha$, $x-\beta$, $x-\gamma$, $x-\delta$, $x-\epsilon$, etc. Formal multiplication of these terms produces a quantity of the given type,

$x^n-(\alpha+\beta+\gamma+\delta+\ldots)x^{n-1}+(\alpha\beta+\alpha\gamma+\beta\gamma+\alpha\delta+\beta\delta+\gamma\delta+\ldots)x^{n-2}-\ldots$ When these terms are equated to corresponding terms of the given equation, there result n independent equations in as many unknown quantities. Therefore the given quantity $x^n-px^{n-1}+qx^{n-2}-rx^{n-3}+sx^{n-4}-tx^{n-5}+\ldots$ can be the result of multiplication of terms $x-\alpha$, $x-\beta$, $x-\gamma$, $x-\delta$, etc.

Example 1.2. The previous quantity $x^n-px^{n-1}+qx^{n-2}-rx^{n-3}+sx^{n-4}-tx^{n-5}+\ldots$ can be the result of multiplication of n factors $\sqrt{x+A}+x+a$, $-\sqrt{x+A}+x+a$, $\sqrt{x+B}+x+b$, $-\sqrt{x+B}+x+b$, ...; the formal multiplication produces a result of the required form, and again there will be n equations in n unknown quantities.

Corollary 1. The roots of the equations $\pm\sqrt{x+A}+x+a = 0$, $\pm\sqrt{x+B}+x+b = 0$, etc., will collectively be the same as the roots of the equation $x^n-px^{n-1}+qx^{n-2}-\ldots = 0$.

Evidently the same method serves to generate an infinity of constructions from which any given quantity may be composed.

(2) *The same method can be used to determine whether or not a given quantity can be formed as the sum of quantities of prescribed forms.*

The assumed quantities are reduced such that the resultant is of the form of the given quantity, and corresponding terms of the given and resultant quantities are equated to each other; then if there are as many or more unknown quantities, as there are independent equations resulting, then the resulting equations will yield up the required values of the unknowns.

Example 2.1. Let the equation be $x^n-px^{n-1}+qx^{n-2}-rx^{n-3}+\ldots = 0$, whose roots are α, β, γ, δ, ϵ, etc.; to determine whether the fraction $\frac{1}{x^n-px^{n-1}+qx^{n-2}-rx^{n-3}+\ldots}$ is the sum of fractions of the form $\frac{a}{x-\alpha}$, $\frac{b}{x-\beta}$, $\frac{c}{x-\gamma}$, $\frac{d}{x-\delta}$, $\frac{e}{x-\epsilon}$, etc.

Now by hypothesis
$$\frac{1}{x^n-px^{n-1}+qx^{n-2}-rx^{n-3}+\ldots} = \frac{a}{x-\alpha}+\frac{b}{x-\beta}+\frac{c}{x-\gamma}+\frac{d}{x-\delta}+\frac{e}{x-\epsilon}+\ldots,$$
and when the assumed fractions are reduced to the form of the given quantity, i.e. to a common denominator, we have

$$\frac{1}{x^n-px^{n-1}+qx^{n-2}-rx^{n-3}+\ldots}=$$

$$\frac{\left\{\begin{array}{r}(a+b+c+d+e+\ldots)x^{n-1}\\ -(a(\beta+\gamma+\delta+\ldots)+b(\alpha+\gamma+\delta+\ldots)+c(\alpha+\beta+\delta+\ldots)+d(\alpha+\beta+\gamma+\ldots)+\ldots)x^{n-2}\\ +(a(\beta\gamma+\beta\delta+\gamma\delta+\ldots)+(b(\alpha\gamma+\alpha\delta+\gamma\delta+\ldots)+\ldots)x^{n-3}\\ -\ldots\end{array}\right\}}{x^n-px^{n-1}+qx^{n-2}-rx^{n-3}+\ldots}.$$

By equating the numerator (= 1) of the given quantity with the numerator of the resulting quantity (since the denominators are the same) we generate n independent linear equations in as many quantities (a, b, c, d, e, ...), from which we easily determine that

$$a=\frac{1}{(\alpha-\beta)(\alpha-\gamma)(\alpha-\delta)(\alpha-\epsilon)\cdots},\quad b=\frac{1}{(\beta-\alpha)(\beta-\gamma)(\beta-\delta)(\beta-\epsilon)\cdots},$$
$$c=\frac{1}{(\gamma-\alpha)(\gamma-\beta)(\gamma-\delta)(\gamma-\epsilon)\cdots},\quad d=\frac{1}{(\delta-\alpha)(\delta-\beta)(\delta-\gamma)(\delta-\epsilon)\cdots},$$
$$e=\frac{1}{(\epsilon-\alpha)(\epsilon-\beta)(\epsilon-\gamma)(\epsilon-\delta)\cdots},\quad \text{etc.}$$

Corollary 2.1. Coefficients constituting a given equation may be found in an infinite variety of ways, whatever may be its construction.

Corollary 2.2. (1) Let α, β, γ, δ, ϵ, ... be the roots of the equation

$$x^n-px^{n-1}+qx^{n-2}-rx^{n-3}+sx^{n-4}-\ldots=0,$$

then $$\frac{1}{x^n-px^{n-1}+qx^{n-2}-\ldots}=\frac{a}{x-\alpha}+\frac{b}{x-\beta}+\frac{c}{x-\gamma}+\frac{d}{x-\delta}+\frac{e}{x-\epsilon}+\ldots$$

where $a=\frac{1}{(\alpha-\beta)(\alpha-\gamma)(\alpha-\delta)\cdots}$, $b=\frac{1}{(\beta-\alpha)(\beta-\gamma)(\beta-\delta)\cdots}$,

$c=\frac{1}{(\gamma-\alpha)(\gamma-\beta)(\alpha-\delta)\cdots}$, etc.

But
$$\begin{aligned}\frac{1}{x^{-n}-px^{-n-1}+qx^{-n-2}-\ldots} &= x^n+p\,x^{-n-1}\\ &+(p^2-q)x^{-n-2}\\ &+(p^3-2pq+r)x^{-n-3}\\ &+(p^4-3p^2q+2pr-s+q^2)x^{-n-4}+\ldots\\ &= 1\cdot x^{-n}\\ &\quad+(\alpha+\beta+\gamma+\delta+\ldots)\ x^{-n-1}\\ &\quad+((\alpha^2+\beta^2+\gamma^2+\delta^2+\ldots)+(\alpha\beta+\alpha\gamma+\beta\gamma+\alpha\delta+\beta\delta+\ldots))\ x^{-n-2}\\ &\quad+((\alpha^3+\beta^3+\gamma^3+\delta^3+\ldots)+(\alpha^2\beta+\alpha^2\gamma+\beta^2\alpha+\beta^2\gamma+\gamma^2\alpha+\gamma^2\beta+\ldots)\\ &\quad+(\alpha\beta\gamma+\alpha\beta\delta+\ldots))\ x^{-n-3}\\ &\quad+\ldots,\end{aligned}$$

where the general m^{th} term is

$$\begin{aligned}
&(\alpha^m+\beta^m+\gamma^m+\dots)\\
&+(\alpha^{m-1}\beta+\beta^{m-1}\alpha+\alpha^{m-1}\gamma+\beta^{m-1}\gamma+\gamma^{m-1}\alpha+\gamma^{m-1}\beta+\dots)\\
&+(\alpha^{m-2}(\beta^2+\beta\gamma+\gamma^2+\dots)\\
&\quad+\beta^{m-2}(\alpha^2+\alpha\gamma+\gamma^2)\\
&\quad+\gamma^{m-2}(\alpha^2+\alpha\beta+\beta^2)+\dots)\\
&+(\alpha^{m-3}(\beta^3+\beta^2\gamma+\beta\gamma^2+\gamma^3+\beta^2\delta+\beta\delta^2+\dots)\\
&\quad+\beta^{m-3}(\alpha^3+\alpha^2\gamma+\alpha\gamma^2+\gamma^3+\alpha^2\delta+\alpha\delta^2+\dots)\\
&\quad+\gamma^{m-3}(\alpha^3+\alpha^2\beta+\alpha\beta^2+\beta^3+\alpha^2\delta+\alpha\delta^2+\dots)+\dots)\\
&+(\alpha^{m-4}(\beta^4+\beta^3\gamma+\beta^2\gamma^2+\beta^2\gamma\delta+\beta\gamma\delta\epsilon+\dots)+\dots)\\
&+(\dots)+(\dots)+\dots
\end{aligned}$$

$= P.$

The dimensions of every term associated with a particular power of x will be the same; this value will be m, if $-n-m$ is the power of x to which the terms are annexed.

All terms whose dimensions are m are collected together.

The coefficient of all of the individual terms will be 1.

Now $\dfrac{a}{x-\alpha}+\dfrac{b}{x-\beta}+\dfrac{c}{x-\gamma}+\dfrac{d}{x-\delta}+\dots=$

$$\begin{aligned}
&(a+b+c+d+\dots)x^{-1}\\
&+(a\alpha+b\beta+c\gamma+d\delta+\dots)x^{-2}\\
&+(a\alpha^2+b\beta^2+c\gamma^2+d\delta^2+\dots)x^{-3}\\
&+\ \dots\\
&+(a\alpha^{n-1}+b\beta^{n-1}+c\gamma^{n-1}+d\delta^{n-1}+\dots)x^{-n}\\
&+(a\alpha^{n}\ \ +b\beta^{n}\ \ +c\gamma^{n}\ \ +d\delta^{n}\ \ +\dots)x^{-n-1}\\
&+(a\alpha^{n+1}+b\beta^{n+1}+c\gamma^{n+1}+d\delta^{n+1}+\dots)x^{-n-2}\\
&+\ \dots\\
&+(a\alpha^{n+m-1}+b\beta^{n+m-1}+c\gamma^{n+m-1}+d\delta^{n+m-1}+\dots)x^{-n-m}\\
&+\ \dots
\end{aligned}$$

$= Q.$

Setting individual corresponding terms in P and Q (i.e., those terms annexed to the same power of x) equal to each other, we have

$$a+b+c+d+\ldots = \frac{1}{(\alpha-\beta)(\alpha-\gamma)(\alpha-\delta)\cdots} + \frac{1}{(\beta-\alpha)(\beta-\gamma)(\beta-\delta)\cdots} + \frac{1}{(\gamma-\alpha)(\gamma-\beta)(\gamma-\delta)\cdots} + \ldots = 0;$$

likewise

$$\frac{\alpha}{(\alpha-\beta)(\alpha-\gamma)(\alpha-\delta)\cdots} + \frac{\beta}{(\beta-\alpha)(\beta-\gamma)(\beta-\delta)\cdots} + \frac{\gamma}{(\gamma-\alpha)(\gamma-\beta)(\gamma-\delta)\cdots} + \ldots = 0;$$

* * *

$$\frac{\alpha^r}{(\alpha-\beta)(\alpha-\gamma)(\alpha-\delta)\cdots} + \frac{\beta^r}{(\beta-\alpha)(\beta-\gamma)(\beta-\delta)\cdots} + \frac{\gamma^r}{(\gamma-\alpha)(\gamma-\beta)(\gamma-\delta)\cdots} + \ldots = S.$$

If $r = n-1$ then $S = 1$; if $r = n$ then $S = \alpha+\beta+\gamma+\delta+\ldots$; if $r = n+1$ then $S = (\alpha^2+\beta^2+\gamma^2+\ldots)+(\alpha\beta+\alpha\gamma+\beta\gamma+\alpha\delta+\beta\delta+\ldots)$; and in general, if $r = n+m-1$, then $S = (\alpha^m+\beta^m+\gamma^m+\ldots)+(\alpha^{n-1}\beta+\beta^{n-1}\alpha+\alpha^{n-1}\gamma+\ldots)+\ldots$, annexed to the power x^{-n-m} in the quantity P.

(2) Let α, β, γ, δ, ... be the roots of the equation

$$p-qx+rx^2-sx^3+tx^4-\ldots\pm x^n = 0.$$

Then $$\frac{a}{\alpha-x}+\frac{b}{\beta-x}+\frac{c}{\gamma-x}+\frac{d}{\delta-x}+\ldots = \pm\frac{1}{p-qx+rx^2-sx^3+tx^4-\ldots},$$

where $$a = \frac{1}{(\alpha-\beta)(\alpha-\gamma)(\alpha-\delta)\cdots},\ b = \frac{1}{(\beta-\alpha)(\beta-\gamma)(\beta-\delta)\cdots},$$
$$c = \frac{1}{(\gamma-\alpha)(\gamma-\beta)(\gamma-\delta)\cdots},\ d = \frac{1}{(\delta-\alpha)(\delta-\beta)(\delta-\gamma)\cdots},\ \text{etc.}$$

Then $$\pm\frac{1}{p-qx+rx^2-sx^3+tx^4-\ldots}$$
$$= \pm\left(\frac{1}{p}+\frac{q}{p^2}x+\frac{q^2-rp}{p^3}x^2+\frac{q^3-2qrp+sp^2}{p^4}x^3+\ldots\right)$$
$$= \frac{1}{\alpha\beta\gamma\delta}\Big(1+\Big(\frac{1}{\alpha}+\frac{1}{\beta}+\frac{1}{\gamma}+\frac{1}{\delta}+\cdots\Big)x$$
$$+\Big(\frac{1}{\alpha^2}+\frac{1}{\beta^2}+\frac{1}{\gamma^2}+\frac{1}{\delta^2}+\cdots+\frac{1}{\alpha\beta}+\frac{1}{\alpha\gamma}+\frac{1}{\beta\gamma}+\frac{1}{\alpha\delta}+\frac{1}{\beta\delta}+\frac{1}{\gamma\delta}+\cdots\Big)x^2$$
$$+\Big(\frac{1}{\alpha^3}+\frac{1}{\beta^3}+\frac{1}{\gamma^3}+\frac{1}{\delta^3}+\cdots+\frac{1}{\alpha^2\beta}+\frac{1}{\beta^2\alpha}+\frac{1}{\alpha^2\gamma}+\frac{1}{\gamma^2\alpha}+\frac{1}{\beta^2\gamma}+\cdots$$
$$+\frac{1}{\alpha\beta\gamma}+\frac{1}{\alpha\beta\delta}+\frac{1}{\alpha\beta\epsilon}+\frac{1}{\beta\gamma\delta}+\cdots\Big)\quad x^3$$
$$+\ldots\Big) = P.$$

This series becomes the same as the one first given, if instead of α, β, γ, δ, ... we write $\frac{1}{\alpha}$, $\frac{1}{\beta}$, $\frac{1}{\gamma}$, $\frac{1}{\delta}$,, etc., and instead of x^m we write x^{n-m}.

Then we also have $\frac{a}{\alpha-x}+\frac{b}{\beta-x}+\frac{c}{\gamma-x}+\cdots$

$= +\frac{a}{\alpha}+\frac{b}{\beta}+\frac{c}{\gamma}+\cdots+\left(+\frac{a}{\alpha^2}+\frac{b}{\beta^2}+\frac{c}{\gamma^2}+\cdots\right)x+\left(+\frac{a}{\alpha^3}+\frac{b}{\beta^3}+\frac{c}{\gamma^3}+\cdots\right)x^2+\cdots$

$= Q$; setting like terms of P and Q equal to each other we see that

$$\frac{1}{\alpha}\cdot\frac{1}{(\alpha-\beta)(\alpha-\gamma)(\alpha-\delta)\cdots}+\frac{1}{\beta}\cdot\frac{1}{(\beta-\alpha)(\beta-\gamma)(\beta-\delta)\cdots}+\frac{1}{\gamma}\cdot\frac{1}{(\gamma-\alpha)(\gamma-\beta)(\gamma-\delta)\cdots}$$
$$+\cdots = \frac{\pm 1}{\alpha\beta\gamma\delta\cdots};$$
$$\frac{1}{\alpha^2}\cdot\frac{1}{(\alpha-\beta)(\alpha-\gamma)(\alpha-\delta)\cdots}+\frac{1}{\beta^2}\cdot\frac{1}{(\beta-\alpha)(\beta-\gamma)(\beta-\delta)\cdots}+\frac{1}{\gamma^2}\cdot\frac{1}{(\gamma-\alpha)(\gamma-\beta)(\gamma-\delta)\cdots}$$
$+\cdots = \frac{\pm 1}{\alpha\beta\gamma\delta\cdots}\left(\frac{1}{\alpha}+\frac{1}{\beta}+\frac{1}{\gamma}+\frac{1}{\delta}+\cdots\right)$; and so on.

Problem XXIX

Given the resolution, to find the equation.

Method 1. With the aid of Problem XXVI or Problem V, the equation is reduced to eliminate irrationalities, and so the problem is solved.

The resulting equation may often be reducible to others having fewer dimensions: to this end, see if it has any divisors; if none are found, it cannot be further reduced.

Example 1. Let x be the unknown quantity, with the given resolution

$$x = \sqrt[3]{-\frac{1}{2}b+\sqrt{\frac{1}{4}b^2+\frac{1}{27}a^3}} + \sqrt[3]{-\frac{1}{2}b-\sqrt{\frac{1}{4}b^2+\frac{1}{27}a^3}}.$$

When this equation is reduced to eliminate irrationalities, the result is the equation $\quad x^9+3bx^6+(3b^2+a^3)x^3+b^3 = 0,$

whose divisors are the cubic equations

$$x^3+ax+b = 0,\ x^3-\left(\frac{1+\sqrt{-3}}{2}\right)ax+b = 0,\ x^3-\left(\frac{1-\sqrt{-3}}{2}\right)ax+b = 0.$$

We have just shown that the resolution

$$x = \sqrt[3]{-\frac{1}{2}b+\sqrt{\frac{1}{4}b^2+\frac{1}{27}a^3}} + \sqrt[3]{-\frac{1}{2}b-\sqrt{\frac{1}{4}b^2+\frac{1}{27}a^3}}$$

is more properly regarded as the solution of an equation of the ninth degree, than simply of a cubic.

Example 2. Let the resolution be $x = \sqrt[3]{-b+\sqrt{b^2-q^3}}+\frac{A}{\sqrt[3]{-b+\sqrt{b^2-q^3}}}$.

After reduction to eliminate irrationalities, this equation becomes

$$\left(x^3-3Ax+\frac{q^3+A^3}{q^3}b\right)^2-(b^2-q^2)\left(\frac{A^3-q^3}{q^3}\right)^2=0.$$

If $A=q$, every root of the resulting equation is a double root.

Example 3. Let the resolution be $x=\sqrt[n]{a+\sqrt{b}}+A\sqrt[n]{a-\sqrt{b}}$. Eliminating all irrationalities except $\sqrt{b}$ produces an equation of degree $n\times n$, whose divisors will be the equations

$$x^n-n(a^2-b)^{1/n}Ax^{n-2}+\frac{n(n-3)}{2}(a^2-b)^{2/n}A^2x^{n-4}$$
$$-\frac{n(n-4)(n-5)}{3\,!}(a^2-b)^{3/n}A^3x^{n-6}+\frac{n(n-5)(n-6)(n-7)}{4\,!}(a^2-b)^{4/n}A^4x^{n-8}$$
$$-\ldots-(A^n+1)a\pm(A^n-1)\sqrt{b}=0;$$

$$x^n-n\pi(a^2-b)^{1/n}Ax^{n-2}+\frac{n(n-3)}{2}\pi^2(a^2-b)^{2/n}A^2x^{n-4}$$
$$-\frac{n(n-4)(n-5)}{3\,!}\pi^3(a^2-b)^{3/n}A^3x^{n-6}+\frac{n(n-5)(n-6)(n-7)}{4\,!}\pi^4(a^2-b)^{4/n}A^4x^{n-8}$$
$$-\ldots-(A^n+1)a\pm(A^n-1)\sqrt{b}=0;$$

$$x^n-n\rho(a^2-b)^{1/n}Ax^{n-2}+\frac{n(n-3)}{2}\rho^2(a^2-b)^{2/n}A^2x^{n-4}$$
$$-\frac{n(n-4)(n-5)}{3\,!}\rho^3(a^2-b)^{3/n}A^3x^{n-6}+\frac{n(n-5)(n-6)(n-7)}{4\,!}\rho^4(a^2-b)^{4/n}A^4x^{n-8}$$
$$-\ldots-(A^n+1)a\pm(A^n-1)\sqrt{b}=0;$$

etc., where 1, π, ρ, ... are the various roots of the equation $x^n-1=0$.

If $A=1$, the irrationality $\sqrt{b}$ vanishes, and the resolution will be of degree $n\times n$; otherwise, to every equation

$$x^n\ -n(a^2-b)^{1/n}Ax^{n-2}+\frac{n(n-3)}{2}(a^2-b)^{2/n}A^2x^{n-4}$$
$$-\frac{n(n-4)(n-5)}{3\,!}(a^2-b)^{3/n}A^3x^{n-6}+\ldots-(A^n+1)a+(A^n-1)\sqrt{b}=0$$

there corresponds another,

$$x^n\ -n(a^2-b)^{1/n}Ax^{n-2}+\frac{n(n-3)}{2}(a^2-b)^{2/n}A^2x^{n-4}$$
$$-\frac{n(n-4)(n-5)}{3\,!}(a^2-b)^{3/n}A^3x^{n-6}+\ldots-(A^n+1)a-(A^n-1)\sqrt{b}=0$$

and so the equation of this resolution has dimensions $2n\times n$.

Example 4. Let the resolution be $\sqrt[n]{a+\sqrt{b}}+\dfrac{A}{\sqrt[n]{a+\sqrt{b}}}$; when all irrationalities are eliminated, the resulting equation is

$$\left(x^n - nAx^{n-2} + \frac{n(n-3)}{2}A^2x^{n-4} - \frac{n(n-4)(n-5)}{3\,!}A^3x^{n-6} + \ldots - \frac{a^n+a^2-b}{a^2-b}\right)^2$$

$$-b\left(\frac{A^n-a^2+b}{(a^2-b)^2}\right)^2 = 0.$$

Corollary. Let $x^n-1 = 0$ have quadratic divisors $z^2-\pi z+1 = 0$, $z^2-\rho z+1 = 0$, $z^2-\sigma z+1 = 0$, $z^2-\tau z+1 = 0$, etc., and let the paired roots of these equations be α and β, γ and δ, ϵ and ζ, η and θ, etc. — i.e., $\beta = \frac{1}{\alpha}$, $\delta = \frac{1}{\gamma}$, $\epsilon = \frac{1}{\zeta}$, $\theta = \frac{1}{\eta}$, etc. Then since the roots of the resolution

$x = \sqrt[n]{a+\sqrt{b}} + \dfrac{A}{\sqrt[n]{a+\sqrt{b}}}$ are

$$\sqrt[n]{a+\sqrt{b}} + \frac{A}{\sqrt[n]{a+\sqrt{b}}},\quad \alpha\sqrt[n]{a+\sqrt{b}} + \frac{A}{\alpha\sqrt[n]{a+\sqrt{b}}},$$

$$\beta\sqrt[n]{a+\sqrt{b}} + \frac{A}{\beta\sqrt[n]{a+\sqrt{b}}},\quad \gamma\sqrt[n]{a+\sqrt{b}} + \frac{A}{\gamma\sqrt[n]{a+\sqrt{b}}},\quad \ldots,$$

the corresponding roots of the equation $x = \sqrt[n]{a+\sqrt{b}}+A\sqrt[n]{a-\sqrt{b}}$ will be $\sqrt[n]{a+\sqrt{b}}+A\sqrt[n]{a-\sqrt{b}}$, $\alpha\sqrt[n]{a+\sqrt{b}}+\beta A\sqrt[n]{a-\sqrt{b}}$, $\gamma\sqrt[n]{a+\sqrt{b}}+\delta A\sqrt[n]{a-\sqrt{b}}$, $\epsilon\sqrt[n]{a+\sqrt{b}}+\zeta A\sqrt[n]{a-\sqrt{b}}$, etc.

Example 5. If the resolution is $x=(a+(b+(c+(d+e^{1/m})^{1/n})^{1/p})^{1/q})^{1/r}+f$, the equation should be transformed to eliminate irrationalities, and the result will be $(((((x-f)^r-a)^q-b)^p-c)^n-d)^m-e = 0$.

And so many similar examples may be treated.

In general, if an equation is of degree n, $x^n-px^{n-1}+qx^{n-2}-rx^{n-3}+\ldots = 0$, its solution will be (as the previous examples have shown) achieved by solving an equation whose degree is much greater than that of the given equation.

Method 2. The following method actually leads to the resolution of equations.

Assume a resolution involving more unknown quantities than the degree (n) of he given equation, and transform the assumed equation so as to eliminate irrational quantities. There will result an equation, having a factor which is the original equation, from which can be extracted n equations in more than n unknowns; on reduction these will lead to a single equation having two or more unknowns.

Then find rational values of the unknown quantities in the resulting equation, which upon substitution into the assumed resolution give the required unknowns.

From the preceding examples and earlier observations, we may compose resolutions appropriate to any given equation; but in equations of the fifth and higher degree the calculations require practically infinite labor. [12]

Lemma

(1) Let α, β, γ, δ, ϵ, ... be roots of the equation $z^n-1 = 0$. Then $\alpha^m+\beta^m+\gamma^m+\delta^m+\ldots = 0$, unless $n = m$ or n is a divisor of m, in which case $\alpha^m+\beta^m+\gamma^m+\delta^m+\ldots = n$.

(2) The sum of all products of the form $\alpha^m\beta^r+\beta^m\alpha^r+\alpha^m\gamma^r+\gamma^m\alpha^r+\beta^m\gamma^r+\gamma^m\beta^r+\ldots$ is 0, unless $n = m+r$ or n is a divisor of $m+r$, in which case $\alpha^m\beta^r+\beta^m\alpha^r+\alpha^m\gamma^r+\ldots = -n$; or if $n = m$ or n divides m, in which case $\alpha^m\beta^r+\beta^m\alpha^r+\alpha^m\gamma^r+\ldots = n^2-n$; or finally if $m = r$, then $\alpha^m\beta^r+\beta^m\alpha^r+\alpha^m\gamma^r+\ldots = \frac{n^2-n}{2}$.

(3.1) The sum of all quantities of the form $\alpha^m\beta^r\gamma^s\delta^t\cdots+\alpha^r\beta^m\gamma^s\delta^t\cdots+\alpha^m\beta^r\gamma^t\delta^s\cdots+\ldots$ is 0, unless n is equal to, or a divisor of, the sum $m+r+s+t+\ldots$. [13]

(3.2) For let π be the number of indices m, r, s, t, ..., and suppose n is equal to, or a divisor of, the sum $m+r+s+t+\ldots$; and for the product $(\pi-1)!\cdot n$ write A.

(3.3) Suppose n is equal to, or a divisor of the sum of ρ of the above indices m, r, s, t, etc. (where $\rho < \pi$), and (since it divides the sum of all of them) also a divisor of the other $\pi-\rho$ indices: find the sum of all quantities of the form $(\rho-1)!(\pi-\rho-1)!n^2$, and call this sum B.

(3.4) Suppose n is equal to, or a divisor of, some number σ of the above indices, and also of some number ρ' of those remaining, so that it will equal or divide the sum of the remaining $\pi-\rho'-\sigma$ indices as well: find the sum of all quantities of the form $(\sigma-1)!(\rho'-1)!(\pi-\rho'-\sigma-1)!n^3$, and call this sum C.

(3.5) In the same way suppose n is equal to, or a divisor of, the sum of some number τ of the indices m, r, s, t, ..., and is likewise equal to or a divisor of the sum of σ' of the remaining ones, and also is equal to, or a divisor of, the sum of ρ'' of those indices not encompassed in the prior two summations; and of course it must be equal to, or a divisor of, the sum of the final $\pi-\tau-\sigma'-\rho''$ subset of the indices: then form the sum of all quantities of the form $(\tau-1)!(\sigma'-1)!(\rho''-1)!(\pi-\tau-\sigma'-\rho''-1)!n^4$, and call it D. Continue in the same manner for five or more partitionings of the indices which are all equal to or divisible by n.

Then the sum of all quantities of the form $\alpha^m\beta^r\gamma^s\delta^t\cdots+\alpha^r\beta^m\gamma^s\delta^t\cdots+\alpha^m\beta^r\gamma^t\delta^s\cdots+\ldots$ is $\pm(A-B+C-D+\ldots)$, where the sign is $+$ if π is odd, and $-$ otherwise.

In the case where a of the indices are equal to m, b are equal to r, c equal to s, d of them equal to t, and so on, the above sum becomes $\pm\dfrac{A-B+C-D+\ldots}{a!b!c!d!\cdots}$.

Theorem 20

(1) Let $x = a\sqrt[3]{p}+b\sqrt[3]{p^2}$; when irrationalities are eliminated, the result is $x^3-3abpx-a^3p-b^3p^2 = 0$. Considering this equation as equivalent to $x^3+qx-r = 0$, we see that $3abp = -q$ and $a^3p+b^3p^2 = r$, whence $p = \frac{-q}{3ab}$, and so r may be rewritten $r = \frac{-a^2q}{3b}+\frac{bq^2}{9a^2}$; or $b = \frac{-q}{3ap}$, whence $r = a^3p-\frac{q^3}{27a^3p}$; or again $a = \frac{-q}{3bp}$ and $r = \frac{-q^3}{27b^3p^2}+b^3p^2$. An arbitrary value may be assumed for any one of a, b, or p, and substituted for the letter in the preceding equations, and the result will be equations which are quadratic, or quadratic in form.

(2) Let the resolution be $x = a\sqrt[4]{p}+b\sqrt[4]{p^2}+c\sqrt[4]{p^3}$, and eliminate the irrationalities; the result will be the equation
$x^4-(2b^2+4ac)px^2-4(a^2bp+bc^2p^2)x-a^4p+b^4p^2-c^4p^3+2a^2c^2p^2-4ab^2cp^2 = 0.$
Matching this equation to the biquadratic $x^4+qx^2-rx+s = 0$, and assuming $p = 1$, we obtain

$$\begin{aligned}
&2b^2+4ac &&= -q\,,\\
&4a^2b+4bc^2 &&= r\,,\\
&a^4+c^4-2a^2c^2+4ab^2c-b^4 &&\\
&= (a^2-c^2)^2+4acb^2-b^4 &&\\
&= (a^2+c^2)^2-4a^2c^2+4acb^2-b^4 &&= -s\,.
\end{aligned}$$

Then from the second equation, $a^2+c^2 = \frac{r}{4b}$ and the first, $4ac = -q-2b^2$, the third becomes

$$(a^2+c^2)^2-4a^2c^2+4acb^2-b^4 = \frac{r^2}{16b^2}-\frac{1}{4}(q+2b^2)^2-b^2q-2b^4-b^4 = -s\,;$$

and it follows that $$4b^6+2qb^4+(\frac{1}{4}q^2-s)b^2-\frac{r^2}{16} = 0,$$

an equation in the form of a cubic, allows us to determine the value of b, and from this the values of a and c readily emerge, and so we have solved the biquadratic equation.

From the same principle different resolutions (of the same structure) of the biquadratic $x^4+qx^2-rx+s = 0$ may be deduced, by making a, c, or p the primary independent variable.

(3.1) Let $x = a\sqrt[2n]{p} + \sqrt[2n]{p^2}$; on eliminating irrationalities we will have an equation

$$\begin{aligned}
&x^{2n}\\
&-2nb^np\ x^n\\
&-\frac{n}{2!}(2nb^{n-1}a^2p)\ x^{n-1}\\
&-\frac{n(n^2-1)}{4!}(2nb^{n-2}a^4p)\ x^{n-2}\\
&-\frac{n(n^2-1)(n^2-4)}{6!}(2nb^{n-3}a^6p)\ x^{n-3}\\
&-\frac{n(n^2-1)(n^2-4)(n^2-9)}{8!}(2nb^{n-4}a^8p)\ x^{n-4}\\
&-\cdots\\
&-\frac{n(n^2-1)(n^2-4)(n^2-9)(n^2-16)\cdots(n^2-(n-2)^2)}{(2n-2)!}(2na^{2n-2}bp)\ x\\
&= a^{2n}p-b^{2n}p^2.
\end{aligned}$$

(3.2) Let the equation be $x = a\sqrt[2n+1]{p}+\sqrt[2n+1]{p^2}$; on eliminating irrationalities we will have an equation

$$\begin{aligned}
&x^{2n+1}\\
&-(2n+1)b^n\ ap\ x^n\\
&-\frac{n(n+1)}{3!}(2n+1)b^{n-1}a^3p\ x^{n-1}\\
&-\frac{n(n^2-1)(n+2)}{5!}(2n+1)b^{n-2}a^5p\ x^{n-2}\\
&-\frac{n(n^2-1)(n^2-4)(n+3)}{7!}(2n+1)b^{n-3}a^7p\ x^{n-3}\\
&-\frac{n(n^2-1)(n^2-4)(n^2-9)(n+4)}{9!}(2n+1)b^{n-4}a^9p\ x^{n-4}\\
&-\cdots\\
&-\frac{n(n^2-1)(n^2-4)(n^2-9)(n^2-16)\cdots(n^2-(n-2)^2)(2n-1)}{(2n-1)!}(2n+1)a^{2n-1}bp\ x\\
&= a^{2n+1}p-b^{2n+1}p^2.
\end{aligned}$$

(3.3) Let $x = a\sqrt[m]{p}+b\sqrt[m]{p^2}$. Then the equation with all irrationalities eliminated will be

$$
\begin{aligned}
&x^m \\
&-ma^{m-2}bp\ x \\
&-\frac{m(m-3)}{2!}a^{m-4}b^2p)x^2 \\
&-\frac{m(m-4)(m-5)}{3!}a^{n-6}b^3p)x^3 \\
&-\frac{m(m-5)(m-6)(m-7)}{4!}a^{m-8}b^4p)x^4 \\
&-\frac{m(m-6)(m-7)(m-8)(m-9)}{5!}a^{m-10}b^5p)x^5 \\
&-\dots \qquad = a^mp\pm b^mp^2,
\end{aligned}
$$

where the last term will be $+b^mp^2$ if m is odd, and $-b^mp^2$ if m is even.

(4.1) Let n be an odd number, and $x = a\sqrt[n]{p}+b\sqrt[n]{p^3}$. Then the equation freed of all irrationalities will be

$$
\begin{aligned}
x^n-p[\ & na^{n-3}bx^2 \\
&+\frac{n(n-5)}{2!}a^{n-6}b^2x^4 \\
&+\frac{n(n-7)(n-8)}{3!}a^{n-9}b^3x^6 \\
&+\frac{n(n-9)(n-10)(n-11)}{4!}a^{n-12}b^4x^8 \\
&+\frac{n(n-11)(n-12)(n-13)(n-14)}{5!}a^{n-15}b^5x^10 \\
&+\frac{n(n-13)(n-14)(n-15)(n-16)(n-17)}{6!}a^{n-18}b^6x^12 \\
&+\dots]
\end{aligned}
$$

$$
\begin{aligned}
\pm p^2\Big[\ & na^{\frac{n-3}{2}}b^{\frac{n+1}{2}}x-\frac{1}{2^2}\frac{n(n-5)(n-7)}{3!}a^{\frac{n-9}{2}}b^{\frac{n+3}{2}}x^3+ \\
&\frac{1}{2^4}\frac{n(n-7)(n-9)(n-11)(n-13)}{5!}a^{\frac{n-15}{2}}b^{\frac{n+5}{2}}x^5- \\
&\frac{1}{2^6}\frac{n(n-9)(n-11)(n-13)(n-15)(n-17)(n-19)}{7!}a^{\frac{n-21}{2}}b^{\frac{n+7}{2}}x^7+ \\
&\frac{1}{2^8}\frac{n(n-11)(n-13)(n-15)(n-17)(n-19)(n-21)(n-23)(n-25)}{9!}a^{\frac{n-27}{2}}b^{\frac{n+9}{2}}x^9 \\
&-\dots\ \Big] = a^np+b^np^3.
\end{aligned}
$$

(The sign of p^2 is taken to be positive, if $\frac{n-3}{4}$ is an integer, otherwise it is negative.)

(4.2) Let n be an even number, and $x = a\sqrt[n]{p}+b\sqrt[n]{p^3}$. Then the equation freed of all irrationalities will be

$$x^n - p(\ na^{n-3}bx^2 + \frac{n(n-5)}{2!}a^{n-6}b^2x^4 + \frac{n(n-7)(n-8)}{3!}a^{n-9}b^3x^6 + \frac{n(n-9)(n-10)(n-11)}{4!}a^{n-12}b^4x^8 + \frac{n(n-11)(n-12)(n-13)(n-14)}{5!}a^{n-15}b^5x^{1}0 + \frac{n(n-13)(n-14)(n-15)(n-16)(n-17)}{6!}a^{n-18}b^6x^{1}2 \ + \dots)$$

$$\pm \underline{p^2}\Big[\ \frac{1}{2}\frac{n(n-4)}{2}a^{\frac{n-6}{2}}b^{\frac{n+2}{2}}x^2 - \frac{1}{2^3}\frac{n(n-6)(n-8)(n-10)}{4!}a^{\frac{n-12}{2}}b^{\frac{n+4}{2}}x^4 + \frac{1}{2^5}\frac{n(n-8)(n-10)(n-12)(n-14)(n-16)}{6!}a^{\frac{n-18}{2}}b^{\frac{n+6}{2}}x^6 - \frac{1}{2^7}\frac{n(n-10)(n-12)(n-14)(n-16)(n-18)(n-20)(n-22)}{8!}a^{\frac{n-24}{2}}b^{\frac{n+8}{2}}x^8 + \frac{1}{2^9}\frac{n(n-12)(n-14)(n-16)(n-18)(n-20)(n-22)(n-24)(n-26)(n-28)}{9!}a^{\frac{n-30}{2}}b^{\frac{n+10}{2}}x^{10} - \dots\ \Big] = a^n p + b^n p^3 \pm \underline{2a^{\frac{n}{2}}b^{\frac{n}{2}}p^2}.$$

(The underlined quantities $\pm p^2$ and $\pm 2a^{\frac{n}{2}}b^{\frac{n}{2}}p^2$ denote $+p^2$ and $+2a^{\frac{n}{2}}b^{\frac{n}{2}}p^2$ if $\frac{n-4}{2}$ is an integer, otherwise $-p^2$ and $-2a^{\frac{n}{2}}b^{\frac{n}{2}}p^2$ respectively.)

This series contains no negative powers of a.

(5.1) Let $x = a\sqrt[n]{p}+b\sqrt[n]{p^{n-1}}$, with n odd; then

$$x^n - nabpx^{n-2} + \frac{n(n-3)}{2!}a^2b^2p^2x^{n-4} - \frac{n(n-4)(n-5)}{3!}a^3b^3p^3x^{n-6} + \frac{n(n-5)(n-6)(n-7)}{4!}a^4b^4p^4x^{n-8} - \dots \quad = a^n p + b^n p^{n-1}.$$

(5.2) Let $x = a\sqrt[n]{p}+b\sqrt[n]{p^{n-1}}$, with n even; then

$$x^n - nabpx^{n-2} + \frac{n(n-3)}{2!}a^2b^2p^2x^{n-4} - \frac{n(n-4)(n-5)}{3!}a^3b^3p^3x^{n-6} + \frac{n(n-5)(n-6)(n-7)}{4!}a^4b^4p^4x^{n-8} - \dots \quad = a^n p \pm \underline{2a^{n/2}b^{n/2}p^{n/2}} + p^{n-1}.$$

(If $n = 4r+2$, where r is an integer, take $+2a^{n/2}b^{n/2}p^{n/2}$ for the underlined term; otherwise it will be $-2a^{n/2}b^{n/2}p^{n/2}$.)

(6.1) Let $x = a\sqrt[n]{p}+b\sqrt[n]{p^{n-2}}$, with n not divisible by 2 or 3. Then the equation freed of all irrationalities will be

$$x^n - na^2bpx^{n-3} + \frac{n(n-5)}{2!}a^4b^2p^2x^{n-6} - \frac{n(n-7)(n-8)}{3!}a^6b^3p^3x^{n-9}$$
$$+\frac{n(n-9)(n-10)(n-11)}{4!}a^8b^4p^4x^{n-12}$$
$$-\frac{n(n-11)(n-12)(n-13)(n-14)}{5!}a^10b^5p^5x^{n-15}$$
$$-\frac{n(n-13)(n-14)(n-15)(n-16)(n-17)}{6!}a^12b^6p^6x^{n-18} - \ldots \text{ (to } m \text{ terms,}$$
where m is the smallest integer greater than $\frac{n}{3}$)
$$-ab^{\frac{n+1}{2}}p^{\frac{n-1}{2}}\Big[nx^{\frac{n-3}{2}} + \frac{1}{2^2}\frac{n(n-5)(n-7)}{3!}a^2bpx^{\frac{n-9}{2}}$$
$$+\frac{1}{2^4}\frac{n(n-7)(n-9)(n-11)(n-13)}{5!}a^4b^2p^2x^{\frac{n-15}{2}}$$
$$+\frac{1}{2^6}\frac{n(n-9)(n-11)(n-13)(n-15)(n-17)(n-19)}{7!}a^6b^3p^3x^{\frac{n-21}{2}}$$
$$+\frac{1}{2^8}\frac{n(n-11)(n-13)(n-15)(n-17)(n-19)(n-21)(n-23)(n-25)}{9!}a^8b^4p^4x^{\frac{n-27}{2}}$$
$$+\ldots\Big] \quad = A = a^np + b^np^{n-2}.$$

If n is an odd multiple of 3, then instead the quantity A above will be $A = a^np + b^np^{n-2} + 3a^{\frac{n}{3}}b^{\frac{2n}{3}}p^{\frac{2n}{3}-1} + 3a^{\frac{2n}{3}}b^{\frac{n}{3}}p^{\frac{n}{3}}$.

(6.2) Let $x = a\sqrt[n]{p} + b\sqrt[n]{p^{n-2}}$, where n is an even number not divisible by 3. The equation free of all irrationalities will be
$$x^n - na^2bpx^{n-3} + \frac{n(n-5)}{2!}a^4b^2p^2x^{n-6} - \frac{n(n-7)(n-8)}{3!}a^6b^3p^3x^{n-9}$$
$$+\frac{n(n-9)(n-10)(n-11)}{4!}a^8b^4p^4x^{n-12}$$
$$-\frac{n(n-11)(n-12)(n-13)(n-14)}{5!}a^10b^5p^5x^{n-15}$$
$$-\frac{n(n-13)(n-14)(n-15)(n-16)(n-17)}{6!}a^12b^6p^6x^{n-18} - \ldots \text{ (to } m \text{ terms,}$$
as specified above)
$$-b^{\frac{n}{2}}p^{\frac{n}{2}-1}\Big[2x^{\frac{n}{2}} + \frac{1}{2}\frac{n(n-4)}{2!}a^2bpx^{\frac{n}{2}-3}$$
$$+\frac{1}{2^3}\frac{n(n-6)(n-8)(n-10)}{4!}a^4b^2p^2x^{\frac{n}{2}-6}$$
$$+\frac{1}{2^5}\frac{n(n-8)(n-10)(n-12)(n-14)(n-16)}{6!}a^6b^3p^3x^{\frac{n}{2}-9}$$
$$+\frac{1}{2^7}\frac{n(n-10)(n-12)(n-14)(n-16)(n-18)(n-20)(n-22)}{8!}a^8b^4p^4x^{\frac{n}{2}-12}$$
$$+\ldots\Big] \quad = A = a^np - b^np^{n-2}.$$

If n is even and divisible by 3, then instead the quantity A above will be $A = a^n p - b^n p^{n-2} - 3a^{\frac{2n}{3}} b^{\frac{n}{3}} p^{\frac{n}{3}} + 3a^{\frac{n}{3}} b^{\frac{2n}{3}} p^{\frac{2n}{3}-1}$.

In all these cases the series stops at the point where the power of a would become negative.

(7) Let α, β, γ, δ, ϵ, ... be the roots of the equation $z^n - 1 = 0$, and let the given equation be $x = a\sqrt[n]{p} + b\sqrt[n]{p^2} + c\sqrt[n]{p^3} + d\sqrt[n]{p^4} + \ldots$

$$+h\sqrt[n]{p^\lambda} + \ldots + k\sqrt[n]{p^\mu} + \ldots + l\sqrt[n]{p^\nu} + \ldots + q\sqrt[n]{p^\xi} + \ldots + r\sqrt[n]{p^\phi} + \ldots$$
$$+s\sqrt[n]{p^{n-4}} + t\sqrt[n]{p^{n-3}} + v\sqrt[n]{p^{n-2}} + u\sqrt[n]{p^{n-1}}.$$

Now the various values of x will be

$$a\sqrt[n]{p}\alpha + b\sqrt[n]{p^2}\alpha^2 + c\sqrt[n]{p^3}\alpha^3 + d\sqrt[n]{p^4}\alpha^4 + \ldots$$
$$+s\sqrt[n]{p^{n-4}}\alpha^{n-4} + t\sqrt[n]{p^{n-3}}\alpha^{n-3} + v\sqrt[n]{p^{n-2}}\alpha^{n-2} + u\sqrt[n]{p^{n-1}}\alpha^{n-1};$$
$$a\sqrt[n]{p}\beta + b\sqrt[n]{p^2}\beta^2 + c\sqrt[n]{p^3}\beta^3 + d\sqrt[n]{p^4}\beta^4 + \ldots$$
$$+s\sqrt[n]{p^{n-4}}\beta^{n-4} + t\sqrt[n]{p^{n-3}}\beta^{n-3} + v\sqrt[n]{p^{n-2}}\beta^{n-2} + u\sqrt[n]{p^{n-1}}\beta^{n-1};$$
$$a\sqrt[n]{p}\gamma + b\sqrt[n]{p^2}\gamma^2 + c\sqrt[n]{p^3}\gamma^3 + d\sqrt[n]{p^4}\gamma^4 + \ldots$$
$$+s\sqrt[n]{p^{n-4}}\gamma^{n-4} + t\sqrt[n]{p^{n-3}}\gamma^{n-3} + v\sqrt[n]{p^{n-2}}\gamma^{n-2} + u\sqrt[n]{p^{n-1}}\gamma^{n-1};$$
$$a\sqrt[n]{p}\delta + b\sqrt[n]{p^2}\delta^2 + c\sqrt[n]{p^3}\delta^3 + d\sqrt[n]{p^4}\delta^4 + \ldots$$
$$+s\sqrt[n]{p^{n-4}}\delta^{n-4} + t\sqrt[n]{p^{n-3}}\delta^{n-3} + v\sqrt[n]{p^{n-2}}\delta^{n-2} + u\sqrt[n]{p^{n-1}}\delta^{n-1};$$

etc. etc. etc. etc.

It follows that the sum of the roots, which is the coefficient of the second term of the required equation, will be

$$a\sqrt[n]{p}(\alpha+\beta+\gamma+\delta+\ldots)(=0) + b\sqrt[n]{p^2}(\alpha^2+\beta^2+\gamma^2+\delta^2+\ldots)(=0)$$
$$+c\sqrt[n]{p^3}(\alpha^3+\beta^3+\gamma^3+\delta^3+\ldots)(=0) + \ldots$$
$$+v\sqrt[n]{p^{n-2}}(\alpha^{n-2}+\beta^{n-2}+\gamma^{n-2}+\delta^{n-2}+\ldots)(=0)$$
$$+u\sqrt[n]{p^{n-1}}(\alpha^{n-1}+\beta^{n-1}+\gamma^{n-1}+\delta^{n-1}+\ldots)(=0).$$

The sum of the product of the roots taken two at a time, which is the coefficient of the third term of the equation, will be

$$a^2\sqrt[n]{p^2}(\alpha\beta+\alpha\gamma+\beta\gamma+\alpha\delta+\ldots)(=0)$$
$$+ab\sqrt[n]{p^3}(\alpha\beta^2+\beta\alpha^2+\alpha\gamma^2+\gamma\alpha^2+\beta\gamma^2+\gamma\beta^2+\alpha\delta^2+\ldots)(=0)$$

— and in general, all the terms will be equal to zero, except for

$$aup\;(\alpha\,\beta^{n-1}+\beta\,\alpha^{n-1}+\alpha\,\gamma^{n-1}+\gamma\,\alpha^{n-1}+\beta\,\gamma^{n-1}+\gamma\,\beta^{n-1}+\alpha\,\delta^{n-1}+\ldots)(=-n)$$
$$+bvp\;(\alpha^2\beta^{n-2}+\beta^2\alpha^{n-2}+\alpha^2\gamma^{n-2}+\gamma^2\alpha^{n-2}+\beta^2\gamma^{n-2}+\gamma^2\beta^{n-2}+\alpha^2\delta^{n-2}+\ldots)(=-n)$$
$$+ctp\;(\alpha^3\beta^{n-3}+\beta^3\alpha^{n-3}+\alpha^3\gamma^{n-3}+\gamma^3\alpha^{n-3}+\beta^3\gamma^{n-3}+\gamma^3\beta^{n-3}+\alpha^3\delta^{n-3}+\ldots)(=-n)$$
$$+dsp\;(\alpha^4\beta^{n-4}+\beta^4\alpha^{n-4}+\alpha^4\gamma^{n-4}+\gamma^4\alpha^{n-4}+\beta^4\gamma^{n-4}+\gamma^4\beta^{n-4}+\alpha^4\delta^{n-4}+\ldots)(=-n)$$
$$+\ldots = -np(au+bv+ct+ds+\ldots).$$

If $n = 2\lambda$, then the coefficient of the product h^2p becomes $-\frac{n}{2}$, and the above sum becomes $-np(au+bv+ct+ds+\ldots+\frac{1}{2}h^2)$.

The sums of the products of the roots taken three at a time, which is the coefficient of the fourth term of the equation, will be

$$a^3\sqrt[n]{p^3}(\alpha\beta\gamma+\alpha\beta\delta+\alpha\gamma\delta+\beta\gamma\delta+\ldots)(=0)$$
$$+a^2b\sqrt[n]{p^4}(\alpha\beta\gamma^2+\alpha\gamma\beta^2+\beta\gamma\alpha^2+\alpha\beta\delta^2+\ldots)(=0)+\ldots$$
$$+a^{n-2}vp(\alpha\beta\gamma^{n-2}+\alpha\gamma\beta^{n-2}+\beta\gamma\alpha^{n-2}+\alpha\beta\delta^{n-2}+\ldots)(=n)$$
$$+abtp(\alpha\beta^2\gamma^{n-3}+\beta\alpha^2\gamma^{n-3}+\gamma\beta^2\alpha^{n-3}+\alpha\beta^2\delta^{n-3}+\ldots)(=1\cdot 2\cdot n)+\ldots.$$

To generalize, all terms contained in the sum vanish, unless they denote quantities of the form $\sqrt[n]{p^n} = p$, or $\sqrt[n]{p^{2n}} = p^2$; that is, the general term may be denoted by the formula

$$hlk\sqrt[n]{p^{\lambda+\mu+\nu}}(\alpha^\lambda\beta^\mu\gamma^\nu+\alpha^\mu\beta^\nu\gamma^\lambda+\alpha^\nu\beta^\mu\gamma^\lambda+\ldots),$$

and where $\lambda+\mu+\nu$ is neither n nor $2n$, the above term vanishes. If $\lambda+\mu+\nu = n$ or $\lambda+\mu+\nu = 2n$, the above term is $1\cdot 2\cdot n(hklp)$ or $1\cdot 2\cdot n(hklp^2)$; but if two of the indices λ, μ, ν are equal to each other the above term must be divided by $1\cdot 2$, and if three indices are equal to each other, i.e. $\lambda=\mu=\nu$, then it must be divided by $1\cdot 2\cdot 3$.

Find all quantities of this kind, where $\lambda+\mu+\nu = n$ or $\lambda+\mu+\nu = 2n$, add them together, and call their sum A.

The sum of all products of four roots multiplied together, which is the coefficient of the fifth term of the required equation, will be

$$a^4\sqrt[n]{p^4}(\alpha\beta\gamma\delta+\alpha\beta\gamma\epsilon+\ldots)(=0)$$
$$+a^3b\sqrt[n]{p^3}(\alpha\beta\gamma\delta^2+\alpha\beta\gamma^2\delta+\alpha\beta^2\gamma\delta+\ldots)(=0)$$
$$+\ldots.$$

We may write the general term as

$$hklq\sqrt[n]{p^{\lambda+\mu+\nu+\xi}}(\alpha^\lambda\beta^\mu\gamma^\nu\delta^\xi+\alpha^\lambda\beta^\mu\gamma^\xi\delta^\nu+\alpha^\mu\beta^\lambda\gamma^\nu\delta^\xi+\alpha^\lambda\beta^\mu\gamma^\nu\epsilon^\xi+\ldots).$$

This term will vanish unless $\lambda+\mu+\nu+\xi = n$ or $2n$ or $3n$, in which case the value of the term will be $-3!(nhklqp)$ or $-3!(nhklqp^2)$ or $-3!(nhklqp^3)$ respectively (unless additionally $\lambda+\mu = \nu+\xi = n$, in which case the said term becomes $-(3!n-n^2)hklqp^2$, and if further $\lambda = \nu$, and consequently also $\mu = \xi$, then it becomes $-(3!n-2n^2)hklqp^2$; lastly if $\lambda = \mu = \nu = \xi = \frac{n}{2}$, this term will be $-(3!n-3n^2)hklqp^2$.)

Finally, in any case where two of the indices λ, μ, ν, or ξ are equal, the specified term must be divided by 2!: if three are equal, it must be divided by 3!, or if four are equal, by 4!, and if two pairs are equal, divide it by $2! \cdot 2!$.

Add together all terms of this sort, and call the sum B.

In like manner determine the sums of products of roots taken five, six, seven, ... at a time, which sums we shall call C, D, E, ... respectively. Then the required equation will be

$$x^n - np(au+bv+ct+ds+\ldots)x^{n-2} - Ax^{n-3} + Bx^{n-4} - Cx^{n-5} + Dx^{n-6} - \ldots = 0.$$

If the resolution is specified as $z = nH + a\sqrt[n]{p} + b\sqrt[n]{p^2} + c\sqrt[n]{p^3} + \ldots$, then by the above principle, writing for x the value $z - nH$ in the equation we derived, we can easily deduce the equation whose root is z.

Corollary 1. Hence an infinity of solvable equations of degree n, involving $n-1$ unknowns, can be constructed in this manner.

Assume a resolution such as $x = H + a\sqrt[n]{p} + b\sqrt[n]{p^2} + c\sqrt[n]{p^3} + \ldots + k\sqrt[n]{p^{n-2}} + l\sqrt[n]{p^{n-1}}$; this resolution involves $n+1$ unknown quantities, so for any two of them we substitute arbitrary values, and eliminate all irrationalities. The result is an equation of degree n in $n-1$ unknowns.

Now say the given equation is $x^n - px^{n-1} + qx^{n-2} - rx^{n-3} + sx^{n-4} - tx^{n-5} + \ldots = 0$. By equating corresponding terms of this equation and the equation resulting from the previous step, we obtain n equations in $n-1$ unknowns. By the method of common divisors the values of all the unknowns may be extracted, if there are any consistent solutions. Further, the technique of reducing the n resulting equations, to eliminate all unknowns, establishes whether or not the given equation admits a resolution of the indicated type.

Corollary 2. Let α, β, γ, δ, ϵ, ... be the roots of an equation of degree n whose solution is required; further let π, ρ, σ, τ, ... be the roots of the equation $x^n - 1 = 0$, and further let $p = 1$. Then suppose

$$\alpha = \pi a + \pi^2 b + \pi^3 c + \pi^4 d + \pi^5 e + \ldots$$
$$\beta = \rho a + \rho^2 b + \rho^3 c + \rho^4 d + \rho^5 e + \ldots$$
$$\gamma = \sigma a + \sigma^2 b + \sigma^3 c + \sigma^4 d + \sigma^5 e + \ldots$$
$$\delta = \tau a + \tau^2 b + \tau^3 c + \tau^4 d + \tau^5 e + \ldots$$
$$\text{etc.} = \text{etc.}$$

where α, β, γ, δ, ... are given, and the n values of π, ρ, σ, τ, ... are determined; then from the n simple equations we have composed, the values of a, b, c, d, e, ... are to be found, and once they are known the required resolution is known, for it is to be $x = a\sqrt[n]{p} + b\sqrt[n]{p^2} + c\sqrt[n]{p^3} + d\sqrt[n]{p^4} + e\sqrt[n]{p^5} + \ldots$.

Problem XXX

To find equations which can be reduced to others of lower degree.

(1) Let $x \equiv x^n-(A+a\sqrt[n]{p}+b\sqrt[n]{p^2}+c\sqrt[n]{p^3}+\ldots+s\sqrt[n]{p^{n-2}}+t\sqrt[n]{p^{n-1}})x^{n-1}$
$+(B+a'\sqrt[n]{p}+b'\sqrt[n]{p^2}+c'\sqrt[n]{p^3}+\ldots+s'\sqrt[n]{p^{n-2}}+t'\sqrt[n]{p^{n-1}})x^{n-2}$
$+(C+a''\sqrt[n]{p}+b''\sqrt[n]{p^2}+c''\sqrt[n]{p^3}+\ldots+s''\sqrt[n]{p^{n-2}}+t''\sqrt[n]{p^{n-1}})x^{n-3}$
$+\ldots=0$;

and let α, β, γ, δ, ... be the various roots of the equation $z^m-1=0$. Then from the principle advanced previously, different versions of the equation $X=0$ may be found, which may be multiplied together, and methods in Chapter 1 show how to derive an equation whose degree is $n\times m$ and which involves $m\times n$ unknowns (A, a, b, c, ...; B, a', b', c', ...; C, a'', b'', c'', ...; etc.) and p. One, two, three, or more of these unknown quantities may be assigned arbitrary values, and so we may produce equations of degree $m\times n$ which can be reduced to equations of degree n.

Corollary. Given m equations of degree n, by means of linear equations an equation $X=0$, of the above form, can be constructed which implicitly contains all of the given equations.

Let the m given equations be

$$x^n-\pi x^{n-1}+\rho x^{n-2}-\sigma x^{n-3}+\tau x^{n-4}-\ldots=0,$$
$$x^n-\pi' x^{n-1}+\rho' x^{n-2}-\sigma' x^{n-3}+\tau' x^{n-4}-\ldots=0,$$
$$x^n-\pi'' x^{n-1}+\rho'' x^{n-2}-\sigma'' x^{n-3}+\tau'' x^{n-4}-\ldots=0,$$

etc. For $\sqrt[n]{p}$ write any quantity (for simplicity, let it be 1); then we assume the m equations

$$A+a\alpha+b\alpha^2+c\alpha^3+\ldots+s\alpha^{n-2}+t\alpha^{n-1}=\pi;$$
$$A+a\beta+b\beta^2+c\beta^3+\ldots+s\beta^{n-2}+t\beta^{n-1}=\pi';$$
$$A+a\gamma+b\gamma^2+c\gamma^3+\ldots+s\gamma^{n-2}+t\gamma^{n-1}=\pi'';$$
etc.

From these m equations we can determine the values of the coefficients a, b, c, ..., s, t; then from the m assumed equations

$$B+a'\alpha+b'\alpha^2+c'\alpha^3+\ldots+s'\alpha^{n-2}+t'\alpha^{n-1}=\rho;$$
$$B+a'\beta+b'\beta^2+c'\beta^3+\ldots+s'\beta^{n-2}+t'\beta^{n-1}=\rho';$$
$$B+a'\gamma+b'\gamma^2+c'\gamma^3+\ldots+s'\gamma^{n-2}+t'\gamma^{n-1}=\rho'';$$
etc.

the values of the coefficients B, a', b', c', ..., s', t' are easily deduced, and so on through all the rest of the coefficients which we seek: once all are found, the required equation will have been determined.

By the same method we can assume equations whose resolution or reduction involves the radical quantities $\sqrt[n]{p}$, $\sqrt[n]{p^2}$, $\sqrt[n]{p^3}$, ..., $\sqrt[n]{q}$, $\sqrt[n]{q^2}$, $\sqrt[n]{q^3}$, ..., $\sqrt[n]{r}$, $\sqrt[n]{r^2}$, $\sqrt[n]{r^3}$, ..., etc. — which, reduced to eliminate irrational quantities, yield equations whose resolution or reduction is known.

(2) In general, any algebraic equation, containing any irrational quantities whatever, may be chosen as the reduced equation; then by the methods of this problem the assumed equation is transformed to remove irrationalities. The result is an equation of higher degree, which can be reduced to the assumed equation of the lower degree.

Theorem 21

Case 1. Let $x = \sqrt[n]{\alpha}+\sqrt[n]{\beta}+\sqrt[n]{\gamma}+\sqrt[n]{\delta}+\ldots$, where $\alpha, \beta, \gamma, \delta, \ldots$ denote roots of the equation $v^{n-1}-pv^{n-2}+qv^{n-3}-\ldots=0$, of degree $n-1$.

The equation whose root is x has n^{n-1} dimensions. If we identify the roots of the equation $x^n-1=0$ respectively as π, ρ, σ, τ, etc., then the quantity $\sqrt[n]{\alpha}$ has the following n values: $\pi\sqrt[n]{\alpha}$, $\rho\sqrt[n]{\alpha}$, $\sigma\sqrt[n]{\alpha}$, etc., and the same holds for $\sqrt[n]{\beta}$, $\sqrt[n]{\gamma}$, $\sqrt[n]{\delta}$, etc.; and consequently the equation whose root is $x = \sqrt[n]{\alpha}+\sqrt[n]{\beta}+\sqrt[n]{\gamma}+\sqrt[n]{\delta}+\ldots$ has $n\times n\times n\times n\times\ldots=n^{n-1}$ roots.

Corollary. The equation whose root is x has the form of an equation of degree n^{n-1}.

Case 2. The letters α, β, γ, δ, ... denoting the same quantities, let $x = \sqrt[2]{\alpha}+\sqrt[2]{\beta}+\sqrt[2]{\gamma}+\sqrt[2]{\delta}+\ldots$; then the resulting equation, cleared of all irrationalities, ascends to 2^{n-1} dimensions, but it has the form of an equation of degree 2^{n-2} since to each positive root corresponds a negative root of equal magnitude.

Case 3. Let $x = \sqrt[rm]{\alpha'}+\sqrt[rm]{\beta'}+\sqrt[rm]{\gamma'}+\sqrt[rm]{\delta'}+\ldots$, where α', β', γ', δ', ... are the rth powers respectively of the roots of an equation of degree $n-1$. Then the equation whose root is x ascends to degree m^{n-1}, but it has the form of an equation of degree m^{n-2}.

This method does not suffice for a general resolution of equations, for in the first case there are only $n-1$ unknown quantities involved in the resolution of an equation whose form is of degree n^{n-2}. Likewise in the second case $n-1$ unknown quantities are produced, to resolve an equation of degree 2^{n-2}, and in the final case $n-1$ arbitrary quantities are contained in an equation whose form is of degree m^{n-2}.

Example 1. Let the given equation be $x^3+qx-r = 0$, and assume $x = \sqrt[3]{\alpha}+\sqrt[3]{\beta}$; it follows that

$$x^3 = \alpha+\beta+3\sqrt[3]{\alpha\beta}(\sqrt[3]{\alpha}+\sqrt[3]{\beta}),$$
$$qx = q(\sqrt[3]{\alpha}+\sqrt[3]{\beta}),$$

and

$$-r = -r.$$

Supposing $\alpha+\beta-r = 0$, and $3\sqrt[3]{\alpha\beta}+q = 0$, we see that $\alpha+\beta = r$, and $\alpha\beta = \frac{-q^3}{27}$, so that α and β will be the two roots of the quadratic equation $z^2-rz-\frac{-q^3}{27}$.

The three roots of the given equation are, respectively,

$$\sqrt[3]{\alpha} + \sqrt[3]{\beta},$$
$$-\frac{1+\sqrt{-3}}{2}\sqrt[3]{\alpha} - -\frac{1-\sqrt{-3}}{2}\sqrt[3]{\beta},$$

and

$$-\frac{1-\sqrt{-3}}{2}\sqrt[3]{\alpha} - -\frac{1+\sqrt{-3}}{2}\sqrt[3]{\beta},$$

and the equation whose roots are the above three numbers will be

$$x^3-3\sqrt[3]{\alpha\beta}x-\alpha-\beta = 0,$$

or, after elimination of the irrational quantity $\sqrt[3]{\alpha\beta}$ it becomes

$$x^9-3(\alpha+\beta)x^6+3(\alpha^2-7\alpha\beta+\beta^2)x^3-(\alpha+\beta)^3 = 0.$$

Example 2. Let the given equation be $x^4+qx^2-rx+s = 0$; for its resolution assume $x = \sqrt[2]{\alpha}+\sqrt[2]{\beta}+\sqrt[2]{\gamma}$. This results in

$$\begin{aligned} x^4 = {} & \alpha^2 + \beta^2 + \gamma^2 + 4(\alpha+\beta+\gamma)(\alpha^{1/2}\beta^{1/2}+\alpha^{1/2}\gamma^{1/2}+\beta^{1/2}\gamma^{1/2}) \\ & + 8\alpha^{1/2}\beta^{1/2}\gamma^{1/2}(\alpha^{1/2}+\beta^{1/2}+\gamma^{1/2}) + 6(\alpha\beta+\alpha\gamma+\beta\gamma), \\ +qx^2 = {} & +2q(\alpha^{1/2}\beta^{1/2}+\alpha^{1/2}\gamma^{1/2}+\beta^{1/2}\gamma^{1/2}) + q(\alpha+\beta+\gamma), \\ -rx = {} & -r(\alpha^{1/2}+\beta^{1/2}+\gamma^{1/2}), \\ +s = {} & s. \end{aligned}$$

Then we assume

$$4(\alpha+\beta+\gamma)+2q = 0;$$
$$\alpha^2+\beta^2+\gamma^2+6(\alpha\beta+\alpha\gamma+\beta\gamma)+q(\alpha+\beta+\gamma)+s = 0;$$

and

$$8\alpha^{1/2}\beta^{1/2}\gamma^{1/2}-r = 0.$$

From the first equation we determine that $\alpha + \beta + \gamma = -\frac{q}{2}$, which, substituted into the second, gives $\frac{q^2}{4}+4(\alpha\beta+\alpha\gamma+\beta\gamma)-\frac{q^2}{2}+s = 0$, whence $4(\alpha\beta+\alpha\gamma+\beta\gamma) = \frac{q^2}{4}-s$; and from the third, $\alpha\beta\gamma = \frac{r^2}{64}$. Consequently the equation whose roots are α, β, and γ will be

$$z^3+\frac{q}{2}z^2+\left(\frac{q^2}{16}-\frac{s}{4}\right)z-\frac{r^2}{64} = 0.$$

The four roots of the given biquadratic will be, respectively,

$$\sqrt[2]{\alpha}+\sqrt[2]{\beta}+\sqrt[2]{\gamma},$$
$$-\sqrt[2]{\alpha}-\sqrt[2]{\beta}+\sqrt[2]{\gamma},$$
$$-\sqrt[2]{\alpha}+\sqrt[2]{\beta}-\sqrt[2]{\gamma},$$

and

$$\sqrt[2]{\alpha}-\sqrt[2]{\beta}-\sqrt[2]{\gamma};$$

and the equation, whose roots are the above quantities, will be

$$x^4-2(\alpha+\beta+\gamma)x^2-8\sqrt[2]{\alpha\beta\gamma}x+\alpha^2+\beta^2+\gamma^2-2\alpha\beta-2\alpha\gamma-2\beta\gamma = 0;$$

or, if the irrational quantity $\sqrt[2]{\alpha\beta\gamma}$ is eliminated, it will become

$$\left(x^4-2(\alpha+\beta+\gamma)x^2+\alpha^2+\beta^2+\gamma^2-2\alpha\beta-2\alpha\gamma-2\beta\gamma\right)^2-64\alpha\beta\gamma x^2 = 0.$$

Indeed, given one resolution, an infinity of others are easily found, all based on the given one although they may appear to be of a different type: for let α', β', γ', δ', ... be such functions of the unknowns of the given equation, that by known algorithms the unknowns themselves can be recovered from the functions α', β', γ', δ', ... — which done, and with the given resolution substituted for the unknown quantities, another resolution results, as desired.

For instance, given the resolution $x = \sqrt[2]{\alpha}+\sqrt[2]{\beta}+\sqrt[2]{\gamma}$, where α, β, and γ are roots of the given equation $z^3 - \frac{q}{2}z^2 + (\frac{q^2}{16}-\frac{s}{4})z - \frac{r^2}{64} = 0$, a resolution $v = \sqrt[2n]{\alpha'}+\sqrt[2n]{\beta'}+\sqrt[2n]{\gamma'}$ is assumed, where $\alpha' = \alpha^n$, $\beta' = \beta^n$, $\gamma' = \gamma^n$; accordingly we find the equation whose roots are the nth powers of the roots of the equation $z^3-\frac{q}{2}z^2+(\frac{q^2}{16}-\frac{s}{4})z-\frac{r^2}{64} = 0$, resulting in an equation whose roots are α', β', γ'. If $n = 2$, the resulting equation will be $v^3 - (\frac{s}{2}+\frac{q^2}{8})v^2 + (\frac{qr^2}{64}+\frac{s^2}{16}-\frac{q^2s}{32}+\frac{q^4}{256})v - \frac{r^4}{4096} = 0$, and $x = \sqrt[4]{\alpha'}+\sqrt[4]{\beta'}+\sqrt[4]{\gamma'}$ will be a root of the equation $x^4+qx^2-rx+s = 0$, whose four roots will be

$$\sqrt[4]{\alpha'}+\sqrt[4]{\beta'}+\sqrt[4]{\gamma'},$$
$$-\sqrt[4]{\alpha'}-\sqrt[4]{\beta'}+\sqrt[4]{\gamma'},$$
$$-\sqrt[4]{\alpha'}+\sqrt[4]{\beta'}-\sqrt[4]{\gamma'},$$

and

$$\sqrt[4]{\alpha'}-\sqrt[4]{\beta'}-\sqrt[4]{\gamma'}.$$

The equation whose root is $x = \sqrt[4]{\alpha'}+\sqrt[4]{\beta'}+\sqrt[4]{\gamma'}$ does not ascend to as high a degree as the equation whose root is $= \sqrt[4]{\alpha}+\sqrt[4]{\beta}+\sqrt[4]{\gamma}$, where α', β', and γ' are respectively equal to α^2, β^2, and γ^2. The former must be referred to the equation whose root is $x = \sqrt[2]{\alpha}+\sqrt[2]{\beta}+\sqrt[2]{\gamma}$; and so for many similar cases.

Example 3. (1) Let the given equation be $x^5+qx^3-rx^2+sx+t = 0$. We assume a resolution $x = \sqrt[5]{\alpha}+\sqrt[5]{\beta}+\sqrt[5]{\gamma}+\sqrt[5]{\delta}$. Carrying out the calculations we see that the general biquadratic equation, whose roots are α, β, γ, and δ, necessarily involves irrationalities in its coefficients, and it follows that the given equation cannot generally be resolved by this method (as could have been established from the foregoing discussion).

(2) The proof of this example is based on the following principle: in the given equation substitute for x its assumed value $\sqrt[5]{\alpha}+\sqrt[5]{\beta}+\sqrt[5]{\gamma}+\sqrt[5]{\delta}$; by comparing the resulting quantities with the coefficients q, r, s, and t, the example is proved.

(3) Another proof utilizes the following principle: the sum of the five roots of the equation, i.e. the sum of five values of the resolution $\sqrt[5]{\alpha}+\sqrt[5]{\beta}+\sqrt[5]{\gamma}+\sqrt[5]{\delta}$, is universally equal to zero and so the sum of the five values of each of the quantities $\sqrt[5]{\alpha}$, $\sqrt[5]{\beta}$, $\sqrt[5]{\gamma}$, $\sqrt[5]{\delta}$, multiplied together, will be zero. Accordingly the five values of the quantity $\sqrt[5]{\alpha}$ must be $\sqrt[5]{\alpha}$, $\pi\sqrt[5]{\alpha}$, $\rho\sqrt[5]{\alpha}$, $\sigma\sqrt[5]{\alpha}$, and $\tau\sqrt[5]{\alpha}$, where 1, π, ρ, σ, and τ are the roots of the equation $x^5-1 = 0$; and likewise for the others.

Then choose any five values of this sort, of the resolution $\sqrt[5]{\alpha}+\sqrt[5]{\beta}+\sqrt[5]{\gamma}+\sqrt[5]{\delta}$.

Each two, three, four, and five values are multiplied together, and the sums of the resulting products (which collected together form the coefficients q, r, s, and t) involve the letters α, β, γ, and δ symmetrically, and may not vanish. But it is easy to establish that from this resolution such values cannot be found.

Likewise the coefficients of the quantities $\sqrt[5]{\alpha}$ and $\sqrt[5]{\beta}$ must be found, whether or not the quantities $\sqrt[5]{\gamma}$ and $\sqrt[5]{\delta}$ vanish. Therefore, if

π is the coefficient of the quantity $\sqrt[5]{\alpha}$, ρ must be the coefficient of both $\sqrt[5]{\beta}$ and $\sqrt[5]{\gamma}$. Then the coefficient of $\sqrt[5]{\gamma}$ must be at the same time π and ρ, which is absurd; and so it follows that this resolution is not sufficiently general to find the roots of the equation $x^5+qx^3-rx^2+sx-t=0$.

It is easy to see, as was indicated earlier, how this may be applied to a proposed resolution $\sqrt[4]{\alpha}+\sqrt[4]{\beta}+\sqrt[4]{\gamma}$ of a biquadratic (multiply the four roots of the equation $x^4-1=0$ by the quantities $\sqrt[4]{\alpha}$, $\sqrt[4]{\beta}$, and $\sqrt[4]{\gamma}$.)

Theorem 22

Case 1. Let the number of positive, negative, and impossible values of the quantity A be respectively a, a', and a''. Then if n is odd, the number of positive, negative, and impossible values of the quantity $\sqrt[n]{A}$ will be a, a', and $na''+(n-1)(a+a')$. If n is an even number, these numbers will be a, a, and $n(a''+a')+(n-2)a$.

Case 2. Let the number of positive, negative, and impossible values of the several quantities A, B, C, ..., P, Q, R, ... be a, a', a''; b, b', b''; c, c', c''; ...; p, p', p''; q, q', q''; r, r', r''; etc. Then the number of real and impossible values of the sum

$$\sqrt[n]{A}+\sqrt[m]{B}+\sqrt[o]{C}+\ldots+\sqrt[h]{P}+\sqrt[k]{Q}+\sqrt[l]{R}+\ldots$$

(where n, m, o, ... are even indices, and h, k, l, ... are odd) will be

$$(2a)(2b)(2c)\cdots(p+p')(q+q')(r+r')\cdots\ (=\sigma),\text{ and}$$

$$n\cdot m\cdot o\cdots h\cdot k\cdot l\cdots(a+a'+a'')(b+b'+b'')(c+c'+c'')\cdots$$
$$(p+p'+p'')(q+q'+q'')(r+r'+r'')\cdots(=\tau)-\sigma,$$

unless any of the values of the forms developed in Example 3.2.3.2.3 (of Corollary 3.2.3.2), Problem XVIII, are contained in the above sum. Let the number of such values be π: then the number of real values will be $\pi+\sigma$, and the number of impossible values will be $\tau-\sigma-\pi$.

Of course, if two or more irrational quantities are identical with respect to two or more of the letters A, B, C, ..., P, Q, R, ..., then the above numbers are diminished accordingly.

Theorem 23

Case 1. The resolution of the general equation

$$x^n+px^{n-1}+qx^{n-2}+\ldots+Rx^2+Qx+P=0$$

encompasses the general resolution of equations of all lower degrees $n-1$, $n-2$, $n-3$, etc.

All we need do is let $P=0$, or $P=Q=0$, etc. Then the given resolution includes all the coefficients of the equations

$$x^{n-1}+px^{n-2}+qx^{n-3}+\ldots+Rx+Q=0,$$
$$x^{n-2}+px^{n-3}+\ldots \qquad \ldots+R=0,$$

etc., with one, two, or more null coefficients.

Example. Let the equation be $x^3+qx+r=0$, whose general resolution is $x=\sqrt[3]{-\frac{1}{2}+\sqrt{\frac{r^2}{4}+\frac{q^3}{27}}}+\sqrt[3]{-\frac{1}{2}-\sqrt{\frac{r^2}{4}-\frac{q^3}{27}}}$.

Therein set $r=0$, and it becomes $x=\sqrt[3]{+\sqrt{+\frac{q^3}{27}}}+\sqrt[3]{+\sqrt{-\frac{q^3}{27}}}$

$$=\sqrt[3]{\frac{q^{3/2}}{3\sqrt{3}}}+\sqrt[3]{\frac{-q^{3/2}}{3\sqrt{3}}}=\begin{cases}\dfrac{q^{1/2}}{\sqrt{3}}-\dfrac{q^{1/2}}{\sqrt{3}} \qquad =0\\ \text{and } \dfrac{-1\pm\sqrt{-3}}{2}\cdot\dfrac{q^{1/2}}{\sqrt{3}}+\dfrac{1\pm\sqrt{-3}}{2}\cdot\dfrac{q^{1/2}}{\sqrt{3}}\\ \qquad =\pm\sqrt{-1}\cdot\sqrt{q}=\pm\sqrt{-q}\,,\end{cases}$$

which latter two are the roots of the equation $x^2+q=0$.

Corollary. This resolution contains the solution of the equation $x=0$, as well as $x=\pm\sqrt{-q}$.

Case 2. (1) Let the equation be

$$ax^n+p'x^{n-1}+q'x^{n-2}+r'x^{n-3}+\ldots+R'x^2+Q'x+P'=0.$$

Its resolution always includes the general solution of all equations of inferior degree ($n-1$, $n-2$, etc.), unless $a=0$. We simply write p, q, r, etc. respectively for $\frac{p'}{a}$, $\frac{q'}{a}$, $\frac{r'}{a}$, etc., leading to a resolution of the equation $x^n+px^{n-2}+qx^{n-2}+\ldots=0$, in whose denominator the letter a does not appear, and which includes the resolutions of all equations of inferior degree.

(2) Alternatively, if $a=0$, in the given resolution of the general equation replace the letters a, p', q', $\ldots$, R', Q', and P' by their counterparts in reverse order, P', Q', R', $\ldots$, q', p', and $a=0$. The

denominator will be a polynomial function in the quantity P'. The resulting resolution consists of the reciprocals of the roots of the equation $p'x^{n-1}+q'x^{n-2}+\ldots+R'x^2+Q'x+P'=0$.

Example. Let the equation be $ax^2+p'x+q'=0$, whose resolution is known, *viz.*, $x=\dfrac{-p'\pm\sqrt{p'^2-4aq'}}{2a}$. Then letting $a=0$ in this resolution does *not* give us the resolution of the equation $p'x+q'=0$. Rather, in the given resolution for the letters a, p', and q', we substitute q', p', and 0. The result is

$$x=\frac{-p'\pm\sqrt{p'^2}}{2q'}=\begin{cases}\dfrac{-p'+p'}{2q'}=0\\[2ex]\dfrac{-p'-p'}{2q'}=\dfrac{-p'}{q'}\end{cases};$$

accordingly the reciprocal of the root of the equation $p'x+q'=0$ will be $\dfrac{-p'}{q'}$, whence its root is $\dfrac{-q'}{p'}$.

A like method may be used to find the roots of equations of inferior degree, from a given resolution of a corresponding equation of higher degree, when consecutive coefficients a, p', ..., P', Q', ... all vanish.

Addendum. (1) Given an equation $A=0$, involving r independent unknown quantities, to decide whether or not it is predicated of another equation involving the same unknowns: first, reduce the equations to equations involving independent quantities only; then reduce the two equations to one, so that one of the above-mentioned quantities may be exterminated, and if there results a self-evident equation, *viz.* $A=A$ or $A-A=0$, then the first equation is justly predicated of the second.

Corollary. From these principles can be demonstrated many propositions given by Pappus and others. [14]

Example. Let $\mathrm{AD}=2\mathrm{AC}=2x$, $\mathrm{DE}=a$, and $\mathrm{EB}=b$ be independent quantities, and prove that if $\mathrm{AB}\times\mathrm{BE}=(2x+a+b)\times b=\mathrm{CD}\times\mathrm{BD}=(x+a+b)(a+b)$, then $\mathrm{CB}(=x+a+b):\mathrm{BD}\,(=a+b)::\mathrm{AC}\times\mathrm{CE}(=x(x+a)):\mathrm{AD}\times\mathrm{DE}(=2x\cdot a)$.

A C D E B

From hence can be deduced the two equations $(b-a)x=a^2+ab$ and $2a(x+a+b)=(a+b)(x+a)$; reduce these two equations to one, so as to exterminate x, and there results the self-evident equation $(a-b)\dfrac{a^2+ab}{b-a}+a^2+ab=0$, and consequently the proposition is true.

(2) If s equations involving $t+r$ unknown and independent quantities be predicated of t equations involving the above-mentioned quantities, reduce the t equations and one of the above-mentioned s equations to one, so that t unknown quantities may be exterminated, and if there results a self-evident equation, then the above-mentioned equation is justly predicated of the t equations; and in the same manner we may reason concerning the remaining $s-1$ equations.

(3) If one equation is justly predicated of another, and both are simple equations, then the latter equation can be predicated of the former. But if the quantity exterminated has more dimensions than one in the equations, then the proposition may not generally be true. These observations may be applied to more equations.

(4) From n given equations $a = 0$, $b = 0$, $c = 0$, etc. can easily be deduced others dependent on them, by finding any direct algebraical functions of the above-mentioned equations, that is, $\phi(a, b, c, \ldots)$, which will always be 0; and in like manner, from the relation between any lines being given, can be deduced innumerable relations between the above-mentioned lines, and other lines dependent on them.

Scholium

The general resolution of an arbitrary equation of degree n is essentially the resolution of an equation of degree $n!$, since it contains roots of index 2, 3, 4, 5, 6, 7, ..., $n-1$, and n. The same resolution applies although two or more of the roots may be equal to each other, so in general the computation required for resolution of equations of high degree will be exceedingly laborious. In vain do we seek any resolutions with roots of higher degree than the degree of the equation $x^n - px^{n-1} + qx^{n-2} - rx^{n-3} + sx^{n-4} - \ldots = 0$. Recent authors, Vieta the first among them, have found many other methods of approximation. All these methods start with a postulated approximate value of one of the roots, from which progressively better approximations to that root are derived. The success of this technique depends, at the least, on the ratio of the assumed value to the true value of the root, if the series is to converge; it may also depend on the property that the assumed quantity is much nearer to one root than to any other root of the given equation.

Example. Let $x^n-px^{n-1}+qx^{n-2}-rx^{n-3}+\ldots+Ax^2-Bx+C = 0$ be an equation whose roots are α, β, γ, δ, ..., and further let a be an assumed value of the root. The given equation is transformed into another whose roots are reduced by the value a, i.e. the roots are $\alpha-a$, $\beta-a$, $\gamma-a$, $\delta-a$, $\epsilon-a$, etc. The resulting equation will be

$$v^n + (na-p)v^{n-1} + \left[\frac{n(n-1)}{2}a^2-(n-1)ap+q\right] v^{n-2} + \ldots + \left[na^{n-1}-(n-1)pa^{n-2}+(n-2)qa^{n-3}-(n-3)ra^{n-4}+\ldots\right]v + \left[a^n-pa^{n-1}+qa^{n-2}-ra^{n-3}+\ldots\right] = 0.$$

Now let $a:\alpha :: 1 : 1+h$, $\alpha : \beta :: 1 : 1+k$, $a:\gamma :: 1 : 1+l$, $a:\delta :: 1 : 1+m$, $a : \epsilon :: 1 : 1+n'$, etc. If a is much nearer to the root α than it is to β, γ, δ, ϵ, ..., then h will be much less than k, l, m, n', etc. The roots of the resulting equation are ha, ka, la, ma, $n'a$, ..., etc. Therefore,

$$ha\cdot ka\cdot la\cdot ma\cdot n'a\cdots = \pm(a^n-pa^{n-1}+qa^{n-2}-ra^{n-3}+sa^{n-4}-\ldots)$$
$$= Q' \text{ (the product of the roots of the resulting equation)}$$

and $ka\cdot la\cdot ma\cdot n'a\cdots+ha\cdot la\cdot ma\cdot n'a\cdots+ha\cdot ka\cdot ma\cdot n'a\cdots+\ldots$

$$= \pm\left[na^{n-1}-(n-1)pa^{n-2}+(n-2)qa^{n-3}-(n-3)ra^{n-4}+\ldots\right]$$
$$= P' \text{ (the sum of the products of each } n-1 \text{ roots).}$$

It follows that the approximation which conventional rules will find,

$$-\frac{a^n-pa^{n-1}+qa^{n-2}-ra^{n-3}+\ldots}{na^{n-1}-(n-1)pa^{n-2}+(n-2)qa^{n-3}-\ldots},$$

is in ratio to the required quantity, $\alpha-a = ha$, as $hklmn'\cdots$ is to $klmn'\cdots+hlmn'\cdots+hkmn'\cdots+\ldots$, i.e. as $\frac{Q'}{ha} : P'$.

Now if the assumed quantity is much nearer to two of the roots, α and β, of the given equation, than it is to any of the others, i.e., if the resulting equation $v^n-\ldots+Tv^4-Sv^3+Rv^2-Qv+P = 0$ has two roots which are much less than any others, then the roots of the quadratic equation $v^2-\frac{Q}{R}v+\frac{P}{R} = 0$ will be a good approximation to those two roots; if three, then the roots of the equation $v^3-\frac{R}{S}v^2+\frac{Q}{S}v-\frac{P}{S} = 0$ will be very near to three of the required roots; and so on. And in like manner we can easily find the ratios of the roots so found to the true roots which we seek.

But discussion of the infinite series, from which the above derived ratios arise, will be deferred until later.

TRANSLATOR'S NOTES

[1] A simpler statement of the relation between the two roots is just that the sum of two of them shall be 7; their product, Q, is not yet determined. (Waring mentions this at the end of this example.)

[2] At this point I have begun numbering corollaries and examples quite differently from Waring. This particular example is numbered simply "Ex. 1", which is hard to reconcile with the previous example being denoted "Ex. 2" (I have called it Example 3.2.2) unless one recognizes that the present one is an example pertaining to the corollary. After a few more lines of text you will see a corollary, which Waring identifies as "Corollary 1"; I call it Corollary 3.2.2.1 because I believe it is subsidiary to the corollary I have numbered 3.2.2, and the examples following it will be called 3.2.2.2.1, 3.2.2.2.2, etc. To add to the confusion, there are several corollaries in the addenda, whose integration into the text of this Chapter requires restructuring of the entire numbering scheme. The occasional references to these items have been correspondingly modified in the text (also a non-trivial exercise).

[3] This corollary comes from the addenda pages (A1.8 - A1.9 in my numbering); the confused wording suggests that some other lines of text were lost or placed elsewhere in the addenda.

[4] While in the previous chapters Waring used Roman numerals for his Problems, and Arabic numerals for his Theorems, here he has begun using Roman numerals for both. I have preserved the distinction between the two numerations. The significance of the asterisk following a Problem or Theorem number remains unexplained.

[5] The citation is in Waring [1762]; it was omitted in the 1782 edition but I have restored it for historical completeness.

[6] Another citation from the 1762 edition, omitted in the 1782 edition.

[7] The most significant aspect of Waring's treatment of this example is the symmetric relation between the roots of the quartic equation and its resolvent cubic. This is, in essence, the first result in the theory of symmetric functions (beyond the basic building blocks which appeared in Chapter 1), a theory whose systematic development was not to appear until the 19th century (Lagrange, Gauss, and others) and was ultimately followed by the theory of permutation groups (Galois, Jordan, ...).

Waring's only fully solved quartic examples are equations whose roots are all integers — a problem easy to compose, and a solution easy to typeset. In the more general case, in this example, he has taken us as far as reducing the quartic equation to a quadratic equation in which the coëfficients are roots of a cubic equation (the traditional Ferrari method, whose history is generally well known and, indeed, is discussed in Mayer's thesis which is translated in the Appendix of this work). However, Waring has neglected to point out that the traditional final stage of solution of the general quartic, i.e. solving the resulting quadratic, can be simplified by applying some of the results of Chapter 1, in a process which I have the audacity to call

Corollary 1.7. *To complete the solution of a quartic equation, having reduced it to a quadratic by Ferrari's method.*

Suppose, for concreteness, that the given equation is $x^4-6x^2-4x-6 = 0$. The beginning steps of the solution are:

(1) Separate into two parts: $x^4 = 6x^2+4x+6$,
and introduce the auxiliary variable y in such a way that the left hand side remains a perfect square:

$$x^4+2x^2y+y^2 = (2y+6)x^2+4x+(y^2+6).$$

(2) The objective is to reduce the problem to two quadratics by taking the square root of each side of the equation, i.e.

$$x^2+y = \pm(\sqrt{2y+6}\,x+\sqrt{y^2+6}).$$

This is possible only if $(\sqrt{2y+6}\,x+\sqrt{y^2+6})^2$

$$= (2y+6)x^2+2\sqrt{(2y+6)(y^2+6)}\,x+(y^2+6)$$
$$= (2y+6)x^2+ \qquad\qquad 4x+(y^2+6),$$

i.e. if $2\sqrt{(2y+6)(y^2+6)} = 4$, which gives us the resolvent cubic,

$$2y^3 + 6y^2 + 12y + 36 - 4 = 0, \text{ or}$$
$$y^3 + 3y^2 + 6y + 16 = 0.$$

We require corresponding values of y and y^2 to complete the solution of the quartic. Unlike all published examples I have seen, this example has not been contrived so that the resolvent cubic has a rational root. The only real root of the cubic is

$$y = \sqrt[3]{\sqrt{37}-6} - \sqrt[3]{\sqrt{37}+6} - 1.$$

(3) Deferring for the moment the evaluation of y^2, by inserting the value of y into the quadratic reduction we now have $x^2 + \sqrt[3]{\sqrt{37}-6} - \sqrt[3]{\sqrt{37}+6} - 1 = \pm \left(x\sqrt{(2\sqrt[3]{\sqrt{37}-6} - 2\sqrt[3]{\sqrt{37}+6} - 1) + 6} + \sqrt{y^2+6} \right)$, which in traditional quadratic form will be written $x^2 \pm$

$$\underbrace{\sqrt{2\sqrt[3]{\sqrt{37}-6} - 2\sqrt[3]{\sqrt{37}+6} + 5}}_{=\pm B}\, x + \underbrace{\sqrt{\sqrt[3]{\sqrt{37}-6} - \sqrt[3]{\sqrt{37}+6} - \sqrt{y^2+6} - 1}}_{= C} = 0.$$

The equation $x^2 - Bx + C = 0$ has the two real roots of the original quartic, so we will focus on it for the rest of this problem.

(4) Now we must evaluate y^2. A "brute force" approach does it just by formally multiplying the algebraic expression for y by itself:

$$y^2 = \sqrt[3]{73+12\sqrt{37}} + \sqrt[3]{73-12\sqrt{37}} - 2\sqrt[3]{\sqrt{37}-6} + 2\sqrt[3]{\sqrt{37}+6} - 1\,.$$

However, Chapter 1 gives us a method to construct an equation whose roots are any algebraic function of the roots of a given equation, so let us find the equation whose roots are the squares of the roots of the resolvent cubic. The simplest way to do this (not mentioned by Waring, oddly enough) uses the power sums: let σ_i represent the sum of the ith powers of the roots of the resolvent cubic, then σ_2 and σ_4 represent the sums of the first two powers of the roots of the required second cubic. Using Newton's identities or Waring's formulas, we have $\sigma_2 = -3$ and $\sigma_4 = 129$, from which the second cubic becomes $z^3+3z^2-60z+w = 0$; w, the final coefficient, is precisely the (negated, because the degree of the equation is an odd number) square of the final coefficient of the resolvent cubic, i.e. $w = -(16^2) = -256$.

Now solve the second cubic, $z^3+3z^2-60z-256=0$:

$$z=y^2=\sqrt[3]{97+2\sqrt{37}}+\sqrt[3]{97-2\sqrt{37}}-1$$

and substitute in the reduced quadratic:

$$x^2-(\sqrt{2y+6})x+(y-\sqrt{y^2+6}) = x^2-\sqrt{2\sqrt[3]{\sqrt{37}-6}-2\sqrt[3]{\sqrt{37}+6}+4}\;x$$
$$+\sqrt[3]{\sqrt{37}-6}-\sqrt[3]{\sqrt{37}+6}-1-\sqrt{6+(\sqrt[3]{97+2\sqrt{37}}+\sqrt[3]{97-2\sqrt{37}}-1)}=0.$$

Then the positive real root of the original equation is $x=$

$$\frac{1}{2}\left(\sqrt{2\sqrt[3]{\sqrt{37}-6}-2\sqrt[3]{\sqrt{37}+6}+4}\right.$$
$$\left.+\sqrt{8-2\sqrt[3]{\sqrt{37}-6}+2\sqrt[3]{\sqrt{37}+6}+4\sqrt{5+\sqrt[3]{97+2\sqrt{37}}+\sqrt[3]{97-2\sqrt{37}}}}\right).$$

Solution by the more conventional method produces $2x=$

$$\sqrt{2\sqrt[3]{\sqrt{37}-6}-2\sqrt[3]{\sqrt{37}+6}+4}$$
$$\pm\sqrt{\begin{array}{l}8-2\sqrt[3]{\sqrt{37}-6}+2\sqrt[3]{\sqrt{37}+6}\\ \quad+4\sqrt{5+\sqrt[3]{73+12\sqrt{37}}+\sqrt[3]{73-12\sqrt{37}}-2\sqrt[3]{\sqrt{37}-6}+2\sqrt[3]{\sqrt{37}+6}}\end{array}}.$$

The depth of radical nesting,

$$\sqrt{\cdots 4\sqrt{\cdots\sqrt[3]{97-2\sqrt{37}}}} \quad vs. \quad \sqrt{\cdots 4\sqrt{\cdots\sqrt[3]{73-12\sqrt{37}}-\sqrt[3]{\sqrt{37}-6}+\cdots}},$$

is the same for the two solutions, but the number of radical subexpressions is greater in the conventional solution — a slight advantage for the new technique, perhaps, although some might argue that solving a second cubic equation is more work than squaring the solution of the resolvent cubic.

As a side effect of this exercise, by comparing the square of the value of y with the solution of the equation for $z=y^2$, we see that

$$\sqrt[3]{97+2\sqrt{37}}+\sqrt[3]{97-2\sqrt{37}}=$$
$$\sqrt[3]{73+12\sqrt{37}}+\sqrt[3]{73-12\sqrt{37}}-2\sqrt[3]{\sqrt{37}-6}+2\sqrt[3]{\sqrt{37}+6}$$

— a non-trivial case of two *algebraic expressions* which have no obvious relation to each other, but both represent the same *algebraic number*. Every root of an algebraic equation that is "solvable by radicals" can in fact be represented by an infinity of different algebraic expressions (cf. Gelfond [1960]), but examples such as this one, more complex than those given by Waring in Problem XX, are not easy to come by.

[8] An early discussion of this example is in Riccati [1771 *a*].

[9] See Riccati [1771 *b*] for discussion of this example.

[10] A problem of this sort, to rationalize the denominator of the fraction $\frac{1}{\sqrt{2}+\sqrt[3]{3}+\sqrt[5]{5}}$, appeared in *Technology Review* (published at the Massachusetts Institute of Technology) in February, 1971, and the solution — a real *tour de force* by Rowe [1971] — appeared three months later.

[11] The concluding paragraph of Problem XXVI actually appeared in Waring's preface (page xxxvii); I made an editorial decision that it more appropriately belongs in the body of Chapter 3. And why did Waring insert these remarks at all? — possibly because there still were those who held that algebra could be called a science only "if we banish from it the ridiculous mysteries arising from the supposition of *negative* quantities, or quantities *less than nothing*..." [Maseres 1780]. A similar cavil appeared in the *Monthly Magazine*, February 1800, page 48:

> If, in the beginning of his career, the Professor could admit such paralogisms into his speculations; and the writings of the mathematicians, for nearly a century before him, may plead his excuse; we are not to be surprised that his discoveries should be built rather on the assumptions of others, than on any new principles of his own. Acquiescing in the strange notion, that nothing could be divided by nothing, and produce a variety of numbers, he as easily adopted the position, that an equation has as many roots, as it has dimensions. Thus 2 and -4 are said to be roots of the equation $x^2+2x = 8$ though 4 can be the root only of the equation $x^2-2x = 8$....

(I have corrected some typographical errors here.) The article was signed "F" (probably the writer's first initial); its similarity to the gratuitous objection raised twenty years earlier convinces me to list it in my bibliography also under the authorship of Francis Maseres. Hutton [1815, 584] says that Maseres too was a candidate for the Lucasian Chair; obviously preferring him to Waring would have done little to advance English mathematics then.

[12] In the retrospective light of Ruffini's [1799] proof that the general quintic could not be solved, even with the remark about "practically infinite labor" this assertion is overly optimistic.

[13] This is where Waring first introduced the "decimal" numbering of sub-paragraphs (3.1, 3.2, ...). From time to time I have had to modify his numbering scheme to reflect the true hierarchical relationship of Examples, Corollaries, and Cases to each other, including adjustments for additions and deletions from the addenda.

[14] Like the "other observations" appended to Problem XV, this Corollary and the preceding and succeeding paragraphs, which are inserted here before the Scholium, came from the third set of Addenda pages. The Example included a marginal note referring to a "Fig.", but no figure appears in any printings that I have seen. (This was not the only Waring publication to omit a figure. The catalogue of the Bibliothèque Nationale, Paris, vol. 218, col. 415, mentions that its copy of the *Proprietates Algebraicarum Curvarum* includes one hand-drawn figure and hand-written corrections to several of the plates.) Substantially the same material, including the missing figure (along with the addendum to Problem XV), appeared in English in the *Philosophical Transactions*, 1787, which I have closely followed in this translation. The Pappian theorems may be found in Jones [1986]; Waring was probably using Robert Simson's restoration of the works Apollonius of Pergæa, based on Pappus's commentary, in which the same figure appears several times (cf. Simson [1776], pp. 8 - 23). Cantor [1907, 571] utilizes essentially the same diagram to demonstrate extensions to the concept that a square has the largest area of all rectangles with a constant perimeter.

Chapter 4

On two or more equations having two or more unknown quantities.

Theorem 24

Let there be n simple equations involving n unknown quantities x, y, z, etc.,

$$\begin{aligned} ax+\ by+\ cz+\ dv+\ldots &= A, \\ a'x+\ b'y+\ c'z+\ d'v+\ldots &= A', \\ a''x+\ b''y+\ c''z+\ d''v+\ldots &= A'', \\ a'''x+b'''y+c'''z+d'''v+\ldots &= A''', \\ \text{etc.,} & \end{aligned}$$

to be reduced to one, so as to exterminate $n-1$ of the unknowns. The result will be $x =$

$$\frac{b'c''d'''\cdots A - bc''d'''\cdots A' - b'cd'''\cdots A'' - b'c''d\cdots A''' - \ldots + bc'd'''\cdots A'' + \ldots}{b'c''d'''\cdots a - bc''d'''\cdots a' - b'cd'''\cdots a'' - b'c''d\cdots a''' - \ldots + bc'd'''\cdots a'' + \ldots}.$$

The numerator of this fraction is the same as the denominator, with the substitution of a, a', a'', a''', $\ldots$ for A, A', A'', A''', $\ldots$. Each term of the denominator contains n factors, *viz.* a, b', c'', d''', $\ldots$; a', b, c'', d''', $\ldots$; etc. Every equation contributes one of the n factors of each term. These n factors will be, severally, the coefficients of all n unknown quantities in the n equations. Every product satisfying these conditions is to be included in the denominator.

Any two products, say $b'c''d'''\cdots$ and $bc''d'''\cdots$, will have identical or opposite signs, according as each has an even or odd number of factors not found in the other term.

Another method: multiply the given equations,

$$\begin{aligned} P &\equiv\ ax+\ by+\ cz+\ dv+\ldots = 0, \\ Q &\equiv a'x+\ b'y+\ c'z+\ d'v+\ldots = 0, \\ R &\equiv a''x+\ b''y+\ c''z+\ d''v+\ldots = 0, \\ &\text{etc.,} \end{aligned}$$

by the quantities 1, α, β, γ, ... respectively, and in the resulting equation $P+\alpha Q+\beta R+\gamma S+\ldots=0$, the coefficients are assumed so as to eliminate the $n-1$ unknowns y, z, v, ...; this gives us $n-1$ equations

$$\begin{aligned}&b+b'\alpha+b''\beta+b'''\gamma+\ldots=0,\\&c+c'\alpha+c''\beta+c'''\gamma+\ldots=0,\\&d+d'\alpha+d''\beta+d'''\gamma+\ldots=0,\\&\text{etc.};\end{aligned}$$

these $n-1$ equations may similarly be depressed to $n-2$ equations, and so on. Working back from this result allows us to obtain the value of x.

Problem XXXI

To transform two equations into one, eliminating an unknown quantity.

Method 1. Arrange the terms of each equation according to their degree in the unknown which is to be eliminated; subtract one of the equations (multiplied by a suitable constant) from the other, then subtract the remainder (again multiplied by a suitable constant if necessary) from the subtrahend of the previous step, and repeat until the selected unknown is entirely eliminated. That is, the unknown quantity is eliminated in the same manner as common divisors are found.

Example. Let the two equations be

$$y^2+(3+2x)y+4x^2+5x+6=0$$

and

$$y^2+(5+2x)y+10x^2+7x+12=0.$$

The operation will proceed thus:

$$\begin{array}{r|l} & 1\\ \hline y^2+(3+2x)y+4x^2+5x+6 & y^2+(5+2x)y+10x^2+7x+12\\ & \underline{y^2+(3+2x)y+\ 4x^2+5x+\ 6}\\ & \qquad 2y+\ 6x^2+2x+\ 6=0\end{array}$$

Continuing,

$$\begin{array}{r|l} & \frac{1}{2}y+\frac{x-3x^2}{2}\\ \hline 2y+6x^2+2x+6 & y^2+\quad(3+2x)y+4x^2+5x+6\\ & \underline{y^2+(3x^2+x+3)y}\\ & \qquad (x-3x^2)y+4x^2+5x+6\\ & \qquad \underline{(x-3x^2)y+\frac{x-3x^2}{2}(6x^2+2x+6)}\\ & \qquad\qquad 4x^2+5x+6-(x-3x^2)(3x^2+x+3)=0\end{array}$$

and so the two equations are transformed into one, in which the unknown quantity y does not appear.

The same method can also be used to reduce three or more (n) equations in three or more (n) unknowns (x, y, z, ...) to m equations such that $n-m$ of the unknown quantities are exterminated: choose one equation of least degree in the unknown now being eliminated, and divide it into all the remaining equations, obtaining new sequences of remainders; and repeat the process for each of the remainders, proceeding until the unknown in question is completely exterminated.

Method 2. (1) To reduce the two equations $ay^2+by+c=0$ and $py^2+qy+r=0$ to one, in order to eliminate y.

The terms of the first equation are mutliplied by p, and those of the second by a, the coefficients of the two terms of highest degree in y (in order to avoid fractions) — the operation produces

$$pay^2+pby+pc=0$$
$$(-)\ pay^2+qay+ra=0.$$

The second equation is subtracted from the first, the residue being $(pb-qa)y+pc-ra=0$. Again to avoid fractions, multiply the equations $py^2+qy+r=0$ and $(pb-qa)y+pc-ra=0$ by $pb-qa$ and py respectively. The result is

$$(p^2b-pqa)y^2+(qpb-q^2a)y+pbr-qar=0$$

and

$$(p^2b-pqa)y^2+(p^2c-pra)y=0.$$

Subtracting the second from the first, we have

$$(qpb-q^2a+pra-p^2c)y+pbr-qar=0.$$

After one more repetition y is completely eliminated; the result is the required expression,

$$(pc-ra)(qpb-q^2a+pra-p^2c)-(pb-qa)(pbr-qar)=0.$$

(2) In like manner let the two equations be $ay^2+by+c=0$ and $py^3+qxy^2+ry+s=0$. Upon eliminating the quantity y, the resultant equation is $(aB-bA)(cB-asb)+(sa^2-cA)^2=0$ where $a=aq-pb$ and $B=ar-pc$.

Also, consider $ax^3+bx^2+cx+d=0$ and $px^3+qx^2+rx+s=0$, and write

$$A=bp-aq,\ B=cp-ar,\ C=dp-as;$$
$$P=ca^2-aAb-bAC+aC^2,$$
$$Q=cAB-aBB-dAC,$$
$$R=dA^2-bAB+aBC;$$

then $PQ+RR=0$.

And so on for more equations involving more unknowns.

Corollary 2.1. Equations so derived are often of higher degree than is apparent from the problem.

(2.1.*a*) Let two given equations be

$$Ay^n + By^{n-1} + Cy^{n-2} + Dy^{n-3} + \ldots = 0$$

and

$$ay^m + by^{m-1} + cy^{m-2} + dy^{m-3} + \ldots = 0$$

where A, B, C, D, ..., a, b, c, d, ... are arbitrary functions of an unknown quantity x. If the first equation is multiplied by $x^r - px^{r-1} + qx^{r-2} - \ldots$, then between the two equations

$$(x^r - px^{r-1} + qx^{r-2} - \ldots)(Ay^n + By^{n-1} + Cy^{n-2} + Dy^{n-3} + \ldots) = 0$$

and

$$ay^m + by^{m-1} + cy^{m-2} + dy^{m-3} + \ldots = 0,$$

there will be r extraneous values of the unknown x, and mr of y, which are not roots of the two given equations. For each root of $x^r - px^{r-1} + qx^{r-2} - \ldots$, substituted into the product

$$(x^r - px^{r-1} + qx^{r-2} - \ldots)(Ay^n + By^{n-1} + Cy^{n-2} + Dy^{n-3} + \ldots) = 0,$$

is also a root of this equation; and if in the equation

$$ay^m + by^{m-1} + cy^{m-2} + dy^{m-3} + \ldots = 0,$$

the same value is substituted for the unknown quantity x, there results an equation of degree m in y. So in general m roots of y correspond to each root of the equation $x^r - px^{r-1} + qx^{r-2} - \ldots = 0$, and it follows that there will be mr values of the root y in the latter pair of equations, which do not satisfy the original given equations.

Now if $x^4 - px^{r-1} + qx^{r-2} - \ldots = 0$ and a have a common divisor, say $x^s - Px^{s-1} + Qx^{s-2} - \ldots$, then at most $(r-s)m + s(m-1)$ extraneous values of the root y will be found; if the three quantities $x^4 - px^{r-1} + qx^{r-2} - \ldots = 0$, a, and b have a common divisor, say $x^t - \pi x^{t-1} + \rho x^{t-2} - \ldots$, then the extraneous roots of y will be at most $(r-s)m + (s-t)(m-1) + t(m-2)$ in number, and so on. The extraneous roots of x will be the same in all of these cases — the roots of $x^r - px^{r-1} + qx^{r-2} - \ldots = 0$, to be precise — unless certain roots of the equation $x^r - px^{r-1} + qx^{r-2} - \ldots = 0$ have no corresponding solutions in the unknown quantity y.

(2.1.*b*) Given the same equations,

$$Ay^n + By^{n-1} + Cy^{n-2} + Dy^{n-3} + \ldots = 0$$

and

$$ay^m + by^{m-1} + cy^{m-2} + dy^{m-3} + \ldots = 0,$$

let these equations be multiplied respectively by V and Z, two algebraic functions of x and y. The resulting equations,

$$(Ay^n + By^{n-1} + Cy^{n-2} + Dy^{n-3} + \dots)V = 0$$

and
$$(ay^m + by^{m-1} + cy^{m-2} + dy^{m-3} + \dots)Z = 0,$$

contain all the roots which are contained in the two original equations, as well as all the roots of the systems

$$V = 0 \text{ and } ay^m + by^{m-1} + cy^{m-2} + dy^{m-3} + \dots = 0,$$
$$Z = 0 \text{ and } Ay^n + By^{n-1} + Cy^{n-2} + Dy^{n-3} + \dots = 0,$$

and $V = 0$ and $Z = 0$ in addition.

(2.1.c) The same reasoning applies to three or more equations, $A = 0$, $B = 0$, $C = 0$, ... in three or more unknowns x, y, z, ...; these equations are multiplied by V, W, Z, ..., which are also algebraic functions of the unknowns x, y, z, ..., and the resulting equations $VA = 0$, $WB = 0$, $ZC = 0$, ... contain all the roots of the original system of equations,

$$A = 0,\ B = 0,\ C = 0,\ \dots$$

as well as those of the systems

$$V = 0,\ B = 0,\ C = 0,\ \dots$$
$$W = 0,\ A = 0,\ C = 0,\ \dots$$
$$Z = 0,\ A = 0,\ B = 0,\ \dots$$
$$\dots,$$
$$V = 0,\ W = 0,\ C = 0,\ \dots$$
$$\dots,$$

and
$$V = 0,\ W = 0,\ Z = 0,\ \dots.$$

Corollary 2.1.1. Hereby we can easily deduce arbitrary numbers of equations, which can be depressed in degree. Functions of the unknown quantities x, y, z, ... — say A, B, C, ..., V, W, Z, ... — are assumed. Then the system of equations

$$VA = 0, \qquad WB = 0, \qquad ZC = 0, \qquad \dots$$

can be depressed into the systems

$$A = 0,\ B = 0,\ C = 0,\ \dots$$
$$A = 0,\ B = 0,\ Z = 0,\ \dots$$
$$V = 0,\ B = 0,\ C = 0,\ \dots$$
$$\dots,$$
$$V = 0,\ B = 0,\ Z = 0,\ \dots$$
$$V = 0,\ W = 0,\ C = 0,\ \dots$$
$$\dots,$$
$$V = 0,\ W = 0,\ Z = 0,\ \dots$$

which latter equations will have all the same roots as the given ones.

Each equation is first reduced so that all have a common denominator, which may then be discarded, leaving in the numerator only quantities which are integral functions of the unknowns x, y, z, $\ldots$.

Example. Let the two equations be

$$y^n+(a+bx)y^{n-1}+(c+dx+ex^2)y^{n-2}+\ldots=0$$

and $$(A+Bx)y^{n-1}+(C+Dx+Ex^2)y^{n-2}+\ldots=0.$$

Multiply the first equation by $A+Bx$ and the second by y, giving

$$(A+Bx)y^n+(A+Bx)(a+bx)y^{n-1}+(A+Bx)(c+dx+ex^2)y^{n-2}+\ldots=0$$

and $$(A+Bx)y^n+(C+Dx+Ex^2)y^{n-1}+\ldots=0,$$

whose difference will be

$$((A+Bx)(a+bx)-C-Dx-Ex^2)y^{n-1}+\ldots=0.$$

Now the two equations

$$((A+Bx)(a+bx)-C-Dx-Ex^2)y^{n-1}+\ldots=0$$

and $$(A+Bx)y^{n-1}+(C+D+Ex^2)y^{n-2}+\ldots=0$$

contain all of the roots of the unknown quantity x that were contained in the two given equations, and (by virtue of having multiplied the first by $A+Bx$) they also contain a value of x that comes from the equation $A+Bx=0$, i.e. $x=-B/a$, which the given equations do not contain. Then the latter system of equations has also $n-2$ values of the quantity y that were not contained in the original equations, unless $A+Bx$ happens to be a divisor of $C+Dx+Ex^2$. Of the latter two equations, if we multiply the first by y, then the two equations

$$((A+Bx)(a+bx)-C-Dx-Ex^2)y^n+\ldots=0$$

and $$y^n+(a+bx)y^{n-1}+\ldots=0$$

have one value ($y=0$) of the unknown y not present in the two given equations, and all values of x as well, that result from setting y to 0 in the equation $(A+Bx)y^{n-1}+(C+Dx+Ex^2)y^{n-2}+\ldots=0$.

This shows how extraneous roots always crop up in the resulting equations; the above arguments, or the method of common divisors, easily show the source of the extraneous roots.

Corollary 2.2. Let the two equations be

$$(x^r+ax^{r-1}+bx^{r-2}+\ldots)y^n+(dx^{r+1}+ex^r+fx^{r-1}+\ldots)y^{n-1}+\ldots=0$$

and $$(x^s+Ax^{s-1}+Bx^{s-2}+\ldots)y^n+(Dx^{s+1}+Ex^s+Fx^{s-1}+\ldots)y^{n-1}+\ldots=0,$$

and reduce them by the method described in this Problem, so that y appears in the two resulting equations to a degree no higher than $n-1$. The degree of x, in the terms where it is paired with n^{n-1}, will be at most $r+s+1$ in one case, and at most $2r+s+2$ or $r+2s+2$ in the other.

If the two given equations are reduced to two others, in which the maximum degree of the unknown quantity y is $n-2$, then the degree of the unknown quantity x, in the terms where y appears to degree $n-2$, will be at most $(r+s+1)+(2r+s+2)+1=3r+2s+4$ in one of them, and $(3r+2s+4)+(r+s+1)+1=4r+3s+6$ in the other; or $2r+3s+4$ in one and $3r+4s+6$ in the other.

Corollary 2.3. Let the two given equations be

$$y^n + ay^{n-1} + by^{n-2} + cy^{n-3} + dy^{n-4} + ey^{n-5} + fy^{n-6} + \ldots = 0$$

and $y^m+Ay^{m-1}+By^{m-2}+Cy^{m-3}+Dy^{m-4}+Ey^{m-5}+Fy^{m-6}+\ldots = 0,$

in which the degree n is greater than m, and a, b, c, ..., A, B, C, ... are functions of a variable x. Divide the first equation by the second: the operation proceeds

$$\begin{array}{r|l} & y^{n-m}+A'y^{n-m-1}+A''y^{n-m-2}+\ldots \\ y^m+Ay^{m-1}+By^{m-2}+\ldots & y^n + ay^{n-1} + by^{n-2} + cy^{n-3} + \ldots \\ & y^n + Ay^{n-1} + By^{n-2} + Cy^{n-3} + \ldots \\ & \quad A'y^{n-1} + B'y^{n-2} + C'y^{n-3} + \ldots \\ & \quad A'y^{n-1} + A'Ay^{n-2} + A'By^{n-3} + \ldots \\ & \qquad A''y^{n-2} + B''y^{n-3} + \ldots \\ & \qquad \text{etc.} \end{array}$$

Then the two equations $y^n + ay^{n-1} + by^{n-2} + cy^{n-3} + \ldots = 0$ and $A'y^{n-1} + B'y^{n-2} + C'y^{n-3} + \ldots = 0$ have all the roots of the unknown quantities x and y that were contained in the equations $y^m+Ay^{m-1}+By^{m-2}+\ldots = 0$ and $y^n+ay^{n-1}+by^{n-2}+\ldots = 0$, and all the roots contained in the equations $y^{n-m} = 0$ and $y^n+ay^{n-1}+by^{n-2}+\ldots = 0$; that is, their roots are the same as those of the system of two equations

$$y^n + ay^{n-1} + by^{n-2} + \ldots = 0$$

and $$(y^m+Ay^{m-1}+By^{m-2}+\ldots)y^{n-m} = 0.$$

Likewise the two equations $A'y^{n-1}+B'y^{n-2}+C'y^{n-3}+\ldots = 0$ and $a''y^{n-2}+B''y^{n-3}+\ldots = 0$ have the same roots as the two equations $A'y^{n-m-1}(y^m+Ay^{m-1}+\ldots) = 0$ and either $A'y^{n-1}+B'y^{n-2}+\ldots = 0$ or $A''y^{n-2}+B''y^{n-3}+\ldots = 0$; and so on.

The same principle can be applied also to larger systems of equations having more unknown quantities.

Corollary 2.4. *To find whether two or more equations together imply anything absurd.*

The equations are reduced until all unknown quantities are eliminated, and if quantities that are by hypothesis equal prove in fact to be unequal, then the given equations are inconsistent; in the contrary case they are consistent. For example, let the equations be $y^2+7x^2+20 = 0$ and $y^2+7x^2+40 = 0$. Subtracting the first equation from the second, we have $40 - 20 = 0$, or $40 = 20$, which is absurd.

This is a consequence of the fact that if there is no absurdity between the divisor, quotient, and remainder, there can be none between the divisor and the dividend; and conversely if there is an absurdity between the divisor and the remainder, then the divisor and dividend are also inconsistent.

All of this can be applied to the case of n equations involving m unknowns.

Corollary 2.5. Evidently the method of reducing two or more equations, of degree 2 or higher, to one, for the purpose of eliminating unknown quantities, is essentially the same as the method whereby two or more simple equations (of degree 1 in each of their unknowns) are reduced to one, also to eliminate unknowns.

Let two such equations be $Ay^m+By^{m-1}-Cy^{m-2}+\ldots = P$, and $ay^n+by^{n-1}+cy^{n-2}+dy^{n-3}+\ldots = Q$, which are to be reduced to one, such that the quantity y is eliminated.

We may regard the quantities y^m, y^{m-1}, y^{m-2}, ..., and y^n, y^{n-1}, y^{n-2}, ... as distinct unknown quantities — that is, we eliminate each power of y in the same way that we would reduce simple equations in several unknowns. And so indeed this method of reducing these equations is equivalent to the preceding method.

If the number of given equations is greater by one than the number of quantities y^m, y^{m-1}, y^{m-2}, ..., and y^n, y^{n-1}, y^{n-2}, ... then by Theorem 24 these equations are easily reduced so that y is entirely eliminated.

In this method of reducing two equations, of degrees n and m, to one, in order to eliminate an unknown quantity, the reduced equation often is of higher degree than the problem demands. But by the method of finding rational divisors the equation can always be depressed to the proper degree.

If this method, or the following one, is used to reduce three or more equations to one, thereby to eliminate one, two, or more unknown quantites, and the reduced equation is of higher degree than the problem calls for, then by the method given in the next Chapter, for finding rational divisors, the degrees of the resulting equations can be lowered.

Method 3. Here is another way to transform two equations into one, so that an unknown quantity is eliminated.

Let the two equations be

$$y^n - py^{n-1} + qy^{n-2} - ry^{n-3} + sy^{n-4} - \ldots = 0$$

and

$$V - Ty + Sy^2 - Ry^3 + Qy^4 - \ldots = 0$$

and further suppose α, β, γ, δ, ... are the roots of y in the first equation, n in number, which substituted into the second give the quantities

$$V - T\alpha + S\alpha^2 - R\alpha^3 + Q\alpha^4 - \ldots$$
$$V - T\beta + S\beta^2 - R\beta^3 + Q\beta^4 - \ldots$$
$$V - T\gamma + S\gamma^2 - R\gamma^3 + Q\gamma^4 - \ldots$$
$$V - T\delta + S\delta^2 - R\delta^3 + Q\delta^4 - \ldots$$

etc.

When these quantities are all multiplied together, the product is

$$V^n - V^{n-1}T\left\{\begin{matrix}\alpha+\\ \beta+\\ \gamma+\\ \delta+\\ \text{etc.}\end{matrix}\right\} + V^{n-1}S\left\{\begin{matrix}\alpha^2+\\ \beta^2+\\ \gamma^2+\\ \delta^2+\\ \text{etc.}\end{matrix}\right\} - V^{n-1}R\left\{\begin{matrix}\alpha^3+\\ \beta^3+\\ \gamma^3+\\ \delta^3+\\ \text{etc.}\end{matrix}\right\} + \ldots$$

$$+V^{n-2}T^2\left\{\begin{matrix}\alpha\beta+\\ \alpha\gamma+\\ \beta\gamma+\\ \alpha\delta+\\ \beta\delta+\\ \gamma\delta+\\ \text{etc.}\end{matrix}\right\} - V^{n-1}TS\left\{\begin{matrix}\alpha^2\beta+\\ \beta^2\alpha+\\ \alpha^2\gamma+\\ \gamma^2\alpha+\\ \beta^2\gamma+\\ \gamma^2\beta+\\ \text{etc.}\end{matrix}\right\} + \ldots$$

By means of Problems III and V this product can be evaluated; it is

$$V^n - V^{n-1}Tp + V^{n-1}S(p^2-2q) - V^{n-1}R(p^3-3pq+3r) + \ldots$$
$$+ V^{n-2}T^2q \qquad - V^{n-2}TS(pq - 3r) \quad + \ldots$$

and by equating it to zero we have transformed the two equations into one, eliminating the unknown quantity y.

Example. Let the two equations be

$$y^2 + (3+2x)y + (4x^2+5x+6) = 0$$

and

$$y^2 + (5+2x)y + (10x^2+7x+12) = 0.$$

Let α and β be the roots of y from the first equation, whose substitution into the second results in the quantities

$$\alpha^2 + (3+2x)\alpha + (\ 4x^2+5x+\ 6) = 0$$

and

$$\beta^2 + (5+2x)\beta + (10x^2+7x+12) = 0,$$

whose product is

$$\alpha^2\beta^2 + (5+2x)\alpha\beta(\alpha+\beta) + (10x^2+7x+12)(\alpha^2+\beta^2) + (5+2x)^2\alpha\beta + (10x^2+7x+12)(5+2x)(\alpha+\beta) + (10x^2+7x+12)^2$$

and by Problem I we may rewrite this product as

$$(4x^2+5x+6)^2 - (5+2x)(4x^2+5x+6)(3+2x) +(10x^2+7x+12)\left[(3+2x)^2-2(4x^2+5x+6)\right] +(5+2x)^2(4x^2+5x+6)-(10x^2+7x+12)(5+2x)(3+2x)+(10x^2+7x+12)^2.$$

Set this product equal to zero; it will be the required equation. [1]

Corollary 3.1. It is readily apparent that the unknown quantities have exactly as many values as this method of reduction discovers; there are no additional ones in the given equations.

Corollary 3.2. When this method is used to transform two equations, of degrees n and m respectively, involving two unknowns, x and y, into one equation, to eliminate one unknown (say y), the resulting equation's degree will be no greater than nm.

Corollary 3.3. Let there be two equations, of degrees n and m,

$$My^n+(\ a+\ bx)y^{n-1} +(\ c+\ dx+\ ex^2)y^{n-2} +(\ f+\ gx+\ hx^2+\ kx^3)y^{n-3} + \ldots =0,$$

$$Ny^m+(A+Bx)y^{m-1}+(C+Dx+Ex^2)y^{m-2}+(F+Gx+Hx^2+Kx^3)y^{m-3}+ \ldots =0$$

in the two unknown quantities x and y. If we ignore all the terms of these equations except those of maximum combined degree in x and y, we have

$$My^n + \ bxy^{n-1} + \ ex^2y^{n-2} + \ kx^3y^{n-3} + \ldots$$

and

$$Ny^m + Bxy^{m-1} + Ex^2y^{m-2} + Kx^3y^{m-3} + \ldots.$$

Now if these resulting quantities have a common divisor, the equation whose root is x or y will be of degree less than nm, and if not, it will be of degree nm.

If the degree of the common divisor is r, then the equation will be of degree $nm-r$ at most.

If there are three or more (h') equations involving as many unknown quantities (x, y, z, v, ...), of degrees m, n, o, l, ... respectively, and we discard all those terms in which the aggregate degrees of all the unknown quantities are less than the respective maxima, there will result quantities A, B, C, D, ... If there is no divisor which is common to all of these resulting quantities, then the equation whose root is x (or y or z or v or ...) is of degree $mnol\ldots$, and otherwise it is of lower degree. That is, we equate to zero h'-1 of the above quantities, $B = 0$, $C = 0$, $D = 0$, etc., and produce another set of h'-1 equations, $A = 0$, $B = 0$, $C = 0$, etc., of which one ($A = 0$) is the one that was omitted from the first set. Then from the equations in the first and second sets we construct ratios of the corresponding values of the unknown quantities x, y, z, ... to each other; and if none of the ratios of the unknown quantities from the first h'-1 equations are the same as in the second set of h'-1 equations, then the resulting equations, whose roots are the values of x (or of y or z or ...) will be of degree $mnol\ldots$, but if two ratios are equal, they will be of lower degree. In the equations $A = 0$, $B = 0$, $C = 0$, etc., for the letters y, z, v, etc. write respectively αx, βx, γx, etc., and there will result h' equations, from which x may be eliminated, leaving only h'-1 unknown quantities α, β, γ, etc. Then by the method of common divisors the corresponding values of α, β, γ, etc. are found (if there are any consistent solutions). If there are no consistent solutions in these variables, then the equation whose root is x (or y or z or ...) must be of degree $mnol\ldots$, otherwise it will be of lower degree.

Now if K' distinct corresponding ratios among the unknown quantities x, y, z, ... are found to be the same in the former as in the latter set of h-1 equations, i.e. if there are K' consistent corresponding values of the several quantities α, β, γ, ..., then the resulting equation must ascend to a degree no higher than $mnol\ldots-k'$.

Corollary 3.4. If the resulting equation, whose roots are the unknown quantities x, and y, is of degree nm, the sum of the roots depends only on those terms of the two given equations in which x and y appear to the nth and $(n-1)^{\text{th}}$, or the mth and $(m-1)^{\text{th}}$, degrees — i.e., it depends only on the terms

$$My^n+bxy^{n-1}+ex^2y^{n-2}+kx^3y^{n-3}+\ldots$$

and

$$ay^{n-1}+dxy^{n-2}+hx^2y^{n-3}+\ldots$$

of the first equation, and

$$Ny^m+Bxy^{m-1}+Ex^2y^{m-2}+Kx^3y^{m-3}+\ldots$$

and

$$Ay^{m-1}+Dxy^{m-2}+Hx^2y^{m-3}+\ldots$$

of the second. The sum of the products in pairs of all the roots depends on the terms of degree n, n-1, and n-2 of the former equation, and those of degree m, m-1, and m-2 of the latter — that is, on the

terms $$My^n+bxy^{n-1}+ex^2y^{n-2}+kx^3y^{n-3}+\dots,$$
$$ay^{n-1}+dxy^{n-2}+hx^2y^{n-3}+\dots,$$
and $$cy^{n-2}+gxy^{n-3}+\dots$$
of the former, as well as the terms
$$Ny^m+Bxy^{m-1}+Ex^2y^{m-2}+Kx^3y^{m-3}+\dots,$$
$$Ay^{m-1}+Dxy^{m-2}+Hx^2y^{m-3}+\dots,$$
and $$Cy^{n-2}+Gxy^{n-3}+\dots$$
of the latter, and similarly for the sum of the products of each three roots, etc.

Likewise, given three or more equations in as many unknowns, x, y, z, v, etc., and the equation whose root is x (or y or z or v or ...) is of degree $mnol\dots$, then the sum of its roots depends on the terms of combined degree n and $n-1$, m and $m-1$, o and $o-1$, l and $l-1$, etc., in the respective equations. The sum of the squares of the roots depends on terms of degree n, -1, and $n-2$; m, $m-1$, and $m-2$; o, $o-1$, and $o-2$; etc. respectively in the given equations, and so on for all higher orders.

If the equation whose root is x (or y or z or ...) is of degree no greater than $mnol-r$, then the sum of its roots depends on the terms of degree n, $n-1$, ..., $n-r-1$; m, $m-1$, ..., $m-r-1$; o, $o-1$, ..., $o-r-1$; l, $l-1$, ..., $l-r-1$; etc. respectively, in the given equations.

Corollary 3.5. Let two equations, of degrees n and m, be
$$P \equiv (ax^r+bx^{r-1}+cx^{r-2}+\dots)y^{n-r} + (dx^{r+1}+ex^r+\dots)y^{n-r-1}$$
$$+(fx^{r+2}+gx^{r+1}+hx^r+\dots)y^{n-r-2} + \dots = 0$$
and
$$Q \equiv (Ax^s+Bx^{s-1}+\dots)y^{m-s} + (Dx^{s+1}+Ex^s+\dots)y^{m-s-1} + \dots = 0.$$
If these equations are reduced to one, eliminating the unknown quantity y, the resulting equation's degree can be no higher than $nm-sr$, and it is easy from the preceding discussions to decide whether this limit is actually attained or not.

Proof. (1) Let the resulting equation be $x^n-px^{n-1}+qx^{n-2}-rx^{n-3}+\dots = 0$, whose roots are α, β, γ, δ, etc.; then Case 3 of Problem III tells us that the degrees of the coefficients p, q, r, s, t, ... in the sum of terms of the form $\alpha^a\beta^b\gamma^c\delta^d\cdots$ (where $a \geq b \geq c \geq d\cdots$ are integers) can be no higher than a. Supposing that each individual coefficient (p, q, r, s, t, ...) is of degree 1, then the degree of the product $p^\mu q^\nu r^\pi s^\rho\cdots$ is $\mu+\nu+\pi+\rho+\dots$. Further, no more than a factors will appear in any term of the abovementioned sum of products $\alpha^a\beta^b\gamma^c\delta^d\cdots$.

(2) Let the equation be $x^n - \frac{p}{h}x^{n-1} + \frac{q}{h}x^{n-2} - \frac{r}{h}x^{n-3} + \ldots = 0$, and since there cannot be more than a factors of the coefficients $(\frac{p}{h}, \frac{q}{h}, \frac{r}{h}, \frac{s}{h}, \ldots)$ in any individual term of the product $\alpha^a\beta^b\gamma^c\delta^d\cdots$, then in the denominator of the sum the quantity h will appear to a degree no higher than a.

(3) In the aforementioned aggregate you will find that the degree of each product $p^\mu q^\nu r^\pi s^\rho\cdots$, that is, the sum $\mu+2\nu+3\pi+\ldots$, is always the same.

(4) If the dimensions of the quantities p, q, r, ... are greater by 1, 2, 3, ... than the dimension of the denominator h, then for all cases in which the sum $\mu+2\nu+3\pi+\ldots$ is the same, the product $\frac{p^\mu}{h^\mu} \times \frac{q^\nu}{h^\nu} \times \frac{r^\pi}{h^\pi}\cdots$, reduced to its common denominator (h^a), will be of the same dimension.

In this context, by the dimensions of the quantities p, q, r, s, ... and of h, I mean the maximum dimensions of the unknown quantity contained in them.

(5) With the dimensions of p, q, r, s, ... and h as stated above (i.e. the dimensions of p, q, r, ... exceed by 1, 2, 3, ... the respective dimensions of h), for example, two equations might be

$$(ax^r+bx^{r-1}+\ldots)y^{n-r}+(dx^{r+1}+ex^r+\ldots)y^{n-r-1}+\ldots+Fx^n+Gx^{n-1}+\ldots=0$$

— whose solutions for y are α, β, γ, δ, ... — and

$$(Ax^s+Bx^{s-1}+\ldots)y^{m-s}+(Dx^{s+1}+Ex^s+\ldots)y^{m-s-1}+\ldots=0.$$

Clearly the dimensions of the individual terms in the product

$$(Ax^s\alpha^{m-s}+Dx^{s+1}\alpha^{m-s-1}+\ldots)\cdot(Ax^s\beta^{m-s}+Dx^{s+1}\beta^{m-s-1}+\ldots)\cdot(Ax^s\gamma^{m-s}+Dx^{s+1}\gamma^{m-s-1}+\ldots)\cdots$$

reduced to its common denominator $h^{m-s} = (ax^r+bx^{r-1}+\ldots)^{m-s}$, will always be the same. Then the product $(Ax^s+Bx^{s-1}+\ldots)\alpha^{m-s}\times(Ax^s+Bx^{s-1}+\ldots)\beta^{m-s}\times(Ax^s+Bx^{s-1}+\ldots)\gamma^{m-s}\cdots$ will be equal to

$$(Ax^s+Bx^{s-1}+\ldots)\alpha^{m-s}\beta^{m-s}\gamma^{m-s}\cdots$$
$$=\pm(Ax^s+Bx^{s-1}+\ldots)^{n-r}\left(\frac{Fx^n+Gx^{n-1}+\ldots}{ax^r+bx^{r-1}+\ldots}\right)^{m-s}$$

in whose numerator the degree of the quantity x, and so of the equation whose root is x, will be no greater than $s(n-r)+n(m-s) = nm-rs$, and this completes the proof.

Corollary 3.5.1. Let $n-r$ and $m-s$ be the maximum dimensions of the quantity y in the two equations $P=0$ and $Q=0$ respectively.

Then the equation whose root is x or y cannot be of degree greater than $nm-rs$.

By a similar principle we can show that if $n-r$ and $m-s$, and $n-t$ and $m-u$ are respectively the highest degrees of the quantities x and y in the equations $P = 0$ and $Q = 0$, then the equation whose root is x or y can be of degree no greater than $nm-rs-tu$.

Corollary 3.6. (1) Let there be three equations $A = 0$, $B = 0$, and $C = 0$, involving the three unknowns x, y, and z, of degrees m, n, and o respectively. Let two of these equations, $A = 0$ and $B = 0$, be reduced to one to eliminate x, the resulting equation $P = 0$ being of degree nm at most and involving the two quantities y and z. Likewise if the two equations $B = 0$ and $C = 0$ are reduced in order to eliminate x, the result is an equation having degree at most no; then if y or z is eliminated from these two equations of degree nm and no, the result is an equation whose degree is no higher than n^2mo. Thus any equation resulting from this process cannot be of degree greater than $nmo \times n$, $nmo \times m$, or $nmo \times o$.

(2) In general, suppose there are α equations of dimensions respectively $n \leq m \leq o \leq r \leq s \leq t \cdots$, in as many unknowns x, y, z, v, w, ..., and W. The α equations may be reduced, in $\frac{\alpha(\alpha-1)}{2}$ distinct ways, to $\alpha-1$ equations from which x has been eliminated, and the degrees of these equations are in all cases no greater than nm or no or mo or mr or ....

The above $\alpha-1$ equations may be reduced to $\alpha-2$ equations, eliminating y, in

$$\alpha(\alpha-1)\frac{\alpha-2}{2}+\alpha \cdot \frac{\alpha-1}{2} \cdot \frac{\alpha-2}{3} \cdot \frac{\alpha-3}{4} = \alpha\frac{\alpha-1}{2} \cdot \frac{\alpha-2}{3} \cdot \frac{\alpha+9}{4}$$

distinct ways, the resulting degrees being no higher than n^2mo, n^2mr, n^2or, ..., m^2no, n^2nr, m^2or, ..., $nmor$, $nmos$, $nros$, etc. respectively.

After this, z may be eliminated from the $\alpha-2$ equations, leaving $\alpha-3$ equations whose degrees will be no greater than n^4m^2or, m^4n^2or, o^4m^2nr, ..., n^3m^3or, o^3r^3nm, ..., n^4mors, m^4nors, ..., n^3morst, ..., etc. Eventually these α equations will be reduced to one, eliminating all but one of the unknowns (say W). The degree of the resulting equation will be no less than $n^{2^{\alpha-2}} \cdot m^{2^{\alpha-3}} \cdot o^{2^{\alpha-4}} \cdot r^{2^{\alpha-5}} \cdot s^{2^{\alpha-6}} \cdots$. The last factor in this expression will look like $k^{2^{-1}}$, for which we will write simply k instead. The coefficients of the terms of the given equations, in which the degrees of the unknowns involved are the greatest, may be so constituted that some of the initial terms of the resulting equations will vanish.

Corollary 3.7. Whenever the resulting equation ascends to a degree higher than that implied by the original given equations, it can always be reduced in degree by finding its rational divisors, until one equation is found having no extraneous roots.

Method 4. If two equations in two unknowns, x and y, are given, we may solve one of them by known rules (if such exist) for one unknown in terms of the other.

Then these values of the unknown (say x) are substituted into the other given equation, and all the resulting quantities are multiplied together, and so the two equations are reduced to one, eliminating the unknown quantity x.

Example 1. Let the two equations be

$$x^2 - 2ax + a^2 - b = 0,$$

and $$x^n - px^{n-1} + qx^{n-2} - rx^{n-3} + sx^{n-4} - \ldots = 0.$$

The two roots of the first equation are $x = a \pm \sqrt{b}$, which are to be substituted for each occurrence of x in the second equation, and the two resulting expressions must be multiplied together. Then if we write

$$A=(a+\sqrt{b})^n+(a-\sqrt{b})^n-p[(a+\sqrt{b})^{n-1}+(a-\sqrt{b})^{n-1}]+q[(a+\sqrt{b})^{n-2}+(a-\sqrt{b})^{n-2}] + \ldots,$$

$$B=(a+\sqrt{b})^n-(a-\sqrt{b})^n-p[(a+\sqrt{b})^{n-1}-(a-\sqrt{b})^{n-1}]+q[(a+\sqrt{b})^{n-2}-(a-\sqrt{b})^{n-2}] + \ldots,$$

the resulting equation will be $A^2 - B^2 = 0$.

Example 2. Let the two equations be

$$x^3 - 3ax^2 + (3a^2 + b)x - a^3 - ba + c = 0,$$

and $$x^n - px^{n-1} + qx^{n-2} - rx^{n-3} + sx^{n-4} - tx^{n-5} + \ldots = 0.$$

Now the three roots of the first equation are

$$a + \sqrt[3]{-\frac{1}{2}c+\sqrt{\frac{1}{4}c^2+\frac{b^3}{27}}}(= \alpha) + \sqrt[3]{-\frac{1}{2}c-\sqrt{\frac{1}{4}c^2+\frac{b^3}{27}}}(= \beta),$$

$$a - \frac{1+\sqrt{-3}}{2}\alpha - \frac{1-\sqrt{-3}}{2}\beta,$$

and $a - \frac{1-\sqrt{-3}}{2}\alpha - \frac{1+\sqrt{-3}}{2}\beta$.

Substitute these three roots for x in the second equation and multiply together the three resulting equations, and write

$$A = (a+\alpha+\beta)^n + (a-\tfrac{1+\sqrt{-3}}{2}\alpha-\tfrac{1-\sqrt{-3}}{2}\beta)^n+(a-\tfrac{1-\sqrt{-3}}{2}\alpha-\tfrac{1+\sqrt{-3}}{2}\beta)^n-$$
$$p\left[(a+\alpha+\beta)^{n-1}+(a-\tfrac{1+\sqrt{-3}}{2}\alpha-\tfrac{1-\sqrt{-3}}{2}\beta)^{n-1}+(a-\tfrac{1-\sqrt{-3}}{2}\alpha-\tfrac{1+\sqrt{-3}}{2}\beta)^{n-1}\right]$$
$$+q\left[(a+\alpha+\beta)^{n-2}+\ldots\right]+\ldots,$$

$$B = (a+\alpha+\beta)^n - \tfrac{1+\sqrt{-3}}{2}(a-\tfrac{1+\sqrt{-3}}{2}\alpha-\tfrac{1-\sqrt{-3}}{2}\beta)^n-\tfrac{1-\sqrt{-3}}{2}(a-\tfrac{1-\sqrt{-3}}{2}\alpha-\tfrac{1+\sqrt{-3}}{2}\beta)^n-$$
$$p\left[(a+\alpha+\beta)^{n-1}-\tfrac{1+\sqrt{-3}}{2}(a-\tfrac{1+\sqrt{-3}}{2}\alpha-\tfrac{1-\sqrt{-3}}{2}\beta)^{n-1}-\tfrac{1-\sqrt{-3}}{2}(a-\tfrac{1-\sqrt{-3}}{2}\alpha-\tfrac{1+\sqrt{-3}}{2}\beta)^{n-1}\right]$$
$$+q\left[(a+\alpha+\beta)^{n-2}+\ldots\right]+\ldots,$$

$$C = (a+\alpha+\beta)^n - \tfrac{1-\sqrt{-3}}{2}(a-\tfrac{1+\sqrt{-3}}{2}\alpha-\tfrac{1-\sqrt{-3}}{2}\beta)^n-\tfrac{1+\sqrt{-3}}{2}(a-\tfrac{1-\sqrt{-3}}{2}\alpha-\tfrac{1+\sqrt{-3}}{2}\beta)^n-$$
$$p\left[(a+\alpha+\beta)^{n-1}-\tfrac{1+\sqrt{-3}}{2}(a-\tfrac{1+\sqrt{-3}}{2}\alpha-\tfrac{1-\sqrt{-3}}{2}\beta)^{n-1}-\tfrac{1+\sqrt{-3}}{2}(a-\tfrac{1-\sqrt{-3}}{2}\alpha-\tfrac{1+\sqrt{-3}}{2}\beta)^{n-1}\right]$$
$$+q\left[(a+\alpha+\beta)^{n-2}+\ldots\right]+\ldots:$$

then the resulting equation will be $A^3 - 3ABC + B^3 + C^3 = 0$, and likewise for higher degrees.

Corollary 4.1. When there are three or more equations in three or more unknowns, this method will produce the same resultant equations as the preceding one.

Corollary 4.2. If two equations $A = 0$ and $B = 0$ both express a relation between x and y, and $A = p+q+r+\ldots = 0$, where p is a homogeneous function of degree n in x and y, q is homogeneous of degree $n-1$ in x and y, r homogeneous of degree $n-2$, etc., then in both equations $A = 0$ and $B = 0$ replace x by the expression zy. The first equation will then be transformed into an equation $\pi-\rho y+\sigma y^2-\tau y^3+\ldots = 0$, where π, ρ, σ, τ, ... are functions of z only. This equation may then be solved for y, and this value of y may be substituted into the other transformed equation, giving rise to a single equation in z only.

And so two or more equations, involving two or more unknown quantities, are by diverse substitutions depending on the forms of the equations transformed into one, eliminating unknown quantities.

Alternatively, suppose we have two equations of the following form:

$$\begin{aligned}&(az^n + bz^{n-1} + cz^{n-2} + dz^{n-3} + \ldots)y^r\\ &+ (a'z^{n-1} + b'z^{n-2} + c'z^{n-3} + d'z^{n-4} + \ldots)y^{r-1}\\ &+ (a''z^{n-2} + b''z^{n-3} + c''z^{n-4} + d''z^{n-5} + \ldots)y^{r-2}\\ &+(a'''z^{n-3} + b'''z^{n-4} + c'''z^{n-5} + d'''z^{n-6} + \ldots)y^{r-3} + \ldots = 0,\end{aligned}$$

and

$$(Az^m + Bz^{m-1} + Cz^{m-2} + Dz^{m-3} + \ldots)y^s$$
$$+ (A'z^{m-1} + B'z^{m-2} + C'z^{m-3} + D'z^{m-4} + \ldots)y^{s-1}$$
$$+(A''z^{m-2} + B''z^{m-3} + C''z^{m-4} + D''z^{m-5} + \ldots)y^{s-2} + \ldots = 0,$$

where $n \geq r$ and $m \geq s$. In these equations for z write $\frac{x}{y}$, and these two equations will be transformed into two others, of degrees n and m respectively, expressing relations between x and y.

Method 4.1. If there are two equations,

$$Ay^m + By^{m-1} + Cy^{m-2} + \ldots = 0 \text{ and } ay^n + by^{n-1} + cy^{n-2} + \ldots = 0,$$

then assume two quantities

$$Py^{m-1} + Qy^{m-2} + Ry^{m-3} + \ldots \text{ and } py^{n-1} + qy^{n-2} + ry^{n-3} + \ldots.$$

Multiply the given equation $Ay^m + By^{m-1} + Cy^{m-2} + \ldots = 0$ by the assumed quantity $py^{n-1} + qy^{n-2} + ry^{n-3} + \ldots$, and equate each term but the last of the resulting product,

$$Apy^{m+n-1} + (Aq + Bp)y^{m+n-2} + (Ar + Bq + Cp)y^{m+n-3} + \ldots = 0,$$

to the corresponding term of the product

$$aPy^{m+n-1} + (aQ + bP)y^{m+n-2} + (aR + bQ + cP)y^{m+n-3} + \ldots = 0$$

obtained by multiplying the given equation $ay^n + by^{n-1} + cy^{n-2} + \ldots = 0$ by the assumed quantity $Py^{m-1} + Qy^{m-2} + Ry^{m-3} + \ldots$; this gives rise to the equations $Ap - aP = 0$, $Aq + Bp - aQ - bP = 0$, etc. Now if these $n + m - 1$ linear equations are reduced to one, to eliminate the unknown quantities P, Q, R, $\ldots$, p, q, r, etc., there results an equation $\pi = 0$ which is the same as that obtained from reducing the two given equations to one to eliminate the unknown quantity y.

If the two equations

$$Ay^m + By^{m-1} + Cy^{m-2} + \ldots = 0 \text{ and } ay^n + by^{n-1} + cy^{n-2} + \ldots = 0$$

are multiplied by the two equations

$$py^n + qy^{n-1} + ry^{n-2} + \ldots = 0 \text{ and } Py^m + Qy^{m-1} + Ry^{m-2} + \ldots = 0,$$

and the two resulting equations are equated term by term to each other, then the system of two equations

$$py^n + qy^{n-1} + ry^{n-2} + \ldots = 0 \text{ and } Py^m + Qy^{m-1} + Ry^{m-2} + \ldots = 0$$

is equivalent to the system of two equations

$$Ay^m + By^{m-1} + Cy^{m-2} + \ldots = 0 \text{ and } ay^n + by^{n-1} + cy^{n-2} + \ldots = 0.$$

The same assertion holds for these multiplications if the assumed equations are of degrees greater than m and n respectively.

Method 4.2. This method is also applicable to the case of three or more (say h) equations in as many unknowns.

It is easy to establish from these two observations what will be the degree of the unknown quantity, i.e. the number of roots in the assumed equations: first, that each product of one equation times the corresponding assumed equation will be of the same degree; and second, that sum of the degrees of all the assumed equations will equal, or exceed by 1, the difference between the said sum and $h-1$, the number of given equations reduced by 1.

In cases of this sort there will be as many or more unknown quantities in the assumed equations, as there are resulting linear equations.

Method 5. There is yet another method of transforming two equations into one, so as to eliminate unknown quantities.

By the method of infinite series, find from one equation the values of the unknown quantity which is to be eliminated, and substitute those values for its occurrences in the other equation. Then equate to zero the product of all the resulting quantities, and the result will be the required equation.

Example. Let the two equations be

$$y^2 - (4x+2)y + 3x^2+4x+4 = 0$$

and

$$y^2 - 3xy + 4x^2 +6 = 0,$$

and we are to transform these into one equation, eliminating y.

By infinite series the two solutions for y in the first equation are found to be

$$y = x + 1 + \frac{3}{2x} + \frac{9}{8x^3} + \dots$$

and

$$y = 3x + 1 - \frac{3}{2x} - \frac{9}{8x^3} - \dots,$$

which, substituted for y in the second equation, produce the expressions

$$2x^2 - x + \frac{11}{2} + \frac{3}{x} + \frac{9}{8x^2} + \dots$$

and

$$4x^2 + 3x + \frac{5}{2} - \frac{9}{8x^3} - \dots.$$

These two expressions are multiplied together, yielding the product

$$8x^4 + 2x^3 + 24x^2 + 20x + 28,$$

which, equated to zero, becomes the equation $8x^4 + 2x^3 + 24x^2 + 20x + 28 = 0$, all of whose roots are imaginary.

All of the corollaries derived from the previously discussed methods may also be discovered from this one.

(5.1) Let there be three or more (h) equations in as many unknowns x, y, z, ..., and w; required to transform all of these equations into one by the method of infinite series, to eliminate all unknowns except w.

From any $h-1$ of these equations find the corresponding values of the quantities x, y, z, ... as series of decreasing degree in w, i.e. let

$$\begin{aligned} x &= aw^{\alpha} + bw^{\alpha-\beta} + cw^{\alpha-2\beta} + \dots \\ y &= a'w^{\alpha'} + b'w^{\alpha'-\beta'} + c'w^{\alpha'-2\beta'} + \dots \\ z &= a''w^{\alpha''} + b''w^{\alpha''-\beta''} + c''w^{\alpha''-2\beta''} + \dots \\ \text{etc.} &= \text{etc.} \end{aligned}$$

and

$$\begin{aligned} x &= Aw^{\pi} + Bw^{\pi-\rho} + Cw^{\pi-2\rho} + \dots \\ y &= A'w^{\pi'} + B'w^{\pi'-\rho'} + C'w^{\pi'-2\rho'} + \dots \\ z &= A''w^{\pi''} + B''w^{\pi''-\rho''} + C''w^{\pi''-2\rho''} + \dots \\ \text{etc.} &= \text{etc.} \end{aligned}$$

and so on.

Now substitute all of the corresponding values of the quantities x, y, z, ... for their occurrences in the one remaining equation; call the resulting series P, Q, R, S, etc. Then multiply these series together, and the resulting product equated to zero, $PQRS \cdots = 0$, will be the required equation whose root is w.

The degree of the equation whose root is w may be deduced from the degrees of the numerators and denominators of the expressions P, Q, R, S, ... that result; the degrees of these numerators and denominators may in turn be found from those terms of the chosen $h-1$ equations in which each occurs to the highest and next lower degree.

But the equation whose root is w may not be finite: then by a method given in the next Chapter we may find a finite fraction which is equal to the infinite series, and the numerator of this fraction (reduced to its lowest terms) will be the required equation.

From the expressions obtained by multiplying together the terms of maximum degree in all the unknowns, the denominator of the required fraction can be found, which denominator, which denominator, divided into the equation whose root is w, results in the required equation in infinite form.

(5.2.1) If there are two or more (h) equations $A = 0$, $B = 0$, $C = 0$, etc., whose degrees are n, m, l, k, etc. respectively, involving two or more unknown quantities x, y, z, etc., and a, b, c, d, etc. are the maximum degrees of the $h-1$ unknown quantities x, z, v, etc. in any of the terms of the equations $A = 0$, $B = 0$, $C = 0$, etc., then the degree of the equation whose root is y will be no greater than the largest of the products $nbcd \cdots$, $macd \cdots$, $labd \cdots$, $kabc \cdots$, etc.

(5.2.2) If the terms of the given equation which is of the highest (n^{th}) degree in the variables x, y, z, ... have a common linear factor $\alpha x+\beta y+\gamma z+\ldots$, then by writing $-\beta y-\gamma z-\ldots+w$ in place of αx we can construct an equation in which none of the quantities y, z, ... appear to a degree as high as n. For example, if the equation is

$$(\alpha x+\beta y+\gamma z+\ldots)^{n-1}(\ ax+\ by+\ c+\ldots+\ du^n+\ ev^m+\ldots)$$
$$+(\alpha x+\beta y+\gamma z+\ldots)^{n-1}(\ a'x+\ b'y+\ c'+\ldots+\ d'u^o+e'v^p+\ldots)$$
$$+(\alpha x+\beta y+\gamma z+\ldots)^{n-1}(a''x+b''y+c''+\ldots+d''u^n+\ldots\qquad)$$
$$+\ldots=0,$$

then by writing $-\beta y-\gamma z-\ldots+w$ for αx we reduce this equation to one of the form $w^{n-1}(\pi y+\rho z+\ldots)+w^{n-2}(\pi' y+\rho' z+\ldots)+\ldots=0$, which involves the quantities y, z, etc. only to the first power.

And likewise for several simultaneous equations. [2]

Method 6. There is yet another way to solve the problem of transforming the two equations

$$A'y^n+(\ a+\ bx)y^{n-1}+(\ c+\ dx+\ ex^2)y^{n-2}+\ldots=0$$

and $$Ny^m+(A+Bx)y^{m-1}+(C+Dx+Ex^2)y^{m-2}+\ldots=0$$

into one, in order to eliminate the unknown quantity x.

Case 1. The two given equations, $A'y^n+(a+bx)y^{n-1}+\ldots=0$ and $Ny^m+(A+Bx)y^{m-1}+\ldots=0$, are multiplied by the two assumed equations $m'y^{mn-n}+(\ p+\ qx)y^{mn-n-1}+(\ r+\ sx+\ tx^2)y^{mn-n-2}+\ldots=0$ and $My^{mn-m}+(P+Qx)y^{nm-m-1}+(R+Sx+Tx^2)y^{nm-m-2}+\ldots=0$ respectively. Equate the resulting products to each other, and then determine the values of the coefficients of the assumed equations such that all terms of the resulting equations vanish, except those involving only powers of the unknown quantity y.

This can be accomplished by solving linear equations. There are always at least as many unknown quantities as there are linear equations to be solved.

Case 2. Let $N(h)^n$ denote the number of terms containing h unknown quantities x, y, z, v, w, etc., in an algebraic equation of degree n.

(1) By an earlier theorem, $N(h)^n=\dfrac{(n+1)(n+2)(n+3)\cdots(n+h)}{h!}$.

(2) The number of terms in an equation of degree n which are divisible by x^p will be the same as the total number of terms in an equation of degree $n-p$, i.e. by the preceding notation this number equals $N(h)^{n-p}$. Then the number of terms, in an equation of degree n, which are

divisible by any of the factors x^p or y^q or z^r or v^s or w^t or ..., will be

$$\begin{array}{llllll}
N(h)^{n-p} - N(h)^{n-p-q} + N(h)^{n-p-q-r} - N(h)^{n-p-q-r-s} + N(h)^{n-p-q-r-s-t} - \dots \\
+N(h)^{n-q} - N(h)^{n-p-r} + N(h)^{n-p-q-s} - N(h)^{n-p-q-r-t} + \dots \quad - \dots \\
+N(h)^{n-r} - N(h)^{n-p-s} + N(h)^{n-p-r-s} - \dots \quad + \dots \quad - \dots \\
+N(h)^{n-s} - N(h)^{n-q-s} + N(h)^{n-q-r-s} - \dots \quad + \dots \quad - \dots \\
+\dots .
\end{array}$$

Case 3. (1) Let there be two equations, of degrees n and m, involving two unknown quantities x and y. Multiply these equations respectively by two equations $A = 0$ and $B = 0$, of degrees $nm-n$ and $nm-m$ respectively, also involving the unknowns x and y (so that each product is an equation of degree mn). Equate corresponding terms of the two resulting equations, except those involving only powers of y. The number of terms so equated will be $nm(\frac{nm+3}{2})-nm$.

But this number must be equal to the number of independent coefficients in the equations $A = 0$ and $B = 0$, which numbers will be (by the preceding case)

$$\begin{aligned}
N(2)^{nm-n} - N(2)^{nm-n-m} &= \frac{(nm-n)(nm-n+3)}{2!} - \frac{(nm-n-m)(nm-n-m+3)}{2!} \\
+N(2)^{nm-m} &= \frac{(nm-m)(nm-m+3)}{2!} \\
&= \frac{nm(nm+3)}{2!},
\end{aligned}$$

whence the equation whose root is x or y cannot be of degree greater than nm.

(2) To generalize, if there are h equations, of degrees n, m, l, k, etc., respectively, in as many unknowns, and if these h equations are multiplied by h assumed equations of degrees $nmlk\cdots-n$, $nmlk\cdots-m$, $nmlk\cdots-l$, $nmlk\cdots-k$, etc. respectively, involving the same unknown quantities, then let the resulting equations be added together, and equate to zero the coefficients of each term of degree $nmlk\cdots$ in the result (barring those involving only powers of y). The number of these terms, and in consequence the number of resulting equations, will be

$$(P+1)\frac{P+2}{2}\cdot\frac{P+3}{3}\cdots\frac{P+h}{h} - P = N(h)^P - P,$$

where $P = nmlk\cdots$; but this number must be equal to the number of distinct independent coefficients from the equations of degrees $P-n$, $P-m$, $P-l$, $P-k$, etc., which arise in the resulting equation. Therefore we see that $(P+1)\frac{P+2}{2}\frac{P+3}{3}\cdots\frac{P+h}{h} - P$ is equal to

$$(P-n+1)\frac{P-n+2}{2}\cdots\frac{P-n+h}{h}-(P-n-m+1)\frac{P-n-m+2}{2}\cdots\frac{P-n-m+h}{h}$$
$$+(P-n-m-l+1)\frac{P-n-m-l+2}{2}\cdots\frac{P-n-m-l+h}{h}$$
$$-(P-n-m-l-k+1)\frac{P-n-m-l-k+2}{2}\cdots\frac{P-n-m-l-k+h}{h}+\cdots$$
$$+(P-m+1)\frac{P-m+2}{2}\cdots\frac{P-m+h}{h}-(P-n-l+1)\frac{P-n-l+2}{2}\cdots\frac{P-n-l+h}{h}$$
$$+(P-n-m-k+1)\frac{P-n-m-k+2}{2}\cdots\frac{P-n-m-k+h}{h}-\cdots$$
$$+(P-l+1)\frac{P-l+2}{2}\cdots\frac{P-l+h}{h}-(P-m-l+1)\frac{P-m-l+2}{2}\cdots\frac{P-m-l+h}{h}$$
$$+(P-n-l-k+1)\frac{P-n-l-k+2}{2}\cdots\frac{P-n-l-k+h}{h}-\cdots$$
$$+(P-k+1)\frac{P-k+2}{2}\cdots\frac{P-k+h}{h}-(P-n-k+1)\frac{P-n-k+2}{2}\cdots\frac{P-n-k+h}{h}$$
$$+(P-m-l-k+1)\frac{P-m-l-k+2}{2}\cdots\frac{P-m-l-k+h}{h}-\cdots$$
$$+\cdots-\cdots+\cdots-\cdots$$
$$=N(h)^{P-n}-N(h)^{P-n-m}+N(h)^{P-n-m-l}-N(h)^{P-n-m-l-k}+\cdots$$
$$+N(h)^{P-m}-N(h)^{P-n-l}+N(h)^{P-n-m-k}-\cdots$$
$$+N(h)^{P-l}-N(h)^{P-m-l}+N(h)^{P-n-l-k}-\cdots$$
$$+N(h)^{P-k}-N(h)^{P-n-k}+N(h)^{P-m-l-k}-\cdots$$
$$+\cdots-\cdots+\cdots-\cdots,$$

whence the equation whose root is x or y or z, etc., cannot be of degree greater than $nmlk\cdots$.

Now if the given degrees (n, m, l, k, etc.) are greater than the degrees $nmlk\cdots-n$, $nmlk\cdots-m$, $nmlk\cdots-l$, $nmlk\cdots-k$, etc., of the general equations by which they are to be multiplied, then the number H, of the abovementioned coefficients, will be greater than the number K, of independent equations; therefore we may assume arbitrary values for $H-K$ of the coefficients. And if the equation whose root is y turns out to have more dimensions than the original problem implies, it may always be depressed by finding its rational divisors.

In many cases it is not necessary to multiply the given equations by the above general equations of degrees $nmlk\cdots-n$, $nmlk\cdots-m$, $nmlk\cdots-l$, $nmlk\cdots-k$, etc.; rather, we may be able to choose certain equations, whose form is dictated by the given equations. Then the equation whose root is x or y, etc., is more easily found, and the degrees of the particular equations may be less than the values here derived.

Multiply the given equations by the above multipliers, and if the equation $\pi = 0$, arising from all of the resulting quantities, contains some multiplier terms independent of the root — or if among the multiplier terms can be found some equations which destroy all terms of the resulting equation $\pi = 0$ except for those containing only one unknown quantity — then all the rest of the terms of the multiplication are to be discarded.

Example 1. Let there be h equations of degrees n, m, l, k, ... respectively, involving n unknown quantities x, y, z, v, etc. Further let $p, q, r, s, \ldots; p', q', r', s', \ldots; p'', q'', r'', s'', \ldots; p''', q''', r''', s''', \ldots$; etc. be the maximum degrees of the quantities x, y, z, v, ... in the equations of degree n, m, l, k, ... respectively. Then the degree of the equation whose root is x or y or z etc. cannot ascend higher than

$$\begin{aligned} nmlk\cdots &- (n-p)(m-p')(l-p'')(k-p''')\cdots - (n-q)(m-q')(l-q'')(k-q''')\cdots \\ &- (n-r)(m-r')(l-r'')(k-r''')\cdots - (n-s)(m-s')(l-s'')(k-s''')\cdots \\ &- \cdots. \end{aligned}$$

Example 2. The same supposed as in the preceding example, if the degrees of x and y together never exceed a, a', a'', a''', ... in the given equations respectively, then if all unknowns but x are eliminated the degree of the resulting equation cannot be greater than

$$\begin{aligned} nmlk\cdots &- (n-p)(m-p')(l-p'')(k-p''')\cdots - (n-q)(m-q')(l-q'')(k-q''')\cdots \\ &- (n-r)(m-r')(l-r'')(k-r''')\cdots - (n-s)(m-s')(l-s'')(k-s''')\cdots \\ &- \cdots \\ +(n-a)(m-a')(l-a'')(k-a''')\cdots &- (p+q-a)(p'-a')(p''-a'')(p'''-a''')\cdots \\ &-(p'+q'-a')(p-a)(p''-a'')(p'''-a''')\cdots \\ &-(p''+q''-a'')(p-a)(p'-a')(p'''-a''')\cdots \\ &-(p'''+q'''-a''')(p-a)(p'-a')(p''-a'')\cdots \\ &- \cdots \end{aligned}$$

Method 7. Let m, l, k, ... be the degrees, respectively of $h-1$ given equations $A = 0$, $B = 0$, $C = 0$, etc., involving h unknown quantities x, y, z, v, etc. Further, let the equation

$$y^n - py^{n-1} + qy^{n-2} - \ldots = 0,$$

whose coefficients are rational functions of $h-1$ of the unknown quantities (x, z, v, ...), be of aggregate degree i, where i is less than any of m, l, k, etc. Let the solutions of this equation for y be α, β, γ, δ, etc. Multiply the $h-1$ given equations $A = 0$, $B = 0$, $C = 0$, ... by general equations of degrees $P-m$, $P-l$, $P-k$, etc., where $P = mlk\cdots$; the resulting polynomials will be referred to as π, ρ, σ, τ, etc. Now substitute the values α, β, γ, δ, ... for y in the polynomial $\pi+\rho+\sigma+\tau+\ldots$, and call the results X, Y, Z, V, etc.; multiply all of these together. Then according to Problem V, Chapter 1, recast the product as $XYZV\cdots \equiv \Phi = 0$, where Φ contains only the unknown quantities x, z, v, ..., and equate to zero each separate term of the equation $\Phi = 0$ except those containing exclusively powers of the unknown quantity x. Often from the resulting linear equations it is possible to reduce the given equations, of degrees m, l, k, ..., and i, to one, such that all the unknowns except x have been eliminated.

It often turns out that the given equations may be multiplied by less general equations of degree $P-m$, $P-l$, $P-k$, etc., or by equations of lower degree whose form is suggested by the form of the given equations.

This method of reducing several (h) equations to one, to eliminate $h-1$ of the unknown quantities, will often be less laborious than the methods that have previously been advanced.

Theorem 25

Let there be h equations $A = 0$, $B = 0$, $C = 0$, $D = 0$, etc. involving the h unknown quantities x, y, z, etc. Let the quantities A, B, C, D, ... be multiplied respectively by quantities a, a', a'', ...; b, b', b'', ...; c, c', c'', ...; d, d', d'', ...; etc.; in which a, a', ... do not involve x; b, b', ... do not involve y; c, c', ... do not involve z; etc. There result h mutually independent equations

$$\begin{aligned} aA + bB + cC + dD + \ldots &= 0,\\ a'A + b'B + c'C + d'D + \ldots &= 0,\\ a''A + b''B + c''C + d''D + \ldots &= 0, \end{aligned}$$

etc. Then these latter quantities, or their terms of highest degree, have no common divisors which are functions of any of the unknowns x, y, z, etc., as long as the h original equations also have none.

If the quantities A, B, C, D, ... have a common divisor, then the quantities

$$\begin{aligned} aA + bB + cC + dD + \ldots &= P,\\ a'A + b'B + c'C + d'D + \ldots &= Q,\\ a''A + b''B + c''C + d''D + \ldots &= R, \end{aligned}$$

etc., have the same common divisor. Conversely, if P, Q, R, S, etc. have a common divisor, then the quantities A, B, C, etc. have the same common divisor. For let P, Q, R, ... be multiplied respectively by α, α', α'', ...; β, β', β'', ...; γ, γ', γ'', ...; etc., in turn, resulting in the quantities

$$\begin{aligned}&\alpha P + \beta Q + \gamma R + \ldots,\\ &\alpha' P + \beta' Q + \gamma' R + \ldots,\\ &\alpha'' P + \beta'' Q + \gamma'' R + \ldots,\\ &\text{etc.}\end{aligned}$$

Then by means of linear equations we can determine the coefficients α, α', α'', ...; β, β', β'', ...; γ, γ', γ'', ...; etc., such that at last

$$\begin{aligned}&\alpha P + \beta Q + \gamma R + \ldots = A,\\ &\alpha' P + \beta' Q + \gamma' R + \ldots = B,\\ &\alpha'' P + \beta'' Q + \gamma'' R + \ldots = C,\\ &\text{etc.}\end{aligned}$$

Q.E.D.

Theorem 26

Let $A = 0$, $B = 0$, $C = 0$, $D = 0$, ... be h mutually independent equations involving the h unknown quantities x, y, z, etc. Then if a, a', a'', ..., b, b', b'', ..., c, c', c'', ..., etc. represent invariable quantities involving none of the unknowns x, y, z, etc.; and further

$$\begin{aligned}aA + bB + cC + dD + \ldots &= 0,\\ a'A + b'B + c'C + d'D + \ldots &= 0,\\ a''A + b''B + c''C + d''D + \ldots &= 0,\end{aligned}$$

etc. are h mutually independent equations, then the values of the unknowns x, y, z, ... which satisfy the latter set of equations will be the same as those values which satisfy the original set.

Problem XXXII

Given h *quantities* A, B, C, D, ... *involving the unknown quantities* x, y, z, ..., *to determine whether or not the sum*

aA + bB + cC + dD + ...

(*where the letters* a, b, c, d, ... *involve none of* x, y, z, ...) *has a divisor.*

Let $aA+bB+cC+dD+\ldots = px^m+qx^{m-1}+rx^{m-2}+\ldots+\gamma x^2+\beta x+\alpha$, and find factors of p and α, which we shall call p' and α' respectively.

Now assume the two quantities

$$p'x^r+q'x^{r-1}+s'x^{r-2}+\ldots+\beta'x+\alpha'$$

and

$$\frac{p}{p'}x^{m-r}+Qx^{m-r-1}+\ldots+Lx+\frac{\alpha}{\alpha'}$$

to be the required divisors. When these two factors are multiplied together the product is

$$px^m+(\frac{p}{p'}q'+p'q)x^{m-1}+\ldots+(\alpha'L+\beta'\frac{\alpha}{\alpha'})x+\alpha\,,$$

of which all terms of the same degree in the unknowns x, y, z, ... must now be equated to the corresponding terms of the quantity $px^m+qx^{m-1}+rx^{m-2}+\ldots+\gamma x^2+\beta x+\alpha$, and the resulting set of equations give us formulas for p', q', β', α', Q, L, etc. Then by the method of common divisors we can determine whether this solution is possible. If it is possible, then the given quantity can be separated into two factors, but if this solution is impossible, then no such factorization exists for the given quantity.

In like manner any given quantities may be factored. [3]

Problem XXXIII

Given two or more (h) *equations* A = 0, B = 0, C = 0, *etc. involving two or more unknown quantities, and another set of* h *equations* P = 0, Q = 0, R = 0, *etc. also involving two or more unknown quantities, to determine whether or not all of the solutions of the first set of equations* (A = 0, B = 0, C = 0, ...) *are equal to the corresponding solutions of the second set of equations.*

Multiply the equations $P=0$, $Q=0$, $R=0$, ... respectively by a, b, c, d, ...; a', b', c', d', ...; a'', b'', c'', d'', ...; etc.; where a, b, c, ...; a', b', c', ...; a'', b'', c'', ...; etc. denote invariable quantities. Now let

$$\begin{aligned} aP+bQ+cR+\ldots&=A,\\ a'P+b'Q+c'R+\ldots&=A,\\ a''P+b''Q+c''R+\ldots&=C, \end{aligned}$$

etc., and by equating corresponding terms on both sides of each of these equations, it is easy to decide whether the resulting equations are mutually contradictory. If these equations are inconsistent then the corresponding roots of the former and latter sets of equations will not be the same. But if none of the above constructed equations are contradictory, then the coefficients a, b, c, $\ldots$; a', b', c', $\ldots$; a'', b'', c'', $\ldots$; etc. are easily found by solving linear equations. [4]

Theorem 27

Let there be two or more (n) independent equations $A = 0$, $B = 0$, $C = 0$, etc., in as many unknown quantities. Then the n independent equations

$$\begin{aligned} PA + QB + RC + \ldots &= 0, \\ P'A + Q'B + R'C + \ldots &= 0, \\ P''A + Q''B + R''C + \ldots &= 0, \\ \text{etc.} \end{aligned}$$

(where the letters P, Q, R, $\ldots$, P', Q', R', $\ldots$, P'', Q'', R'', $\ldots$, etc. denote functions of the unknown quantities x, y, z, $\ldots$) will include all of the roots of the quantities x, y, z, $\ldots$ which were contained in the original equations $A = 0$, $B = 0$, $C = 0$, etc.; unless the quantities P, Q, R, $\ldots$, P', Q', R', $\ldots$, etc. have factors in common with their corresponding multipliers A, B, C, etc.

We substitute the corresponding values of the quantities x, y, z, etc. for the quantities themselves in the above equations, so that the expressions A, B, C, $\ldots$ vanish; and therefore the quantities $PA+QB+RC+\ldots$, $P'A+Q'B+R'C+\ldots$, $P''A+Q''B+R''C+\ldots$, etc. also vanish, and it follows that all the solutions of the equations $A = 0$, $B = 0$, $C = 0$, etc. for the unknowns x, y, z, $\ldots$ will also be found among the solutions of the latter set of equations for the same unknown quantities.

Corollary 1. Given n equations

$$\begin{aligned} PA + QB + RC + \ldots &= 0, \\ P'A + Q'B + R'C + \ldots &= 0, \\ P''A + Q''B + R''C + \ldots &= 0, \end{aligned}$$

etc., in the n unknown quantities x, y, z, etc., then the sets of corresponding values of x, y, z, etc. which satisfy the equations $A = 0$, $B = 0$, $C = 0$, etc. will also satisfy the given set of equations.

Corollary 2. By Problem XXXI we can often say what will be the degree of the equation whose root is x, deducing this from the equations $PA+QB+RC+\ldots=0$, $P'A+Q'B+R'C+\ldots=0$, $P''A+Q''B+R''C+\ldots=0$, etc., and separately from the equations $A=0$, $B=0$, $C=0$, etc. From these numbers we can determine the number of values of x which satisfy the first set of equations but are not found among the roots of the second set. For example, if there are two equations

$$P \equiv y^n + (a + bx)y^{n-1} + (c + dx + ex^2)y^{n-2} + \ldots = 0$$

and

$$Q \equiv y^m + (A + Bx)y^{m-1} + (C + Dx + Ex^2)y^{m-2} + \ldots = 0,$$

whose unknown quantities x and y have mn roots respectively, then the two equations

$$\alpha x \cdot P + \beta y \cdot Q = 0 \text{ and } \gamma x \cdot P + \delta y \cdot Q = 0$$

can have no more than $(n+1)(m+1)-nm = n+m+1$ solutions which are not also solutions of the given equations $P=0$ and $Q=0$.

Theorem 28

(1) Let there be algebraic equations $A=0$, $B=0$, $C=0$, etc., which are functions of the unknown quantities x, y, z, v, etc., and let the quantities A, B, C, D, ... have a common divisor π. Then $\pi=0$ is an equation whose solutions are some of the values of the indeterminates x, y, z, etc.

(2) Let $\frac{A}{\pi}=a$, $\frac{B}{\pi}=b$, $\frac{C}{\pi}=c$, etc.; then $a=0$, $b=0$, $c=0$, etc. are equations whose roots are identical to some of the roots of the equations $A=0$, $B=0$, $C=0$, etc.

Problem XXXIV

Given an algebraic quantity σ, to find its value.

(1) Let σ be a given function of the quantities x, y, z, ..., and in this function substitute for x, y, z, ... their assigned values a, b, c, ... Then from the various irrational quantities within the resulting expression, the several values of the function σ may be extracted.

(2) If σ is a fraction whose numerator vanishes on one, two, or more (π) occasions when the quantities x, y, z, ... are set to the values a, b, c, d, ..., and the corresponding denominators do not vanish, then if n is the number of different values of the irrationalities in the denominator which are not found in the numerator, there will be $\pi \times n$ values of the function σ which are zero.

(3) If the denominator of the said fraction σ vanishes on one, two, or more (n) occasions, when the corresponding value of the numerator is not zero, then let m be the number of different values of the irrationalities in the numerator, which are not found in the denominator: then $n \times m$ values of the fraction σ are absurd, i.e. their value is not finite.

(4.1) If σ is a rational function of a quantity x,

$$\sigma = \frac{(x-\alpha)^m}{(x-\alpha)^n} \times \frac{P}{Q}$$

whose numerator and denominator both vanish when $x=\alpha$, and further if $x-\alpha$ is not a divisor of either P or Q, then if m is greater than n, the value of the function σ will be zero when $x=\alpha$; if m is less than n, the said value for $x=\alpha$ will be infinite, that is, absurd; and if $n=m$ the value of the function σ will be $\frac{P}{Q}$ when $x=\alpha$.

(4.2) Let $\sigma = \frac{A}{B}$ be an irrational algebraic fraction whose numerator and denominator both vanish when $x=\alpha$. By Problem XXVI we can find factors (A', A'', A''', ..., B', B'', B''', ...) which produce rational quantities when multiplied respectively by A and B. If in the factors A', A'', ...; B', B'', ... there is no factor equal to zero, then upon multiplication of both the numerator and the denominator by $A'A''A''' \cdots$ and $B'B''B''' \cdots$ respectively, the resulting fraction will be a quantity of the form

$$\frac{(x-\alpha)^m}{(x-\alpha)^n} \cdot \frac{P}{Q} \cdot \frac{B'B''B'''\ldots}{A'A''A'''\ldots},$$

in which $(x-\alpha)^m P = AA'A''A''' \ldots$ and $(x-\alpha)^n Q = BB'B''B''' \ldots$, with P and Q not divisible by $x-\alpha$. Then if m is greater than n, $\frac{A}{B}$ will be zero; if m is less than n, $\frac{A}{B}$ will be an infinite quantity, which is to say, absurd; and if $M=n$ then $\frac{A}{B} = \frac{PB'B''B'''\ldots}{QA'A''A'''\ldots}$.

If some factor if the quantity A', A'', A''', ..., B', B'', B''', ... is seen to vanish, the problem cannot be resolved by this method.

Corollary. From the number of factors of the quantities A and B which vanish when $x=\alpha$, the indices m and n are easily determined, and so it follows that the value of the fraction $\frac{A}{B}$ will be $\frac{A'}{B'}$ or $\frac{A''}{B''}$ or $\frac{A'''}{B'''}$ or ..., according as α is a root of multiplicity one, two, three, or ....

Problem XXXV

Given an equation having one or more unknown quantities, to determine the constitution of its coefficients.

Case 1. Given an equation $x^n-px^{n-1}+qx^{n-2}-rx^{n-3}+sx^{n-4}-\ldots=0$, to determine the constitution of its coefficients p, q, r, s, etc.

Let α, β, γ, δ, ... be the roots of the given equation (of which there are n altogether): substituting each of these for x produces the equations

$$\alpha^n-p\alpha^{n-1}+q\alpha^{n-2}-r\alpha^{n-3}+s\alpha^{n-4}-\ldots=0,$$
$$\beta^n-p\beta^{n-1}+q\beta^{n-2}-r\beta^{n-3}+s\beta^{n-4}-\ldots=0,$$
$$\gamma^n-p\gamma^{n-1}+q\gamma^{n-2}-r\gamma^{n-3}+s\gamma^{n-4}-\ldots=0,$$
$$\delta^n-p\delta^{n-1}+q\delta^{n-2}-r\delta^{n-3}+s\delta^{n-4}-\ldots=0,$$

etc.

When these equations are reduced to one, to eliminate all of the coefficients but one (p, whose composition we seek), we find that $p=\alpha+\beta+\gamma+\delta+\ldots$; in the same manner we see that $q=\alpha\beta+\alpha\gamma+\beta\gamma+\alpha\delta+\ldots$, $r=\alpha\beta\gamma+\alpha\beta\delta+\ldots$, and so on.

Case 2. Given an equation

$$y^n+(a+bx)y^{n-1}+(c+dx+ex^2)y^{n-2}+(f+gx+hx^2+kx^3)y^{n-3}+\ldots=0,$$

involving the two unknown quantities x and y, to determine the composition of its coefficients.

Let the roots of the unknown quantity y be α, β, γ, δ, ..., and let the corresponding roots of the unknown quantity x be π, ρ, σ, τ, ...; the total number of pairs of corresponding roots x and y will be $\frac{n(n+3)}{2}$. When these corresponding roots are substituted for the respective occurrences of x and y there result the equations

$$\alpha^n+(a+b\pi)\alpha^{n-1}+(c+d\pi+e\pi^2)\alpha^{n-2}+(f+g\pi+h\pi^2+k\pi^3)\alpha^{n-3}+\ldots=0,$$
$$\beta^n+(a+b\rho)\beta^{n-1}+(c+d\rho+e\rho^2)\beta^{n-2}+(f+g\rho+h\rho^2+k\rho^3)\beta^{n-3}+\ldots=0,$$

$$\gamma^n+(a+b\sigma)\gamma^{n-1}+(c+d\sigma+e\sigma^2)\gamma^{n-2}+(f+g\sigma+h\sigma^2+k\sigma^3)\gamma^{n-3}+\ldots=0,$$
$$\delta^n+(a+b\tau)\delta^{n-1}+(c+d\tau+e\tau^2)\delta^{n-2}+(f+g\tau+h\tau^2+k\tau^3)\delta^{n-3}+\ldots=0,$$
etc.

By hypothesis there are $\frac{n(n+3)}{2}$ of these equations, and the number of unknown coefficients a, b, c, d, e, etc. is equal to the same fraction. Reduce all of these $\frac{n(n+3)}{2}$ equations to one, to eliminate all but one of the coefficients (beginning with a), and this will establish how many of which members, i.e., what function of the corresponding roots of the unknown quantities, compose this particular coefficient; and similarly for the rest of them.

Corollary 1. If the number of corresponding roots of the unknown quantities x and y, in an equation of degree n, is greater than $\frac{n(n+3)}{2}$, then you may choose any $\frac{n(n+3)}{2}$ of the pairs of corresponding roots to determine the constitution of the equation's coefficients.

Corollary 2. If the number (m) of corresponding roots of the unknown quantities x and y is less than $\frac{n(n+3)}{2}$, then the equation is constituted in an infinity of ways; in other words any values whatever may be chosen for $\frac{n(n+3)}{2}-m$ of the coefficients.

Corollary 3. Let there be two equations, of degrees n and m respectively, in the two unknowns x and y, and suppose $n<m$. Then the coefficients of the equation of degree n can be assigned arbitrary values to produce $\frac{n(n+3)}{2}$ values of x and $\frac{n(n+3)}{2}$ values of y; in the equation of degree m the coefficients may be varied in an infinite number of ways.

But if $n=m$, then we cannot arbitrarily choose values of the coefficients of either equation to produce the $\frac{n(n+3)}{2}$ values of x and $\frac{n(n+3)}{2}$ values of y: if this were possible, the two equations would have to be equivalent. But it is possible, in an infinity of ways, to assume values for the coefficients such that any $\frac{n(n+3)}{2}-1$ quantities are respectively the corresponding values of x and y.

Example. Let there be two quadratic equations in two unknowns,

$$y^2 + (\,a + \,bx)y + \,\,c + \,dx + \,ex^2 = 0$$

and
$$y^2 + (A + Bx)y + C + Dx + Ex^2 = 0.$$

In either equation there are $\dfrac{2(2+3)}{2} = 5$ unknown coefficients, and substituting for the unknown quantities x and y any five quantities in either equation produces in either case the same five linear equations, which shows that the two equations are equivalent.

Therefore we should substitute into either equation only four pairs of values for the unknown quantities x and y, which being done, from either quadratic are derived four linear equations involving five unknown coefficients, so in each of them any arbitrary value may be assumed for one of the unknown coefficients. However we must be careful not to assume the same value for the same unknown coefficient in both equations.

Let the four values chosen for the unknown quantity y be 0, 2, 4, and 2; and choose for x the four corresponding values 0, 1, 1, and 2 respectively. Substitute these into each equation, and then for one coefficient (b and B, in this example) choose the numbers 4 and -2 respectively. The first quadratic then gives forth the four linear equations

$$c = 0,$$
$$4 + 2a + 8 + e + d = 0,$$
$$16 + 4a + 16 + e + d = 0,$$

and
$$4 + 2a + 16 + 4e + 2d = 0,$$

from which we find that the first equation is

$$y^2 + (4x - 10)y - 8x^2 + 16x = 0,$$

and in like manner the second is found to be

$$y^2 - (2x + 4)y - 2x^2 + 10x = 0.$$

In all cases the number of corresponding roots of the unknown quantities x and y that may be arbitrarily chosen, is equal to the number of coefficients in the given equation to which we did not assign an arbitrary value.

The method of reasoning is the same for more equations involving more unknown quantities, but observe that the equation of degree n involving m unknown quantities has $\dfrac{(n+1)(n+2)(n+3)\cdots(n+m)}{m!} - 1$ unknown coefficients.

It is easily established that an infinity of similar constitutions can be given for these equations.

Theorem 29

(1) Let there be n independent equations involving m unknown quantities, then $m-n$ of the said quantities may be assumed *ad libitum*.

(2) Let there be n independent equations involving m unknown quantities, and if the said n equations may be reduced such that there result r mutually independent equations involving only known quantities, then in the said n equations $m-n-r$ unknown quantities may be assumed *ad libitum*.

Instances of this Theorem are utilized in Examples 2 and 3 of Problem XXII.

Problem XXXVI

Let there be two equations in two unknown quantities (x, y)*; let* $\alpha, \beta, \gamma, \delta, \epsilon, \ldots$ *be roots of the unknown quantity* y, *and let* $\pi, \rho, \sigma, \tau, \upsilon, \ldots$ *be the corresponding roots of* x; *and let* m *and* n *be given indices: to find the sum*

$$\alpha^{\mathrm{m}}\pi^{\mathrm{n}} + \beta^{\mathrm{m}}\rho^{\mathrm{n}} + \gamma^{\mathrm{m}}\sigma^{\mathrm{n}} + \delta^{\mathrm{m}}\tau^{\mathrm{n}} + \ldots$$

Suppose $y^m x^n = z$, so in the given equation for y we may substitute the value $\frac{z^{1/m}}{x^{n/m}}$; then transform the two equations into one, eliminating x, and there results an equation whose root is z, the sum of whose roots is the sum required.

Corollary. The sum of these roots is often easier to find, in that not uncommonly it depends only on those terms of highest degree in the given equations, or those of the highest and second highest degrees, etc. [5]

Problem XXXVII

Case 1. *The same letters having the same significance, supposing* m, n, o, r, s, ... *and* a, b, c, d, e, ... *to be given indices: to find the sum of all quantities of the form*
$\alpha^{\mathrm{m}}\pi^{\mathrm{a}}\beta^{\mathrm{n}}\rho^{\mathrm{b}}\gamma^{\mathrm{o}}\sigma^{\mathrm{c}}\cdots+\alpha^{\mathrm{m}}\pi^{\mathrm{a}}\beta^{\mathrm{n}}\rho^{\mathrm{b}}\delta^{\mathrm{o}}\tau^{\mathrm{c}}\cdots+\beta^{\mathrm{m}}\rho^{\mathrm{a}}\alpha^{\mathrm{n}}\pi^{\mathrm{b}}\gamma^{\mathrm{o}}\sigma^{\mathrm{c}}\cdots+\ldots$

(1) Write first

$$p^m q^a = \alpha^m\pi^a + \beta^m\rho^b + \gamma^m\sigma^a + \delta^m\tau^a + \epsilon^m\nu^a + \ldots$$

$$p^n q^b = \alpha^n \pi^b + \beta^n \rho^b + \gamma^n \sigma^b + \delta^n \tau^b + \epsilon^n \nu^b + \dots$$
$$p^o q^c = \alpha^o \pi^c + \beta^o \rho^c + \gamma^o \sigma^c + \delta^o \tau^c + \epsilon^o \nu^c + \dots$$
$$p^r q^d = \alpha^r \pi^d + \beta^r \rho^d + \gamma^r \sigma^d + \delta^r \tau^d + \epsilon^r \nu^d + \dots$$
$$etc. = \qquad \text{etc.}$$
$$p^{m+n} q^{a+b} = \alpha^{m+n} \pi^{a+b} + \beta^{m+n} \rho^{a+b} + \gamma^{m+n} \sigma^{a+b} + \dots$$
$$p^{m+o} q^{a+c} = \alpha^{m+o} \pi^{a+c} + \beta^{m+o} \rho^{a+c} + \gamma^{m+o} \sigma^{a+c} + \dots$$
$$p^{n+o} q^{b+c} = \alpha^{n+o} \pi^{b+c} + \beta^{n+o} \rho^{b+c} + \gamma^{n+o} \sigma^{b+c} + \dots$$
$$p^{m+r} q^{a+d} = \alpha^{m+r} \pi^{a+d} + \beta^{m+r} \rho^{a+d} + \gamma^{m+r} \sigma^{a+d} + \dots$$
$$etc. = \qquad \text{etc.}$$
$$p^{m+n+o} q^{a+b+c} = \alpha^{m+n+o} \pi^{a+b+c} + \beta^{m+n+o} \rho^{a+b+c} + \gamma^{m+n+o} \sigma^{a+b+c} + \dots$$
$$p^{n+o+r} q^{a+b+d} = \alpha^{m+n+r} \pi^{a+b+d} + \beta^{m+n+r} \rho^{a+b+d} + \gamma^{m+n+r} \sigma^{a+b+d} + \dots$$
$$p^{n+o+r} q^{b+c+d} = \alpha^{n+o+r} \pi^{b+c+d} + \beta^{n+o+r} \rho^{b+c+d} + \gamma^{n+o+r} \sigma^{b+c+d} + \dots$$
$$etc. = \qquad \text{etc.}$$
$$p^{m+n+o+r} q^{a+b+c+d} = \alpha^{m+n+o+r} \pi^{a+b+c+d} + \beta^{m+n+o+r} \rho^{a+b+c+d} + \gamma^{m+n+o+r} \sigma^{a+b+c+d} + \dots$$
$$p^{m+n+o+s} q^{a+b+c+e} = \alpha^{m+n+o+s} \pi^{a+b+c+e} + \beta^{m+n+o+s} \rho^{a+b+c+e} + \gamma^{m+n+o+s} \sigma^{a+b+c+e} + \dots$$
$$\text{etc.} = \qquad \text{etc.}$$

And so proceed, observing the law of substitution of Problem III.

Further, write

$$A = p^m q^a + p^n q^b + p^o q^c + p^r q^d + \dots$$
$$B = \frac{p^{m+n} q^{a+b} \cdot A}{p^m q^a \cdot p^n q^b} + \frac{p^{m+o} q^{a+c} \cdot A}{p^m q^a \cdot p^o q^c} + \frac{p^{n+o} q^{b+c} \cdot A}{p^n q^b \cdot p^o q^c} + \dots$$
$$C = \frac{p^{m+n+o} q^{a+b+c} \cdot A}{p^m q^a \cdot p^n q^b \cdot p^o q^c} + \frac{p^{m+n+r} q^{a+b+d} \cdot A}{p^m q^a \cdot p^n q^b \cdot p^r q^d} + \frac{p^{n+o+r} q^{b+c+d} \cdot A}{p^n q^b \cdot p^o q^c \cdot p^r q^d} + \dots$$
$$D = \frac{p^{m+n+o+r} q^{a+b+c+d} \cdot A}{p^m q^a \cdot p^n q^b \cdot p^o q^c \cdot p^r q^d} + \frac{p^{m+n+o+s} q^{a+b+c+e} \cdot A}{p^m q^a \cdot p^n q^b \cdot p^o q^c \cdot p^s q^e} + \frac{p^{n+o+r+s} q^{b+c+d+e} \cdot A}{p^n q^b \cdot p^o q^c \cdot p^r q^d \cdot p^s q^e} + \dots$$
$$BB = \frac{p^{m+n} q^{a+b} \cdot p^{o+r} q^{c+d} \cdot A}{p^m q^a \cdot p^n q^b \cdot p^o q^c \cdot p^r q^d} + \frac{p^{m+o} q^{a+c} \cdot p^{n+r} q^{b+d} \cdot A}{p^m q^a \cdot p^o q^c \cdot p^n q^b \cdot p^r q^d} + \dots$$
$$\text{etc.} = \qquad \text{etc.} \qquad \text{etc.},$$

following the law of substitution of the same Problem. The required sum will be

$$A - B + 2C - 1 \cdot 2 \cdot 3 \cdot D + 1 \cdot 2 \cdot 3 \cdot 4 \cdot E - \dots$$
$$+ 1 \cdot 2 \cdot BB - 1 \cdot 1 \cdot 2 \cdot BC + \dots$$

This is essentially the same series that was derived in Problem III.

(2) Already in Problems III and IV was taught a procedure which leads to the method given here of finding the sum of the values of any function, whether rational or irrational, of the roots of the unknown quantities x and y, and of infinite series in their powers as well.

But simpler progressions for the coefficients may often exist, and in the same context we may easily determine the maximum degree to which any unknown quantity will be involved in the resulting aggregate.

> Case 2. *Let there be more equations, with the same number of unknown quantities* x, y, z, v, ... *as equations; and let* α, β, γ, δ, ... *be the roots of the unknown quantity* x; π, ρ, σ, τ, ... *the corresponding roots, respectively, of the unknown quantity* y; κ, λ, μ, ν, ... *the corresponding roots, respectively, of the unknown quantity* z; Γ, Δ, E, Z, ... *the corresponding values of the unknown quantity* v; *and so on; lastly let* a, f, l, p, ... *be given indices: to find the sum* $\alpha^{\mathrm{a}}\pi^{\mathrm{f}}\kappa^{\mathrm{l}}\Gamma^{\mathrm{p}}\cdots+\beta^{\mathrm{a}}\rho^{\mathrm{f}}\lambda^{\mathrm{l}}\Delta^{\mathrm{p}}\cdots+\gamma^{\mathrm{a}}\sigma^{\mathrm{f}}\mu^{\mathrm{l}}\mathrm{E}^{\mathrm{p}}\cdots+\delta^{\mathrm{a}}\tau^{\mathrm{f}}\nu^{\mathrm{l}}\mathrm{Z}^{\mathrm{p}}\cdots+\ldots$

(1) Suppose $x^a y^f z^l v^p = w$, and transform these equations into one, eliminating all the unknowns x, y, z, v, ... except the assumed one (w). The result is an equation whose root is w, and the sum of its roots is the required sum.

This also can often be achieved by other methods.

(2) By a series observing the same law as that derived in the preceding series of this problem, and by the aforementioned methods from Problems III and IV, we can investigate the sum of the values of any functions whatever, be they rational or irrational, of the roots of the unknown quantities x, y, z, v, etc.

(3) Moreover, the same may be said of functions of the several roots together with adventitious unknown quantities.

If α is a true root of the unknown quantity x, this root α may be substituted in place of x in all of the equations; by adventitious roots I mean, on the other hand, roots which have crept into some of the equations through this substitution, but which are not common to all of the equations.

Many other propositions of this kind may be found by transposing terms from one side to the other in a given equation; by adding, subtracting, or multiplying given equations together, etc.; and by substituting in the resulting equations for x, y, z, ... their corresponding roots α, π, κ, ..., β, ρ, λ, ..., γ, σ, μ, ..., etc.; and so on.

Problem XXXVIII

To transform two equations, with two unknowns x *and* y, *into one whose unknown quantity shall have a particular algebraic relation to the roots of the unknown quantities of the two equations.*

(1) The roots of the unknown quantities are treated as if they are known, i.e., the roots of the unknown quantity x are assumed to be α, β, γ, δ, etc., and the corresponding roots of the unknown quantity y respectively π, ρ, σ, τ, etc. Then from the given relation (if it is not so complex as to require expression by equations) the roots of the required quantity are found, and consequently by the previous Problem the sum of its roots, which is the coefficient of the second term of the equation.

Every two, three, etc. roots of the required quantity are multiplied together, and with the help of the two preceding Problems we can find the sums of the resulting products, which are respectively the third, fourth, etc. coefficients of the required equation. Thus all of the equation's coefficients, and so the equation itself, are produced.

(2) If certain given equations express the relation between the roots of the unknown quantities of the two given equations and certain other assumed unknown quantities, then once again we regard the roots of the given equations as known, and insert them in all possible combinations into the equations of relation; then all of the corresponding equations so produced are multiplied together, and by the two preceding Problems is found the aggregate of all of them. This aggregate, which no longer involves the unknown quantities of the two given equations, is what we require.

Corollary 1. If three or more equations with three or more unknowns are given, by the same method we may find equations whose roots have any specified algebraic relation to the roots of the given equations.

Corollary 2. If the relation between the unknown quantities of the given and required equations involves only corresponding values of the unknown quantities, then the required equation may always be found by reduction of several equations to one, eliminating unknown quantities of the given equations.

Theorem 30

If there are two, three, or more equations $A = 0$, $B = 0$, $C = 0$, $D = 0$, etc., involving two, three, or more unknowns x, y, z, etc., and if by some combination of additions, subtractions, and multiplications by rational quantities we can construct an equation $Z = 0$, where Z is the sum of several real squares, i.e. $Z = P^2+Q^2+R^2+\ldots$, where P, Q, R, ... denote rational functions of the quantities x, y, z, etc., then the given equations $A = 0$, $B = 0$, $C = 0$, $D = 0$, etc. have no solution in which all corresponding values of the unknown quantities x, y, z, etc. are real numbers, unless the equations $P = 0$, $Q = 0$, $R = 0$, etc. have solutions for the unknown quantities x, y, z, etc. which are also solutions of $A = 0$, $B = 0$, $C = 0$, etc. for the same unknowns.

Corollary. Hence each of the unknowns (e.g. x) has no real root (e.g. α) in the said equations $A = 0$, $B = 0$, $C = 0$, $D = 0$, etc., unless α is a root of multiplicity 2 or more.

Hence it is easy to generate an infinity of equations of this sort, which have no real roots.

Theorem 31

If there are two or more equations $A = 0$, $B = 0$, $C = 0$, etc. involving two or more unknown quantities x, y, etc., for each equation we may identify the conditions under which none, or two, or four, etc. or all of the roots of one of the unknowns (x or y or ...) will be impossible. From these cases we can determine the limits between which real solutions of the given set of equations must lie.

Example. Let the two equations be

$$y^2 + (a+bx) + cx^2+dx+e = 0$$

and

$$y^2 + (p+qx)y + rx^2+sx+t = 0.$$

First, if b^2-4c is positive then the roots of y in the first equation will be real, provided that x does not lie between the two roots (π and ρ) of the equation

$$(b^2-4c)x^2+(2ab-4d)x+a^2-4e = 0.$$

Second, if b^2-4c is negative then the roots of y will be real if x does lie between π and ρ. Likewise if q^2-4p is positive, then the roots of y will be real if x is not found between the two roots (α and β) of the equation

$$(q^2-4r)x^2+(2pq-4s)x+p^2-4t = 0,$$

while if q^2-4r is negative, the roots of y will be real if x lies in the stated interval between α and β.

Thus if both b^2-4c and q^2-4r are positive, then y will be real provided that x lies neither between π and ρ, nor between α and β. If b^2-4c is positive and q^2-4r is negative, y will be real if x is between α and β, but not between π and ρ. The reader can supply the remaining cases. Hence we can find limits on the values of y corresponding to all values of x lying between the limits just deduced.

By the same method first we can find limits between which y must lie, in order that the values of x shall be real, and then we can find the limits between which lie real values of x, which correspond to all values of y between the previously stated limits.

Then by comparing the limits so discovered, not uncommonly we can determine whether or not the roots of all the unknown quantities of the given equations can ever simultaneously be real.

Theorem 32

Given two or more equations involving two or more unknown quantities (x, y, z, ...), these equations being transformed into others whose roots are the squares, or some other algebraic function, of the roots of the given equations; from the principles given in Chapter 2 the number of impossible roots (or other types) in the given equations can be deduced from the number of impossible roots (or other types) in the resulting equation.

Theorem 33

(1) Given a quantity expressing the value of a series which progresses by even powers in the quantities x, y, z, etc. Let α, β, γ, δ, ... be roots of the equation $\kappa^m-1=0$; λ, μ, ν, ... the roots of the equation $\kappa^n-1=0$; and π, ρ, σ, τ, ... the roots of the equation $\kappa^r-1=0$. In the given equation for x write αx, for y write λy, for z write πz, and so on; then substitute αx, μy, πz, ...; αx, μy, ρz, ...; βx, λy, πz, ...; etc. There will result $nmr\cdots$ distinct quantities A, B, C, D, E, etc. Then

$$\frac{\left\{\begin{matrix}(\alpha^{n-d}\lambda^{m-e}\pi^{r-f}\cdots)A+(\alpha^{n-d}\mu^{m-e}\pi^{r-f}\cdots)B+\\(\alpha^{n-d}\mu^{m-e}\rho^{r-f}\cdots)C+(\beta^{n-d}\lambda^{m-e}\pi^{r-f}\cdots)D+\cdots\end{matrix}\right\}}{nmr\cdots}$$

will be the sum of the terms, in the above series, that are of the form

$$x^{pn+d}\times y^{p'm+e}\times z^{p''r+f}\times\cdots$$

where p, p', p'', ... denote positive whole numbers.

(2) Let there be an equation $P=0$, where P is a function of the unknown quantities x, y, z, etc. As before, for y, z, etc. substitute respectively αy, λz, ...; then αy, μz, ...; then βy, μz, ...; etc., where α, β, γ, ... are roots of the equation $x^n-1=0$ and λ, μ, ν, ... are roots of the equation $x^m-1=0$, and so on. There result thereby $nm\cdots$ equations, from which values corresponding to the unknown quantity x are found, which, arranged according to increasing powers of the quantities y, z, etc., shall be identified as A, B, C, D, etc. Then

$$\frac{\alpha^{n-d}\lambda^{m-e}\cdots A+\alpha^{n-d}\mu^{m-e}\cdots B+\beta^{n-d}\mu^{m-e}\cdots C+\cdots}{n\cdot m\cdots}$$

is equal to the sum of all terms, in the above series, of the form $y^{pn+d}\cdot z^{p'm+e}\cdots$, where p, p', etc. are integers.

Problem XXXIX

Given two or more equations in two or more unknown quantities, to find those values of the roots of the unknown quantities which have a fixed relation (either given, or to be determined) among themselves.

Case 1: where formal representations of the roots of the unknown quantities x, y, etc. are given.

Upon substitution of the expressions for the roots in place of the unknown quantities x, y, etc. themselves, the required roots, if they exist, can be found by the method of finding common divisors.

Example 1. Let the two equations be

$$y^n + (a + bx)y^{n-1} + (c + dx + ex^2)y^{n-2} + \ldots = 0$$

and

$$y^m + (A + Bx)y^{m-1} + (C + Dx + Ex^2)y^{m-2} + \ldots = 0,$$

and be it required to find three roots of y in an arithmetic progression.

Let the arithmetic progression be α, $\alpha+\beta$, $\alpha+2\beta$, which substituted into the given equations for y results in six equations involving five unknown quantities. If we refer to the values of x as, say, π, ρ, and σ, then the six equations will be

$$\alpha^n + (a+b\pi)\alpha^{n-1} + (c+d\pi+e\pi^2)\alpha^{n-2} + \ldots = 0,$$
$$(\alpha+\beta)^n + (a+b\rho)(\alpha+\beta)^{n-1} + (c+d\rho+e\rho^2)(\alpha+\beta)^{n-2} + \ldots = 0,$$
$$\alpha^m + (A+B\pi)\alpha^{m-1} + (C+D\pi+E\pi^2)\alpha^{m-2} + \ldots = 0,$$
$$(\alpha+\beta)^m + (A+B\rho)(\alpha+\beta)^{m-1} + (C+D\rho+E\rho^2)(\alpha+\beta)^{m-2} + \ldots = 0,$$
$$(\alpha+2\beta)^n + (a+b\sigma)(\alpha+2\beta)^{n-1} + (c+d\sigma+e\sigma^2)(\alpha+2\beta)^{n-2} + \ldots = 0,$$

and $(\alpha+2\beta)^m + (A+B\sigma)(\alpha+2\beta)^{m-1} + (C+D\sigma+E\sigma^2)(\alpha+2\beta)^{m-2} + \ldots = 0.$

Here the five unknown quantities are α, β, π, ρ, and σ. We must determine whether or not these equations are mutually contradictory, and so the example is completed.

Example 2. (1) To find whether any root of one unknown quantity (x) is equal to the corresponding root of the other unknown quantity (y).

Let the required root be α. In the given equations substitute α for both x and y, and determine whether or not the resulting equations are consistent with this hypothesis.

(2) To find whether any root of the unknown quantity x is equal to any root of the unknown quantity y.

Substitute α and π for x and y in one equation (or more), and substitute ρ and α for x and y in another (or others); then there result equations in which, by the method of common divisors, α will be found to be a root, if the proposition is true.

Example 3. Required to find the number of pairs (x, y) both equal to zero, in the simultaneous solution of the two equations

$$a+bx+cy+dx^2+exy+fy^2+gx^3+hx^2y+kxy^2+Ly^3+\ldots=0$$

and
$$A+Bx+Cy+Dx^2+Exy+Fy^2+Gx^3+Hx^2y+Kxy^2+Ly^3+\ldots=0.$$

If a and A, being of degree zero with respect to both x and y, both vanish, the corresponding values of x and y will be equal to zero. If these terms vanish, and the terms bx, cy, Bx, and Cy also vanish, then four values of the root x will be equal to zero, as will be four corresponding values of the root y. If all of the terms a, A, bx, cy, Bx, Cy, dx^2, exy, fy^2, Dx^2, Exy, and Fy^2 vanish, nine pairs of corresponding values of x and y will be equal to zero, and so on.

If the terms of degree $r-1$, $r-2$, $r-3$, ..., 0 of one equation, and the terms of degree $s-1$, $s-2$, $s-3$, ..., 0 of the other all vanish, then sr roots of the equation, whose roots are x or y, will be equal to zero.

If there are three or more equations in three or more unknown quantities (x, y, z, ...) in which the first equation's terms of degree $r-1$, $r-2$, $r-3$, ..., 0, the second equation's terms of degree $s-1$, $s-2$, $s-3$, ..., 0, the third equation's terms of degree $t-1$, $t-2$, $t-3$, ..., etc. all vanish, then the equation whose roots are x or y or z or ... will have a total of $rst\cdots$ roots equal to zero. These results can be deduced from the principle, discussed in my *Proprietates Algebraicarum Curvarum*, that the curve designated by these equations has a double or triple or higher-order point whenever $x=y=\ldots=0$.

Now if the terms of lowest degree are divisible by xy, as many or more corresponding values of the unknown quantities can be made equal to zero, as there are terms of lowest degree; their exact number depends on the terms of next larger degree, and so on.

Example 4. The same equations being proposed, to determine in general whether or not they can be satisfied by x=0: write 0 in place of x in all of them, and find out if the resulting equations have a common divisor; if they do, the multiplicity of the root $x=0$ is the degree of the common divisor.

Example 5. If the two equations are

$$y^{2n}+bxy^{2n-1}+(c+ex^2)y^{2n-2}+(gx+kx^3)y^{2n-3}+\ldots=A,$$

and
$$y^{2n}-bxy^{2n-1}-(c+ex^2)y^{2n-2}-(gx+kx^3)y^{2n-3}-\ldots=B;$$

or alternatively,

$$y^{2n}+(c+ex^2)y^{2n-2}+(l+mx^2+n'x^4)y^{2n-4}+\ldots=0$$

and
$$y^{2m}+(C+Ex^2)y^{2m-2}+(L+Mx^2+Nx^4)y^{2m-4}+\ldots=0,$$

then the values of the unknown quantities x and y will be equal in magnitude and opposite in sign.

For if we substitute α and $-\alpha$ for x, and β and $-\beta$ for y, in the two equations, the same values result, whence if the positive quantities α and β are true values of x and y respectively, then so will be the negative quantities $-\alpha$ and $-\beta$.

The same reasoning applies to three or more equations in three or more unknowns.

Example 6. Given the two equations

$$My^n+(a+bx)y^{n-2}+(c+dx+ex^2)y^{n-2}+(f+gx+hx^2+kx^3)y^{n-3}+\ldots=0$$

and

$$Ny^m+(A+Bx)y^{m-1}+(C+Dx+Ex^2)y^{m-2}+(F+Gx+Hx^2+Kx^3)y^{m-3}+\ldots=0$$

having two unknowns, x and y; to determine whether or not two roots of x, and the two corresponding roots of y, are equal.

Multiply successive terms of these equations (arranged by powers of either x or y) by the successive terms of an arithmetic progression; this produces quantities which will vanish when two values of x, and two values of y, are respectively equal. Thus,

$$nMy^{n-1}+((n-1)a+bx)y^{n-2}$$
$$+(n-2)(c+dx+ex^2)y^{n-3}+(n-3)(f+gx+hx^2+kx^3)y^{n-4}+\ldots=0$$

and $mNy^{m-1}+(m-1)(A+Bx)y^{m-2}$

$$+(m-2)(C+Dx+Ex^2)y^{m-3}+(m-3)(F+Gx+Hx^2+Kx^3)y^{m-4}+\ldots=0;$$

moreover $\quad by^{n-1}+(d+2ex)y^{n-2}+(g+2hx+3kx^2)y^{n-3}+\ldots=0$

and $\quad by^{n-1}+(D+2Ex)y^{m-2}+(G+2Hx+3Kx^2)y^{m-3}+\ldots=0;$

finally $\quad (n-1)by^{n-2}+[(n-2)d+2(n-2)ex]y^{n-3}+\ldots=0$

and $\quad (m-1)By^{m-2}+[(m-2)D+2(m-2)Ex]y^{m-3}+\ldots=0:$

these are true when two values of the unknown quantity x are equal, and so are two corresponding values of the unknown quantity y.

This example is easily deduced from the principles given in Problem VI and Theorems 3 and 4.

Also, if the roots are not equal, these resulting equations will show the limits of the roots.

Example 7. Let the two equations be

$$y^n+(a+bx)y^{n-1}+(c+dx+ex^2)y^{n-2}+\ldots+Ax^m+Bx^{m-1}+Cx^{m-2}+Dx^{m-3}+\ldots=0$$

and

$$y^n+(a+bx)y^{n-1}+(c+dx+ex^2)y^{n-2}+\ldots+Px^r+Qx^{r-1}+Rx^{r-2}+Sx^{r-3}+\ldots=0,$$

that is, all corresponding terms involving y are the same in the two equations, then each value of the unknown quantity x will occur with multiplicity n. For take any value $(\alpha, \beta, \gamma, \ldots)$ of x as a root of the equation $\quad Ax^m + Bx^{m-1} + \ldots - Px^4 - Qx^{r-1} - \ldots = 0$
and substitute this value for x in the given equation or equations; there will result various equations of the nth degree in y, and so n values of y correspond to each root x, or in other words, each value of x occurs with multiplicity n.

All of this can be extended to the case of more equations in more unknown quantities.

Example 8. Given two equations in the two unknown quantities x and y, to determine whether there are at least two (a total of m) values of y which correspond to each value of the quantity x.

Write a for x in the given equations, and reduce the two given equations $A = 0$ and $B = 0$, by the method of common divisors, such that there result the two quadratic equations

$$\alpha y^2+\beta y+\gamma = 0 \text{ and } \pi y^2+\rho y+\sigma = 0$$

or, if $m>2$, two equations of degree m,

$$\alpha y^m+\beta y^{m-1}+\gamma y^{m-2}+\ldots = 0 \text{ and } \pi y^m+\rho y^{m-1}+\sigma y^{m-2}+\ldots = 0,$$

in which equations $\alpha, \beta, \gamma, \delta, \ldots, \pi, \rho, \sigma, \tau, \ldots$ are functions of the quantity a. Then equate corresponding coefficients of these equations to each other, i.e.

$$\frac{\beta}{\alpha} = \frac{\rho}{\pi}, \frac{\gamma}{\alpha} = \frac{\sigma}{\pi}, \frac{\delta}{\alpha} = \frac{\tau}{\pi}, \text{ etc.}$$

Then from these resulting equations, again using the method of common divisors, that value a of x, if there be any such in the two equations $A = 0$ and $B = 0$, can be found to which correspond two (or m) values of the unknown quantity y.

Case 2. If the relation is implicit (expressed by equations), then by reducing the several equations to one, so as to eliminate unknowns, the desired roots can be found, if they exist.

Case 3. If an equation is given which expresses the relation between various roots of the unknown quantity x only, then the given equations are reduced so as to eliminate all but the unknown x; then simply test to see whether the given equation is a divisor of the resulting equation or not.

Case 4. Given any functions whatever of the roots of unknown quantities, having to each other a given relation, these functions are to be found from the given equations; and then from the resulting and given equations the required roots, if any exist, may be found.

Corollary 1. Given any quantity composed in an arbitrary way of diverse irrational quantities ($A+B+C+D+E+\ldots$), to find whether or not it admits of any real values.

Assume $a+p\sqrt{-1}$ as a root of the irrational quantity A, $b+q\sqrt{-1}$ as a root of the irrational quantity B, $c+r\sqrt{-1}$ as a root of the irrational quantity C, $d+s\sqrt{-1}$ as a root of the irrational quantity D, etc.

Now assume $p+q+r+s+\ldots = 0$; the *quæsita* of the problem are satsified by determining whether or not the equations resulting from the assumption $p+q+r+s+\ldots = 0$ have any consistent solutions.

Corollary 2. Given two or more equations having two or more unknown quantities (x, y, z, etc.), if the impossible quantity $a+\sqrt{-b^2}$ is a root of the unknown quantity x, then so is $a-\sqrt{-b^2}$.

This can be established simply by substituting each of these values in place of the unknown quantity x.

The following points may be added:

(1) If there are two or more equations in two or more unknowns, the same irrationalities will be found in the corresponding values of all of the unknown quantities, unless two or more values of an unknown quantity are equal to each other.

(2) Given two equations in two unknowns, x and y, if one root ($a+\sqrt{-b^2}$) is impossible, then the corresponding value of y will also be impossible, unless two, three, or more values of the unknown quantity y are equal to each other and correspond to the said root of the unknown quantity x. And so for more unknown quantities contained in more equations.

(3) If there are n equations in the n unknown quantities x, y, v, w, z, etc., then to find the number of sets of real values corresponding to the roots x, y, z, v, etc. we must first find the number of real values which one of the roots (say x) has. There will be the same number of real corresponding values of the other unknowns y, z, v, etc., unless two, three, or more values of an unknown quantity are equal to each other. In that event we must determine, by the preceding Problem, how many real values of the unknown quantities y, z, v, etc. correspond to two or more equal or impossible values of the unknown quantity x, and then the number we seek can easily be calculated.

(4) Rules for detecting impossible roots in given simultaneous equations can be deduced:

first, by substituting quantities for the roots in the given equations;

second, by means of quantities deduced from the given equations;

third, by transforming the given equations to others whose roots have an assignable relation to the roots of the given equations; or

fourth, by determining the circumstances under which the roots of any two unknown quantities are equal to each other once, twice, thrice, or with higher multiplicity.

When the given equations are transformed into others whose roots have an assignable relation to the roots of the given equations, the number of impossible roots of the resulting equations is often directly equal to the number of impossible roots of the resulting equations.

(5) Similar techniques are applicable for the investigation of the number of positive and negative roots of the unknown quantities.

Problem XL

Given two or more equations in two or more unknown quantities, to find how many pairs (x, y) *have like, and how many unlike, signs.*

Find the equation whose root is $\frac{x}{y}$: for as many of its roots will be positive, as there are pairs of roots $(x,\ y)$ of the original equations having identical signs, and it will have as many negative roots as there are pairs of roots in the given equations having opposite signs; an exception is the case where x has an impossible root $a+b\sqrt{-1}$ and the corresponding value of y is $ma+mb\sqrt{-1}$.

It is similarly easy to find equations whose roots are all negative or all positive.

For every equation of degree n,

$$x^n - px^{n-1} + qx^{n-2} - rx^{n-3} + \ldots = 0,$$

will be the difference between two equations of degree n or a higher degree, both having exclusively real roots, or having m real positive roots, r negative roots, and the rest impossible; and these equations can be varied in an infinity of ways to achieve the desired end.

Similar reasoning applies to the sum of two equations of degree n.

Likewise it may be extended to two or more equations involving two or more unknown quantities.

Problem XLI

Case 1. (1) *Given two or more equations* A $= 0$ *and* B $= 0$, *in two unknown quantities* x *and* y, *having factors (i.e.* $\mathrm{A} = \alpha \cdot \beta \cdot \gamma \cdot \delta \cdots$, $\mathrm{B} = \pi \cdot \rho \cdot \sigma \cdot \tau \cdots$*), to depress these equations into others of lesser degree.*

Equate each factor of one equation in turn to each factor of the other, and these will be the reduced equations, i.e., the reduced equations will be

$$(\alpha = 0, \pi = 0),\ (\alpha = 0, \rho = 0),\ (\alpha = 0, \sigma = 0),\ (\alpha = 0, \tau = 0),\ \ldots$$
$$(\beta = 0, \pi = 0),\ (\beta = 0, \rho = 0),\ (\beta = 0, \sigma = 0),\ (\beta = 0, \tau = 0),\ \ldots$$
$$(\gamma = 0, \pi = 0),\ (\gamma = 0, \rho = 0),\ (\gamma = 0, \sigma = 0),\ (\gamma = 0, \tau = 0),\ \ldots$$
$$(\delta = 0, \pi = 0),\ (\delta = 0, \rho = 0),\ (\delta = 0, \sigma = 0),\ (\delta = 0, \tau = 0),\ \ldots$$

etc. In the next chapter will be shown how to determine whether or not the given equations have factors.

Corollary 1.1.1. Let the divisors of the two given equations,

$$y^n + (a+bx)y^{n-1} + \ldots = 0 \text{ and } y^m + (A+Bx)y^{m-1} + \ldots = 0$$

be respectively

$$y+\alpha x+\beta,\ y+\gamma x+\delta,\ y+\epsilon x+\zeta,\ \ldots\ (n \text{ factors of the first equation})$$

and $y+\pi x+\rho,\ y+\sigma x+\tau,\ y+\upsilon x+\mu,\ \ldots$ (m factors of the second equation). If we equate a factor of one equation, and a factor of the other, to zero, i.e. $y+\alpha x+\beta = 0,\ y+\pi x+\rho = 0$ (whence $x = \frac{\rho-\beta}{\alpha-\pi},\ y = \frac{\pi\beta-\rho\alpha}{\alpha-\pi}$)

and likewise

$$y+\alpha x+\beta = 0,\ y+\sigma x+\tau = 0 \text{ (whence } x = \frac{\tau-\beta}{\alpha-\sigma},\ y = \frac{\sigma\beta-\tau\alpha}{\alpha-\sigma})$$

and so on, we have in all nm distinct pairs $(x,\ y)$ which satisfy the given equations. But in the most general case we may not be able to assume as many as nm pairs of corresponding values: only as many can be assumed, as the number of unknowns and independent coefficients. Corresponding values sometimes have a connection between them, that is, for instance, all the values resulting from one of the given equations may be proportional to all the values resulting from another of the given equations; accordingly, if some of the values are given, the others arise directly from them.

The same may be said of equations that are not separable into factors; in this case the individual values of a certain unknown quantity from one of the given equations may be expressible by irrarional expressions or by infinite series of terms; the values of this quantity may be proportional to the values similarly resulting from a different one of the equations.

Corollary 1.1.2. Let there be two equations, each of degree n, in two unknowns. It was shown earlier that we cannot assume as many as $\frac{n(n+3)}{2}$ distinct pairs of corresponding values of the two unknown quantities, otherwise the two equations are identical. However, if corresponding values are assumed, having the abovementioned sort of affinity among themselves, then n^2 corresponding values can be assumed, and there will be also n^2 different equations of the above type; but if we can find n^2+1 distinct corresponding values, this will prove either that the two equations are identical, or that they have a common divisor.

(2) If there are two equations having no divisors, these equations may be multiplied by assumed quantities, and then we take the sum or difference of the resulting equations; often equations with factors result. (Whether or not they can be factored, can be determined according to the following Chapter.)

Example. Let the two equations be

$$6y^2 - 5y - 4x^2 + 34x - 31 = 0$$

and

$$y^2 - (11x - 1)y + 14x^2 = 9x + 4 = 0.$$

These equations have no divisors. Now subtract the latter equation from twice the former, and divide the result by 11; the result is the equation

$$y^2 + (x-1)y - 2x^2 + 7x - 6 = (y-x+2)(y+2x-3) = 0.$$

Then add the first equation to five times the second, and the result is

$$y^2 + 5xy + 6x^2 - x - 1 = (y-2x+1)(y-3x-1) = 0,$$

whence the corresponding values are respectively

$$x = -1,\ 1,\ -\frac{3}{2},\ \frac{2}{5};\quad y = -3,\ 1,\ -\frac{7}{2},\ \frac{11}{5}.$$

Corollary 1.2.1. Hereby it is easy to construct an infinity of such equations. Begin with two equations which have factors (but no factors in common); multiply these equations by suitably chosen quantities, and from the sum or difference, etc., of the resulting equations we can compose equations which have no divisors.

And so for even more equations having also more unknown quantities.

Case 2. There being three or more (h) equations $A = 0$, $B = 0$, $C = 0$, $D = 0$, etc., involving three or more (h) unknown quantities x, y, z, v, etc., each having divisors, i.e. $A = \alpha\beta\gamma\delta\cdots$, $B = \alpha'\beta'\gamma'\delta'\cdots$, $C = \alpha''\beta''\gamma''\delta''\cdots$, etc., to depress these equations into others of lower degree.

Equate to zero the respective factors of each of the equations $A = 0$, $B = 0$, $C = 0$, ..., i.e. ($\alpha = 0$, $\alpha' = 0$, $\alpha'' = 0, \ldots$), ($\alpha = 0$, $\beta' = 0$, $\alpha'' = 0, \ldots$), ($\alpha = 0$, $\alpha' = 0$, $\beta'' = 0, \ldots$), etc. The roots of the unknown quantities in the resulting simultaneous equations will also be the roots of the unknown quantities in the given equations.

Corollary 2.1. If there are three or more equations, of degrees n, m, l, k, ..., and

the first has n factors $y{+}\alpha x{+}\beta z{+}\gamma v{+}\ldots{+}a$,
$y{+}\alpha' x{+}\beta' z{+}\gamma' v{+}\ldots{+}a'$, $y{+}\alpha'' x{+}\beta'' z{+}\gamma'' v{+}\ldots{+}a''$, etc.;

the second has m factors $y{+}\kappa x{+}\lambda z{+}\mu v{+}\ldots{+}b$
$y{+}\kappa' x{+}\lambda' z{+}\mu' v{+}\ldots{+}b'$, $y{+}\kappa'' x{+}\lambda'' z{+}\mu'' v{+}\ldots{+}b''$, etc.;

the third has l factors $y{+}\pi x{+}\rho z{+}\sigma v{+}\ldots{+}c$,
$y{+}\pi' x{+}\rho' z{+}\sigma' v{+}\ldots{+}c'$, $y{+}\pi'' x{+}\rho'' z{+}\sigma'' v{+}\ldots{+}c''$, etc.;

and so on, we suppose each divisor of each equation equal to zero, i.e.

$$(y{+}\alpha x{+}\beta z{+}\ldots{+}a{=}0,\ y{+}\kappa\, x{+}\lambda\, z{+}\ldots{+}b{=}0,\ y{+}\pi\, x{+}\rho\, z{+}\ldots{+}c{=}0,\ \ldots)$$
$$(y{+}\alpha x{+}\beta z{+}\ldots{+}a{=}0,\ y{+}\kappa' x{+}\lambda' z{+}\ldots{+}b'{=}0,\ y{+}\pi x{+}\rho z{+}\ldots{+}c{=}0,\ \ldots)$$
$$(y{+}\alpha x{+}\beta z{+}\ldots{+}a{=}0,\ y{+}\kappa x{+}\lambda z{+}\ldots{+}b{=}0,\ y{+}\pi' x{+}\rho' z{+}\ldots{+}c'{=}0,\ \ldots)$$

etc. Then from each resulting set of simultaneous equations we can deduce values corresponding to the various unknown quantities y, x, z, v, etc.

Hence there can be $n\cdot m\cdot l\cdot k\cdots$ distinct sets of values corresponding to the unknown quantities x, y, z, v, etc., and the equation whose root is x or y or z or ... may be of degree $n\cdot m\cdot l\cdot k\cdots$. Whence, since the maximum degree of the equation whose root is x or y or z or ... depends only on the terms of highest degree in the various equations, we see that the resulting equation cannot be of degree higher than

$h \times m \cdot l \cdot k \cdots$ when the total number of given equations is h in as many unknowns (their degrees being n, m, l, k, etc. as noted before) and we reduce them to one, eliminating all unknowns but the quantity y.

These equations cannot have as many as $n \cdot m \cdot l \cdot k \cdots$ freely assigned values corresponding to the unknown quantities x, y, z, etc.

If we are given $n \cdot m \cdot l \cdot k \cdots +1$ distinct corresponding values of the separate unknown quantities x, y, z, v, etc., then by multiplying each given equation by any chosen number such that h new equations result, whose roots in all their unknowns are identical to those of the given equations, we can deduce the existence of a common divisor in two or more of the resulting equations.

Case 3. Let there be three or more (h) equations $A = 0$, $B = 0$, $C = 0$, etc., whose degrees are n, m, l, k, etc., in three or more (h) unknowns x, y, z, v, etc., in which y appears to a maximum degree of a, a', a'', etc. respectively. Suppose further that the degree of the equation whose root is y is at most

$$n \cdot m \cdot l \cdot k \cdots - (n-a)(m-a')(l-a'') \cdots;$$

and suppose $h+1$ other equations $A' = 0$, $B' = 0$, $C' = 0$, $\ldots$, $H = 0$, are given, of degrees n, m, l, k, $\ldots$, N in the $h+1$ unknown quantities x, y, z, v, $\ldots$, V, in which y appears to the maximum degree a, a', a'', a''', $\ldots$, $a^{(h)}$. Then in this case the degree of the equation whose root is y cannot exceed

$$n \cdot m \cdot l \cdot k \cdots N - (n-a)(m-a')(l-a'') \cdots (N-a^{(h)}).$$

Assume that we may write the equation $H = 0$ as

$$(V+\alpha x+\beta z+\ldots)(V+\alpha' x+\beta' z+\ldots)(V+\alpha'' x+\beta'' z+\ldots)\cdots$$
$$\times[y^{a^{(h)}} + (\pi x+\rho z+\sigma v+\ldots+\tau)y^{a^{(h)}-1} + \ldots]$$

where the number of factors of the form $V+\alpha x+\beta z+\ldots$, $V+\alpha' x+\beta' z+\ldots$, etc. is $n-a^{(h)}$. Consider $V+\alpha x+\beta z+\ldots$, $V+\alpha' x+\beta' z+\ldots$, etc., and substitute for V the values resulting in the first h of the equations $A' = 0$, $B' = 0$, $C' = 0$, etc. Now the equation whose root is y, that results from the h equations determined by this hypothesis, has at most

$$n \cdot m \cdot l \cdot k \cdots - (n-a)(m-a')(l-a'')(k-a''') \cdots$$

roots. Therefore as a consequence of the $N-a^{(h)}$ substitutions for V in the above equations there can result not more than

$$(N-a^{(h)})[n \cdot m \cdot l \cdot k \cdots - (n-a)(m-a')(l-a'')(k-a''') \cdots]$$

distinct roots of y. But from the h equations $A' = 0$, $B' = 0$, $C' = 0$, etc., combined with the equation $y^{a^{(h)}} + (\pi x+\rho z+\sigma v+\ldots+\tau)y^{a^{(h)}-1} +$

$\ldots = 0$, there can result at most $a^{(h)} \cdot n \cdot m \cdot l \cdot k \cdots$ distinct roots of y; it follows that the equation whose root is y is of degree

$$(N-a^{(h)})[n \cdot m \cdot l \cdot k \cdots - (n-a)(m-a')(l-a'')(k-a''') \cdots]$$
$$a^{(h)} \cdot n \cdot m \cdot l \cdot k \cdots$$
$$= N \cdot n \cdot m \cdot l \cdot k \cdots - (N-a^{(h)})(n-a)(m-a')(l-a'')(k-a''') \cdots$$

at most.

Or consider the two equations

$(x^p - ax^{p-1} + bx^{p-2} - \ldots)[y^{n-p} + (a'+b'x)y^{n-p-1} + (c'+d'x+e'x^2)y^{n-p-2} + \ldots] = 0$

and

$(x^q + Ax^{q-1} + Bx^{q-2} + \ldots)[y^{m-q} + (A'+B'x)y^{m-q-1} + (C'+D'x+E'x^2)y^{m-q-2} + \ldots]$ $=0$, of degrees n and m. Let α, β, γ, $\ldots$ be the roots of the equation $x^p - ax^{p-1} + bx^{p-2} - \ldots = 0$, and substitute these values for x in the equation $y^{m-q} + (A'+B'x)y^{m-q-1} + (C'+D'x+E'x^2)y^{m-q-2} + \ldots = 0$. Then for each value of x there will be $m-q$ resulting values of y, so for the p values of x (α, β, γ, $\ldots$) there will be altogether $p(m-q)$ values of y. Likewise substitute each of the q roots of the equation $x^q + Ax^{q-1} + Bx^{q-2} + \ldots = 0$ for x in the equation

$$y^{n-p} + (a'+b'x)y^{n-p-1} + (c'+d'x+e'x^2)y^{n-p-2} + \ldots = 0$$

and there will be $q(n-p)$ corresponding values of the quantity y. Also from the two equations

$$y^{n-p} + (a'+b'x)y^{n-p-1} + (c'+d'x+e'x^2)y^{n-p-2} + \ldots = 0$$

and $\quad y^{m-q} + (A'+B'x)y^{m-q-1} + (C'+D'x+E'x^2)y^{m-q-2} + \ldots = 0$

there may result up to $(n-p)(m-q)$ roots or values of the quantity y, which combined with the others gives, in all,

$$p(m-q) + q(n-p) + (n-p)(m-q) = nm - pq$$

values of the quantity y.

It does not matter, in this investigation, whether or not $x^p - ax^{p-1} + bx^{p-2} - \ldots$ or $x^q + Ax^{q-1} + Bx^{q-2} + \ldots$ divides the given equations; but it does matter whether the degrees of the quantities x and y in the terms of the given equations do, or do not, exceed the degrees of the corresponding quantities in these latter equations: for this determines whether the degree of the equation whose root is x or y (of the original equations) does, or does not, exceed the degree of the equation whose root is x or y (of the latter equations).

In the above equations $A' = 0$, $B' = 0$, $C' = 0$, $D' = 0$, $\ldots$, $H' = 0$, let a, a', a'', a''', $\ldots$ be the maximum degrees of the quantity y; then by so much the degree of the equation whose root is y can be lowered, i.e. by $(n-a)(m-a')(l-a'')(k-a''')\cdots$; likewise, let b, b', b'', b''', ..., c, c', c'', c''', ..., d, d', d'', d''', ..., etc. be the maximum degrees of the quantities x, z, v, ... in the same equations respectively, so that the degree of the equation whose root is y can further be lowered by the amounts $(n-b)(m-b')(l-b'')(k-b''')\cdots$, $(n-c)(m-c')(l-c'')(k-c''')\cdots$, $(n-d)(m-d')(l-d'')(k-d''')\cdots$, etc. Therefore the equation whose root is y or ... cannot be of degree higher than

$$\begin{aligned} n\cdot m\cdot l\cdot k\cdots &- (n-a)(m-a')(l-a'')(k-a''')\cdots \\ &- (n-b)(m-b')(l-b'')(k-b''')\cdots \\ &- (n-c)(m-c')(l-c'')(k-c''')\cdots \\ &- (n-d)(m-d')(l-d'')(k-d''')\cdots \\ &- \text{etc.} \end{aligned}$$

And so it is easy, in many cases, to derive the maximum degree of the equation whose root is y from the given equations, from the terms of highest degree in the given equations, using the principle that was laid down earlier.

Now if the terms of highest degree, in the given equations, have a common divisor, or if independent equations having a common divisor of degree r can be formed from them by addition, subtraction, or multiplication by unknown quantities, then the maximum degree of the equation whose root is the unknown quantity y will be reduced by at least the amount r from the amount that was stated before. The degrees of equations, whose roots are unknown quantities, can also be lowered when two or more of their roots turn out to be equal to each other. By similar reasoning it is possible to deduce the number of values of any unknown quantity, in n simultaneous equations involving as many unknowns.

By the method of infinite series given earlier, we can always find the amount by which the degree of the unknown quantities may be lowered.

The equations may be reduced, in order to eliminate unknown quantities, and divisors of the equations may be found by separating the equation so found into two parts, such that we can extract a root of the same order of both parts. This may also be accomplished by transforming the given equations into others, whose roots have a given assignable relation to the roots of the given equations. By this method also two or more terms may be deleted from these equations, exactly as we delete them from an equation having only one unknown quantity. We can indeed find equations which can be reduced in these ways: the given equations are transformed into others of higher degree, and the inverse process transforms the resulting equations back to the given ones, etc. And by the same method which was presented for investigation of the nature of the coefficients of a transformed equation involving only one unknown quantity, the coefficients of transformed simultaneous equations involving several unknown quantities may also be investigated.

Problem XLII

To find two, three, or (h) *equations, having two, three, or* (h) *unknown quantities* (x, y, z, ...) *which can be resolved.*

Case 1. Arbitrary algebraic expressions are assumed for the resolutions of x, y, z, etc., and by the principle given in the preceding chapter the number of distinct values of each irrational quantity in these supposed resolutions is found, which numbers we shall represent by π, ρ, σ, τ, etc. Then h equations are assumed, of the form
$y^n+(a+bx+cz+\ldots)y^{n-1}+(A+Bx+Cz+\ldots+Ex^2+Fxz+Gz^2+\ldots)y^{n-2}+\ldots=0.$
In these equations for the unknown quantities x, y, z, etc. their supposed values are written; then in each resulting equation the coefficients of each irrationality, and of all the rational terms together, are separately equated to zero, and from the simple equations thus derived, the coefficients a, b, c, d, ..., A, B, C, D, ..., etc. emerge.

If there are as many independent resulting linear equations as there are coefficients a, b, c, ..., A, B, C, ..., etc., included among them, then the coefficients can be found from the given equations; if the number of coefficients exceeds the number of independent equations, then as many may be arbitrarily chosen, as the difference between the number of coefficients and the number of independent equations.

These resolutions are often not general solutions of the resulting equations, that is, they will not include all of the sets of values of their unknown quantities.

In these cases it is essential to be certain that no two, three, or more of the h given equations are the same, or in general, that there be no equation which is dependent on the rest of them.

The number of distinct irrationalities contained in any one equation cannot be greater than $\pi\cdot\rho\cdot\sigma\cdot\tau\cdots$.

Case 2. If the values proposed for the different letters x, y, z, ... do not contain the same irrational quantities, then not infrequently it is easier to construct equations of the form

$$Ay^n + ay^{n-1} + \ldots + Bx^n + bx^{n-1} + \ldots + Cz^n + cz^{n-1} + \ldots + \ldots = 0$$

among whose terms are none of the form $y^\alpha x^\beta z^\gamma \cdots$ (where α, β, γ, ... are whole numbers), i.e. if $y = \sqrt{p}+q$, $x = \sqrt{p'}+q'$, $z = \sqrt{p''}+q''$, etc., then the assumed equations should be of the form

$$y^2 + ay + bx^2 + cx + dz^2 + ez + f = 0.$$

In these assumed equations substitute the values that were initially assumed for y, x, and z; the result will be

$$p+q^2+aq+b(p'+q'^2)+cq'+d(p''+q''^2)+eq''+f$$
$$+(2q+a)\sqrt{p}+(2bq'+c)\sqrt{p'}+(2dq''+e)\sqrt{p''} = 0.$$

Equating to zero every term containing an irrationality ($\sqrt{p}$, $\sqrt{p'}$, or $\sqrt{p''}$) produces the equations

$$2q + a = 0 \text{ (whence } a = -2q)$$
$$2bq' + c = 0 \text{ (whence } c = -2bq'\text{, and so } b \text{ may be freely chosen)}$$
$$2bq''+e = 0 \text{ (whence } e = -2dq''\text{, and } e \text{ may be freely chosen)}$$

and finally $$p+q^2+aq+b(p'+q'^2)+cq'+d(p''+q''^2)+eq''+f = 0,$$

from which the coefficient f may be deduced, completing the assumed equation. And similarly two other equations of the same form, different from and independent of the first and of each other, may be assumed, and so there will result three equations in three unknowns, some of whose roots (sets of values for x, y, and z) are the originally given quantities.

If the given values for two or more of the unknown quantities x, y, z, ... involve the same irrationalities, then it is usually necessary to assume equations containing terms of the form $x^\alpha y^\beta \cdots$, where α, β, ... are whole numbers.

Theorem 34

Let

$$y = a\sqrt[n]{p}+b\sqrt[n]{p^2}+c\sqrt[n]{p^3}+\ldots+\sqrt[n]{p^{n-1}}+A\sqrt[m]{q}+B\sqrt[m]{q^2}+\ldots+\sqrt[m]{q^{m-1}}$$
$$+H\sqrt[n]{p}\sqrt[m]{q}+K\sqrt[n]{p}\sqrt[m]{q^2}+L\sqrt[n]{p^2}\sqrt[m]{q}+\ldots$$
$$+\alpha\sqrt[k]{r}+\beta\sqrt[k]{r^2}+\ldots+\pi\sqrt[n]{p^i}\sqrt[m]{q^o}\sqrt[k]{r^u}+\ldots,$$

$$x = a'\sqrt[n]{p} + b'\sqrt[n]{p^2} + c'\sqrt[n]{p^3} + \ldots + \sqrt[n]{p^{n-1}} + A'\sqrt[m]{q} + B'\sqrt[m]{q^2} + \ldots + \sqrt[m]{q^{m-1}}$$
$$+ \alpha'\sqrt[n]{p}\sqrt[m]{q} \quad + K'\sqrt[n]{p}\sqrt[m]{q^2} \qquad + L'\sqrt[n]{p^2}\sqrt[m]{q} \qquad + \ldots$$
$$+ \alpha'\sqrt[k]{r} + \beta'\sqrt[k]{r^2} + \ldots,$$
$$z = a''\sqrt[n]{p} + \ldots$$

etc.,

where the letters a, b, c, $\ldots$, A, B, $\ldots$, H, K, L, $\ldots$, α, β, $\ldots$, π, $\ldots$, etc., and a', b', c', $\ldots$, A', B', $\ldots$, H', K', L', $\ldots$, α', $\ldots$, a'', $\ldots$, etc. and p, q, r, $\ldots$ are given quantities; x, y, z, $\ldots$ are unknown quantities, h in number, and the letters n, m, k, l, o, u, $\ldots$ denote integers.

Compute the sum of the individual degrees $m-1$, $n-1$, $k-1$, $\ldots$, also the sum of all products of two of these numbers, and the sums of all products of three, four, five, etc. of them; call these sums respectively P, Q, R, S, T, etc. Now assume h equations of the form

$$\Gamma y^i + (\delta + \epsilon x + \ldots)y^{i-1} + (\zeta + \eta x + \theta x^2 + \ldots)y^{i-2} + \ldots = 0,$$
$$\Gamma' y^i + (\delta' + \epsilon' x + \ldots)y^{i-1} + \ldots \qquad = 0,$$

etc.,

choosing i sufficiently large that $(i+1)(\frac{i+2}{2})(\frac{i+3}{3})\cdots(\frac{i+h}{h}) - 1$ exceeds or is equal to the sum $P+Q+R+S+\ldots$; then in all of these equations substitute for x, y, z, $\ldots$ their prescribed values. From the resulting equations we can deduce suitable values for the coefficients Γ, δ, ϵ, $\ldots$, Γ', δ', ϵ', $\ldots$, etc., of all h independent equations; thus we can find h independent equations whose unknown quantities x, y, z, $\ldots$ have the values assigned above.

In most cases the quantities x, y, z, $\ldots$ will have more values than just the pre-assigned set.

This method affords a formal solution of two or more equations involving two or more unknowns, that is more general than any other which is presently known.

Theorem 35

Let there be two, three, or more (i.e., h) equations involving two, three, or more (h) unknown quantities and containing irrational expressions. Then by Problem XXVI these equations may be transformed so as to eliminate all irrationalities, and the result will be equations whose resolution is the same as that of the original equations.

Corollary. Conversely, the resulting equations can be depressed back to the given equations, and by substitutions similar to those given earlier in Problem XXX, we can deduce the reduction of two or more equations, in two or more unknowns, to equations of lower degree, in a manner more general than any heretofore known.

Several further points about resolving or reducing equations deserve mention here.

(1) 1. Suppose there are two or more equations having two or more unknown quantities (x, y, z, ...) among which are some from which quantities can be derived, which are either directly contained in the others, or from which quantities in the others can be easily deduced. Then by substituting these quantities for their values in the remaining equations, the desired reduction sometimes results.

2. Often by such substitutions some of the unknowns may be eliminated from the rest of the equations.

(2) By equating different values of the same quantities, selected from different equations, the desired resolution or reduction may often be deduced.

(3) From the addition, subtraction, multiplication, etc. of different equations it is often possible to deduce the value of some quantity which, once found, permits one to deduce the required reduction.

(4) Determine whether or not any of the given equations have divisors, or whether equations having divisors can be generated by the addition, subtraction, multiplication, etc. of some of the given equations (possibly multiplied by arbitrary quantities). If the resulting equations have factors, the degree of the problem can be lowered.

(5) Let there be an equation in two or more (n) unknown quantities x, y, z, ... in which only terms of degrees n and $n-1$, or n, $n-1$, and $n-2$, etc., are present; or suppose such an equation can be deduced

through the addition, subtraction, etc. of given equations. In this equation for x, z, ... write πy, ρy, etc. There results an equation from which the quantity y is easily deduced.

(6) The reduction or resolution of an equation may sometimes be established through various transformations.

(7) From n given equations having n unknown quantities x, y, z, ... the form of the equation whose root is x or y or z or ..., or some function thereof, may sometimes be inferred. If you believe it possible that the solution is of a form in which two or more roots stand in a given relation to each other, then two assumed general quantities, having the stated relation, should be substituted for the respective unknowns in the given equations; and if from both such substitutions the same equations result, then the resolution does indeed have the proposed formula.

Example 1. Let the equations $y^2+xz = a$, $x^2+yz = b$, and $z^2+yx = c$ be given; to determine whether the negative of any solution for x, y, z is also a solution.

In the given equations, for y, x, and z substitute respectively α, λ, and π, as well as their negatives $-\alpha$, $-\lambda$, and $-\pi$. The same equations, $\alpha^2+\lambda\pi = a$, $\lambda^2+\alpha\pi = b$, $\pi^2+\alpha\lambda = c$, result from either substitution, confirming that for each positive value of each unknown quantity there is another value, of equal magnitude but negative in sign, which also satisfies the equation.

Whatever may be the form of the equation whose root is x or y or z, it is independent of the above substitutions.

Example 2. Given the two equations $x^2 - 3y^2 + 7y = a$ and $1-3y^2x^2+7yx^2 = ax^2$, to determine whether the equation whose root is x is a recurrent equation.

When an equation is recurrent, if α is a root then so is $\frac{1}{\alpha}$. Accordingly we write first α, and then $\frac{1}{\alpha}$, for x in the two given equations. From the first substitution result the equations $\alpha^2-3y^2+7y = a$ and $1-3\alpha^2y^2+7\alpha^2y = a\alpha^2$, and from the second, $\frac{1}{\alpha^2}-3y^2+7y = a$ (whence $1-3\alpha^2y^2+7\alpha^2y = a\alpha^2$) and $1-\frac{3y^2}{\alpha^2}+\frac{7y}{\alpha^2} = \frac{a}{\alpha^2}$ (whence

$\alpha^2-3y^2+7y = a$); it follows that the equation whose root is x is recurrent, since the latter two equations are the same as the former two.

(8) If in the given equations are given irrational quantities which are independent of each other, and new assumed quantities are substituted for them, the resolution may often be more easily deduced.

Example. Let there be four equations,

$$(x^\alpha+y^\alpha+z^\alpha+w^\alpha)^m(x^\alpha+y^\alpha+z^\alpha)^n = a,$$
$$(x^\beta+y^\beta+w^\beta)^r(x^\beta+y^\beta+z^\beta+w^\beta)^s = b,$$
$$(x^\alpha+y^\alpha+z^\alpha+w^\alpha)^m(x^\alpha+y^\alpha+z^\alpha)^r = c,$$
$$(x^\alpha+y^\alpha+z^\alpha+w^\alpha)^s(x^\alpha+y^\alpha+z^\alpha)^n = d;$$

then write A, B, C, and D respectively for the irrationalities $(x^\alpha+y^\alpha+z^\alpha+w^\alpha)^m$, $(x^\alpha+y^\alpha+z^\alpha)^n$, $(x^\beta+y^\beta+z^\beta+w^\beta)^s$, and $(x^\beta+y^\beta+z^\beta)^r$, with the results $AB = a$, $CD = b$, $A+D = c$, $B+C = d$, whence A, B, C, and D are readily found, and so ultimately y, x, z, and w.

(9) To find equations which can be depressed: having assumed some equations, transform them into others having higher degrees than the given equations; then the transformed equations are easily depressed to the given ones.

These methods are not significantly different from those given in the previous Chapter, in which is treated the case of a single equation having a single unknown quantity.

(10) On occasion the given equations can be reduced to the form of perfect n^{th} powers, which are then further reduced by extraction of the nth roots.

If it is possible at all, it should be easy to reduce them to perfect powers by addition, subtraction, etc.

(11) The degrees of the unknown quantities in the given equations may be depressed, if the given equations can be transformed into others of the same degree as the given equations, in which

- the terms of highest degree, in those unknown quantities which belong to two or more of the equations, have a common divisor; or,
- terms of degree zero with respect to all unknown quantities vanish; or,
- more generally, some roots become equal to zero; or,
- the given equations can be transformed into others of the same degree, some of whose roots are equal.

So much for methods of depressing given equations, having two or more unknown quantities, into others of lower degree, or into others whose form is of lower degree, or into others of a form whose resolution or further reduction is known.

Example 1. Let the equations be

$$x^n + n(\tfrac{n-1}{2})x^{n-2}y^2 + n(\tfrac{n-1}{2})(\tfrac{n-2}{3})(\tfrac{n-3}{4})x^{n-4}y^4 + \ldots = A,$$

and $x^{n-1}y + (\tfrac{n-1}{2})(\tfrac{n-2}{3})x^{n-3}y^3 + (\tfrac{n-1}{2})(\tfrac{n-2}{3})(\tfrac{n-3}{4})(\tfrac{n-4}{5})x^{n-5}y^5 + \ldots = B;$

then $x+y = \sqrt[n]{A+nB}$, $x-y = \sqrt[n]{A-nB}$. The n values of the sum $x+y$ will always be equal in some order to the n values of the difference $x-y$, so the equation whose root is $x+y$ or $x-y$ cannot be of degree higher than n; but the equation whose root is x or y will be of degree n^2.

Example 2. Let the equations be

$$(x+y)\left[x^{n-1} - ax^{n-2}y + (n-2)(\tfrac{n-1}{2})x^{n-3}y^2 - (\tfrac{n-3}{2})(\tfrac{n-2}{3})ax^{n-4}y^3 + \ldots\right] = A,$$

and $(x+y)\left[x^{n-2}y + (\tfrac{n-3}{2})(\tfrac{n-2}{3})x^{n-4}y^3 + \ldots\right] = B;$

then $x+y = \sqrt[n]{A+(a+(n-1))B}$.

Example 3. Let the equations be

$$(x-y)\left[x^{n-1} - ax^{n-2}y + (n-2)(\tfrac{n-1}{2})x^{n-3}y^2 - (\tfrac{n-3}{2})(\tfrac{n-2}{3})ax^{n-4}y^3 + \ldots\right] = A,$$

and $(x-y)\left[x^{n-2}y + (\tfrac{n-3}{2})(\tfrac{n-2}{3})x^{n-4}y^3 + \ldots\right] = B;$

then $x+y = \sqrt[n]{A+(a-(n-1))B}$.

And so for similar equations in infinite variety.

Example 4. Let the given equations be

$$x^m y^r z^s v^t \cdots = A, \qquad \dot{x}^m \dot{y}^r \dot{z}^s \dot{v}^t \cdots = B,$$

$$\ddot{x}^m \ddot{y}^r \ddot{z}^s \ddot{v}^t \cdots = C, \qquad \dddot{x}^m \dddot{y}^r \dddot{z}^s \dddot{v}^t \cdots = D,$$

etc.

The equations whose roots are x, y, z, v, ... are easily seen to be simple in form, and by adding them, subtracting them, etc., we can formulate an infinity of equations which can then easily be reduced back to these equations. [6]

Theorem 36

(1) Let there be h equations,

$p + q + r + s + t + \ldots = 0,$ $\quad p' + q' + r' + s' + t' + \ldots = 0,$

$p'' + q'' + r'' + s'' + t'' + \ldots = 0,$ $\quad p''' + q''' + r''' + s''' + t''' + \ldots = 0,$

etc.

in the same number of unknowns x, y, z, v, w, etc. Further, in these equations let p, q, r, s, t, ... be homogeneous functions of the unknown quantities x, y, z, ..., of degrees n, $n-m$, $n-2m$, $n-3m$, etc. respectively; and in like manner p', q', r', s', ..., p'', q'', r'', ... p''', q''', r''', ..., etc. are homogeneous functions of the same unknown quantities x, y, z, ..., of degrees n', $n'-m'$, $n'-2m'$, $n'-3m'$, ..., n'', $n''-m''$, $n''-2m''$, $n''-3m''$, ..., n''', $n'''-m'''$, $n'''-2m'''$, $n'''-3m'''$, ..., etc. Then the equation whose root is x will have the form $x^\alpha + Px^{\alpha-m} + Qx^{\alpha-2m} + Rx^{\alpha-3m} + \ldots = 0$. The equations whose roots are y, z, ... will have similar formulas. (By homogeneous functions of the quantities x, y, z, ... I mean functions in which each term is of the same total degree in all of the unknown quantities.)

(2) Let there be two equations, $x^n - px^{n-1} + qx^{n-2} - \ldots = 0$ and $y^m - Py^{m-1} + Qy^{m-2} - \ldots = 0$, where p, q, ... and P, Q, ... are given quantities. If from these two equations can be derived two equations, each involving both x and y, whose roots are the same as the roots of the given equations, then in the resulting equations there will always be n identical values of y corresponding to each single value of x, and m identical values of x corresponding to each single value of y.

Theorem 37

Case 1. Let there be two (or more) equations, $A = 0$ and $B = 0$, involving symmetrically their two unknown quantities x and y; and these two equations are to be reduced to one, to eliminate the second unknown, y. Then the degree of the resulting equation, whose root is x, will be twice the degree of an equation whose root is $x+y$, x^2+y^2, x^3+y^3, xy, or indeed any rational function in which the quantities x and y are symmetrically involved.

Proof. Let the roots of the unknown quantity x be respectively α, β, γ, δ, ... π, ρ, σ, τ; then the corresponding roots of the unknown quantity y will be τ, σ, ρ, π, ... δ, γ, β, α. The values of the unknown quantity $z = x+y$ will be $\alpha+\tau$, $\beta+\sigma$, $\gamma+\rho$, $\delta+\pi$, ... $\pi+\delta$, $\rho+\gamma$, $\sigma+\beta$, and $\tau+\alpha$. Evidently where α is a chosen value of x and τ is the value of y that corresponds to this choice of x, so will α be the value of y corresponding to the choice of τ for x. Because of the symmetry, if in the equations $A = 0$ and $B = 0$ we substitute α for x and τ for y, the result will be the same as if we substituted τ for x and α for y. Hence for every value of the compound unknown quantity $z = x+y$ there is another one equal to it ($\alpha+\tau = \tau+\alpha$, $\beta+\sigma = \sigma+\beta$, etc.) And so either unknown quantity (x or y) will have twice as many values as will the unknown quantity $x+y$, and the degree of the equation whose root is x (or equivalently y) will be twice as great as the degree of the equation whose root is $x+y$.

Example 1. Let the two equations be $(x^2+y^2)(x^4+y^4) = b$ and $(x^2+y^2)(x+y) = a$. Now x^2+y^2 is a common divisor of the terms, in x and y, of highest degree in both equations, so it follows that the degree of the equation whose root is x or y will not be $6\times 3 = 18$: rather it will be the same as the degree of the equation whose root is x or y, derived from the two equations $(x^2+y^2)(x+y) = a$ and $a(x^4+y^4) = b(x+y)$ by substituting for x^2+y^2, in the first equation, its value $\frac{a}{x+y}$ deduced from the second — and that degree is 12. But also x and y are symmetrically involved in these two equations, so the equation whose root is $z = x+y$ will be only of degree 6, and by the preceding theorem evidently it will be quadratic in form, *viz.* $az^6+(2b-2a^2)z^3-a^3 = 0$.

Example 2. Let the two equations be

$$x^4+x^3y+x^2y^2+xy^3+y^4 = a$$

and $(x^4+x^3y+x^2y^2+xy^3+y^4)(x^4-x^3y+x^2y^2-xy^3+y^4) = b$;

the terms of highest degree in x and y together, in these equations, have the common divisor $x^4+x^3y+x^2y^2+xy^3+y^4$, and so the equation whose root is x (or whose root is y) has not $4 \times 8 = 32$, but $4 \times 4 = 16$, for its degree. Now let α, β, γ, and δ be the roots of x in the equation $x^4+x^3y+x^2y^2+xy^3+y^4 = a$, and substitute α, β, γ, δ in turn for x in the equation

$$(x^4+x^3y+x^2y^2+xy^3+y^4)(x^4-x^3y+x^2y^2-xy^3+y^4) = b,$$

and there result the equations

$$a(\alpha^4-\alpha^3y+\alpha^2y^2-\alpha y^3+y^4)-b = 0,$$
$$a(\beta^4-\beta^3y+\beta^2y^2-\beta y^3+y^4)-b = 0,$$
etc.,

and it follows that the degree of the equation whose root is y must be $4 \times 4 = 16$. Stated otherwise, in the latter equation we substitute for $x^4+x^3y+x^2y^2+xy^3+y^4$ the value a deduced from the former equation. Similar to what we showed before, the equation whose root is x (or y) has the form of an equation of the eighth degree, i.e. its negative roots are equal in magnitude to its positive roots. At the same time, these two equations both involve the unknown quantities x and y symmetrically, and so the equation whose root is $z = x^2+y^2$ cannot be of degree greater than $\frac{16}{4} = 4$, and furthermore it will be quadratic in form, *viz.*

$$4a^2z^4 - (2a^2+2b)az^2 = (a^2-b)^2.$$

Example 3. Let the two equations be

$$(x^4+y^4)(x^2+y^2) = bx^2y^2 \quad \text{and} \quad (x^2+y^2)(x+y) = axy.$$

Since in these two equations x^2+y^2 will be a quadratic common divisor of the terms of highest degree (six and three respectively), twelve roots will turn out to be zero (because, where the terms of degree zero or one in the lower-degree equation vanish, the terms of degrees zero, one, two, and three in the other must also vanish). Also, xy is a divisor of the terms of degree two in one of the equations, and of degree four in the other; therefore the equation whose root is x or y will be of degree $6 \times 3-12-2 = 4$. Hence, since x and y are symmetrically involved, the equation whose root is $x+y = z$ will be only of degree two: specifically, it will be $z^2+\frac{b}{a}z+\frac{b-a^2}{2} = 0$.

In other words, these two equations could be written as

$$a(x^4+y^4) = bxy(x+y) \quad \text{and} \quad (x^2+y^2)(x+y) = axy,$$

in which eight roots of the quantities x and y are found to vanish.

Example 4. Let the two equations be $x^3+x^2y+xy^2+y^3 = axy$, and $(x^3+x^2y+xy^2+y^3)(x^6-x^5y+x^3y^3-xy^5+y^6) = bx^3y^3$. By the method of reasoning just described, we see that the equation whose root is x or y can be of degree $9 \times 3 - 9 \times 2 - 3 = 6$, and so the equation whose root is $x+y = z$ will be of degree 3; in fact it is

$$z^3 - az^2 - \left(\frac{a^2}{3}+\frac{2b}{3a}\right) z + \frac{a^3}{3} - \frac{b}{3} = 0.$$

Example 5. Let the two equations be $x^3+x^2y+xy^2+y^3 = axy$, and $x^{3n}+x^{2n}y^n+x^ny^{2n}+y^{3n} = bx^ny^n$. Then, where m denotes an integer,

first, if $n = 6m+3$, the equation whose root is x or y will be of degree $3n \times 3 - 3n \times 2 - 3 = 3n - 3 = 18m + 6$, the terms of highest degree in the two equations have a common divisor of degree 3, and the equation whose root is $x+y$ is of degree $9m + 3$;

second, if $n = 2m$, the equation whose root is x or y is of degree $3n \times 3 - 3n \times 2 - 2 = 3n - 2 = 6m - 2$, the most significant terms have a quadratic common divisor, and the equation whose root is $x+y$ is of degree $3m - 1$; and

third, if $n = 6m\pm1$, the equation whose root is x or y is of degree $3n - 1$, and the equation whose root is x or y is of degree $\frac{3n-1}{2} = \frac{18m \pm 3 - 1}{2} = (9m - 2 \text{ or } 9m + 1)$.

And in general the equation whose root is $x+y$ will be

$$z^n+(a-z)^n-\frac{nz^3}{2z+a}[z^{n-2}+(a-z)^{n-2}]+\frac{n(n-3)}{2}\cdot\frac{z^6}{(2z+a)^2}[z^{n-4}-(a-z)^{n-4}]+\ldots = b.$$

The rule for this series is developed in Problem I.

Example 6. Let the two equations be $x^2+zx+z^2+zy+y^2 = az$, where $z^2=xy$, and $x^4+x^3y+x^2y^2+xy^3+y^4 = bxy$. It is easy to see that the equation whose root is z is only of degree 2; it will be

$$z^2 + \left(\frac{a}{2}+\frac{b}{2a}\right) z - \left(\frac{a}{2}-\frac{b}{2a}\right)^2 = 0.$$

And the equation $v = \frac{a}{2} - \frac{b}{2a}$, whose root is $v = x+y$, has two equal roots.

Similarly let the equations be $x^5+x^4y+x^3y^2+x^2y^3+xy^4+y^5 = ax^2y^2$, and $x^{10}+x^8y^2+x^6y^4+x^4y^6+x^2y^8+y^{10} = bx^4y^4$; the equation whose root is x or y has degree $5 \times 10 - 5 \times 8 = 10$, and the equation whose root is $x+y$ has degree $\frac{10}{2} = 5$.

Also let the equations be

$$x^7+x^6y+x^5y^2+x^4y^3+x^3y^4+x^2y^5+xy^6+y^7 = ax^3y^3,$$

and $x^{14}+x^{12}y^2+x^{10}y^4+x^8y^6+x^6y^8+x^4y^{10}+x^2y^{12}+y^{14} = bx^6y^6$;

the equation whose root is x or y has degree 14, and therefore the equation whose root is $x+y$ is of degree 7.

And so *ad infinitum* for successively higher degrees — but enough of this.

Case 2. Given two equations $A = 0$ and $B = 0$, in which x and y appear symmetrically, and again with $\alpha, \beta, \gamma, \delta, \ldots \pi, \rho, \sigma, \tau$ represent the roots of the quantity x, and corresponding to them the roots of the unknown y are $\tau, \sigma, \rho, \pi, \ldots \delta, \gamma, \beta, \alpha$ respectively. Then the corresponding values of $x-y$ will be respectively $\alpha-\tau$, $\beta-\sigma$, $\gamma-\rho$, $\delta-\pi$, $\ldots$ $\pi-\delta$, $\rho-\gamma$, $\sigma-\beta$, $\tau-\alpha$; hence the number of roots of the quantity $x-y$ will be equal to the number of roots of the quantity x or y. But to every positive root of the quantity $x-y$ corresponds a negative root of equal magnitude, i.e. the positive root $\alpha-\tau$ corresponds to the negative root $\tau-\alpha$, which has the same absolute value — and likewise $\beta-\sigma$ and $\sigma-\beta$, $\gamma-\rho$ and $\rho-\gamma$, $\delta-\pi$ and $\pi-\delta$, etc. It follows that the equation whose root is $x-y$, of which the magnitudes of the positive and negative roots are all equal in pairs, has the form of an equation whose degree is half that of the equation whose root is x (or y).

The same may be asserted, in general, of any rational function $P(x,y)$, in whose square, P^2, x and y are symmetrically involved. In such a case there will be pairs of values P and $-P$ which are equal in magnitude but opposite in sign; therefore the equation whose root is P has the form of an equation whose degree is half that of the equation whose root is either x or y.

Now if the given function consists of several fractions, let them be reduced to a common denominator, and then if in the numerator and the denominator (or in their squares) the quantities x and y are symmetrically involved, the equation whose root is either x or y has twice the degree of the equation whose root is the given function. For instance, if the function is $\frac{x}{y}+\frac{y}{x}$, reduced to a common denominator it becomes $\frac{x^2+y^2}{yx}$, a fraction in which x and y are symmetrically involved.

The equation whose root is x, deduced from an equation in which x and y are symmetrically involved, contains an even number of roots, equal or unequal, unless some value of x has a corresponding identical value of y.

The principle may be extended to more equations in which more unknowns are symmetrically involved.

Case 3. Given the two equations $A = 0$ and $B = 0$ as before, let C be any rational algebraic function of the letters x and y. In C for x everywhere write y, and for y write x, and call the resulting expression D. Assume $\frac{C}{D} = z$: then the equation whose root is z is recurrent, and consequently the equation whose root is $v = \frac{C}{D}+\frac{D}{C}$ has only half the degree of the equation whose root is either x or y.

This is established by the reasoning employed in the preceding examples.

Theorem 38

(1) $x^n+y^n = z^n-nz^{n-2}v+\frac{n(n-3)}{2!}z^{n-4}v^2-\frac{n(n-4)(n-5)}{3!}z^{n-6}v^3$
$+\frac{n(n-5)(n-6)(n-7)}{4!}z^{n-8}v^4-\ldots$, where $z = x+y$ and $v = xy$. [7]

(2) Let there be any irrational expression which symmetrically involves the quantities x, y, z, etc. If this quantity is reduced, by the rules of Problem XXVI, so as to eliminate irrationalities, the result is a rational quantity which remains symmetric in the said quantities x, y, z, etc.

(3) Every integral power, of a quantity in which the letters x, y, z, etc. are symmetrically involved, will also symmetrically involve those quantities.

Theorem 39

If there are any n equations in n unknown quantities, and x, y, z, v, ... are any m of the unknown quantities appearing symmetrically to the same powers in all of the equations, while X, Y, Z, V, ... are r unknowns appearing symmetrically to the same powers in one of the equations, and α, β, γ, δ, ... are s unknowns appearing symmetrically to the same powers in another single equation, etc. Then the degree of the equation whose root is

$$x+y+z+v+\ldots+X+Y+Z+V+\ldots+\alpha+\beta+\gamma+\delta+\ldots+\text{etc.}$$

— or any rational function which symmetrically involves all of x, y, z, v, ... and X, Y, Z, V, ... and α, β, γ, δ, ..., etc. — will be in ratio to the degree of the equation whose root is

$$x+y+\ldots+X+Y+\ldots+\alpha+\beta+\ldots+\text{etc.}$$

— or any rational function in which the m' quantities x, y, ... and the r' quantities X, Y, ... and the s' quantities α, β, ..., etc., are symmetrically involved — as 1 is in ratio to

$$m\cdot\frac{m-1}{2}\cdot\frac{m-2}{3}\cdots\frac{m-m'+1}{m'}\times r\cdot\frac{r-1}{2}\cdot\frac{r-2}{3}\cdots\frac{r-r'+1}{r'}$$
$$\times s\cdot\frac{s-1}{2}\cdot\frac{s-2}{3}\cdots\frac{s-s'+1}{s'}\times\cdots$$

(but if $m'=0$, for $m\cdot\frac{m-1}{2}\cdot\frac{m-2}{3}\cdots\frac{m-m'+1}{m'}$ write 1, and likewise for r', s', etc.)

Example. Let the number of unknown quantities α, β, γ, δ, ϵ, ... be n, and let the n given equations be

$$\alpha+\beta+\gamma+\delta+\epsilon+\ldots=a,$$
$$\alpha^2+\beta^2+\gamma^2+\delta^2+\epsilon^2+\ldots=b,$$
$$\alpha^3+\beta^3+\gamma^3+\delta^3+\epsilon^3+\ldots=c,$$
$$\alpha^4+\beta^4+\gamma^4+\delta^4+\epsilon^4+\ldots=d,\text{ etc.}$$

Then the degree of the equation whose root is $\alpha+\beta+\gamma+\delta+\ldots=a$ will be, to the degree of the equation whose root is α, as $1{:}n$. Therefore the equation whose root is α will be of degree n; in fact it will be

$$\alpha^n-a\alpha^{n-1}+\frac{a^2-b}{2!}\alpha^{n-2}-\frac{a^3-3ab+2c}{3!}\alpha^{n-3}$$
$$+\frac{a^4-6a^2b+8ac-6d+3b^2}{4!}\alpha^{n-4}-\ldots=0.$$

(The law of this series comes from Example 1, Problem V.)

And so an infinity of examples of this sort may be analyzed, since an infinity of examples similar to this may be constructed.

The same, *mutatis mutandis*, may be asserted of any rational function P, in whose square the quantities x, y, ..., X, Y, ..., α, β, ...

etc. are within each group symmetrically involved, insofar as for every positive root of the equation whose root is P there is a negative root of the same magnitude.

Theorem 40

Case 1. Let $A = 0$ be an equation containing the two unknown quantities x and y. In it write $\frac{1}{x}$ for x, and the result will be the equation $B = 0$; from the two equations $A = 0$ and $B = 0$ an equation whose root is x is to be deduced. Then the equation so deduced will have two roots $x = +1$ and $x = -1$, while the rest of the roots will all be contained in the recurrent equation $x^n+ax^{n-1}+bx^{n-2}+\ldots+bx^2+ax+1 = 0$.

Proof. (1) Let α be a root, or a value of the root x, in the equations $A = 0$ and $B = 0$. Then $\frac{1}{\alpha}$ is another value of the root x. For by substituting α for x in the equations $A = 0$ and $B = 0$ we obtain the results $\pi = 0$ and $\sigma = 0$; then by substituting $\frac{1}{\alpha}$ instead of α in the same two equations we obtain the results $\sigma = 0$ and $\pi = 0$. Therefore, if α is a value of x that satisfies the two equations $A = 0$ and $B = 0$, the value $\frac{1}{\alpha}$ also satisfies the same two equations, and thus the equation whose root is x is equal to the product

$$(x-\alpha)(x-\frac{1}{\alpha})(x-\beta)(x-\frac{1}{\beta})(x-\gamma)(x-\frac{1}{\gamma})\cdots = 0$$

which is a recurrent equation.

(2) To show that in addition x takes the values $+1$ and -1, write $x = 1$ in both $A = 0$ and $B = 0$, and the result will be the same equation $\sigma = 0$; then from the number of roots of the quantity y contained in the resulting equation the multiplicity of the root $x = 1$ can be determined, for each such value of y takes the corresponding value $x = 1$. In consequence, there will be as many roots of x having the value $+1$ as there are values of y contained in the equation $\sigma = 0$.

In like manner the multiplicity of the value $x = -1$ can be found.

(3) Let $x^n + ax^{n-1} + bx^{n-2} + \ldots + bx^2 + ax + 1 = 0$ be a recurrent equation. Then,

first, if n is an even number, the equation whose root is $z = x+\frac{1}{x}$ is of degree $\frac{1}{2}n$: for if α is one such value of x, then $\frac{1}{\alpha}$ is another, but if $x = \alpha$ then $z = \alpha+\frac{1}{\alpha}$, and if $x = \frac{1}{\alpha}$ then $z = \frac{1}{\alpha}+\frac{1}{\frac{1}{\alpha}} = \frac{1}{\alpha}+\alpha$, so the same value $(\alpha+\frac{1}{\alpha})$ corresponds to the two values α and $\frac{1}{\alpha}$ of x;

second, if n is an odd number then one value of the quantity x is -1, and the corresponding value of $z = x+\frac{1}{x}$ is $-1-\frac{1}{1} = -2$; and since in all other cases one value of $z(= \alpha+\frac{1}{\alpha})$ corresponds to two values $(\alpha$ and $\frac{1}{\alpha})$ of x, the total number of values of z will be $\frac{n-1}{2}$ in addition to the value $z = -2$ that was just derived.

The same assertion holds for the degree of any equation whose root z is any rational function of x, the root of a recurrent equation, provided that when $\frac{1}{x}$ is written in place of x in the given function, the resulting expression is the same.

Case 2. (1) In an equation $A = 0$ which contains two unknown quantities x and y, write $a-x$ in place of x to produce an equation $B = 0$. Then find the equation whose root is x from the two equations $A = 0$ and $B = 0$. One value of x which satisfies the equation so deduced will be $x = \frac{a}{2}$; the rest of the roots are contained in an equation of the form

$$(x-\alpha)(x-a+\alpha)(x-\beta)(x-a+\beta)(x-\gamma)(x-a+\gamma)(x-\delta)(x-a+\delta)\cdots = 0.$$

It is easy to show, by writing α and $a-\alpha$ for x in the two equations $A = 0$ and $B = 0$, that $a-\alpha$ is a root or value of x in the said equations whenever α is a root of the same quantity x.

To show that one value of x will be $\frac{a}{2}$ simply write $\frac{a}{2}$ in place of x in both $A = 0$ and $B = 0$: the same equation $\sigma = 0$ results in both cases. As before, the multiplicity of the value $x = \frac{a}{2}$ is equal to the degree of y in the equation $\sigma = 0$.

(2) Let v be any rational function of the quantities x and $a-x$, in which x and $a-x$ are symmetrically involved; then the equation whose root is v has at most half as many roots as the equation whose root is x or $a-x$.

If in any such function v, in place of x we write either α or $a-\alpha$, the same quantity results; consequently one value of the function v corresponds to two values, α and $a-\alpha$, of the quantity x.

Let u be a rational function of the quantities x and $a-x$, in which x and $a-x$ are symmetrically involved to a factor of -1 *pres*, e.g., $x^m(a-x)^r$ and $x^r(a-x)m$ have opposite signs (i.e. if the $+$ sign is affixed to one, the $-$ sign will be affixed to the other): then the equation whose root is u consists entirely of pairs of roots equal in magnitude and opposite in sign.

This easily emerges, when α and $a-\alpha$ are written in place of x.

(3) In the function v let $a-x$ be written for x, and x for $a-x$, and call the result w. Then the equation whose root is $u = \frac{v}{w^2}$ is recurrent, and the degree of the equation whose root is $\frac{v}{w}+\frac{w}{v}$ is only half that of the equation whose root is x or y.

Case 3. More generally, let x be any function of the quantity z, *viz.* $x = \phi(z)$, such that z is the same function of x (i.e. $z = \phi(x)$.) Then in a given equation $A = 0$, substitute for x its value $\phi(z)$, and in the resulting equation for z write $\phi(x)$ and call the result $B = 0$. Then the values of the quantity x in the assumed equation $z = \phi(x)$ will be the same as the values of x that satisfy the two assumed equations $A = 0$ and $B = 0$, and so if α is one value of the quantity x, then $\phi(\alpha)$ will be another value of x.

Let v be a rational function of x and $\phi(x)$ in which x and $\phi(x)$ enter symmetrically. Then the equation whose root is v can have at most half as many roots as the equation whose root is x or $\phi(x)$.

If in the stated function v the signs of the quantities $\alpha^m(\phi(\alpha))^r$ and $\alpha^r(\phi(\alpha))^m$ are consistently opposite, then the equation whose root is v has pairs of roots of equal magnitude, one positive and the other negative.

If in any rational function v of the quantities x and $\phi(x)$, $\phi(x)$ is substituted for x and x is written in place of $\phi(x)$, the resulting quantity being w, then the equation whose root is $\frac{w}{v}$ will be recurrent; and so on.

Case 4. Let $A = 0$ be any equation expressing the relationship between the two unknown quantities x and y. In the equation $A = 0$ for x y is written, and x is written for y; call the resulting equation $B = 0$. Then the equation whose root is $x+y$ (or any rational function

π, in which x and y are symmetrically involved) can never have more than half as many roots as the equation whose root is the unknown quantity x or y.

If in such a function π the signs of the quantities $x^m y^r$ and $x^r y^m$ are always contrary, then the equation whose root is π has pairs of roots of equal magnitude with contrary signs.

If for x we write y, and for y we write x, in the rational function $\pi(x,y)$, resulting in a quantity ρ, then the equation whose root is $\frac{\pi}{\rho}$ will be recurrent.

Case 5. Let $A = 0$, where A is a function of the quantities x and y; and assume $x = \phi(x,y)$. In the given equation $A = 0$ for x substitute the assumed function $\phi(x,y)$, and let the resulting equation be $B = 0$. Then the values of the unknown quantities x and y in the equations $A = 0$ and $x = \phi(x,y)$ will be a subset of the values of the same quantities x and y that satisfy the two equations $A = 0$ and $B = 0$.

Case 6. Let $A = 0$ as before, and let an equation $P = 0$ be assumed, expressing the relation between the two unknown quantities x and y of the equation $A = 0$ and a new quantity z as well. Reduce the two equations $A = 0$ and $P = 0$ to one, eliminating the unknown quantity x, and call the resulting equation $B = 0$. Now write z in place of x in the two equations $P = 0$ and $B = 0$, and call the two resulting equations $Q = 0$ and $C = 0$. Then the roots of the unknown quantities x and y in the equations $A = 0$ and $Q = 0$ will be a subset of the roots of the same quantities in the equations $A = 0$ and $C = 0$.

Case 7. Let $A = 0$ be an equation containing the unknown quantities x and y, and let $Q = 0$ be an assumed equation expressing the relation between the quantities x and z, in which these two quantities enter symmetrically. Then let the two equations $A = 0$ and $Q = 0$ be reduced to one, $L = 0$, whereby the unknown quantity x is eliminated, and in the equation $L = 0$ let x be written in place of z, calling the resulting equation $B = 0$. Reduce the equation $B = 0$ in order to eliminate any irrationalities, the equation resulting therefrom being $C = 0$. Then in the equation $Q = 0$, for z write x; the roots of x in the resulting equation will be a subset of the values of x that satisfy the two equations $A = 0$ and $C = 0$. If in the equation $Q = 0$ α and β are corresponding values of the quantities x and z, and if α is a value of x

that satisfies the equations $A = 0$ and $B = 0$, then β is another root of x satisfying these same two equations. Moreover, similar propositions may also be deduced about more than two equations having more than two unknown quantities.

Case 8. If in a given equation $A = 0$ involving two unknown quantities x and y we write x for y and y for x to produce an equation $B = 0$, then the equation whose root is x or y has twice the degree of the equation whose root is $x+y$ or any other rational function in which the quantities x and y are symmetrically involved. For if, in the equations $A = 0$ and $B = 0$, the corresponding values α and π, of x and y respectively, are written in place of x and y, and then conversely π and α are written in place of x and y, the equations $P = 0$ and $Q = 0$ result from the first substitution, while $Q = 0$ and $P = 0$ result from the second (the same two equations, in different order). Consequently if α is a root of x, π will be another root, and so the two values $\alpha+\pi$ and $\pi+\alpha$ are seen to be equal, proving the proposition.

The same applies to three or more equations involving symmetrically three or more unknowns, and also, *mutatis mutandis*, to the corresponding aspects of the preceding cases.

If in the square of the equation whose root is a rational function of x and y, x and y are symmetrically involved, then the equation whose root is the square of the said functions in the preceding cases will have half as many roots as the equation whose root is x or y.

Problem XLIII

(1) *Let* A *and* B *be functions of the letters* x, y, *etc.; to determine whether* A *is a function of the quantity* B.

Assume $A = v$ and $B = w$, and reduce the two resulting equations to one in order to eliminate x. If in this process $y, \ldots$ all vanish, then A will be a function of B; in any other case it will not.

(2) *Let* A *and* B *be two quantities, and let* P = 0, Q = 0, R = 0, ... *be equations, where* A, B, P, Q, R, *etc. are functions of the unknown quantities* x, y, z, *etc.; to determine whether the quantity* A *is a function of the quantity* B.

Assume $A = v$, $B = w$, and $P = 0$, $Q = 0$, $R = 0$, etc. Reduce all of these equations to one, so that the unknown quantities x, y, z, etc. vanish: if all the unknown quantities vanish at once, except for the quantities v and w, then v will be a function of w, and otherwise it will not.

Theorem 41*

Let the equation be $x^n - px^{n-1} + qx^{n-2} - rx^{n-3} + sx^{n-4} - tx^{n-5} + \ldots = 0$, and further let $p' = P$, $pP - q' = Q$, $pQ - qP + r' = R$, $pR - qQ + rP - s' = S$, $pS - qR + rQ - sP + t' = T$, and so on. If a, b, c, d, e, ... are sums of the given equation's roots, the squares of its roots, their cubes, their quadrato-quadrates, etc., then $\dot{a} = P$, $\frac{1}{2}\dot{b} = Q$, $\frac{1}{3}\dot{c} = R$, $\frac{1}{4}\dot{d} = S$, $\frac{1}{5}\dot{e} = T$, etc., provided that in the fluxions $\dot{a}$, $\dot{b}$, $\dot{c}$, $\dot{d}$, $\dot{e}$, ... for $\dot{p}$, $\dot{q}$, $\dot{r}$, $\dot{s}$, $\dot{t}$, etc. we write respectively p', q', r', s', t', etc.

The proof is elementary.

Scholium

In studying the roots of several equations having two or more unknowns x, y, etc., very often all of the preceding methods are futile; then we must have recourse to approximations.

(1) For example let $y^n + (a + bx)y^{n-1} + \ldots = 0$ and $y^m + (A + Bx)y^{m-1} + \ldots = 0$ be two given equations in two unknown quantities x and y, and let the corresponding roots of x and y be respectively α, β, γ, δ, ... and π, ρ, σ, τ, ... In order through approximation to find the corresponding values α and π of the unknown quantities x and y, it is necessary first that two quantities k and l be given, much nearer to the particular corresponding roots α and π than they are to any other of the roots (β, γ, δ, ...; ρ, σ, τ, ...) of the quantities x and y. In the given equations for x and y write respectively $z - k$ and $v - l$, and let the resulting equations be

$$\Gamma + \Delta z + \mathrm{E}v + \mathrm{Z}z^2 + \mathrm{H}zv + \Theta v^2 + \ldots = 0,$$

and $$\Pi + \mathrm{P}z + \Sigma v + \mathrm{T}z^2 + \Upsilon zv + \Phi v^2 + \ldots = 0,$$

and from the constitution of the coefficients we can see that $\Gamma+\Delta z+\mathrm{E}v$ is a very near approximation to 0, as is $\Pi+\mathrm{P}z+\Sigma v$, whence the approximation to z will be $\frac{\mathrm{E}\Pi-\Sigma\Gamma}{\Delta\Sigma-\mathrm{EP}}$, and the approximation to v will be $\frac{\Delta\Pi-\mathrm{P}\Gamma}{\mathrm{PE}-\Delta\Sigma}$.

Now if k and l are much better approximations to two corresponding roots of the unknown quantities x and y respectively, than to any others, and we substitute $z-k$ and $v-l$ for x and y in the given equations, then from the constitution of the coefficients we can take either

$$\Gamma + \Delta z + \mathrm{E}v = 0 \qquad \text{(nearly)}$$

$$\text{and } \Pi + \mathrm{P}z + \Sigma v + \mathrm{T}z^2 + \Upsilon zv + \Phi v^2 = 0 \qquad \text{(nearly)},$$

or

$$\Pi + \mathrm{P}z + \Sigma v = 0 \qquad \text{(nearly)}$$

$$\text{and } \Gamma + \Delta z + \mathrm{E}v + \mathrm{Z}z^2 + \mathrm{H}zv + \Theta v^2 = 0 \qquad \text{(nearly)},$$

from which equations approximations to the quantities z and v can be found. And so we can progress, having values of k and l that are much closer to three or more of the corresponding pairs of roots of the unknown quantities than to any of the rest. And if these continuing approximations can be shown to become perpetually much better, then from this we can establish that at last they will differ from the true roots by less than any prescribed amount.

This also establishes a method of finding the ratio which the approximation has to the true root, expressed in terms of the different roots of the unknown quantities.

(2) To find successive approximations to an impossible root $a+b\sqrt{-1}$ in one equation with only one unknown quantity, or to find impossible corresponding roots of the unknowns in two or more equations having two or more unknown quantities, it is necessary to have an approximation that is much closer to the real part, and another that is much closer to the imaginary part of one root, than to the real or imaginary part of any other root. And by essentially the same method that was just given, successively better aproximations to both the real and imaginary parts can be generated. In this sense everything that was previously said regarding real roots applies also to impossible roots; equally, all of the preceding remarks apply to n equations involving n unknowns.

Example. Let the equation be $x^n-px^{n-1}+qx^{n-2} - \ldots = 0$, and let $a+b\sqrt{-1}$ be an approximation to an impossible root of the given equation. In the given equation for x write $a+a'+(b+b')\sqrt{-1}$; there results

$$
\begin{aligned}
&a^n - \frac{n(n-1)}{2}a^{n-2}b^2 + \frac{n(n-1)(n-2)(n-3)}{4\,!}a^{n-4}b^4 - \dots \\
&\quad -pa^{n-1} + \frac{(n-1)(n-2)}{2\,!}pa^{n-3}b^2 - \frac{(n-1)(n-2)(n-3)(n-4)}{4\,!}pa^{n-5}b^4 + \dots \\
&\quad +qa^{n-2} - \frac{(n-2)(n-3)}{2\,!}qa^{n-4}b^2 + \dots \qquad = A \\
&+ \Bigg(na^{n-1}b - \frac{n(n-1)(n-2)}{3\,!}a^{n-3}b^3 + \frac{n(n-1)(n-2)(n-3)(n-4)}{5\,!}a^{n-5}b^5 - \dots \\
&\qquad -(n-1)pa^{n-2}b + \frac{(n-1)(n-2)(n-3)}{3\,!}pa^{n-4}b^3 \\
&\qquad - \frac{(n-1)(n-2)(n-3)(n-4)(n-5)}{5\,!}pa^{n-6}b^5 + \dots \\
&\quad +(n-2)qa^{n-3}b - \frac{(n-2)(n-3)(n-4)}{3\,!}qa^{n-5}b^3 + \dots - \dots \Bigg)\sqrt{-1} \qquad = B\sqrt{-1} \\
&+ \Bigg(na^{n-1} - \frac{n(n-1)(n-2)}{2\,!}a^{n-3}b^2 + \frac{n(n-1)(n-2)(n-3)(n-4)}{4\,!}a^{n-5}b^4 - \dots \\
&\qquad -(n-1)pa^{n-2} + \frac{(n-1)(n-2)(n-3)}{2\,!}pa^{n-4}b^2 \\
&\qquad - \frac{(n-1)(n-2)(n-3)(n-4)(n-5)}{4\,!}pa^{n-6}b^4 + \dots \\
&\quad +(n-2)qa^{n-3} - \frac{(n-2)(n-3)(n-4)}{2\,!}qa^{n-5}b^2 + \dots - \dots \Bigg)a' \qquad = Ca' \\
&- \Bigg(n(n-1)a^{n-2}b - \frac{n(n-1)(n-2)(n-3)}{3\,!}a^{n-4}b^3 \\
&\qquad + \frac{n(n-1)(n-2)(n-3)(n-4)(n-5)}{5\,!}a^{n-6}b^5 - \dots \\
&\quad -(n-1)(n-2)pa^{n-3}b + \frac{(n-1)(n-2)(n-3)(n-4)}{3\,!}pa^{n-5}b^3 - \dots \\
&\quad +(n-2)(n-3)qa^{n-4}b - \frac{(n-2)(n-3)(n-4)(n-5)}{3\,!}qa^{n-6}b^3 + \dots - \dots \Bigg)b' = Eb' \\
&+ \Bigg(n(n-1)a^{n-2}b - \frac{n(n-1)(n-2)(n-3)}{3\,!}a^{n-4}b^3 \\
&\qquad + \frac{n(n-1)(n-2)(n-3)(n-4)(n-5)}{5\,!}a^{n-6}b^5 - \dots \\
&\quad -(n-1)(n-2)pa^{n-3}b + \frac{(n-1)(n-2)(n-3)(n-4)}{3\,!}pa^{n-5}b^3 - \dots \\
&\quad +(n-2)(n-3)qa^{n-4}b - \frac{(n-2)(n-3)(n-4)(n-5)}{3\,!}qa^{n-6}b^3 + \dots
\end{aligned}
$$

$$-\ldots\Big)a'\sqrt{-1} \qquad = Fa'\sqrt{-1}$$

$$+\Big(na^{n-1} - \frac{n(n-1)(n-2)}{3!}3a^{n-3}b^2 + \frac{n(n-1)(n-2)(n-3)(n-4)}{5!}5a^{n-5}b^4 - \ldots$$

$$-(n-1)pa^{n-2} + \frac{(n-1)(n-2)(n-3)}{3!}3pa^{n-4}b^2$$

$$- \frac{(n-1)(n-2)(n-3)(n-4)(n-5)}{5!}5pa^{n-6}b^4 + \ldots$$

$$+(n-2)qa^{n-3} - \frac{(n-2)(n-3)(n-4)}{3!}3qa^{n-5}b^2 + \ldots - \ldots\Big)b'\sqrt{-1} = Db'\sqrt{-1}$$

$$+\ldots = 0.$$

Now set $A+Ca'-Eb' = 0$ and $(B+Db'+Fa')\sqrt{-1} = 0$, and consequently the approximation is $a' = \dfrac{EB+DA}{FE+CD}$, $b' = \dfrac{FA-CB}{FE+CD}$.

By the same method new approximations can perpetually be found.

If two or more values of the quantities a or b are very nearly equal to each other, then from the principle given earlier their approximations are easily extracted.

The same principle may also be applied to find similar approximations in two or more equations having two or more unknown quantities.

(3) Let two equations be

$$x^n - p\,x^{n-1} + q\,x^{n-2} - r\,x^{n-3} + s\,x^{n-4} - t\,x^{n-5} + \ldots = 0$$

and $x^n-(p+p')x^{n-1}+(q+q')x^{n-2}-(r+r')x^{n-3}+(s+s')x^{n-4}-(t+t')x^{n-5}+\ldots=0$ whose roots are a, b, c, d, e, $\ldots$ and $a+a'$, $b+b'$, $c+c'$, $d+d'$, $e+e'$, $\ldots$ respectively. Then

$$x^n-(p+p')x^{n-1}+(q+q')x^{n-2}-(r+r')x^{n-3}+\ldots$$
$$= (x-a-a')(x-b-b')(x-c-c')(x-d-d')\cdots$$

which very nearly is

$$x^n - (a+b+c+d+\ldots+a'+b'+c'+d'+\ldots)x^{n-1}$$
$$+ [(ab+ac+bc+\ldots)+a(b'+c'+d'+\ldots)$$
$$+b(a'+c'+d'+\ldots)+c(a'+b'+d'+\ldots)+\ldots]x^{n-2}$$
$$- [(abc+abd+acd+bcd+\ldots)+ab(c'+d'+\ldots)$$
$$+ac(b'+d'+\ldots)+bc(a'+d'+\ldots)+ad(b'+c'+\ldots)+\ldots]x^{n-3}$$
$$+\ldots = 0$$

provided a', b', c', d', $\ldots$ are much nearer to zero than are a, b, c, d, $\ldots$.

Hence, approximately,

$$\begin{aligned}
&a'+b'+c'+d'+\dots &&= p' = P,\\
&aa'+bb'+cc'+dd'+\dots &&= pp'-q' = Q,\\
&a^2a'+b^2b'+c^2c'+d^2d'+\dots &&= pQ-qP+r' = R,\\
&a^3a'+b^3b'+c^3c'+d^3d'+\dots &&= pR-qQ+rP-s' = S,\\
&\dots,\\
&a^{n-5}a'+b^{n-5}b'+c^{n-5}c'+\dots &&= N,\\
&a^{n-4}a'+b^{n-4}b'+c^{n-4}c'+\dots &&= M,\\
&a^{n-3}a'+b^{n-3}b'+c^{n-3}c'+\dots &&= L,\\
&a^{n-2}a'+b^{n-2}b'+c^{n-2}c'+\dots &&= K, \text{ and}\\
&a^{n-1}a'+b^{n-1}b'+c^{n-1}c'+\dots &&= I; \text{ whence}
\end{aligned}$$

$$\begin{aligned}
a' &= \frac{I-(b+c+d+\dots)K+(bc+bd+cd+be+\dots)L-(bcd+bce+\dots)M+\dots}{(a-b)(a-c)(a-d)(a-e)\cdots}\\
&= \frac{a^{n-1}p'-a^{n-2}q'+a^{n-3}r'-a^{n-4}s'+\dots}{na^{n-1}-(n-1)pa^{n-2}+(n-2)qa^{n-3}-(n-3)ra^{n-4}+\dots}, \text{ and}\\
b' &= \frac{I-(a+c+d+\dots)K+(ac+ad+cd+ae+\dots)L-(acd+ace+\dots)M+\dots}{(b-a)(b-c)(b-d)(b-e)\cdots}\\
&= \frac{b^{n-1}p'-b^{n-2}q'+b^{n-3}r'-b^{n-4}s'+\dots}{nb^{n-1}-(n-1)pb^{n-2}+(n-2)qb^{n-3}-(n-3)rb^{n-4}+\dots}.
\end{aligned}$$

Now if two roots of the equation $x^n-px^{n-1}+qx^{n-2}-rx^{n-3}+\dots = 0$ are nearly equal to each other, then from the quadratic equation

$$\begin{aligned}
&\left(\frac{n(n-1)}{2}a^{n-2}-\frac{(n-1)(n-2)}{2}pa^{n-3}+\frac{(n-2)(n-3)}{2}qa^{n-4}-\dots\right)a'^2\\
&\quad+\left(na^{n-1}-(n-1)(p+p')a^{n-2}+(n-2)(q+q')a^{n-3}-\dots\right)a'\\
&\quad-p'a^{n-1}+q'a^{n-2}-r'a^{n-3}+\dots = 0,
\end{aligned}$$

two values of the quantity a' can be found suitably near to them, i.e., slightly greater or less than the roots a and b respectively.

And if three or more roots a, b, c, etc. of the equation $x^n-px^{n-1}+qx^{n-2}-rx^{n-3}+\dots = 0$ are nearly equal to each other, by a similar principle we can derive three values of a' that are close enough to the said roots a, b, c, etc., differing only by a small increment or decrement.

(4) Let $a>b>c>d>\ldots$ be the roots of the equation $x^n - px^{n-1} + qx^{n-2} - rx^{n-3} + sx^{n-4} - \ldots = 0$, and likewise let $A>B>C>D>\ldots$ be the roots of the equation $x^n - Px^{n-1} + Qx^{n-2} - Rx^{n-3} + Sx^{n-4} - \ldots = 0$ (all roots of the two equations being real). Then the equation $x^n - px^{n-1} + qx^{n-2} - \ldots = 0$ can be transformed, by repeated additions of very small quantities $-p'x^{n-1} + q'x^{n-2} - r'x^{n-3} + s'x^{n-4} - \ldots$, into the equation $x^n - Px^{n-1} + Qx^{n-2} - \ldots = 0$, so that at no stage of the transformations are two roots of the transformed equation equal to each other. For in this manner the root a is transformed to A, b to B, c to C, etc. by progressive incrementation, and consequently there is no cause for two roots to become equal at any step.

Suppose in the equation $x^n - px^{n-1} + qx^{n-2} - r''x^{n-2} + \ldots = 0$ there are $2r$ impossible roots, and suppose there are $2h+2r$ impossible roots in the equation $x^n - Px^{n-1} + Qx^{n-2} - Rx^{n-3} + Sx^{n-4} - \ldots = 0$. Then by repeated additions of quantities of the form $-p'x^{n-1} + q'x^{n-2} - r'x^{n-3} + \ldots$, where p', q', r', s', $\ldots$ are infinitesimally small quantities, the equation $x^n - px^{n-1} + qx^{n-2} - r''x^{n-3} + \ldots = 0$ cannot be transformed into the equation $x^n - Px^{n-1} + Qx^{n-2} - Rx^{n-3} + \ldots = 0$, unless in h cases two roots of the intermediate equation become equal to each other.

Every equation $x^n - px^{n-1} + qx^{n-2} - \ldots = 0$ can be transformed into any other of the form $x^n - Px^{n-1} + Qx^{n-2} - \ldots = 0$ by repeated additions of the quantity $-p'x^{n-1} + q'x^{n-2} - r'x^{n-3} + \ldots$, where p', q', r', s', $\ldots$ are infinitesimally small quantities, such that never are more than two roots equal to each other, and in consequence every root of any equation $x^n - Px^{n-1} + Qx^{n-2} - \ldots = 0$ is of the form $\alpha + \beta\sqrt{-1}$; and by the methods previously disclosed, for reducing equations such that unknown quantities are eliminated, we can see that every unknown quantity in two or more equations involving two or more unknowns also is of the stated form.

If in fact $n-1$ roots are nearly equal to each other, then by the resolution of an equation of lower degree $(n-1)$ we can obtain approximations to those $n-1$ roots.

The same observations may also be applied to the case of two, three, or more equations with two, three, or more unknown quantities, being transformed into the same number of other equations in a similar manner.

TRANSLATOR'S NOTES

[1] Although the only concrete example for Method 3 is a quadratic, which could be "solved" to express α and β as irrational quadratic expressions involving x, algebraic solvability is not a requirement for this method; as Waring's reference to Problem I indicates, each of the coefficients of the resultant equation is a symmetric function of the solutions of the first equation for y, and therefore is expressible in terms of the coefficients of y in the first equation, which are polynomials or rational functions of x.

[2] The text labeled "(5.2.2)" comes from Addenda page A3.8, where it was assigned to "pag. 229" and numbered "2.2". The page assignment is clearly incorrect, as there is no paragraph or subsection labeled 2 or 2.1 in the vicinity of page 229; the theme of the addendum is clearly related to Problem XXXI, and my best guess is that here (page 209-210 of the Latin text) is the referent Waring intended for this addendum.

[3] For a different treatment of a similar matter, see problem LVI, Chapter 5.

[4] The question of determining whether two systems of equations are equivalent is resolvable as indicated in this Problem, but the amount of work increases prohibitively with the degree and the number of variables in the equations. Waring's procedure is mathematically equivalent to the "Gröbner basis" reduction which has recently received a great deal of attention from computer algebra specialists, but the algorithm, which is conceptually analogous to the Gaussian elimination process for triangularizing sets of linear equations, is much more efficient computationally. Evidently Gröbner himself never published the algorithm; a remark by his former student Buchberger [1976] indicates that Gröbner suggested it to him in 1964, and it was first published in Buchberger's [1965] doctoral thesis. Subsequent refinements of the algorithm have appeared frequently in *SIGSAM Bulletin*, the quarterly publication of the ACM's Special Interest Group on Symbolic &

Algebraic Manipulation; a particular Fortran implementation is given by Winkler *et al.* [1985]. (Of course, when Waring wrote this work even the method of Gaussian elimination was not known, and for Waring's contemporaries solution of a system of linear equations involved work which increased in proportion to the factorial of the order of the problem. Clearly when Waring says something like this is "easily" done, he means that the algorithm is known, and he is not concerned that the practical time for the computation might be measured on a geologic scale.)

[5] An extensive exposition of the topics of this and the following Problem appeared in the next century in a paper by Méray[1867], a series of three papers by Junker[1891; 1893; 1894], and one by Brill[1893].

[6] As printed, this example uses the (now obsolete) Newtonian raised dot notation, where $\dot{x}$ is a derivative, $\ddot{x}$ is a second derivative, etc. I think it more likely that Waring intended to have x', x'', x''', etc. in these formulas, since partial derivatives in this context make little sense (and so simple a case as $x^2y^2 = A$, $\dot{x}^2\dot{y}^2 = B$, interpreting $\dot{x}^2$ as $(\frac{\partial f}{\partial x})^2$ where $f(x) = \sqrt{\frac{A}{y^2}}$, etc., has no solution unless $A = B = 1$). In other examples a few pages later the raised dot notation appears again where clearly primes are called for; hereafter I will make the correction without comment.

[7] This formula also appears in Waring's *Meditationes Analyticæ* (ed. II, 1785, p. 137).

Chapter 5

On rational and integral quantities.

Problem XLIV

Given an equation

$$x^n+px^{n-1}+qx^{n-2}+rx^{n-3}+\ldots+Px^3+Qx^2+Rx+S=0,$$

having only one unknown quantity x, *to find its rational divisors.*

Case 1. *First*: let p, q, r, ..., P, Q, R, S be integers, and since

$$(x^{n-1}+px^{n-2}+qx^{n-3}+rx^{n-4}+\ldots+Px^2+Qx+R)x=S,$$

the root x, if it is an integer, will be a divisor of the final term, S, of the given equation.

Transform the given equation into another, whose roots are greater or less than those of the original by a given number a; then $x\pm a$ will be a divisor of the final term of the resulting equation, which will be $S=a^n\mp pa^{n-1}+qa^{n-2}\mp ra^{n-3}+\ldots$.

For example, let a be successively 0, 1, -1, 10, -10, 100, -100, etc., and suppose z is a root of the given equation; then in effect we substitute for x in the given equation the numbers 0, 1, -1, 10, -10, etc., and of the resulting numbers z, $z-1$, $z+1$, $z-10$, $z+10$, etc. will be the respective divisors.

If a is respectively 0, 1, -1, 2, -2, 3, -3, etc., and z is a root of the given equation, then z, $z-1$, $z+1$, $z-2$, $z+2$, $z-3$, $z+3$, etc. will be the respective divisors of the numbers resulting from substitution of the quantities 0, 1, -1, 2, -2, 3, -3, etc., for the root x in the given equation.

From the said divisors accordingly may be found the rational roots of the given equation.

This problem was solved by Jacob van Wassenaer. [1]

Second: If fractional quantities are contained in the given equation, it can be changed into another which has no fractional terms, e.g. if the equation is

$$Nx^n+px^{n-1}+qx^{n-2}+rx^{n-3}+\ldots+Qx^2+Rx+S=0,$$

where n, p, q, r, ..., Q, R, S denote integers, then write $\frac{z}{N}$ for x, producing the equation

$$z^n+pz^{n-1}+qNz^{n-2}+rN^2z^{n-3}+\ldots+QN^{n-3}z^2+RN^{n-2}z+SN^{n-1}=0,$$

whose root shall be α, say; then $Nx-\alpha$ is the required factor.

Thus every root which is a rational fraction will have for its denominator some number which evenly divides N. As above, let α be a divisor of S, and now also let m divide N. Then when we substitute 0, 1, -1, 10, -10, etc. for x in the equation

$$Nx^n+px^{n-1}+qx^{n-2}+rx^{n-3}+\ldots+Qx^2+Rx+S=0,$$

the resulting quantities will be multiples of α, $\alpha-m$, $\alpha+m$, $\alpha-10m$, $\alpha+10m$, etc.; or if instead we substitute 0, 1, -1, 2, -2, etc., the resulting quantities will be divisors of α, $\alpha-m$, $\alpha+m$, $\alpha-2m$, $\alpha+2m$, etc.; then $mx-\alpha$ becomes the trial factor of the given equation.

By the same token the given equation may be transformed into another whose roots are any rational function whatever of the given equation's roots; and then π will be the same rational function of α, where α and π are corresponding rational roots of the given and resulting equations.

Case 2. *First*: here is another method of finding rational divisors of an equation

$$S+Rx+Qx^2+Px^3+Ox^4+\ldots+x^n=0$$

in which S, R, Q, P, ... are integers.

If a is the required root (an integer) then $\frac{S}{a}$, $\frac{\frac{S}{a}+R}{a}$, $\frac{\frac{\frac{S}{a}+R}{a}+Q}{a}$, ... will be integers; that is, provided that A, B, C, D, E, ... are the aforesaid quotients, then $\frac{S}{a}$, $\frac{Ra+S}{a^2}$, $\frac{Qa^2+Ra+S}{a^3}$, $\frac{Pa^3+Qa^2+Ra+S}{a^4}$, ... will be integers.

For example, let the given equation be $-35-44x+7x^2-7x^3+x^4=0$. For S, R, Q, P, and O write respectively -35, -44, 7, -7, and 1. The divisors of the coefficient -35 are 1, 5, 7, -1, -5, and -7; for the rational root a we assume the divisor 7, then

$$\frac{S}{a} = -\frac{35}{7} = -5,$$

$$\frac{A+R}{a} = \frac{-5-44}{7} = -7,$$

$$\frac{B+Q}{a} = \frac{-7+7}{7} = 0,$$

$$\frac{C+P}{a} = \frac{0-7}{7} = -1,$$

$$\frac{D+O}{a} = \frac{-1+1}{7} = 0,$$

whence it follows that the number 7 is a rational root of the given equation.

Second: If fractional quantities are contained in the given equation, it can be changed, as above, into another which has no fractional terms, and then by this or some other method we can determine whether or not it has integral divisors.

Let the coefficient of the term x^n be any integer N, one of whose factors is m, with $\frac{a}{m}$ the required root of the given equation

$$S+Rx+Qx^2+Px^3+Ox^4+\ldots+Nx^n = 0;$$

then

$$\frac{S}{a},\ \frac{\frac{mS}{a}+R}{a},\ \frac{\frac{\frac{mS}{a}+R}{a}+Q}{a},\ \text{etc.}$$

— or, letting A, B, C, D, ... denote the same quantities as before,

$$\frac{S}{a},\ \frac{mA+R}{a},\ \frac{mB+Q}{a},\ \frac{mC+P}{a},\ \frac{mD+O}{a},\ \text{etc.}$$

or, what is the same,

$$\frac{S}{a},\ \frac{Ra+mS}{a^2},\ \frac{Qa^2+mRa+m^2S}{a^3},\ \frac{Pa^3+mQa^2+m^2Ra+m^3S}{a^4},\ \text{etc.}$$

will be integral numbers.

Case 3. If no divisor occurs by this method, we are to conclude that the equation admits of no linear factors. But it may have a quadratic divisor; suppose it to be mx^2-ax+b, in which case b is a divisor of S, the final term of the given equation, and m is a divisor of the coefficient N of the given equation's leading term x^n. Then $m-a+b$ will be a divisor of the quantity $S+R+Q+P+O+\ldots$, and $m+a+b$ will be a divisor of $S-R+Q-P+O-\ldots$; and so

$$\frac{R+\frac{aS}{b}}{b} = \frac{Rb+aS}{b^2}, \quad \frac{Q-\frac{mS}{b}+a\left(\frac{R+\frac{aS}{b}}{b}\right)}{b} = \frac{Qb^2-(mS-aR)b+a^2S}{b^3}, \quad \text{etc.}$$

will be integers — or, letting A, B, C, D, E, ... denote the preceding numbers, then $\frac{S}{b}=A$, $\frac{r+aA}{b}=B$, $\frac{Q+aB-mA}{b}=C$, $\frac{P+aC-mB}{b}=D$, etc. will be integers.

And in the same way we can easily derive a rule by which divisors of the third or higher degrees can be detected.

There is yet another way to find the quadratic divisors, mx^2-ax+b, of a given equation. In the given equation substitute for x three, four, or more terms of the progression 3, 2, 1, 0, -1, -2, -3, etc. Let all the divisors of the numbers that result be singly added to and subtracted from the squares of the correspondent terms of that progression, multiplied by some numeral divisor m of the coefficient N, and set the sums or differences beside each term of the progression. Among these numbers note all collateral progressions, which run through those sums and differences. Then suppose $\pm b$ to be a term of such like progressions, that stands against the term of the first progression which arises from having substituted 0 for x, and $\pm a$ the difference which arises by subtracting $\pm b$ from the next superior term, which stands against the term 1 of the first progression, and this gives the quadratic factor to be tried. [2]

Now if the given equation has no quadratic divisors, it may yet have a cubic, biquadratic, or one of even higher degree, which may be found by seeking among the abovementioned sums and differences progressions which, though not truly arithmetic progressions themselves, have higher-order (second, third, ...) differences whose terms do form an arithmetic progression.

In the same way the given equation may be transformed into another whose roots are any rational function of the given equation's roots, and thence may be found the required rational divisors of quadratic, cubic, or higher order.

Finding a factor of a quantity involving only one literal term is done in the same way as we find the divisors of an equation involving one unknown quantity.

Case 4. Given a quantity involving two or more unknown quantities, to determine whether it can be factored or not.

First: All quantities but one must be supposed equal to zero, and the divisors of the resulting quantity must be found. Then suppose all but two of the unknowns equal to zero, and find divisors of the resulting quantity, and so on. In other words, group the terms of the given equation together according to their degree in some one of the unknowns, say y, and find the divisors of the final term; then substitute for y terms of the sequence -2, -1, 0, 1, 2, etc., and thus by the preceding method the required divisors may be found.

If there are more than two unknowns, the operation must be repeated more times; each stage serves to eliminate one of the unknown quantities.

Second: If fractional quantities are contained in the given quantity, then reduce all terms to their common denominator, and find the divisors of both the numerators and denominator; if the given quantities contain irrationalities, then for each distinct irrationality substitute a distinct letter, and by the preceding method of finding divisors of polynomials in two or more unknowns the required factors may often be extracted.

Case 5. A formula containing fractional or irrational quantities in the given equation or equations may often be turned into a formula involving only integers or rational fractions. For instance, suppose the equation is $x^n+ax^{n-1}+\sqrt[m]{c+dx^r}+\ldots=0$; here, if x is a rational fraction, $\sqrt[m]{c+dx^r}$ will also be a rational fraction (unless the equation contains other irrational quantities, which added to $\sqrt[m]{c+dx^r}$ destroy it, or reduce it to a rational fraction); and obviously if we can determine that $\sqrt[m]{c+dx^r}$ is a rational or integral quantity, an expression for x will be easy to determine.

Case 6. In certain cases it follows easily from other observations that an equation will not have any rational divisors. For example, consider $x^{2n}+3ax^m+3bx^r+\ldots=3d+2$, where a, b, $\ldots$, d are integers. Then since no square (x^{2n}) has a remainder of 2 when divided by 3, this equation admits of no integral divisors; and similarly for other residues after division by any numbers, which do not conform to the numeric nature of the problem. Also by root extraction, and consideration of the limits between which roots must lie, an infinity of such impossibility results may be derived.

Case 7. From the difference between certain quantities, or their ratio, or some other relation between them, it can often be proved that no quantities of a given sort can exist, or limits result between which the numbers of the given sort, if they exist, must be found.

Example 7.1. The difference between two successive squares, x^2 and $(x+1)^2$, is $(x+1)^2-x^2 = 2x+1$; no number greater than x^2, by an amount less than $2x+1$, can be a square. That is, if the quantity is x^2+x+b we take $2x+1 = x+b$, whence $x = b-1$, and it follows that x^2+x+b cannot be a square if x is greater than $b-1$.

In a more general case, if the quantity is $x^2+ax+b = x^2+2ex+e^2$, then $x = \frac{b-e^2}{2e-a}$, whence e must be less than $\frac{a}{2}$ and not greater than $\sqrt{b}$.

Example 7.2. The quantity x^3+ax^2+bx+c, where a, b, and c are integers, cannot be a cube if ax^2+bx+c is less than $3x^2+3x+1$. This leads to limits on the value of x, and such reasoning may be extended to higher powers and the corresponding limits.

Example 7.3. If $x^2+2ax = 4z^{2n}$, then $x = -a+\sqrt{4z^{2n}+a^2}$; but $\sqrt{4z^{2n}+a^2}$ cannot be less than $2z^n+1$, hence $4z^n+1$ will never be greater than a^2, etc.

Case 8. Let any quantity be given, having a simple divisor which is not equal to any other of its divisors; then the given quantity is not a square, cube, biquadrate, or any higher power of an integer.

Example 8.1. If the quantity x has a simple divisor a, and a^2 is not a divisor of x, then a will be a divisor of $ex+fx^2+gx^3+\ldots$, where e, f, g, ... are all integers; and so $x(\pm d+ex+fx^2+gx^3+\ldots)$, where d is not divisible by a, cannot be a square, cube, biquadrate, or higher power.

Example 8.2. If a^m is the highest power of a which is a factor of x, and a^r $(r \leq m)$ is not a divisor of d, then $x(d+ex+fx^2+\ldots)$ is not the $(r+m)^{\text{th}}$ power of any integer.

If $x = \alpha^{r+m}$, then for the product $x(d+ex+fx^2+\ldots)$ to be a power of the prescribed order, it is necessary that $d+ex+fx^2+\ldots = \beta^{r+m}$, where both α and β are integers.

Theorem 42

Consider two equations in two unknowns x and y: if x is rational, then y will also be rational; and if they are not rational, then the expressions for x and y must always involve the same irrationalities, i.e., the quantity x must be expressed in terms of the same irrationality as y (unless two or more of the values of one of the quantities, x or y, are equal to each other). But if α and β are two different values of x that correspond to one value, call it π, of y, then the quadratic equation whose roots are α and β has no irrationalities except those found in the corresponding value π of the unknown quantity y; or if three values α, β, and γ, of the unknown x, correspond to one value π of the unknown quantity y, then the cubic equation whose roots are α, β, and γ will have no irrational quantities, other than irrationalities contained in π; and likewise for all higher degrees.

The same holds for more equations involving more unknown quantities.

Corollary. Let there be two equations, in two unknowns,

$$Nx^n+(a+by)x^{n+1}+(c+dy+ey^2)x^{n-2}+\ldots=0$$

and $$Hx^m+(A+By)x^{m-1}+\ldots=0,$$

all of whose coefficients N, a, b, c, …, H, A, B, … are (positive or negative) integers. In the two equations, as N and H are the coefficients of the terms of maximum degree in the unknown quantity x, let S and s be the coefficients of the terms of maximum degree in the unknown quantity y. Where x and y are both rational quantities, if y is actually an integer, then x will be a fraction whose denominator is the common divisor of the coefficients N and H; and reciprocally if x is an integer, y must be a fraction whose denominator is the common divisor of the coefficients S and s. But if the quantity x is an arbitrary fraction whose denominator is λ, and if the maximum degree of x in the two equations is q and r respectively, then the quantity y will be a fraction whose denominator is a common divisor of the quantities $\lambda^q S$ and $\lambda^r s$. Let

$$Sy^n+A'xy^{n-1}+B'x^2y^{n-2}+\ldots$$

and $$sy^m+a'xy^{m-1}+b'x^2y^{m-2}+\ldots$$

be the terms of the two equations, respectively, of the highest degree with respect to the two unknown quantities x and y. Then $y-\alpha x$, $y-\beta x$, $y-\gamma x$, $y-\delta x$, … are the simple divisors of the first quantity, and $y-\pi x$,

$y-\rho x$, $y-\sigma x$, ... are the second quantity's divisors. The quantity x will thus be a fraction whose denominator is a divisor of the quantity $S^m \cdot s^n \cdot (\alpha-\pi)(\alpha-\rho)(\alpha-\sigma)\cdots(\beta-\pi)(\beta-\rho)(\beta-\sigma)\cdots(\gamma-\pi)(\gamma-\rho)(\gamma-\sigma)\cdots$.

The above rule is easily adapted to cases where some of the terms are equal to zero, and extended to more equations involving more unknown quantities.

Problem XLV

Given two equations involving two unknown quantities, to find the integral corresponding roots, if there are any, of the unknown quantities x *and* y.

Method 1. In the given equations for the unknown quantity y substitute three or more terms in the arithmetic series -1, 0, 1, 2, ..., and the series -1, 0, 1, 10, ... Arrange the terms of the two resulting quantities according to descending powers of x, and then by repeated subtraction of the smaller from the larger, and the remainder from the previous subtrahend and minuend, eliminate the quantity x. Then if all the terms resulting from the first substitutions are divisible by the respective terms of the arithmetic series $z+1$, z, $z-1$, $z-2$, etc., and those resulting from the latter substitution are multiples of the respective terms of the arithmetic series $z+1$, z, $z-1$, $z-10$, etc., then the divisor z will be the required integral root of y.

Example. Let the two equations be

$$y^2 + (1+3x)y + 2x^2+4x-18 = 0,$$

and $$y^2 + 3x\ y + 3x^2+4x-17 = 0.$$

Substituting 0 for y we have the quantities $2x^2+4x-18$ and $3x^2+4x-17$. The transformation of these quantities to eliminate x proceeds thus:

$$(2x^2+4x-18)\cdot 3 = 6x^2+12x-54$$
$$(3x^2+4x-17)\cdot 2 = 6x^2+8x-34.$$

Subtract the smaller from the larger: the remainder will be $4x-20$.

Continuing,

$$(2x^2+4x-18)\cdot 2 = 4x^2+8x-36$$
$$(4x-20)\cdot x = \underline{4x^2-20x\qquad}$$
$$\text{remainder} = 28x-36$$
$$(4x-20)\cdot 7 = \underline{28x-140}$$
$$\text{remainder} = 104.$$

And likewise from	$(3x^2+4x-17)\cdot 4$	$= 12x^2-16x-\ 68$
we subtract	$(4x-20)\cdot 3x$	$= \underline{12x^2-60x\qquad}$
whose remainder will be		$76x-\ 68$
from which we subtract	$(4x-20)\cdot 19$	$= \underline{76x-380}$
leaving a remainder of		$312;$

any integral root of the quantity y must be a divisor of these two numbers, 104 and 312, meaning in this particular instance that it must divide 104. Likewise on substitution of 1 and -1 for the quantity y we obtain the resulting quantities

$$2x^2+7x-16, \quad 3x^2+7x-16$$

and

$$2x^2+\ x-18, \quad 3x^2+\ x-16$$

respectively; the numbers resulting from these quantities will be

$$(512,\ 768) \text{ and } (972,\ 1458)$$

whose greatest common divisors are 256 and 486, respectively. We determine the divisors of the numbers 256, 104, 486, ..., among which there is only one arithmetic progression, *viz.* 1, 2, 3; and the common difference is 1, as it must be. Therefore the number 2, which was the divisor of the number 104 (which we obtained by substituting 0 for the unknown quantity y) is the number we will try as the integral value of the quantity y.

From the preceding Problem we can easily extend this to a method of finding fractional corresponding roots of the quantities x and y.

Also by the preceding Problem those rational quadratic, cubic, and higher degree equations, whose roots are the roots of an unknown quantity y among two or more unknowns x, y, etc., can be found in the cases where two or more corresponding values of x are equal. Otherwise the reduction of these equations is necessarily more complicated.

And thus as for one equation involving one unknown quantity, for more equations with more unknowns, the given equations are transformed into others whose roots are some functions of the roots of the given equations, from which methods may be deduced to find the given equations' roots.

Method 2. There is another way to solve the problem: let two equations involving two unknowns be written as

$$S - Ry + Qy^2 - Py^3 + \ldots = 0$$

and

$$s - ry + qy^2 - py^3 + \ldots = 0$$

in which the letters S, R, Q, P, ..., s, r, q, p, ... each denote an integral function of the letter x, and let z be the required rational and integral root of y. Then

$$\frac{S}{z}, \frac{Rz-S}{z^2} = \frac{R-A}{z}, \frac{Qz^2-Rz+S}{z^3} = \frac{Q-B}{z}, \ldots,$$

and
$$\frac{s}{z}, \frac{rz-s}{z^2} = \frac{r-a}{z}, \frac{qz^2-rz+s}{z^3} = \frac{q-b}{z}, \ldots,$$

(where A, B, C, ... and a, b, c, ... in each case denote the preceding fraction) will all be integers.

This method also allows us to make use of what was previously stated concerning fractional roots, and concerning the finding of quadratic, cubic, and higher degree equations whose roots are solutions for one of the unknown quantities of the given equation.

But when it is required that the roots be integers, the integral values may often be easily deduced by considering the limits between which the roots are known to lie, or their divisors, or the given form of the numbers.

All of the above may also, *mutatis mutandis*, be extended to three or more equations involving three or more unknown quantities. [3]

Problem XLVI

Given an equation, to determine whether it can be solved by extracting square roots, cube roots, etc.

When the equation is given, necessarily so are its dimensions; then assume a general resolution which involves only roots of index two, or three, etc., such that, when surds are eliminated, the result will be an equation of the degree of the given equation. The solution is completed by equating the terms of the given and resulting equations.

Example. To find out whether a given equation, of degree 2, 4, 8, ... can or cannot be resolved with the aid of the extraction of square roots.

Assume the most general resolution involving only square roots, such that the resulting equation is of degree 2, 4, 8, etc.; these will be

$x = \sqrt{a}+b$

or $x = \sqrt{a+m\sqrt{b}} + n\sqrt{b} + c$

or $x = \sqrt{a+m\sqrt{b+n\sqrt{c}}+o\sqrt{c}} + r\sqrt{b+n\sqrt{c}} + s\sqrt{c} + d,$
etc.; and if these equations are reduced, to eliminate irrational quantities, the results will be

$$x^2 - 2bx + b^2 - a = 0,$$
$$x^4 - 4cx^3 - (2n^2b - 6c^2 + 2a)x^2$$
$$-(4c^3 - 4ac + 4mnb - 4n^2cb)x + (n^2b + c^2 - a)^2 - (m - 2nc)\,b = 0,$$
$$x^8 - 8dx^7 + \ldots = 0,$$

etc. Set the resolutions equal to the given equations; there will be as many or more equations as there are unknown quantities, whence by the preceding problems we can discover whether or not integer or rational values exist for the quantities a, m, b, n, c; or for a, m, b, n, c, o, r, s, d; etc., and this completes the problem. [4]

An analogous method allows us to attack equations of higher degree, with the aid of square, cube, and higher-order roots. That is, general expressions are assumed for the roots, necessarily including all quantities of a given sort, and by their reduction it is possible to decide whether or not the given equations are susceptible to solution by such extractions.

Lemma

(1) If $\frac{A}{a}$ and $\frac{B}{b}$ are two fractions, both reduced to lowest terms, and their denominators are not equal, then $\frac{A}{a} \pm \frac{B}{b}$ cannot be an integer.

(2) If $\frac{A}{a}$, $\frac{B}{b}$, $\frac{C}{c}$, and $\frac{D}{d}$, ... are fractions, all reduced to lowest terms, where A, B, C, D, ... and a, b, c, d, ... are integers, and if some prime number π or a power of it, π^n, divides the denominator a but divides none of the other denominators, then $\frac{A}{a} \pm \frac{B}{b} \pm \frac{C}{c} \pm \frac{D}{d} \pm \ldots$ cannot be an integer.

Problem XLVII

Given any n *quantities which are rational and integral functions of the quantity* x, *to find a number or expression which contains as a factor all of the common divisors of the given quantities.*

(1) Reduce the quantities, by the ordinary process for finding common divisors, so as to eliminate the unknown quantity x. There result $n-1$ expressions, whose greatest common divisor contains all the common divisors of the given quantities. In this operation fractions are to be avoided at all times.

Example 1. Let the two given quantities be $5x+1$ and $7x+4$. First, to avoid fractions, multiply $7x+4$ by 5, making it $35x+20$. Then the operation proceeds

$$\begin{array}{r|l} 5x+1 & 35x+20\,(\,7 \\ & \underline{35x+\ 7} \\ & \qquad\ \ 13 \end{array}$$

Hence if $5x+1$ and $7x+4$, where x is any integer, have a common divisor other than 1, it can only be 13.

Example 2. Let the two quantities be $3x^2+2x+8$ and $4x^2+7x+20$. We find their greatest common divisor, first multiplying $3x^2+2x+8$ by 4, for the sake of avoiding fractions, and it becomes $12x^2+8x+32$. Then

$$\begin{array}{r|l} 4x^2+7x+20 & 12x^2+\ 8x+32\,(\,3 \\ & \underline{12x^2+21x+60} \\ & \qquad\quad 13x+28 \end{array}$$

Now divide the divisor, $4x^2+7x+20$, by the remainder, $13x+28$, first multiplying $4x^2+7x+20$ by 13 to keep the results integral, obtaining $52x^2+91x+260$; thus

$$\begin{array}{r|l} 13x+28 & 52x^2+\ 91x+260\,(\,4x \\ & \underline{52x^2+112x\qquad\ } \\ & \qquad\ -\ 21x+260; \end{array}$$

now (to avoid fractions) multiply $21x-260$ by 13, obtaining $273x-3380$, whence

$$\begin{array}{r|l} 13x+28 & 273x-3380\,(\,21 \\ & \underline{273x+\ 588} \\ & \qquad\ -3968 \end{array}$$

Therefore $3968 = 31 \times 128$ is a number in which will be contained as factors all common divisors of the given quantities, $4x^2+7x+20$ and $3x^2+2x+8$, for any integer values of x; and for the quantities $4x^2+7x+20$ and $13x+28$, and even for $13x+28$ and $21x-260$.

In fact the quantities $4x^2+7x+20$ and $3x^2+2x+8$ never have the common divisor 31×128; at most they are multiples of 16×31. The rest of the divisors arose out of the multiplications to avoid fractions.

Example 3. Let the two quantities be x^2-5x+6 and x^3+x^2-x-10; the operation begins

$$
\begin{array}{rl}
x^2-5x+6 \,) & x^3+\ x^2-\ \ x-10\,(\ x+6 \\
& \underline{x^3-5x^2+\ 6x\qquad} \\
& \quad +6x^2-\ 7x-10 \\
& \quad \underline{+6x^2-30x+36} \\
& \qquad\quad +23x-46
\end{array}
$$

In order to avoid fractions, we multiply x^2-5x+6 by 23 before we divide it by $23x-46$, i.e. we proceed thus:

$$
\begin{array}{rl}
23x-46\,) & 23x^2-115x+138\ [= 23(x^2-5x+6)]\ (\ x-3 \\
& \underline{23x^2-\ 46x} \\
& \qquad -69x+138 \\
& \qquad \underline{-69x+138}
\end{array}
$$

Hence $23x-46 = 23(x-2)$ will be the quantity containing all the common divisors. Now if we divide x^2-5x+6 and x^3-x^2-x-10 by $x-2$, the results are $x-3$ and x^2+3x+5, of which the unique divisor will be 23. Thus the divisors of the given quantities are found in their common divisor $x-2$ and in the divisors of the quantities $x-3$ and x^2+3x+5 resulting from the division of each original quantity by the common divisor $x-2$. And so for many other similar cases.

It may happen that the resulting quantities have more divisors than the problem entails; these arise from, and are factors of, the multipliers which are used to multiply the given quantities, or the remainders, for the purpose of avoiding fractions.

(2) If some of the quantities A, B, C, ... have factors, i.e. if $A = a \times \alpha$, then we should find the quantities which contain all of the divisors of the quantities a, B, C, ..., and of the quantities α, B, C, etc.

If two or more of A, B, etc. have a common divisor, i.e, $A = a \times \alpha^m$, $B = b \times \alpha^n$, etc., then we need to find all divisors of the set of quantities α, C, etc.; of the set a, α; C, etc.; of the set b, α, C, etc.; of the set a, b, C, etc.; and so on, to complete the problem.

(3) The same principle can also be applied to three or more given quantities, or to m quantities containing two or more (but fewer than m) unknown quantities, as well as to propositions of impossibility.

Example. Let the two given quantities be $a^r - b^r$ and $a^s - b^s$, in which a and b are mutually prime, as are r and s. The above method would show the expression containing all of their common factors to be of the form $a^t(a-b)$; but since a and b are mutually prime, neither a nor a^t will be a divisor of either b^r or b^s — nor, it follows, of either $a^r - b^r$ or $a^s - b^s$. Therefore $a-b$ will be the unique common divisor of the quantities $a^r - b^r$ and $a^s - b^s$. When these quantities are divided by their common divisor $a-b$, there result two mutually prime numerical quantities $a^{r-1}+a^{r-2}b+a^{r-3}b^2+\ldots+b^{r-1}$ and $a^{s-1}+a^{s-2}b+a^{s-3}b^2+\ldots+b^{s-1}$.

If the common divisors of quantities such as the above are given, we can easily deduce formulæ for the unknown quantities which admit the said common divisors.

Theorem 43

Case 1. Two quantities,

$$a^n + \frac{n(n-1)}{2}a^{n-2}b^2 + \frac{n(n-1)(n-2)(n-3)}{2 \cdot 3 \cdot 4}a^{n-4}b^4$$
$$+\frac{n(n-1)(n-2)(n-3)(n-4)(n-5)}{2 \cdot 3 \cdot 4 \cdot 5 \cdot 6}a^{n-6}b^6 + \ldots = \frac{(a+b)^n + (a-b)^n}{2} = P, \text{ and}$$

$$na^{n-1} + \frac{n(n-1)(n-2)}{2 \cdot 3}a^{n-3}b^2 + \frac{n(n-1)(n-2)(n-3)(n-4)}{2 \cdot 3 \cdot 4 \cdot 5}a^{n-5}b^4$$
$$+\frac{n(n-1)(n-2)(n-3)(n-4)(n-5)(n-6)}{2 \cdot 3 \cdot 4 \cdot 5 \cdot 6 \cdot 7}a^{n-7}b^6 + \ldots = \frac{(a+b)^n - (a-b)^n}{2b} = Q,$$

(where a and b are mutually prime) can have no common divisor, except 2, 4, 8, 16, 32, ... 2^n.

Proof. Divide $2P = (a+b)^n + (a-b)^n$ by $(a+b)^n - (a-b)^n$; the operation will proceed

$$(a+b)^n - (a-b)^n \,\big)\, (a+b)^n + (a-b)^n \,\big(\, 1$$
$$\underline{(a+b)^n - (a-b)^n}$$
$$2(a-b)^n$$

Therefore if the quantities P and Q have a common divisor, the quantities $(a+b)^n \pm (a-b)^n$ and $2(a-b)^n$ must have the same common divisor. So divide $(a+b)^n + (a-b)^n$ by $2(a-b)^n$; the operation will proceed

$$2(a-b)^n \overline{\big)\, 2(a+b)^n+2(a-b)^n \big(}\, 1$$
$$\underline{2(a-b)^n}$$
$$\text{remainder} = 2(a+b)^n$$

Hence, if the above quantities have a common divisor, $(a+b)^n$ and $(a-b)^n$ must have the same common divisor. Dividing $a+b$ by $a-b$, we have

$$a-b \overline{\big)\, a+b \big(}\, 1$$
$$\underline{a-b}$$
$$\text{remainder} = 2b$$

So $2b$ and $a \pm b$ share the same common divisor as the quantites $a+b$ and $a-b$; and it follows, since a and b are mutually prime, 2 will be the unique common divisor of the quantities $a+b$ and $a-b$, whence 2, 4, 8, 16, ... 2^n covers the set of all common divisors of the quantities $(a+b)^n$ and $(a-b)^n$, and hence of the quantities

$$\frac{(a+b)^n + (a-b)^n}{2} \quad \text{and} \quad \frac{(a+b)^n - (a-b)^n}{2} .$$

Q.E.D.

Case 2. Let a and c be mutually prime numbers, and in the above expressions P and Q substitute c, c^2, c^3, etc. for b^2, b^4, b^6, etc. Then the two resulting quantities will have no common factor, except, as before, 2, 4, 8, 16, ... 2^n.

This is easily established by the preceding case and the operation in Problem XLVII.

Similar propositions may also be asserted regarding the sums of each product of two, three, four, or more terms of those series which arise from the expansion of binomial, trinomial, etc. quantities in terms of successive even powers of the letter x.

And we can easily detect those cases in which the above quantities are actually divisible by 2^n or 2^{n-1} or 2^{n-2} or ... or 8 or 4 or 2. For instance, let $n = 3$ and let the given quantity be $a+\sqrt{c}$, where a and c are mutually prime. Then the given quantities, deduced by the above rules (*viz.* a^3+3ac and $3a^2+c$), have the common divisor $2^3 = 8$ if and only if $a = 2x+1$ and $c = 4z+1$, where x is any integer and z is an odd number; they have the common factor $2^2 = 4$ when $a = 2x+1$ and $c = 4v+1$, where x and v are integers; they have only the common factor 2 when $a = 2x+1$ and $b = 2v+1$; and in any other case (i.e. if a or c is an even number) they admit no common divisor.

Problem XLVIII

To reduce the equation

$$x^{2n}+px^{2n-1}+qx^{2n-2}+rx^{2n-3}+sx^{2n-4}+tx^{2n-5}+\ldots=0$$

to two others,

$$P\equiv x^n+ax^{n-1}+bx^{n-2}+cx^{n-3}+dx^{n-4}+ex^{n-5}+fx^{n-6}+\ldots$$
$$=\sqrt{l}(\alpha x^{n-1}+\beta x^{n-2}+\gamma x^{n-3}+\delta x^{n-4}+\ldots)$$

through the extraction of square roots. [5]

When the equation P is reduced, to eliminate the irrational quantity l, the result is

$$x^{2n}+2ax^{2n-1}+(2b+a^2-l\alpha^2)x^{2n-2}+(2c+2ba-2l\alpha\beta)x^{2n-3}$$
$$+(2d+2ac+b^2-l[2\alpha\gamma-\beta^2])x^{2n-4}+(2e+2da+2bc-2l[\alpha\delta+\beta\gamma])x^{2n-5}$$
$$+\ldots=0.$$

Equating corresponding terms of the given and resulting equations, we have the $2n$ equations

$$2a=p,\qquad 2b+a^2-l\alpha^2=q,\ 2(c+ab-l\alpha\beta)=r,$$
$$2d+2ac+b^2-l(2\alpha\gamma+\beta^2)=s,\ 2(e+ad+bc-l[\alpha\delta+\beta\gamma])=t,$$

etc.

These $2n$ equations have $2n+1$ unknown quantities, one of which (l) always appears in a product with some of the n unknown quantities α, β, γ, δ, etc. Accordingly, we reduce these $2n$ equations to n equations, eliminating the n unknown quantities a, b, c, etc. The result will be n equations having no unknown quantities except those (α, β, γ, δ, ϵ, etc.) that are always multiplied by the unknown quantity l. Thus there result n numbers, of which each is divisible by l. To wit, from the first equation we have $a=\frac{p}{2}$;

from the second (where $\pi=q-\frac{1}{4}p^2$), $\qquad 2b=\pi+l\alpha^2$;

from the third (where $\rho=r-\frac{1}{2}p\pi$), $\qquad 2c=\rho+l(2\alpha\beta-\frac{p\alpha^2}{2})$;

from the fourth (where $\sigma=s-\frac{1}{2}p\rho-\frac{1}{4}\pi^2$),

$$2d=c-l\left(p\alpha\beta-\frac{p^2\alpha^2-2\pi\alpha^2-l\alpha^4}{4}-2\alpha\gamma-\beta^2\right);$$

and so on. What we seek now is a common divisor which is an integer but not the square of an integer, which has a remainder of 1 upon division by 4 when the alternate terms p, r, t, ... are odd multiples of the common divisor. I allow divisors in this context to be fractions as well as integers, provided their denominators are powers of 2. Now if a is half of an odd number, then so is α; if b, so is β; if c, so is γ; and so on. And if the n numbers indeed have the common divisor l, then the quæsitum is a square (Γ^2) which, multiplied by l and added, with proper choice of sign, to the final term of the equation, composes a perfect square. Then writing the given equation as

$$P+Qx+Rx^2+\ldots = 0$$

we assume the equation

$$A+Bx+Cx^2+\ldots = \sqrt{l}(\Gamma+\Delta x+\mathrm{E}x^2+\ldots);$$

this allows us to restate the given equation as

$$A^2-l\Gamma^2+2(AB-l\Gamma\Delta)x+\ldots = 0,$$

whence $P = A^2-l\Gamma^2$, $Q = 2AB-2l\Gamma\Delta$, etc.

This observation is limited by the following restriction: the quantities $A\pm\sqrt{l}\,\Gamma$, $B\pm\sqrt{l}\Delta$, $C\pm\sqrt{l}\,\mathrm{E}$, etc. must not be larger than four times the largest term of the given equation; an infinity of similar restrictions apply.

Corollary. (1) Let the equation be $x^4+px^3+qx^2+rx+s = 0$, and suppose the reduction is $x^2+\frac{p}{2}x+b = \sqrt{l}(\alpha x+\beta)$: let $q-\frac{1}{4}p^2 = \pi$, $r-\frac{1}{2}p\pi = \rho$, $s-\frac{1}{2}p\rho-\frac{1}{4}\pi^2 = \sigma$. From the above rule we know that l must be a common divisor of the terms ρ and 2σ, and it must be odd, not a square, and greater by 1 than the nearest multiple of 4, if either p or r is odd. The candidate for α must be a divisor of the quantity $\frac{\rho}{l}$, if p is even, or half of an odd divisor, if p is odd; from this number subtract $\frac{1}{2}p\alpha$, and call the remainder β; then for b we propose $\frac{\pi+l\alpha^2}{2}$, and try, if l divides b^2-s, whether the root is rational and equal to β; if these conditions are met, the given equation can be so reduced, otherwise it does not admit such a reduction.

The above requirements are all logical consequences of the equations $a = \frac{p}{2}$, $2b = \pi+l\alpha^2$, $l\left(\frac{p\alpha^2}{2}-2\alpha\beta\right) = \rho$, and $s = b^2-l\beta^2$, considered together.

(2) Another method may be deduced to resolve this corollary: if the second term of the given equation is deleted, the result may be written $x^4+Qx^2+Rx+S=0$. Then we must determine whether or not the equation $P^3+2QP^2+(Q^2-4x)P-R^2=0$ has a rational root which is not a square; if so, the corollary is solved. The same and many similar methods may be extended easily to equations of higher degree.

(3) Given n equations having m unknowns (where m is less than n) that are not multiplied by the quantity l, and r other unknown quantities which are multiplied by l, we reduce the n given equations so as to eliminate the unknown quantities that are not multiplied by l. Thence we can deduce different numbers which are all divisible by l, and from any $m+1$ of the equations can be investigated the number whose divisor is l.

Problem XLIX

Given an equation of degree $4n$,
$$x^{4n}+px^{4n-1}+qx^{4n-2}+rx^{4n-3}+sx^{4n-4}+tx^{4n-5}+\ldots=0,$$
to find whether or not it can be depressed to an equation of degree n by successive extraction of square roots.

Assume an equation
$$x^n+ax^{n-1}+bx^{n-2}+\ldots+\sqrt{l}(\alpha x^{n-1}+\beta x^{n-2}+\ldots)$$
$$=\sqrt{r+s\sqrt{l}}(\pi x^{n-1}+\rho x^{n-2}+\ldots)$$
which, reduced to eliminate the irrational expressions, will be
$$x^{4n}+4ax^{4n-1}+(4b-2r\pi^2+6a^2-2l\alpha^2)x^{4n-2}+\ldots=0.$$
Now collect together the equations
$$4a=p,$$
$$4b-2r\pi^2+6a^2-2l\alpha^2=q,$$
etc.,

which gives us $4n$ equations with $3n+3$ unknown rational quantities, $n+1$ of which (α, β, ... and s) always appear in a term which also includes the multiplier l, so by eliminating all the other unknowns we will be left with a set of numbers which must all be divisible by l.

Every unknown quantity will be either an integer or an odd number divided by 2, 4, etc. The possible denominators for each of the unknown quantities can be deduced by the principles laid down in Problem XLVII and Theorem 43.

These unknown quantities may be found by the method of common divisors, since there are more equations than unknowns; or following the principles contained in the preceding Problem. [6]

Generalizations of this approach permit us to determine whether or not an equation can be reduced by the aid not only of quadratic, but of cubic or higher order root extractions as well.

The same reasoning is applicable to multiple equations involving multiple unknown quantities.

Theorem 44

Case 1. In the following, let n and m represent integers, while e, f, g, and k are rational numbers, and $\sqrt{f}$, $\sqrt{k}$, and $\sqrt[2n+1]{g}$ are irrational quantities. Then,

Case 1.1. The expansion of $(e+\sqrt{f})^m$ will be

$$e^m+\frac{m(m-1)}{2!}e^{m-2}f+\frac{m(m-1)(m-2)(m-3)}{4!}e^{m-4}f^2+\dots$$

$$+\left[me^{m-1}+\frac{m(m-1)(m-2)}{3!}e^{m-3}f+\frac{m(m-1)(m-2)(m-3)(m-4)}{5!}e^{m-5}f^2+\dots\right]\sqrt{f}$$

which we may separate into a rational part, $A =$

$$e^m+\frac{m(m-1)}{2!}e^{m-2}f+\frac{m(m-1)(m-2)(m-3)}{4!}e^{m-4}f^2+\dots$$

and an irrational part, $B =$

$$\left[me^{m-1}+\frac{m(m-1)(m-2)}{3!}e^{m-3}f+\frac{m(m-1)(m-2)(m-3)(m-4)}{5!}e^{m-5}f^2+\dots\right]\sqrt{f}$$

whence $A^2-B^2 = (A+B)(A-B) = (e+\sqrt{f})^m(e-\sqrt{f})^m$

$$= e^{2m} - me^{2m-2}f + \frac{m(m-1)}{2!}e^{2m-4}f^2-\frac{m(m-1)(m-2)}{3!}e^{2m-6}f^3 + \dots$$

$$= (e^2 - f)^m.$$

Case 1.2. Let $\sqrt{k}(e+\sqrt{f}) = x$, then $x^m =$

$$(e^m+\frac{m(m-1)}{2!}e^{m-2}f+\dots)k^{\frac{m}{2}}+(me^{m-1}+\frac{m(m-1)(m-2)}{3!}e^{m-3}f+\dots)k^{\frac{m}{2}}f^{\frac{1}{2}}.$$

Now write
$$(e^m+\frac{m-1)}{2!}e^{m-2}f+\ldots)k^{\frac{m}{2}} = A,$$
$$(me^{m-1}+\frac{m(m-1)(m-2)}{3!}e^{m-3}f+\ldots)k^{\frac{m}{2}}f^{\frac{1}{2}} = B.$$
Then $a^2-B^2 = (e^2-f)^m \cdot k^m$.

Case 1.3. Let $x = (e+\sqrt{f})(\sqrt[2n+1]{g})$, or $x = \sqrt{k}(e+\sqrt{f})(\sqrt[2n+1]{g})$; then
$$x^{2n+1} = \Big(e^{2n+1} + \frac{(2n+1)(2n)}{2!} + \frac{(2n+1)(2n)(2n-1)(2n-2)}{4!}e^{2n-3}f^2+\ldots\Big)g$$
$$+\Big((2n+1)e^{2n}+\frac{(2n+1)(2n)(2n-1)}{3!}e^{2n-2}f+\ldots\Big)g\sqrt{f} \qquad \text{— or,}$$
$$x^{2n+1} = \Big(e^{2n+1}+\frac{(2n+1)(2n)}{2!}+\frac{(2n+1)(2n)(2n-1)(2n-2)}{4!}e^{2n-3}f^2+\ldots\Big)gk^{\frac{2n+1}{2}}$$
$$+\Big((2n+1)e^{2n}+\frac{(2n+1)(2n)(2n-1)}{3!}e^{2n-2}f+\ldots\Big)g(k^{\frac{2n+1}{2}})\sqrt{f}.$$
Now for the rational quantity write
$$(e^{2n+1} + \frac{(2n+1)(2n)}{2!}e^{2n-1}f + \ldots)g = A,$$
and for the irrational quantity write
$$\Big((2n+1)e^{2n} + \frac{(2n+1)(2n)(2n-1)}{3!}e^{2n-2}f + \ldots\Big)g\sqrt{f} = B,$$
or the analogous formulas incorporating k; then
in the former case $A^2-B^2 = (e^2-f)^{2n+1}g^2$, and
in the latter case $\Big(Ak^{\frac{2n+1}{2}}\Big)^2 - \Big(Bk^{\frac{2n+1}{2}}\Big)^2 = (e^2-f)^{2n+1}k^{2n+1}g^2$, whence
$$\sqrt[2n+1]{A^2-B^2} = (e^2-f)(\sqrt[2n+1]{g^2}) \quad \text{(in the former case), or}$$
$$\sqrt[2n+1]{A^2-B^2}\cdot k = (e^2-f)k(\sqrt[2n+1]{g^2}) \quad \text{(in the latter case);}$$
but $\sqrt[2n+1]{g^2}$ is an irrational quantity, and therefore in both cases $\sqrt[2n+1]{A^2-B^2}$ is also irrational (as long as $\sqrt[2n+1]{g}$ is irrational).

Corollary 1. Hence, if A and B represent integers, or the square roots of integer quantities, with A greater than B, and if $\sqrt[2n+1]{A^2-B^2}$ is an integer, then the root $\sqrt[2n+1]{A+B}$ will never appear in the formula for $(e+\sqrt{f})\cdot\sqrt[2n+1]{g}$, but it may appear in the formula for $e+\sqrt{f}$ or $(e+\sqrt{f})\sqrt{k}$.

Case 1.4. (1) Let $(a^2-b)N^2 = (A^2-B^2)N^2 = \pi^{2n+1}$, where the letters a, b, N, and π denote integer quantities. If we multiply N by $a+\sqrt{b}$ the result is $Na+N\sqrt{b}$; the $(2n+1)^{\text{th}}$ root of this quantity, $\sqrt[2n+1]{Na+N\sqrt{b}}$, may appear in the formula for $e+\sqrt{f}$, but not in the formula for $(e+\sqrt{f})\sqrt[2n+1]{g}$.

(2) Let $(a^2-b)N^2 = (A^2-B^2)N^2 = \pi^{2n+1}$; then $\sqrt[2n+1]{N^{\frac{1}{2}}a+N^{\frac{1}{2}}\sqrt{b}}$ may be contained in the formula for $(e+\sqrt{f})\sqrt{k}$, but not in the formula for $(e+\sqrt{f})(\sqrt[2n+1]{g})\sqrt{k}$.

For in the first case $Na = A$ and $N\sqrt{b} = B$; in the second, $N^{\frac{1}{2}}a = A$ and $N^{\frac{1}{2}}\sqrt{b} = B$, whence the difference of their squares will be A^2-B^2 $(= N^2(a^2-b)$ in the former case, and $N(a^2-b)$ in the latter); and consequently $\sqrt[2n+1]{Na+N\sqrt{b}}$ and $\sqrt[2n+1]{N^{\frac{1}{2}}a+N^{\frac{1}{2}}\sqrt{b}}$ may occur in the formulas for $e+\sqrt{f}$ and $(e+\sqrt{f})k$ respectively, but never in the above formulas which contain $\sqrt[2n+1]{g}$.

Corollary 2. Hence, let a quantity $a+\sqrt{b}$ be given, and multiply it by N, where N is chosen such that $N^2(a^2-b) = \pi^{2n+1}$ or $N(a^2-b) = \pi^{2n+1}$. Then the $(2n+1)^{\text{th}}$ root of the resulting quantity $N(a+\sqrt{b})$ or $N^{\frac{1}{2}}(a+\sqrt{b})$ may be of the form $e+\sqrt{f}$ or $\sqrt{k}+\sqrt{f}$, etc.

Case 2. If the root is $\sqrt[n]{e}+\sqrt[n]{f} = \pi$, then we can find the power
$(\sqrt[n]{e}+\sqrt[n]{f})^m = e^{\frac{m}{n}}+f^{\frac{m}{n}}+me^{\frac{m-1}{n}}f^{\frac{1}{n}}+\dfrac{m(m-1)}{2!}e^{\frac{m-2}{n}}f^{\frac{2}{n}}+\ldots = P.$
Then by Problem XXVI we can find a quantity which, multiplied by P, produces a rational product Q of degree m; Q in fact will be $(e\pm f)^m$, *viz.* $(e+f)^m$ if n is odd, otherwise $(e-f)^m$. For example, if $n = 3$, then
$(e^{\frac{1}{3}}+f^{\frac{1}{3}})^m =$

$$\left(e^{\frac{m}{3}}+\frac{m(m-1)(m-2)}{3!}e^{\frac{m-3}{3}}f+\frac{m(m-1)\cdots(m-4)(m-5)}{6!}e^{\frac{m-6}{3}}f^2+\ldots\right)(=A)$$
$$+\left(me^{\frac{m-1}{3}}f^{\frac{1}{3}}+\frac{m(m-1)(m-2)(m-3)}{4!}e^{\frac{m-4}{3}}f^{\frac{4}{3}}+\ldots\right)(=B)$$
$$+\left(\frac{m(m-1)}{2!}e^{\frac{m-2}{3}}f^{\frac{2}{3}}+\frac{m(m-1)(m-2)(m-3)(m-4)}{5!}e^{\frac{m-5}{3}}f^{\frac{5}{3}}+\ldots\right)(=C)$$

And thus
$(A+\alpha B+\alpha^2C)(A+\beta B+\beta^2C)(A+\gamma B+\gamma^2C) = A^3-3ABC+B^3+C^3 = (e+f)^3$
where α, β, and γ are the three roots of the equation $z^3-1 = 0$.

If π is a trinomial, or higher, quantity, the same method can be used to deduce π^m or other powers.

Case 3. Let m and n be two mutually prime numbers, and let
$$A+B\sqrt[n]{p}+C\sqrt[n]{p^2}+D\sqrt[n]{p^3}+\ldots+H\sqrt[n]{p^{n-2}}+L\sqrt[n]{p^{n-1}} = \sigma$$
be a quantity for whose m^{th} root we require a formula.

Find a quantity P, which on multiplication with σ gives a product $P\sigma = \pi^m$; then the formula for the required root will be

$$\lambda = \frac{a+b\sqrt[n]{p}+c\sqrt[n]{p^2}+\ldots+k\sqrt[n]{p^{n-2}}+h\sqrt[n]{p^{n-1}}}{Q\sqrt[n]{P}}$$

where A, B, C, D, ... H, L; a, b, c, ... k, h; and Q, P, and π are all integers. The number Q is easily deduced from Problem XLVII and Theorem 43.

Compute the formal expansion of λ^n, and by equating the corresponding terms of λ^n and σ, the integral values of a, b, c, d, etc. can be worked out.

And so for an infinity of similar propositions.

Problem L

Given an equation, to find whether or not it has any roots of given form(s), the literal expressions of the formulas always denoting rational quantities.

By the method developed in Chapter 3, determine whether or not the equation has any roots of a particular form; then by the two previous Problems determine whether or not the literal quantities of the formula can represent integer or rational quantities.

In the majority of cases the solution is most easily effected by direct comparison of corresponding terms.

Example 1. Let the formula be $\sqrt{x}+\sqrt{y}$, where x and y represent rational quantities, x the larger of the two. And let the given equation be $w^2 = a+\sqrt{b} = x+y+2\sqrt{xy}$, whence $x+y = a$ and $2\sqrt{xy} = \sqrt{b}$, and consequently

$$x^2-2xy+y^2 = a^2-b, \qquad x-y = \sqrt{a^2-b}.$$

Therefore $$x = \frac{a+\sqrt{a^2-b}}{2} \quad\text{and}\quad y = \frac{a-\sqrt{a^2-b}}{2},$$

which shows that the roots of w cannot be of the stated form, unless a^2-b is a square or zero.

If $w^2 = \sqrt{a}+\sqrt{b}$, then it can easily be reduced to the form

$$w^2 = \sqrt{a}\left(1+\sqrt{\tfrac{b}{a}}\right) \quad\text{or}\quad w^2 = \sqrt{b}\left(1+\sqrt{\tfrac{a}{b}}\right).$$

The same method may also be applied to extraction of fourth, eighth, sixteenth, ... roots, or to extracting the square or higher root of three or more quantities of the general formula $a+\sqrt{b}+\sqrt{c}+\sqrt{d}+\ldots.$ Or, the root of more than two quantities may rather be found by the process of dividing any two roots by a factor of a third radical, which produces a rational and integral expression; for the root of that expression will in effect supply two terms of the required root.

Example 2. Given a quantity $N(a+\sqrt{b})$, where $N^2a^2-N^b = \pi^{2n+1}$, to find an expression for the root $\sqrt[2n]{Na+N\sqrt{b}}$ in the form $e+\sqrt{f}$, if such exists (where N, a, b, n, e, and f are integers).

(1) Let $x+\sqrt{y} = \sqrt[2n+1]{N(a+\sqrt{b}}$, thus $x-\sqrt{y} = \sqrt[2n+1]{N(a-\sqrt{b}}$, and so $(x+\sqrt{y})(x-\sqrt{y}) = x^2-y = \sqrt[2n+1]{N^2a^2-N^2b} = \pi$. Assuming $a>\sqrt{b}$, compute the sum

$$\frac{1}{2}\left(\sqrt[2n+1]{N(a+\sqrt{b})}\right)+\frac{1}{2}\left(\sqrt[2n+1]{N(a+\sqrt{b})}\right),$$

correct to the first decimal place, and let z be the integer nearest to this sum. Then $z = x$ and $y = z^2-\pi$. If on the other hand $a<\sqrt{b}$, compute

$$\frac{1}{2}\left(\sqrt[2n+1]{N(a+\sqrt{b})}\right)-\frac{1}{2}\left(\sqrt[2n+1]{N(a+\sqrt{b})}\right),$$

correct to the first decimal place, and let z be the integer nearest to this sum. Then again $z = x$ and $y = z^2-\pi$. This is a straightforward consequence of the preceding discussion.

(2) If simply $a^2-b = \pi^{2n+1}$, select a rational number v slightly larger than the required root $\sqrt[2n+1]{a+\sqrt{b}}$, the difference being less than $\frac{1}{2}$. Then if $a^2>b$, and z is the integer nearest to, and not greater than, the sum $v+\frac{\pi}{v}$, half of its value will be the rational part of the required root, i.e. $\frac{1}{2}z = x$ and $y = \frac{1}{4}z^2-\pi$.

If $a^2<b$, and z is the integer nearest to, and not less than, the difference $v-\frac{\pi}{v}$, then $\frac{1}{2}z = x$ and $y = \frac{1}{4}z^2-\pi$.

Proof. Let $v = \sqrt[2n+1]{a+\sqrt{b}}+\rho$, where by hypothesis ρ is not greater than $\frac{1}{2}$. But $\sqrt[2n+1]{a+\sqrt{b}}$ is always greater than $\sqrt[2n+1]{a-\sqrt{b}}$, and greater than 1 also. Then

$$\frac{\pi}{v} = \frac{\sqrt[2n+1]{a^2-b}}{\sqrt[2n+1]{a+\sqrt{b}}+\rho} = \sqrt[2n+1]{a-\sqrt{b}}-\frac{\rho\left(\sqrt[2n+1]{a-\sqrt{b}}\right)}{\sqrt[2n+1]{a+\sqrt{b}}}+\frac{\rho^2\left(\sqrt[2n+1]{a-\sqrt{b}}\right)}{\left(\sqrt[2n+1]{a+\sqrt{b}}\right)^2}-\ldots$$

so that $$v+\frac{\pi}{v} = \sqrt[2n+1]{a+\sqrt{b}} + \sqrt[2n+1]{a-\sqrt{b}} + \rho - \frac{\rho\left(\sqrt[2n+1]{a-\sqrt{b}}\right)}{\sqrt[2n+1]{a+\sqrt{b}}} + \dots$$

But $$\sqrt[2n+1]{a+\sqrt{b}} + \sqrt[2n+1]{a-\sqrt{b}} = 2x;$$
therefore $v+\frac{\pi}{v}$ is greater than $2x$ by the amount
$$\rho - \frac{\rho\left(\sqrt[2n+1]{a-\sqrt{b}}\right)}{\sqrt[2n+1]{a+\sqrt{b}}} + \dots,$$
which must always be less than $\frac{1}{2}$. Moreover,
$$v-\frac{\pi}{v} = \sqrt[2n+1]{a+\sqrt{b}} - \sqrt[2n+1]{a-\sqrt{b}} + \rho + \frac{\rho\left(\sqrt[2n+1]{a-\sqrt{b}}\right)}{\sqrt[2n+1]{a+\sqrt{b}}} - \dots$$
Now if a^2 is less than b, then $\sqrt[2n+1]{a+\sqrt{b}} - \sqrt[2n+1]{a-\sqrt{b}} = 2x$. Therefore $2x$ is less than $v-\frac{\pi}{v}$ by an amount which is less than
$$\rho + \frac{\rho\left(\sqrt[2n+1]{a-\sqrt{b}}\right)}{\sqrt[2n+1]{a+\sqrt{b}}},$$
which (since ρ is not greater than $\frac{1}{2}$) is an amount less than 1.

(3) Given a quantity $N^{\frac{1}{2}}(a+\sqrt{b}) = N^{\frac{1}{2}}(A+B)$, where $N(a^2-b) = \pi^{2n+1}$, and A is greater than B, to express the root $\sqrt[2n+1]{N^{\frac{1}{2}}(a+\sqrt{b})}$ in the form $\sqrt{x}+\sqrt{y}$, where $\sqrt{x}$ is greater than $\sqrt{y}$.

Compute $\sqrt[2n+1]{N^{\frac{1}{2}}(A+B)}$ numerically, rounded to the nearest integer, and call it r, i.e. $r = \sqrt[2n+1]{N^{\frac{1}{2}}(a+\sqrt{b})} \pm \rho$, where ρ is less than $\frac{1}{2}$; in consequence
$$\begin{aligned} r+\frac{\pi}{r} = {} & \sqrt[2n+1]{N^{\frac{1}{2}}(a+\sqrt{b})} + \sqrt[2n+1]{N^{\frac{1}{2}}(a-\sqrt{b})} \pm \rho \mp \rho \sqrt[2n+1]{\frac{a-\sqrt{b}}{a+\sqrt{b}}} \\ & + \rho^2 \sqrt[2n+1]{\frac{a-\sqrt{b}}{(a+\sqrt{b})^2}} \pm \rho^3 \sqrt[2n+1]{\frac{a-\sqrt{b}}{(a+\sqrt{b})^3}} + \dots \\ = {} & \sqrt{x} \pm \text{ some quantity less than } \rho \text{ (and so less than } \tfrac{1}{2}). \end{aligned}$$

Now divide $A\sqrt{N}$ by its largest rational integral divisor, and call the quotient s; then s will be an irrational factor of the larger component $\sqrt{x}$, and will not be less than 1. When we divide the sum $r+\frac{\pi}{r}$ (which differs from the required solution by no more than $\frac{1}{2}$) by $2s$, i.e. by twice the irrational quantity just produced, then the quotient differs by less than $\frac{1}{2}$ from a rational numerical factor t; therefore if $\dfrac{r+\frac{\pi}{r}}{2s}$ is approximated most nearly by the integer t, then ts will be the greater part of the required root, whence the lesser part will be $\sqrt{t^2s^2-\pi}$, whence the required root will be $ts+\sqrt{t^2s^2-\pi}$.

From the lemmas we see that if $c+\sqrt{\frac{e}{f}} = x+\sqrt{y}$, or $\sqrt{\frac{c}{d}}+\sqrt{\frac{e}{f}} = \sqrt{x}+\sqrt{y}$, where $\frac{c}{d}$ and $\frac{e}{f}$ are fractions reduced to lowest terms, and the denominators d and f are unequal, the quantities N, a, and b in the expressions

$$N(a+\sqrt{b}) = (x+\sqrt{y})^n \quad \text{or} \quad N^{\frac{1}{2}}(a+\sqrt{b}) = (\sqrt{x}+\sqrt{y})^n$$

cannot all be integers simultaneously.

In the quantity $x+\sqrt{y}$ or $\sqrt{x}+\sqrt{y}$, suppose that x and y are integers and mutually prime. Then if

$$(x+\sqrt{y})^n = A+B\sqrt{y}, \quad \text{or} \quad (\sqrt{x}+\sqrt{y})^n = A\sqrt{x}+B\sqrt{y},$$

the quantities A and B will have no common divisor except 2, 4, 8, 16, ... 2^{n-2}, 2^{n-1}, or 2^n; this is proved easily by Theorem 43. Hence the above quantity $A+B\sqrt{y}$ or $A\sqrt{x}+B\sqrt{y}$ (where $A^2-B^2y = \pi^n$ or $A^2x-B^2y = \pi^n$) is multiplied by 2^n; by this device the root of the given quantity may always be extracted, provided an extraction of this sort is possible at all.

And for the remaining steps in this operation, if the given quantity is a fraction, or if some of its parts have a common divisor, then extract the roots separately of the denominator and the factors.

(4) Let the given quantity be $A+B$, whose n^{th} root is required. Let the required root be $\dfrac{x+y}{\sqrt[2n]{p}} = \sqrt{A+B}$, whence $x^2-y^2 = \sqrt[n]{(A^2-B^2)p}$, where x, y, and p are integers, and $(x^2-y^2)^n = (A^2-B^2)p = r^n$, where r is the smallest n^{th} power by which A^2-B^2 can be divided, in order to reduce the calculation to the lowest possible terms. We know that the difference of the squares $x^2-y^2 = r$; now find the sum of the squares,

$$x^2+y^2 = \frac{1}{2}\sqrt[n]{(A+B)^2 \cdot p} + \frac{1}{2}\sqrt[n]{(A-B)^2 \cdot p}.$$

Now if the root can indeed be exhibited, it is necessary that the sum of the two irrational quantities be an integer. This can occur if, of the required integers that are closest to the irrational values $\sqrt[n]{(A+B)^2 \cdot p}$ and $\sqrt[n]{(A-B)^2 \cdot p}$, one is greater than the true value and the other is less. Accordingly, if $\sqrt[n]{(A+B)^2 \cdot p}$ is nearly equal to $s \pm$ a fraction,

and $\sqrt[n]{(A-B)^2 \cdot p}$ is nearly equal to $t \mp$ a fraction, then $x^2+y^2 = \frac{s+t}{2}$, $x^2 = \frac{2r+s+t}{4}$, $y^2 = \frac{s+t-2r}{4}$,

so that the required root will be $\dfrac{\sqrt{2r+s+t}+\sqrt{s+t-2r}}{2\sqrt[2n]{p}}$.

This rule always finds the required roots, if any such exist; but a divisor 2, not contained in the previous methods, has been introduced. If the given quantity $A+B$ is multiplied by 2^n, then from the quantity resulting from the above rules the required root can always be extracted, provided that it can be found by this rule: for instance, let the given quantity be $2+\sqrt{5}$; its cube root is $\frac{1+\sqrt{5}}{2}$, which the earlier methods cannot find, but if $2+\sqrt{5}$ is multiplied by 2^3, then the cube root of the quantity $16+8\sqrt{5}$ can be identified by the present method.

(5) But all of the above rules fail, when one of the quantities A or B is impossible. In this case suppose $(A^2-B^2)p = r^n$, and let the required root be $\dfrac{x+y}{2\sqrt[2n]{p}}$. Assuming the equation

$$z^n - nr^2z^{n-2} + \frac{n(n-3)}{2!}r^4z^{n-4} - \frac{n(n-4)(n-5)}{3!}r^6z^{n-6} + \ldots = 2p(A^2+B^2),$$

from which the value of z is determined, then $x^2+y^2 = \frac{1}{2}z$, $x^2-y^2 = r$; whence $x = \dfrac{\sqrt{z+2r}}{2}$ and $y = \dfrac{\sqrt{z-2r}}{2}$, and the nth root of the binomial $A+B$ will be $\dfrac{\sqrt{z+2r} \pm \sqrt{z-2r}}{2\sqrt[2n]{p}}$.

In the above equations for z substitute ru, making the resulting equation $$u^n - nu^{n-2} + \frac{n(n-3)}{2!}u^{n-4} - \ldots = \frac{A^2+B^2}{A^2-B^2},$$

so that the required root will be $\frac{\sqrt{u+2}\pm\sqrt{u-2}}{2}(\sqrt{A^2-B^2})$.

Another method: first reduce the given quantity, such that $A^2-B^2 = \pi^n$, and then find the roots of the equation

$$z^n - n\pi z^{n-2} + \frac{n(n-3)}{2}\pi^2 z^{n-4} - \ldots = 2A;$$

then $z = 2x$ and $y = \sqrt{\frac{z^2-4\pi}{4}}$, whence the required root will be $x+y = \frac{z+\sqrt{z^2-4\pi}}{2}$.

Problem LI

Given a quantity composed of an infinite number of parts, to determine whether or not it can be expressed by a rational fraction involving some of its assumed quantities.

(1) Divide the given quantity by the assumed quantity, if the given is the greater of the two (otherwise divide the assumed quantity by the given quantity). Divide the divisor by the remainder, and continuing divide this divisor by the new remainder, and if the remainder vanishes at some point (the operation does not continue forever) then the given quantity is a rational function of the assumed quantity, or else it is not.

Example 1. To determine whether or not 1.1212121212... can be expressed as a rational fraction.

I assume the integer number 1, and divide the (larger) given quantity by the (smaller) assumed quantity as the first step of the operation, which progresses:

1
1)1.12121212 . . . 8
remainder = .12121212 . . .)1 (divisor)
.96969696. . . 4
2nd remainder = .03030303 . . .).12121212 . . . (2nd divisor)
.12121212. . .
0 0 0 0

in which the final remainder vanishes, confirming that the given quantity may be expressed as a rational fraction.

The same method can be applied to literal quantities and the like.

Example 2a. Given the series

$$2-\frac{5x}{3}-\frac{35x^2}{9}+\frac{40x^3}{27}+\frac{565x^4}{81}-\frac{35x^5}{243}+\ldots;$$

to find whether or not it is a rational algebraic function of the letter x. Divide 1 by the given series, the series itself by the first remainder, and so on — i.e., proceed as one would using the conventional method of finding common divisors:

$$\begin{array}{r|l}
 & \frac{1}{2} \\
\hline
2-\frac{5}{3}x-\frac{35}{9}x^2+\frac{40}{27}x^3+\frac{565}{81}x^4-\frac{35}{243}x^5+\ldots & 1 \\
\text{remainder} = & \frac{5}{6}x+\frac{35}{18}x^2-\frac{40}{54}x^3-\frac{565}{162}x^4+\ldots;
\end{array}$$

$$\begin{array}{r|l}
 & \frac{12}{5x}-\frac{38}{5} \\
\hline
\frac{5}{6}x+\frac{35}{18}x^2-\frac{40}{54}x^3-\frac{565}{162}x^4+\ldots & 2-\frac{5}{3}x-\frac{35}{9}x^2+\frac{40}{27}x^3+\frac{565}{81}x^4-\ldots \\
 & 2+\frac{14}{3}x-\frac{16}{9}x^2-\frac{226}{27}x^3+\ldots \\
\hline
 & -\frac{19}{3}x-\frac{19}{9}x^2+\frac{266}{27}x^3+\ldots \\
 & -\frac{19}{3}x-\frac{133}{9}x^2+\frac{152}{27}x^3+\ldots \\
\hline
\text{remainder} = & \frac{114}{9}x^2+\frac{114}{27}x^3+\ldots;
\end{array}$$

$$\begin{array}{r|l}
 & \frac{5}{76x}+\frac{5}{38} \\
\hline
\frac{114}{9}x^2+\frac{114}{27}x^3+\ldots & \frac{5}{6}x+\frac{35}{18}x^2+\ldots \\
 & \frac{5}{6}x+\frac{5}{18}x^2+\ldots \\
\hline
 & \frac{5}{3}x^2+\ldots \\
 & \frac{5}{3}x^2+\ldots; \\
\hline
\end{array}$$

and as we shall show in Corollary 2 of the following Theorem,

$$2-\frac{5}{3}x-\frac{35}{9}x^2+\ldots=\cfrac{1}{\frac{1}{2}+\cfrac{1}{\frac{12}{5x}-\frac{38}{5}+\cfrac{1}{\cfrac{1}{\frac{5}{76x}+\frac{5}{38}}}}}=\cfrac{1}{\frac{1}{2}+\cfrac{1}{\frac{12}{5x}-\frac{38}{5}+\frac{76x}{5+10x}}}$$

$$=\cfrac{1}{\frac{1}{2}+\frac{5x+10x^2}{12-14x}}=\frac{12-14x}{6-2x+10x^2}=\frac{6-7x}{3-x+5x^2}.$$

Example 2*b*. Given the fraction

$$\frac{a}{c}+\frac{cb-da}{c^2}x+\frac{d^2a-dcb}{c^3}x^2+\frac{d^2cb-d^3a}{c^4}x^3+\ldots,$$

then the operations would proceed as follows:

$$\frac{a}{c}+\frac{cb-da}{c^2}x+\frac{d^2a-dcb}{c^3}x^2+\frac{d^2cb-d^3a}{c^4}x^3+\ldots\;)\,1\,\left(\frac{c}{a}\right.$$

$$\begin{array}{l}
1+\dfrac{cb-da}{ca}x+\dfrac{d^2a-dcb}{c^2a}x^2+\ldots\\
\hline
+\dfrac{da-cb}{ca}x+\dfrac{dcb-d^2a}{c^2a}x^2+\dfrac{d^3a-d^2cb}{c^3a}x^3+\ldots
\end{array}$$

$$\frac{da-cb}{ca}x+\frac{dcb-d^2a}{c^2a}x^2+\frac{d^3a-d^2cb}{c^3a}x^3+\ldots\;\Bigg)\;\frac{a}{c}+\frac{cb-da}{c^2}x+\frac{d^2a-dcb}{c^3}x^2+\ldots\;\left(\frac{a^2}{(da-cb)x}+\frac{ab}{da-cb}\right.$$

$$\begin{array}{ccccc}
\dfrac{a}{c} & + & \dfrac{cb-da}{c^2}x & + & \dfrac{d^2a-dcb}{c^3}x^2+\ldots\\
\dfrac{a}{c} & - & \dfrac{da}{c^2}x & + & \dfrac{d^2a}{c^3}x^2+\ldots\\
\hline
 & & \dfrac{b}{c}x & - & \dfrac{db}{c^2}x^2+\ldots\\
 & & \dfrac{b}{c}x & - & \dfrac{db}{c^2}x^2+\ldots\\
\hline
\end{array}$$

Since the remainder now vanishes, the operation terminates and the series can be expressed as a rational fraction involving the quantity x. By Corollary 2 of the following Theorem, this will be

$$\frac{a}{c}+\frac{cb-da}{c^2}x+\frac{d^2a-dcb}{c^3}x^2+\ldots=\cfrac{1}{\frac{c}{a}+\cfrac{1}{\frac{a^2}{(da-cb)x}+\frac{ab}{da-cb}}}$$

$$= \frac{1}{\frac{c}{a}+\frac{(da-cb)x}{a^2+abx}} = \frac{a+bx}{c+dx}\,.$$

But if the operation had not terminated, then the series could not have been expressed as a rational function of the quantity x.

Corollary 1. If the given quantity is a numerical fraction, then its digits, as we easily see by the process of division, are repeated in periods of constant length *ad infinitum* — thus, in the first example, they will be 12, 12, 12, 12, 12, ... repeated indefinitely; and conversely, if the digits are repeated in an infinity of identical periods, then the given quantity is a numerical fraction. Let A be the first sequence of digits in that example (.12); the first sequence multiplied by $\alpha = .01$ in the same example produces the second sequence (.0012); the second sequence times α produces the third, and so on (in decimal notation α will always be .1 or .01 or .001, etc.) Then the given quantity will be

$$1+A+\alpha A+\alpha^2 A+\alpha^3 A+\alpha^4 A+\ldots\text{(to infinity)} = 1+\frac{A}{1-\alpha} = 1+\frac{.12}{1-.01} = \frac{111}{99}\,.$$

And so for all sequences of this sort.

Corollary 2. Let 47619 be the sequence of digits in a decimal fraction, i.e. the fraction is .47619 47619 47619 ... to infinity. If we divide unity by 47619 the quotient pattern is 210000 210000 21..., and reciprocally if we divide unity by 21 the result is 476190476190....
And so it is for any other fraction.

Corollary 3. From the quotients acquired by the method described in this Problem, we can acquire fractions which approximate the given fraction itself, such that no other fraction with terms no larger than these can approach it more closely. If the series is 1. 1546 1546 1546..., the process will be, as directed in this Problem,

$$\begin{array}{llll}
 & 1 & & \\
1) & 1.1546\ 1546\ \ldots & & \\
 & \underline{1} & 6 & \\
 & 1546\ 1546\ \ldots &)\,1 & \\
 & & \underline{.9276\ 9276\ \ldots} & 2 \\
 & & .0723\ 0723\ \ldots &)\,.1546\ 1546\ \ldots \\
 & & & \underline{.1446\ 1446\ \ldots} \\
 & & & .0100\ 0100\ \ldots
\end{array}$$

and so on; the successive quotients are seen to be 1, 6, 2, 7, 4, 2, 1, 7. Then since dividing the numerator by the denominator gives the first quotient (1), $\frac{1}{1}$ will be less than the required fraction. But the denominator divided by the first remainder gives the second quotient, 6, and consequently the remainder will be less than one-sixth of the denominator, and so the fraction $1+\frac{1}{6}=\frac{7}{6}$ will be larger than the true value. And since the first remainder divided by the second gives the quotient 2, and the two immediately preceding fractions were $\frac{1}{1}$ and $\frac{7}{6}$, we see that $\frac{7\times 2+1}{6\times 2+1}=\frac{15}{13}$ will be slightly less than the correct fraction; and, 7 being the next quotient, $\frac{15\times 7+7}{13\times 7+6}=\frac{112}{97}$ will be slightly greater than the correct fraction. And based on the succession 1, 6, 2, 7, 4, 2, 1, 7 of quotients, the fractions will be

$$\frac{1}{1},\quad \frac{7\times 2+1}{6\times 2+1}=\frac{15}{13},\quad \frac{112\times 4+15}{97\times 4+13}=\frac{463}{401},\quad \frac{1038\times 1+463}{899\times 1+401}=\frac{1501}{1300};$$

$$\frac{6\times 1+1}{6}=\frac{7}{6},\quad \frac{15\times 7+7}{13\times 7+6}=\frac{112}{97},\quad \frac{463\times 2+112}{401\times 2+97}=\frac{1038}{899},\quad \frac{1501\times 7+1038}{1300\times 7+899}=\frac{11545}{9999}.$$

Here the fractions in the upper row denote approximations that are less than the true value, and those in the lower row denote approximations that are greater (except the last one, which is exactly the fraction's value). Between these we can interpolate others, e.g. between the fractions $\frac{c}{d}$ and $\frac{na+c}{nb+d}$ we can interpolate the fractions

$$\frac{a+c}{b+d},\ \frac{2a+c}{2b+d},\ \frac{3a+c}{3b+d},\ \frac{4a+c}{4b+d},\ \ldots,\ \frac{(n-1)a+c}{(n-1)b+d}.$$

This method of approximation was first discovered by Wallis.

Let a, b, c, d, e, f, g, ... be the successive quotients. Then the corresponding fractions will be

$$\frac{1}{1},\ \frac{ab+1}{b},\ \frac{abc+c+a}{bc+1},\ \frac{abcd+cd+ad+ab+1}{bcd+b+d},$$
$$\frac{abcde+cde+ade+abe+abc+a+c+e}{bcde+de+be+bc+1},\ \text{etc.}$$

The numerators and denominators of this series are both generated according to the same rule, except that one letter is always present in one and is always absent in the other; this is the initial quotient a. This letter is present in the numerator or denominator, according as the initial quotient results from division of the numerator by the denominator, or of the denominator by the numerator.

In this series the leading term will be *abcdefg*...; all the others are derived from it. First each two successive quotients are deleted in turn. To make this more explicit, consider that a and b are successive quotients, so by deleting them from the first term we are left with *cdefg*..., the second of the required terms. Then b and c are successive quotients, which must be deleted from the first term, leaving *adefg*..., which is the third required term. Also c and d are successive quotients: deleting c and d from the first term produces the fourth required term, *abefg*...; and so on. Then two pairs of successive quotients are simultaneously removed from the first term, and this produces a further one of the required terms; after this in turn three, four, ... pairs of successive quotients are simultaneously removed, until the resulting terms have only one quotient apiece (or none, in which case we supply a term 1). The sum of all terms which can be formed according to the foregoing method will be the required series. [7]

(2) Let there be two fractions, one of which has α digits in its denominator and β digits in its numerator, the other having n and m digits in its denominator and numerator respectively. Divide the two numerators by their respective denominators. Suppose that $n+m$ is not less than $\alpha+\beta$. Then there cannot be more than $n+m$ digits in the quotient of these two fractions, unless both fractions have the same value. Hence a fraction which has $n+m+1$ digits, in the quotient which results upon division of its numerator by its denominator, the same number as are present in a given series, will be as close as or closer to the value of the series, as any fraction having only $n+m$ digits in its numerator and denominator together.

Let there be two unknowns, N and M, having n digits, m of which are prime numbers, and let r be an s-digit number. Then in the rth powers of M and N there will be at least $m-s-1$ prime digits.

By a not very different method we can determine whether there are two or more rational fractions, such that after multiplication by given infinite series the sum of the products will be equal to a rational fraction; this allows us to investigate whether or not a given quantity which has a non-terminating decimal expansion es equal in value to a rational function. But this is basically an issue related to the summation of infinite series.

Theorem 45

If the fraction
$$\cfrac{\alpha}{a+\cfrac{\beta}{b+\cfrac{\gamma}{c+\cfrac{\delta}{d+\cfrac{\epsilon}{e+\cfrac{\zeta}{f+\cfrac{\eta}{\cdots}}}}}}}$$
is reduced to a common fraction, there will be the sequence

$\frac{\alpha}{a}$ greater than the true value;

$\cfrac{\alpha}{a+\cfrac{\beta}{b}}$ less than the true value;

etc., each fraction being alternately greater or less than the true value:

the third is $\cfrac{\alpha}{a+\cfrac{\beta}{b+\cfrac{\gamma}{c}}} = \alpha\left(\frac{bc+\gamma}{abc+a\gamma+a\beta}\right)$;

the fourth is $\alpha\left(\frac{bcd+b\delta+d\gamma}{abcd+ab\delta+ad\gamma+dc\beta+\delta\beta}\right)$;

the fifth is $\alpha\left(\frac{bcde+bc\epsilon+be\delta+ed\gamma+\gamma\epsilon}{abcde+abc\epsilon+abe\delta+ade\gamma+cde\beta+a\gamma\epsilon+e\beta\delta+c\beta\epsilon}\right)$;

and so on. If we write 1 in place of α, β, γ, δ, ϵ, ..., the series will be the same as that derived in Corollary 3 of the preceding Problem. The rule governing the letters α, β, γ, δ, ϵ, etc. may be enunciated this way: write the terms of the series

$a,\ b,\ c,\ d,\ e,\ f,\ \ldots\ l,\ m,$ etc.

and below them $\alpha,\ \beta,\ \gamma,\ \delta,\ \epsilon,\ \zeta,\ \ldots\ \lambda,\ \mu,$ etc.

Then, as instructed in the previous Problem, strike out two adjacent letters of the upper sequence, i.e. either a and b, or b and c, or c and d, or d and e, etc. Then each time two letters are deleted, the single Greek letter below the second of the two Roman letters (e.g. μ below l, m) is inserted in their place. For example, in the fifth fraction the first term of the numerator is $bcde$ and the second is $bc\epsilon$: in this case the two adjacent letters d and e were deleted, so in their place ϵ was written, since its position in the sequence α, β, γ, δ, ... is below the second term (e) of those being deleted from the upper sequence.

These fractions may also be constructed in the following way: write in columns,

$$\begin{array}{ccccccc} & a, & b, & c, & d, & e, & \text{etc.} \\ & \frac{0}{1}, & \frac{\alpha}{a}, & \frac{b\alpha}{ab+\beta}, & \frac{bc\alpha+\gamma\alpha}{abc+a\gamma+c\beta}, & \frac{bcd\alpha+b\delta\alpha+d\gamma\alpha}{abcd+ab\delta+ad\gamma+cd\beta+\beta\delta}, & \text{etc.} \\ \alpha, & \beta, & \gamma, & \delta, & \epsilon, & \zeta, & \text{etc.} \end{array}$$

Then (in the middle row) multiplying one fraction's numerator by the letter written above it, and adding to this the product of the preceding fraction's numerator times the letter below that fraction, gives the numerator of the next fraction. Likewise multiply each denominator by the upper letter in the same column, and add to this the product of the preceding denominator multiplied by the lower letter in the preceding column, to derive the next denominator in the sequence.

Corollary 1. The differences between successive fractions of this series, i.e. subtraction of each fraction from the preceding one, form the series

$$\frac{\alpha}{a}, \quad \frac{\alpha\beta}{a(ab+\beta)}, \quad \frac{\alpha\beta\gamma}{(ab+\beta)(abc+a\gamma+c\beta)}, \text{ etc.}$$

and the rule for generating successive terms of this series is immediately evident.

Corollary 2. Required to transform the common fraction $\frac{A}{B}$ into continued fraction form: divide A by B and call the quotient a and the remainder C. Then divide the previous divisor by the new remainder (C), calling the new quotient b and the new remainder D, etc.; i.e., let the operation be

$$\begin{array}{l} B)\underline{A(}\ a \\ \quad C)\underline{B(}\ b \\ \quad\quad D)\underline{C(}\ c \\ \quad\quad\quad E)\underline{D(}\ d \\ \quad\quad\quad\quad \text{etc.} \end{array}, \qquad \text{then } a+\cfrac{1}{b+\cfrac{1}{c+\cfrac{1}{d+\cfrac{1}{e+\cfrac{1}{\dots}}}}} = \frac{A}{B};$$

and its reciprocal will be $\frac{B}{A} = \cfrac{1}{a+\cfrac{1}{b+\cfrac{1}{c+\cfrac{1}{d+\dots}}}}$

This technique was first invented by Viscount Brounker. A series of this sort that runs to infinity is beyond the scope of this work; for that the reader is referred to tracts on infinite series.

Problem LII

Given an equation ax+by+cz+dv+ . . . +f = 0, *in which each of the unknown quantities* x, y, z, v, . . . *is of the first degree, and* a, b, c, d, . . . f *are integers, to find its integral solutions.*

Case 1. Let a, the coefficient of the term ax, be the smallest in the given equation. Divide the rest of the terms by, cz, dv, . . . f by this coefficient a; and then let $\dot{b}y$ be the remainder of the division of the term by by the coefficient a, let $\dot{c}z$ be the remainder of the division of the term cz by the coefficient a, and similarly for the others, i.e. the sum of the remainders will be $\dot{b}y+\dot{c}z+\dot{d}v+\ldots+\dot{f}$. Assume a new equation $au+\dot{b}y+\dot{c}z+\dot{d}v+\ldots+\dot{f} = 0$. Now let b be the smallest coefficient in the new equation, and divide the other terms au, $\dot{c}z$, $\dot{d}v$, . . . $\dot{f}$ by this coefficient $\dot{b}$, and denote the remainders of the division of the terms au, $\dot{c}z$, $\dot{d}v$, . . . by this coefficient, respectively, as $\dot{a}u$, $\ddot{c}$, $\ddot{d}v$, . . ., and their sum as $\dot{a}u+\ddot{c}+\ddot{d}v+\ldots+\ddot{f}$. Assume a new equation $\dot{b}\phi+\dot{a}u+\ddot{c}+\ddot{d}v+\ldots+\ddot{f} = 0$; and repeat this process until you reach the point where the remainder is 0. Then in the final equation, arbitrary integer values may be assumed for all of its unknowns, and with the help of the previous equations in the sequence the corresponding integral values of the unknown quantities x, y, z, v, . . . may be found.

Some examples may make this resolution more intelligible.

Example 1. Find the integral values corresponding to x and y, when $16y+20 = 7x$.

Divide $16y+20$ by 7, 7 being the smallest coefficient; the operation is

$$7)\overline{16y+20}(2y+2$$
$$\underline{14y+14}$$

remainder = $2y+6$; now set $2y+6 = 7v$ and divide this new equation by the new smallest coefficient, 2, the operation proceeding

$$2)\overline{7v-6}(3v-3$$
$$\underline{6v-6}$$

remainder = v ; assume the new equation $v = 2w$, and then from the second equation we have $2y+6 = 7v = 14w$, whence $y = 7w-3$, and from the original equation $16y+20 = 112w-28 = 7x$ it follows that $x = 16w-4$. Thus the corresponding values of x and y will be $16w-4$ and $7w-3$ respectively, where w is any integer number.

Example 2. $8x+12y+20 = 15v$.

Divide by the least coefficient, 8:

$$8\overline{)12y+20-15v}(y+2-v$$
$$\underline{8y+16-\ 8v}$$

remainder = $4y+\ 4-\ 7v$. Assume the second equation, $8z+4y+4 = 7v$; dividing this by 4, its least coefficient, proceeds

$$4\overline{)8z+4-7v}(2z+1-v$$
$$\underline{8z+1-4v}$$

remainder = $-3v$. Now we take $4w = 3v$, and divide terms of this third equation by their lesser coefficient, 3:

$$3\overline{)4w}(w$$
$$\underline{3w}$$

remainder = w; for the fourth equation assume $1w = 3u$, and then from the third equation, $4w = 12u = 3v$, i.e. $v = 4u$;
from the second equation, $8z+4y+4 = 7v = 28u$, therefore $y = 7u-1-2z$; and from the first (i.e., the given) equation,

$$8x+12y+20 = 8x+84u-24z-12+20 = 15v = 60u,$$

whence $x = 3z-1-3u$,
the corresponding value of y is $y = 7u-2z-1$,
and $v = 4u$,
where the values of z and u may be any desired integers.

Case 2. Let $px'+qy'+rz'+\ldots = l$; and if $\frac{l}{p} = \alpha$, where α, p, q, r, ... and l, m, n, ... are all integers, then $\alpha-mq-nr-\ldots$, mp, np, ... will be integral corresponding values of the quantities x', y', z', ... respectively. Hence in the previous method if it happens that an equation is proposed in which l is divisible by p, one integer value corresponding to each of the unknowns x, y, z, ... may be discovered immediately, without the tedium of the rest of the procedure.

Corollary 1. Let the number of unknown quantities in the given equation be m; then there may be as many as $m-1$ quantities whose values can be arbitrarily assigned.

Corollary 2. Let $ax+by+cz+dv+\ldots = A$; if the given coefficients a, b, c, d, ... have a common divisor, which however is not a factor of the number A, then the given equation has no solution in which all of x, y, z, v, ... are integers.

Corollary 3. Given limits, between which some of the unknowns are to be confined, these limits may be substituted for their respective unknowns in the given and resulting equations, which permits us to display all those limits which follow from the stated limits. Thus if it is required that some quantity be a positive whole number, we write 1 (the smallest positive integer) in its place in the given and resulting equations, and thus we establish limits which are consequences of the original limit of 1. We can also in this way determine the number of values contained between two limits.

Corollary 4. Let the equation be $ax+by = P$, where π is the greatest common divisor of the coefficients a and b, and let p and q be two corresponding values of the unknown quantities x and y. Then the corresponding values of x and y will be

$$p,\ p\pm\frac{b}{\pi},\ p\pm\frac{2b}{\pi},\ p\pm\frac{3b}{\pi},\ p\pm\frac{4b}{\pi},\ \text{etc.}$$

and

$$q,\ q\pm\frac{a}{\pi},\ q\pm\frac{2a}{\pi},\ q\pm\frac{3a}{\pi},\ q\pm\frac{4a}{\pi},\ \text{etc.}$$

respectively. Accordingly, given a satisfactory value for each of the unknown quantities x and y, all correspondent values are easily deduced.

(1) One value may be found by the method previously given. Alternatively, if a and b have no common divisor, the greater coefficient (say a) may be divided by the lesser (b), the divisor by the remainder, and so on, until we reach a remainder π which divides P. With α, β, γ, δ, ϵ, ζ, ... representing the quotients from these divisions, we then write

$$\begin{array}{cccccccc} \alpha, & \beta, & \gamma, & \delta, & \epsilon, & \zeta, & & \text{etc.} \\ +1, & -\alpha, & \beta A+1, & -\gamma B-A, & \delta C+B, & -\epsilon D-C, & \zeta E+D, & \text{etc.} \end{array}$$

where A, B, C, D, E, ... denote the absolute values respectively of the preceding terms. Now multiply the last term of the lower row by $\frac{P}{\pi}$, and the product will be the required value of y. If in fact we require a smaller value, we have but to divide the above product by the quantity a, and the remainder will be an equal or smaller value of y.

Example. Suppose $256x-87y = 50$. Divide 256 by 87, the divisor 87 by the first remainder, and so on. The first remainder that divides

$P = 50$ turns out to be 5, and the corresponding quotients will be 2 and 1. Hence the sequence is

2, 1

1, −2, $2\times 1+1 = 3$;

whence the value of y that emerges will be $3\times\frac{P}{\pi} = \pm 3\times\frac{50}{5} = \pm 30$; it will be +30 if P is −50, otherwise −30. The second remainder that divides P will be 2. The corresponding quotients, three in number, are 2, 1, and 16, hence

2, 1, 16

1, −2, $2\times 1+1 = 3$, $-(16\times 3+2) = -50$;

whence the second value of y discovered by this method will be $50\times\frac{P}{2} = 50\times\frac{50}{2} = 1250$, which will be negative if the preceding quantity (30) is positive (as 3 and −50 are of opposite signs), and conversely will be positive if the preceding value (−30) is negative. The third and final remainder, which divides 50, will be 1, and the respective quotients are 2, 1, 16, and 2; hence

2, 1, 16, 2

1, −2, $2\times 1+1 = 3$, $-(16\times 3+2) = -50$, $50\times 2+3 = 103$;

so we multiply 103 by $P = -50$, and have the value $50\times 103 = 5150$ for the unknown quantity y, and its corresponding value of x is 1750. The smallest positive integer values of the quantities y and x are thus found to be 30 and 10, respectively.

(2) The quantities A, B, C, D, E, ... will be respectively α, $\alpha\beta+1$, $\alpha\beta\gamma+\alpha+\gamma$, $\alpha\beta\gamma\delta+\alpha\beta+\alpha\delta+\gamma\delta+1$, $\alpha\beta\gamma\delta\epsilon+\alpha\beta\gamma+\alpha\beta\epsilon+\alpha\delta\epsilon+\gamma\delta\epsilon+\alpha+\gamma+\epsilon$, etc., whose rule of formation will be the same as that given in Problem LI, Corollary 3.

(3) Given the sequence

$$\begin{array}{lllllll} \beta, & \gamma, & \delta, & \epsilon, & \zeta, & & \text{etc.} \\ 1, & -\beta, & \gamma H+1, & -\delta I-H, & \epsilon K+I, & \zeta L+K, & \text{etc.} \end{array}$$

where H, I, K, L, ... denote the absolute values of the preceding terms, if the last term of the lower row is multiplied by $\frac{P}{\pi}$, the product will be the required value of x, corresponding to the value of y that has already been found — e.g., from the preceding example β, γ, and δ will be respectively 1, 16, and 2, whence

1 16 2
1, -1, 16×1+1 = 17, -2×17-1 = -35;

and so $-1\times\frac{-50}{5} = 10$, $17\times\frac{-50}{2} = -425$, and $-35\times-50 = 1750$ will be the three values of x corresponding to the three values 30, 1250, and 5150 already found for y.

Corollary 5. Let c, d, e, f, g, ... be the remainders respectively of the division of the coefficient a by b, of the divisor b by the first remainder, of the first remainder by the second, and so on, to the remainder 1 (which divides P); then one value of y will be

$$y = \pm aP\left(\frac{1}{ab}-\frac{1}{bc}+\frac{1}{cd}-\frac{1}{de}+\frac{1}{ef}-\dots\right),$$

this series being continued to that denominator, one of whose factors divides P. Corresponding to this the value of x will be

$$x = \pm bP\left(\frac{1}{bc}-\frac{1}{cd}+\frac{1}{de}-\frac{1}{ef}+\dots\right),$$

this series being continued to the same term as the preceding series for the value of y. Then

$$\begin{aligned} ax+by &= abP\left(\frac{1}{ab}-\frac{1}{bc}+\frac{1}{cd}-\frac{1}{de}+\frac{1}{ef}-\dots\right)+abP\left(\frac{1}{bc}-\frac{1}{cd}+\frac{1}{de}-\frac{1}{ef}+\dots\right) \\ &= abP\left(\frac{1}{ab}\right) = P; \end{aligned}$$

whence, if the first series is a value of the quantity y, the second will be a value of the quantity x. We may prove in this way that the stated value of y is an integer: in the quantity

$$ab\left(\frac{1}{ab}-\frac{1}{bc}+\frac{1}{cd}-\frac{1}{de}+\frac{1}{ef}-\dots\right),$$

for the quotients $\frac{a-c}{b}$, $\frac{b-d}{c}$, $\frac{c-e}{d}$, $\frac{d-f}{e}$, ... respectively write α, β, γ, δ, etc. Then $ab\left(\frac{1}{ab}\right)$, $ac\left(\frac{1}{ab}-\frac{1}{bc}\right)$, $ad\left(\frac{1}{ab}-\frac{1}{bc}+\frac{1}{cd}\right)$, $ae\left(\frac{1}{ab}-\frac{1}{bc}+\frac{1}{cd}-\frac{1}{de}\right)$, $ad\left(\frac{1}{ab}-\frac{1}{bc}+\frac{1}{cd}-\frac{1}{de}+\frac{1}{ef}\right)$, will be the above quantities 1, $-\alpha$, $\beta A+1$, $-\gamma B-A$, $\delta C+B$, etc. The value of x may be investigated similarly.

Corollary 6. Let α, β, γ, δ, ϵ, ... represent the above quotients from the division of the quantity a by the quantity b, the division of the divisor b by the remainder c, etc. respectively. Then the remainders c, d, e, f, g, ... will be, respectively,

$$c = a - b\alpha,$$
$$d = b(\alpha\beta+1) - a(\beta),$$
$$e = a(\beta\gamma+1) - b(\alpha\beta\gamma+\alpha+\gamma),$$
$$f = b(\alpha\beta\gamma\delta+\alpha\beta+\alpha\delta+\gamma\delta+1) - a(\beta\gamma\delta+\beta+\delta),$$

and in general it will be

$$\pm b(\alpha\beta\gamma\delta\epsilon\zeta\cdots+\gamma\delta\epsilon\zeta\cdots+\alpha\delta\epsilon\zeta\cdots+\alpha\beta\epsilon\zeta\cdots+\alpha\beta\gamma\zeta\cdots+\alpha\beta\gamma\delta\cdots$$
$$+\ldots+\alpha\beta\cdots+\gamma\delta\cdots+\epsilon\zeta\cdots+\ldots+\text{etc.})$$
$$\mp a(\beta\gamma\delta\epsilon\zeta\cdots+\delta\epsilon\zeta\cdots+\beta\epsilon\zeta\cdots+\beta\gamma\zeta\cdots+\beta\gamma\delta\cdots+\beta\cdots+\delta\cdots+\zeta\cdots$$
$$+\ldots).$$

These two terms follow the rule of Corollary 3, Problem LI, but the latter term (multiplied by a) excludes the first quotient, α.

Problem LIII

Let there be two equations haveing three or more unknown quantities, all of the first degree; to transform them into one, to eliminate one of the unknown quantities (say x*), such that, if the unknowns of the resulting equation are integers the eliminated quantity* x *will also be an integer.*

(1) Let the two equations be

$$\alpha x+\beta y+\ldots = A$$

and

$$\pi x+\rho y+\ldots = B.$$

Assume the equation $\alpha m-\pi n = 1$, and by the preceding Problem we can find integer values of m and n, which we shall call r and s respectively. The first of the given equations is then multiplied by r, the second by s, producing

$$r\alpha x+r\beta y+\ldots = rA, \quad s\pi x+s\rho y+\ldots = sB,$$

whose difference will be $x+(r\beta-s\rho)y+\ldots = rA-sB$, whence $x = rA-sB+(s\rho-r\beta)y+\ldots$; this expression is substituted as the value of x, and the problem is complete.

Example 1. Let the two equations be

$$7x+6y+4z = 12 \text{ and } 5x+10y+12z = 20,$$

to transform these into one, to eliminate x according to the stated conditions. Assume the equation $7m-5n = 1$; 3 and 4 emerge as correspondent integer values for the unknown quantities m and n. Then the former equation is multiplied by 3, the latter by 4, resulting in two equations

$$21x+18y+12z = 36 \text{ and } 20x+40y+48z = 80,$$

whose difference will be

$$x - 22y - 36z = -44,$$

whence$x = 22y + 36z - 44$, whose value being substituted into the first equation in place of x results in

$$160y+256z-308 = 12, \text{ i.e. } 5y+8z = 10.$$

(2) If the quantities α and π have a common divisor, then the preceding method may fail to solve the problem.

In this case use the methods of the previous Problem to find integral values for one of the unknowns in one of the equations, which substituted in the other equation for the occurrences of that unknown quantity, yields an equation, which can be used to find the corresponding values of all of the unknown quantities.

Example 2. Let the two equations be

$$6x+20y-8z+9v = 33 \text{ and } 9x+14y+12z-21v = 99,$$

to transform these into one in order to eliminate x, etc.

From the first equation, by the method of the earlier Problem, we find the integral correspondent values of the quantities x, y, z, and v; these will be

$$v = 1 - 2w$$
$$y = 3w - 3u + z$$
$$x = 10u - 7w - 2z + 4$$

where u, z, and w are any arbitrary integers — which values, substituted into the second equation, give us

$$21w + 48u + 8z = 84;$$

and so the two equations have been transformed into one, eliminating x under the stated conditions.

Corollary 1. From the resulting equation we see that $w = -4+8a$ and $z = 21-21a-6u$, where a and u are any whole numbers; thus $v = 9-16a$, $y = 3a-9u+9$, and $x = 22u-14a-10$.

Problem LIV

(1) *To find an equation in two unknown quantities,* x *and* y, *whose corresponding values for the unknowns* x *and* y *are respectively*
$\mathrm{x} = \mathrm{Az^n+Bz^{n-1}+Cz^{n-2}}+\ldots,\quad \mathrm{y} = \mathrm{az^{\alpha}+bz^{\alpha-1}+cz^{\alpha-2}}+\ldots,$
where $\mathrm{n} \geq \alpha$. [8]

Assume the equation $Py^m+(p+qx)y^{m-1}+(r+sx+tx^2)y^{m-2}+\ldots = 0$, choosing m such that $\frac{m(m+3)}{2} \geq nm+1$. In the assumed equation, for x and y substitute their required values $Az^n+Bz^{n-1}+Cz^{n-2}+\ldots$ and $az^{\alpha}+bz^{\alpha-1}+cz^{\alpha-2}+\ldots$, and equate to zero each term of the equation of degree mn that results. Now if $\frac{n(m+3)}{2} > nm+1$, then an infinity of such equations may be deduced; if equal, then only one may be deduced, in which the values corresponding to the unknown quantities x and y have the specified formulas.

Corollary. An infinity of equations may be assumed, which resolve this problem; the only requirement is that the number of unknown coefficients must exceed by at least one the degree of the equation obtained from substituting the postulated values of the unknown quantities x and y for the letters themselves in the assumed equation.

(2) In much the same way equations may be found having two unknown quantities x and y, whose corresponding values are

$$x = \frac{az^n+bz^{n-1}+cz^{n-2}+\ldots}{a'z^r+b'z^{r-1}+c'z^{r-2}+\ldots},\quad y = \frac{Az^m+Bz^{m-1}+Cz^{m-2}+\ldots}{A'z^s+B'z^{s-1}+C'z^{s-2}+\ldots},$$

and so also for an infinite variety of similar cases.

Corollary 2.1. Hence an infinite variety of equations may be found, in two unknowns x and y, whose corresponding values may be expressed in general by any formula involving only a single unknown quantity z.

Corollary 2.2. If there are $n-1$ unknown quantities in the given expressions for the corresponding values of the unknown quantities x, y, ..., an equation involving n unknown quantities x, y, ... must be assumed, such that, when the given values are substituted for each corresponding unknown quantity, every term of the resulting equation vanishes. In most instances such equations can be assumed, provided only that the number of unknown coefficients in the assumed equation is at least as large as the number of terms in the resulting equation.

And similarly for the deduction of larger numbers of equations in which the unknown quantities are specified.

Problem LV

Given an equation having two unknown quantities, x *and* y, *to find the rational corresponding values of the unknown quantities, provided that these corresponding values can be expressed in general by an algebraic function of an assumed unknown quantity* z.

Let the equation be

$$Py^n + (p+qx)y^{n-1} + (r+sx+tx^2)y^{n-2} + \ldots = 0.$$

For the unknown quantities x and y substitute, respectively,

$$x = Az^n+Bz^{n-1}+Cz^{n-2}+\ldots \text{ and } y = az^n+bz^{n-1}+cz^{n-2}+\ldots$$

(if integer values are required for the unknown quantities x and y), or

$$x = \frac{Az^n+Bz^{n-1}+Cz^{n-2}+\ldots}{Pz^n+Qz^{n-1}+Rz^{n-2}+\ldots} \text{ and } y = \frac{az^n+bz^{n-1}+cz^{n-2}+\ldots}{Pz^n+Qz^{n-1}+Rz^{n-2}+\ldots}$$

(if rational, not necessarily integer, values are desired); by equating to zero each term of the resulting equation, we can determine the required values of the unknown coefficients.

Example. Let the equation be

$$4y^2 - 4xy + x^2 + 7y - 4x - 4 = 0.$$

For the unknown quantities x and y write, respectively,

$$y = z^2 + bz + c, \quad x = az^2 + bz + c;$$

the resulting equation is

$$(4-4a+a^2)z^4 + (8B-4aB-4b+2ab)z^3$$
$$+(8C-4c-4aC+2ac+4B^2-4Bb+b^2+7-4a)z^2$$
$$+(8BC-4bC-4Bc+2cb+7B-4b)z + 4C^2-4Cc+c^2+7C-4c-4=0.$$

Equate each coefficient separately to zero, and we find that the values of the coefficients a, b, c, B, and C are respectively 2, 11, 1, 6, and 1.

In this instance, when the coefficient of z^4 vanishes, so does the coefficient of z^3, and we are left with more unknowns than equations, so that the given equation admits of an infinity of such resolutions.

Similar reasoning lets us determine the rational (non-integral) values of the unknown quantities — thus also we may assume either

$$y = z^2 + bz + c, \quad x = az^2 + bz + c;$$

or $x = \dfrac{Az^n+Bz^{n-1}+Cz^{n-2}+\dots}{Pz^n+Qz^{n-1}+Rz^{n-2}+\dots}$ and $y = \dfrac{az^n+bz^{n-1}+cz^{n-2}+\dots}{Pz^n+Qz^{n-1}+Rz^{n-2}+\dots}$.

Corollary. Let the given equation be

$$py^n + (a+bx)y^{n-1} + (c+dx+ex^2)y^{n-2} + (\dots+kx^3)y^{n-3} + \dots = 0.$$

If $py^n+bxy^{n-1}+ex^2y^{n-2}+\dots$, the sum of the terms of highest degree, has no rational factor of the first degree, then the corresponding values of the unknown quantities x and y cannot generally be expressed by an integral or rational algebraic function of the assumed quantity z.

If these terms do have a rational divisor of the first degree, say $y-nx$, assume $x = az^n+bz^{n-1}+\dots$, and then y will be $y = \alpha az^n+\dots$. Or, if for x and y we assume respectively

$$x = \frac{az^r+bz^{r-1}+\dots}{pz^s+qz^{s-1}+\dots} \quad \text{and} \quad y = \frac{Az^r+Bz^{r-1}+\dots}{Pz^s+Qz^{s-1}+\dots},$$

then $\frac{A}{P} = \frac{a}{p} \cdot \alpha$.

If there is only one term in the given equation representing the highest degree, then the method of convergent series must be used to determine the relationship between the two terms of highest degree in z. For example, if $x = az^n+bz^{n-1}+\dots$, $y = Az^m+Bz^{m-1}+\dots$, by means of finding convergent series we can establish the relationship between a and A, n and m.

And likewise for the relations between the terms of maximum degree respectively in z, when the required values of the unknown quantities x and y may be rational instead of strictly integer numbers.

The same reasoning applies to m equations in $m+1$ or more unknowns.

Problem LVI

Given two equations having three or more unknown quantities, x, y, *and* z, *to reduce them to one in order to eliminate* x, *in such a way that when the remaining unknown quantities in the resulting equation are rational, the eliminated quantity will also be a rational quantity.*

Method 1. The two equations are reduced by the first of the methods given in Problem XXXII; and if together they yield an equation $Ax = B$, where A and B denote rational functions of the unknown quantities y, z, v, ... (but not x), then substitute $\frac{B}{A}$ for x in one of the given equations. Then reduce this and the other given equation to one, so as to eliminate x; and then if the remaining unknown quantities y, z, v, ... are rational, so also will be the eliminated quantity x.

Method 2. If the above method of reduction does not produce an equation of the form $Ax = B$, but rather gives you two resulting equations

$$Ax^2 + Bx = C \quad \text{and} \quad Ax^2 + Bx = D,$$

where A, B, C, and D denote rational functions of the unknown quantities y, v, z, ... from the given equation, then to every value of each of the unknown quantities y, v, z, ... correspond two values of x. And if the square root of the quantity B^2+4AC or B^2+4AD cannot be extracted, then we must eliminate each of the remaining unknown quantites y r v or z or ... which satisfy the condition that they do not have two or more correspondent values of any of the further remaining unknowns. If in the given equations we thus find some quantity, say v, having this property, then the process leads to an equation $Av = B$. If no such quantity can be found, then the stated method fails to reduce the two equations to one in such a way that the eliminated unknown quantity will be rational whenever the values of the unknown quantities in the resulting equation are rational as well.

Method 3. If the preceding method fails to reduce $n+1$ equations to n equations, then try to find a formula expressing one of the unknown quantities rationally in terms of the rest of them. This formula (or these formulæ), substituted in place of the corresponding unknown quantities, will afford the desired reduction of $n+1$ equations to n or fewer.

Or, if two (or n) unknown quantities can be expressed rationally in terms of the remaining unknown quantities together with one (or $n-1$) assumed new unknown quantity (quantities), then from these rational expressions the given equations may be reduced to a smaller number, as stipulated.

Corollary 1. If there are $n+1$ equations, and n of their unknown quantities appear only to the first degree in n of the equations, then those n equations may be solved for those n unknowns, which solutions may be substituted for their respective variables in the one remaining equation. Thus the $n+1$ equations may be reduced to one, in which n unknown quantities have been eliminated, such that if the remaining unknown quantities in the remaining equation are rational, so will be the eliminated quantities.

(1) Let equations $A = 0$, $B = 0$, $C = 0$, ... be given, where A, B, C, ... are functions of the quantities x, y, z, etc. Then assume u, v, w, ... as rational functions of the quantities x, y, z, ..., and reduce all of these equations so as to eliminate x, y, z, etc. The resultant equations will be $P = 0$, $Q = 0$, $R = 0$, ..., in which if x, y, z, ... of the given equations are rational, the quantities u, v, w, ... of the resultant equations will be rational.

(2) Conversely, let x, y, z, ... be rational functions of u, v, w, ...; then if u, v, w, ... in the resultant equations $P = 0$, $Q = 0$, $R = 0$, etc. are rational, the quantities x, y, z, ... will be rational in the given equations.

(3) If the assumed functions are not rational, then from these functions and formulas either rational or irrational expressions in the resultant equations may be deduced from the formulas in the given equations; and reciprocally, from the formulas of the quantities in the resultant equations, may be deduced the formulas of the quantities in the given equations. And thus also for integral quantities.

Corollary 1.1. Thus numerous equations may be constructed, which yield a given solution. [9]

Supplement to Problem LVI

Here it seems appropriate to add some rules for particular methods to discover rational values of the unknown quantities in a single given equation.

Case 1. When an equation is given involving the unknown quantities x, y, z, v, ..., and each term of the given equation is of degree n or $n-1$ in whatever combination of the unknowns x, y, z, etc., if all of the degrees are in fact n, then for the unknown quantities y, z, ...

in the given equation substitute respectively αx, βx, ...; and so the quantity x is eliminated from the given equation. If some of the terms are of degree n and others are of degree $n-1$, then this method produces an equation in which x appears only to the first degree, and accordingly the value of the unknown quantity x may be written as a rational expression involving the other unknown quantities y, z, v, etc.

Corollary. This method is often useful also when two or more equations of this sort are given.

Example. Let $zy+az+by = v^2$ and $zy+ez+\frac{be}{a}y = w^2$ be given. For y, v, and w by the above rule, we write αz, βz, and γz. From the first equation we have $z = \frac{a+b\alpha}{\beta^2-\alpha}$, and from the second, $z = \frac{e+\frac{be}{a}\alpha}{\gamma^2-a}$, whence $\frac{a}{\beta^2-\alpha} + \frac{e}{\gamma^2-\alpha}$, by which α may be expressed in terms of the quantities β and γ. And so for an infinity of propositions of this sort.

Case 2. Let the equation be

$$ax^2 + by^2 + cz^2 + \ldots + A = pv^2 + qw^2 + ru^2 + \ldots + B,$$

where A and B denote arbitrary rational functions of the unknown quantities, whose terms are all of degree one or zero with respect to the unknown quantities x, y, z, v, w, u, etc. Further let

$$a + b\beta^2 + c\gamma^2 + \ldots = p\pi^2 + q\rho^2 + r\sigma^2 + \ldots$$

Substitute $\beta x+y' = y$, $\gamma x+z' = z$, etc., and $\pi x+v' = v$, $\rho x+w' = w$, $\sigma x+u' = u$, etc., and write these expressions in place of the stated quantities in the given equation. There results an equation which is of the first degree at most in x, whence the unknown quantities x, y, z, v, ... can all be expressed in terms of the unknown quantities y', z', ... v', u', w', etc.

Example 2.1. In the equation $x^2+y^2 = z^2$ for z write $x+z'$, and there results the equation $y^2 -z'^2$, whence

$$x = \frac{y^2-z'^2}{2z'}, \quad z = \frac{y^2+z'^2}{2z'}.$$

Example 2.2. In the equation $a^2+b^2 = x^2+y^2$ assume $x = a+x'$ and $y = b-y'$, upon substitution of which results the equation

$$2ax'+x'^2-2by'+y'^2 = 0$$

which is susceptible to solution by the above method. The terms in question do not go above the second degree.

Another way to solve this example is to multiply the given equation by z^2, the result being

$$(a^2+b^2)z^2 = \frac{z^2x^2}{(=v^2)} + \frac{z^2y^2}{(=w^2)}\,.$$

Now for v and w respectively write $az+v'$ and $bz-w'$, which produces the equation

$$2azv' + v'^2 - 2bzw' + w'^2 = 0\,,$$

therefore

$$z = \frac{v'^2+w'^2}{2(bw'-av')}\,.$$

And similar methods work for equations of higher degree.

Corollary. When two or more equations are given, this method of substitution, which transforms quantities in such a way that some of them are destroyed, is frequently performed to advantage. That is, its virtues are most evident in cases of simple, double, or multiple concurrent equalities, where the terms of greatest (or least) degree are rendered equal by the first substitution, and from this the required resolution may often be deduced.

Example 2.3. Let $ax^3+bx^2+cx+d = v^2$; to find rational expressions for v and x. Put $v = d+\frac{cx}{2d}$ (in order to cancel out the two terms d^2+cx); then

$$d^2 + cx + \frac{c^2x^2}{4d^2} = d^2+cx+bx^2+ax^3,$$

whence $b+ax = \frac{c^2}{4d^2}$, and thus a rational value of x is established.

Example 2.4. Let $x^3 + y^3 = x^3 - z^3$; assume $x = p+q$, $y = p-q$, $v = r+s$, and $z = r-s$, then

$$x^3 + y^3 = 2p^3 + 6pq^2 = v^3 - z^3 = 2s^3 + 6r^2s,$$

and thus $p(p^2 + 3q^2) = s(s^2 + 3r^2)$.

For p, q, r, and s in this equation write, respectively, $ax+3by$, $bx-ay$, $dy-cx$, and $3cy-dx$; the result is

$$(ax + 3by)(a^2 + 3b^2) = (3c6 - dx)(d^2 + 3c^2),$$

whence

$$x = \frac{3c(d^2+3c^2)-3b(a^2+3b^2)}{a(a^2+3b^2)+d(d^2+3c^2)}\,y\,.$$

Case 3. It often helps to find the value of some quantity, using the ordinary methods of resolution of equations, and then to equate the formally irrational part of the resulting expression to some rational quantity, and similarly for the irrational parts of all of the other quantities, either separately or together. And so from simple or double equalities resolutions may often be achieved.

Example. Let $a^2z^2 + bz + c = v^2$ and $az^2 + dz + e = w^2$; suppose $az+\alpha$ and $az+\beta$ respectively equal to v and w. Then by the preceding case the two resulting equations are

$$bz + c = 2a\alpha z + \alpha^2 \text{ and } dz + e = 2\alpha\beta z + \beta^2,$$

from which
$$z = \frac{\alpha^2-c}{b-2a\alpha} = \beta^2-ed-2a\beta\ .$$

Now solve for the unknown quantity α, and equate its irrational part to a chosen rational quantity, by any available means, from which the required values of α and β are deduced.

Case 4. Given n equations having $n+m$ unknown quantities, all of which are required to be rational, choose arbitrary rational values for $m-1$ of the unknown quantities, which will render the complete solution much easier, for thereby rational values of all $n+m$ unknown quantities may often be deduced.

Example 4.1. Let $x^4 + y^4 + z^4 = v^2$. Set $v = \alpha-x^2$, whereby $x^2 = \frac{\alpha^2-y^4-z^4}{2\alpha}$. To cancel out y^4 and z^4, rendering the solution even easier, set $\alpha = z^2+y^2$, which results in the equation $x^2 = \frac{z^2y^2}{y^2+z^2}$; then choose $y^2+z^2 = w^2$, after which values of the unknowns y, z, and w may be found to complete the solution.

Example 4.2. Let $z^2 + a = v^2 = (z + \alpha)^2$, $zy + a = \beta^2$, and $y^2 + a = w^2 = (y + \gamma)^2$. By making the assumption $\beta=\alpha$, we get $y = \frac{\alpha-\gamma^2}{2\gamma} = -2\alpha$, from which we can now deduce values corresponding to the quantities α, γ, y, etc.

Example 4.3. Let $zy+a^2 = w^2$, $zv+a^2 = u^2$, and $yv+a^2 = x^2$. Assign $u = a+tz$, $x = vyt+s$, and, in order to make the solution much more tractable, assume $w = a+vy$. Then the result is

$$y = \frac{s^2-a^2}{2avt^2+2at-2svt}\ .$$

Corollary. In order that the resolution of n equations, containing $n+m$ unknown quantities $x, y, z, \ldots$, be fully general, it must be possible to express individually each of the values of the unknown quantities x, y, z, ... by rational functions which involve m independent quantities that may be arbitrarily chosen.

These resolutions cannot be regarded as general resolutions of the given problem; rather they are solutions of a particular problem within the given problem — that is, they are enumerated as a particular case of the given problem. It is of course possible that the general solution may be revealed by reasoning forward from resolutions of this sort pertaining to certain known cases.

Case 5. (1) One value of each unknown quantity having been found by whatever means, the chosen value of some unknown, increased or diminished (or otherwise altered), may be substituted in place of the same unknown in each of the equations; this gives rise to new equations, from which by the preceding methods new values of each of the unknown quantities may be sought, and so on.

(2) If the given equations are $P = 0$, $Q = 0$, $R = 0$, ..., each of which is of the form

$$Ay^n+(a+bx+cz+\dots)y^{n-1}+(e+fx+gz+hx^2+kxz+lz^2+\dots)y^{n-2}+\dots = 0,$$

where n is an integer; and if a corresponding value α, β, γ, δ, ... is known for each of the unknown quantities x, y, z, ... in the given equations; then in those equations for x, y, z, ... write $x'+\alpha$, $y'+\beta$, $z'+\gamma$, ... respectively, and equations will result in which each term vanishes if its dimensions in the quantities x', y', z', ... are all zero. [10]

It is also worth noting that equations which appear to involve the same unknown quantities may often be decoupled, e.g. consider $2x^2 = y^3$ and $(z+x)^2 = w^3$, both of which equations involve the same unknown quantity x. But writing v for $z+x$ gives rise to two independent equations, $2x^2 = y^3$ and $v^2 = w^3$, whose solutions are the resolutions of two independent problems.

Case 6. Equations may easily be found, having several unknowns, with known rational or integral values for each unknown quantity.

Assume resolutions, from which rational values can be found corresponding to each of the unknown quantities. Then form equations which the chosen resolutions satisfy; and what was required, is done.

Example 6.1. Let $xy+x+y = v^2$ and $z = 1+x+y\pm 2v$ be the assumed resolutions, whose rational solutions are known. Then from these it will follow that $xz+x+z = (v\pm x\pm 1)^2$, $yz+y+z = (v\pm y\pm 1)^2$, $xy+z$, $xz+y$, $yz+x$, $xy+xz+yz$, and $xy+xz+yz+x+y+z$ are all squares.

Example 6.2. Suppose $P = 0$ is an equation which may be satisfied by rational values of its unknowns. Then $q^2+\alpha P+\beta P^2+\gamma P^3+\dots$ will be a square, $r^3+\pi P+\rho P^2+\sigma P^3+\dots$ will be a cube, and so on, whatever be the values assumed for q, α, β, γ, ..., r, π, ρ, σ, ..., etc. Thus can be formed as many equations as we wish, the rational values of whose corresponding unknowns may easily be deduced. Or two or more equations $P = 0$, $Q = 0$, etc., may be assumed, and then

$$p^2 + \alpha P + \beta Q + \ldots + \gamma P^2 + \delta PQ + \epsilon Q^2 + \ldots$$

will be a square, and

$$q^3 + \alpha P + \beta Q + \ldots + \gamma P^2 + \delta PQ + \epsilon Q^2 + \ldots$$

will be a cube — and so on for higher powers also.

Example 3. If it is required that a given quantity be a square, cube, etc., add to it or subtract from it arbitrary quantities, so that the expression is in the form of a square, cube, etc. Then the added or subtracted quantities are equated to zero (if possible), producing equations which will render the given quantity a square or a cube, etc. The same strategy may be used for several simultaneous unknown quantities. For instance, let the given quantity be

$$x^2 + xy + y^2 + ax + by + e;$$

add to it the quantity

$$c^2 - e + (2c-b)y + (2c-a)x + xy,$$

producing the square $(x+y+c)^2$. Accordingly we assert

$$c^2 - e + (2c-b)y + (2c-a)x + xy = 0,$$

which will make the given quantity a square.

Case 7. Not infrequently propositions of this sort can be resolved according to the following method: reduce the given equations, if possible, to other forms whose relations are known, and then their resolutions may be deduced.

Example. Let $x^3 + y^3 = w^2$, where x and y are mutually prime. Then this can be reduced to the form $(x + y)(x^2 - xy + y^2) = w^2$, and it is easy to show that $x+y$ and x^2-xy+y^2 must be co-prime unless they have a common divisor 3. Whence $x+y$ = either u^2 or $3u^2$, and x^2-xy+y^2 = either v^2 or $3v^2$.

Case 8. In resolving several problems of this sort it is often worthwhile to choose quantities which afford general or particular solutions to a certain case of the problems; then from the assumed quantities and the remaining cases of the problems their general resolutions may often be derived.

Example 8.1. To find three numbers, the sum or difference of any two of which is to be a square. For the three numbers assume either

$a^2x^2+b^2y^2$, $2axy$, $a^2y^2+b^2x^2$;
or f^2g^2+1, $2fg$, f^2+g^2;
or $\frac{1}{2}(f^4g^4+g^4+f^4+1)$, $\frac{1}{2}(f^4g^4-g^4-f^4+1)+2f^2g^2$, $\frac{1}{2}(f^4g^4-g^4-f^4+1)-2f^2g^2$;
etc.: each of these triplets satisfies four cases of the given problem. Additionally, the three expressions

$$6a^{18}b^2+24a^{14}b^6-92a^{10}b^{10}+24a^6b^{14}+6a^2b^{18},$$
$$a^{20}+21a^{16}b^4-6a^{12}b^8-6a^8b^{12}+21a^4b^{16}+b^{20},$$
and $$10a^{18}b^2-24a^{14}b^6+60a^{10}b^{10}-24a^6b^{14}+10a^2b^{18}$$
(where a and b are integers, and $6a^4b^4+b^8<3a^8$) are three numbers which will satisfy the problem; but this formula does not generate all numbers which satisfy the problem, as the general expressions for the individual numbers we seek must include three quantities that can be chosen at will.

Example 8.2. (1) Required that x^2-xy+y^2 and $x+y$ be squares: the values $x = 2pq-p^2$ and $y = q^2-p^2$ satisfy the first condition, making x^2-xy+y^2 a square; then the second condition, that $x+y = q^2+2pq-2p^2 = (q+p)^2-3p^2$ be a square, is satisfied by $p = 2mn$, $q = 3m^2-2mn+n^2$.

(2) Required that x^2-xy+y^2 and $x+y$ each be three times a square: the numbers

$x = 2p^2-2pq-q^2$, $y = p^2+2pq-2q^2$ (whence $x+y=3p^2-3q^2$)
or $x = 2p^2-2pq-q^2$, $y = p^2-4pq+q^2$ (whence $x+y=3p^2+6pq$)
or $x = p^2+2pq-2q^2$, $y = -p^2+4pq-q^2$ (whence $x+y=6pq-3q^2$)

satisfy the first requirement; then the appropriate value of $x+y$ will be $3v^2$, say, and we derive respectively

$p = m^2+n^2$, $q = 2mn$
or $p = 2m^2$, $q = m^2-n^2$
or $p = m^2+n^2$, $q = 2m^2$.

Case 9. From the known rules of extraction of roots new equations may often be deduced such that, if the corresponding values of these equations are rational, those of the unknowns of the given equation will also be rational.

The irrationalities contained in the root are assigned to be rational fractions or integers, and this is how the new equations may be deduced.

Example 9.1. Let the equation be $ay^2+(bx+c)y+dx^2+ex+f = 0$; then by the standard procedure $y = \sqrt{\frac{(bx+c)^2-4a(dx^2+ex+f)}{4a^2}} - \frac{bx+c}{2a}$, whence $(bx+c)^2-4a(dx^2+ex+f) = Ax^2+Bx+C$ will be a square (w^2, say). For x write $z-\frac{B}{A}$, and we will have $Az^2+\frac{B^2}{4A}+C = v^2$ to be solved for rational z.

Example 9.2. Let the equation be $y^3-(a+bx)y-\frac{1}{2}(cx^2+dx+e) = 0$;

then $$y = \sqrt[3]{cx^2+dx+e+\sqrt{\frac{(cx^2+dx+e)^2}{4}+\frac{(a+bx)^3}{27}}}+\ldots;$$

so make $\frac{(cx^2+dx+e)^2}{4}+\frac{(a+bx)^3}{27} = (Ax^2+Bx+C)^2$ equal to a square, v^2, and $-\frac{1}{2}(cx^2+dx+e) \pm (Ax^2+Bx+C)$ equal to a cube, w^3.

Further remarks on Problem LVI

1. To discover whether or not a given method is a general resolution of a given problem.

This is often determined through substitutions in the constructed resolution; if all substitutions for an unknown quantity or quantities, contained in the given resolution, are completely general, then the resolution is general.

2.1. In the following let x, y, z, v, $\ldots$ all denote integers. Then in particular x will denote all integers, the fraction $\frac{x}{y}$ willdenote all rational quantities, the root $\sqrt[m]{x}$ will denote the mth roots of all integers, and $\sqrt[m]{\frac{x}{y}}$ will denote the mth roots of all rational quantities, etc.

2.2. The quantity α denoting any rational quantity $\frac{x}{y}$, then $\sqrt[m]{\alpha}$ will denote the mth roots of all rational quantities, etc.

2.3. If a quantity α appears only to the first degree in a given resolution, then the resolution is general and generates all rational numbers.

2.4. In the numerator let x be any integer number, not appearing above the first degree, and having no coefficient (except unity): then the resolution generates all integers.

2.5. Let α, β, m, n be any integers. Then there are more values of the power α^n in a given range, than there are of β^m, if m is greater than n. In particular, if H is a given quantity, to which the numbers α^n and β^m are nearly equal, then the number of integer values of the powers α^n and β^m between 0 and H are roughly in the ratio of $H^{\frac{1}{n}} : H^{\frac{1}{m}}$ $(= H^{\frac{m-n}{nm}} : 1)$.

If $a\alpha^n$ and $b\beta^m$ are nearly equal to H, where a and b are integers, then the number of integer values of the quantities $a\alpha^n$ and $b\beta^m$ between 0 and H will be in the ratio $b^{\frac{1}{m}} H^{\frac{m-n}{mn}} : a^{\frac{1}{n}}$.

2.6. This allows us to proceed to the determination of approximate ratios of he number of values in one given resolution, to the number of values in another given resolution — as when ax^m, by^m, cz^l, ... and $p\pi^h$, $q\rho^i$, $r\sigma^k$, ... appear in the respective solutions, where a, b, c, ..., p, q, r, ..., m, n, l, ..., and h, i, k, ... all denote integers.

3. Given an equation $L = 0$, involving the n unknown quantities x, y, z, ..., where x, y, z, ... denote arbitrary integral or rational quantities, to find a general formula (if one exists) for the unknowns of this equation.

3.1. First, find the factors A, B, C, ... of the given equation, if it has any. Equate each factor separately to zero, and we reduce the given equation to two or more other equations $A = 0$, $B = 0$, $C = 0$, etc. If $x = P$, $x = Q$, $x = R$, ... are general resolutions of the equations $A = 0$, $B = 0$, $C = 0$, ..., then $x = P$, $x = Q$, $x = R$, ... will be general resolutions of the equation $L = 0$.

3.2. If the quantity has no factors, then the general solution for the quantity x will involve the n-1 quantities y, z, w, etc. Now these n-1 distinct quantities may often be reduced to a smaller number; e.g. if the n expressions involving the quantities y, z, v, ... are $A(ay+bz+cv+\ldots)^n+B(ay+bz+cv\ldots)^{n-1}+C(ay+bz+cv+\ldots)^{n-2}+\ldots$, we may write W in place of $ay+bz+cv+\ldots$ in the given equation $L = 0$, and it will be reduced to another equation involving n-m+1 unknown quantities. Or in general we may write the most general possible rational expressions π, ρ, σ, τ, ... for the unknown quantities y, z, v, ..., such that from given assumed quantities the unknowns y, z, v, ... are easily recovered. And if by this method we obtain a resultant equation which involves fewer unknown quantities than the original, then in the general expression for x the number of assumed quantities π, ρ, σ, τ, ... will be fewer than n-1, the number of other unknown quantities contained in the given equation.

The highest degree of such an unknown quantity in the required resolution may always be obtained by comparing the degrees and terms contained in the given equation.

4. These equations may be made independent through substitutions, provided that equations, expressing relationships between the unknown quantities in tow or more of the equations on the one hand, and in the new equations on the other, can be assumed such that each unknown quantity is eliminated from all but one of the equations in which it occurs, and the same new assumed quantity is not introduced into two or more equations; and provided that the assumed equations are so chosen that rational correspondent values of the unknown quantities in the assumed equations can easily be inferred from given values of the unknown quantities of the independent equations so deduced.

Problem LVII

Given an equation having two or more unknowns x *and* y, *to find integral corresponding values for its unknowns.*

Find an approximate value of one unknown (x) in terms of the other unknown (y) and the given quantities, which value, augmented or diminished by some assumed quantity, is substituted for x. From the resulting equation, in the same manner, we find an approximate value of the assumed quantity, and substitute this value (augmented or diminished by yet another assumed quantity) in place of the original assumed quantity. And by repeating this process indefinitely, until the true value of one or the other of the unknown quantities emerges, one may find the required values of the unknown quantities of the given equation from the preceding substitutions.

Example. Let $13a^2+1 = x^2$. Here the approximate value of x will be $3a$. Suppose $x = 3a+b$, which is substituted for x in the given equation, the result being $4a^2+1 = 6ab+b^2$. Then assume $a = b+c$, from which $4b^2+8bc+4c^2+1 = 6b^2+6bc+b^2$, which reduces to $2bc+4c^2+1 = 3b^2$;

then assume	$b = c+d$,	with the result	$3c^2+1 = 4cd+3d^2$;
"	$c = d+e$,	" " "	$2de+3e^2+1 = 4d^2$;
"	$d = e+f$,	" " "	$e^2+1 = 6ef+4f^2$;
"	$e = 6f+g$,	" " "	$6fg+g^2+1 = 4f^2$;
"	$f = g+h$,	" " "	$3g^2+1 = 2gh+4h^2$;
lastly, suppose	$g = h+k$,	" " "	$4hk+3k^2+1 = 3h^2$.

From this last equation we arrive at the value $k = 1$, and so $h = 2$. Then $g = h+k = 3$, $f = g+h = 5$, $e = 6f+g = 33$, $d = e+f = 38$, $c = d+e = 71$, $b = c+d = 109$, $a = b+c = 180$, so $x = 3a+b = 649$.

In like fashion we may find corresponding integer values of the unknown quantities in equations of degree higher than the quadratic.

The same method may also be applied to equations having three or more unknown quantities.

This method, slightly modified, can be used to find fractional corresponding values of the unknown quantities.

But in many cases this method offers little or no advantage over trial and error: for example, to solve $xy = 399$, there is no quantity that is suitable for assumption as an approximate value, and so it will be in many other cases.

Note 1. These equations often have neither integral solutions, nor rational, nor irrational of any prescribed type, corresponding to all of the unknown quantities which they involve.

(1) By the methods previously given for finding impossible roots, we can compose certain equations of this type, as well as upper and lower bounds for the real solutions of the corresponding unknown quantities.

(2) If some real roots are given, then various equations can be found which have neither integral, nor rational, nor irrational solutions of a prescribed type, for the other unknown corresponding quantities.

Note 2. (1) We can in some cases easily perceive whether or not the unknowns of certain equations can be integers, or rational numbers, etc., through solution of the given equations.

Example 1.

Let the given equation be $ax^my^n+bx^{n+1}yn-1+cx^{n+2}y^{n-2}+\dots = 0$; substitute πy for x in the equation, the result being $a\pi^m+b\pi^{m+1}+c\pi^{m+2}+\dots = 0$. Then if the resolution of this equation gives no rational roots for π, there can be no corresponding roots of x and y that are both rational.

Example 2.
Let $(x+y+z+v+\ldots)^n = A$; then if the nth root of A is not a rational quantity, the corresponding values of x, y, z, v, ... cannot all be rational at once.

(2) We may also recognize this possibility or impossiblity through division of variables in the given equations by divisors from any source whatever, as well as by noting the form of the factors which divide the said quantities in the given equations, which will be discussed in greater detail later in this Chapter.

(3) This may also be established by observing that no integral corresponding values of some of the unknown quantities can be found between their upper and lower bounds, whence it can be proved that no wholly integral sets of values can be given. Limits of this type should always be given, as they usually cannot be deduced without great difficulty. And so for many more methods.

Problem LVIII

Given two or more (n) *equations in more than* n *unknown quantities,* x, y, ..., *to find integral values corresponding to these unknowns.*

An approximate value of one or more of the unknown quantities is found in terms of the remaining unknown and given quantities, and then we represent this value as increased or diminished by some assumed quantity, or by one assumed quantity for each separate value if we choose to approximate several unknowns. Substitute these values, with their increments or decrements, in their respective places in each equation, and so on, until all the unknowns are found, or until some unknown quantity's value becomes 1, 0, or a known quantity, etc.

Example. Let the two equations be

$$x^2 = y^2 + 5z^2 - 12 \quad \text{and} \quad 7x^2 = 10y^2 + 28z^2 - 69.$$

Assume $x = y+2z-v$, and substitute this for the value of x, resulting in

$$y^2 + 4yz + 4z^2 - 2yv - 4zv + v^2 = y^2 + 5z^2 - 12;$$

thus $\quad v^2 - 4zv - 2yv = z^2 - 4yz - 12,$

and $\quad 7v^2 - 28z\,v - 14yv + 28yz = 3y^2 - 69.$

Now at this point we suppose in the first equation $v = z{+}w$, and in the second $v = y{-}u$, whence, if we write in the two preceding equations $z{+}w$ for v and $z{+}w{+}u$ for y, there result two equations

$$2z^2 + 2zw + w^2 = 12 + 2zu - 2wu$$

and $$10z^2 + 20zw + 10w^2 - 8zu + 20wu = 69 - 3u^2.$$

Then assume $w = 0$, giving $u = 1$, $z = 3$, $v = z{+}w = 3$; so $y = z{+}w{+}u = 4$, and $x = y{+}2z{-}v = 7$.

Corollary. The same method may be applied to n equations in $n{+}1$ or more unknowns. But the amount of work is much greater, and often no method can be given, except repeated trial and error.

This method can also find sets of corresponding fractional values.

Problem LIX

Given one, two, or more integral values of each of the unknown quantities in a given equation involving two or more unknowns, and given further a method of deducing another integral value from each unknown, to deduce an infinity of integral values for each unknown quantity.

Case 1. Repeat continuously the operation that deduces other values from one or more already known, and the objective is attained.

Example 1. Let the equation be $ax^2{+}1 = y^2$; having found r and s, a pair of correspondent integer values for x and y, another integer value of x will be found to be $2rs$, and the corresponding value of y will be $2ar^2{+}1 = 2s^2{-}1$. Any expression

$$2^m s^m - (m{-}1)\cdot 2^{m-2}s^{m-2} + \frac{(m{-}2)(m{-}3)}{2!}\cdot 2^{m-4}s^{m-4} - \frac{(m{-}3)(m{-}4)(m{-}t)}{3!}\cdot 2^{m-6}s^{m-6} + \frac{(m{-}4)(m{-}5)(m{-}6)(m{-}7)}{4!}\cdot 2^{m-8}s^{m-8} - \dots,$$

multiplied by r, will be an integer root of x, to which corresponds

$$y = 2^m s^{m+1} - (m{+}1)2^{m-2}s^{m-1} + \frac{(m{+}1)(m{-}2)}{2!}\cdot 2^{m-4}s^{m-3} - \frac{(m{+}1)(m{-}3)(m{-}4)}{3!}\cdot 2^{m-6}s^{m-5} + \frac{(m{+}1)(m{-}4)(m{-}5)(m{-}6)}{4!}\cdot 2^{m-8}s^{m-7} - \frac{(m{+}1)(m{-}5)(m{-}6)(m{-}7)(m{-}8)}{5!}\cdot 2^{m-10}s^{m-9} + \dots.$$

If we write $ar^2{+}1$ for s^2 in these series, we have equivalent series in terms of the quantity r. These series terminate when the exponent of s becomes negative.

The same values of the quantity x may be designated by the following product: $r \times \left(2s+1\frac{1}{1}\right)\left(2s-1+\frac{2s-1}{2s}\right)\left(2s-1+\frac{4s^2-2s-1}{4s^2-1}\right)$

$$\times\left(2s-1+\frac{8s^3-4s^2-4s+1}{8s^3-4s}\right)\cdots$$

i.e., if $r = 2$, $2s = 6 = 5+\frac{1}{1}$, the value of the quantity x will be

$$2\left(5+\frac{1}{1}\right)\left(5+\frac{29}{35}\right)\left(5+\frac{169}{204}\right)\cdots$$

where the numerator of each fraction is equal to the same fraction's denominator, diminished by the denominator of the preceding fraction; and the denominator is equal to the numerator of the preceding term when the latter is expressed as an improper fraction. This solution was first given by Viscount Brounker.

These methods do not find all possible values: from any given values others may be sought in many different ways. But since by hypothesis no rule can be given which generates every possible solution, it follows that no rule can be found which encompasses another less general than itself, and no method of reasoning sufficies to deduce new rules that are not limited in the same way as the original rule.

Example 2. Let $ax^2+bx+c = y^2$, and let p and $q = \sqrt{ap^2+bp+c}$ be corresponding values of the unknown quantities x and y; and suppose we wish to discover further values corresponding to these quantities. Assume $dp+eq+f = x$ and $gp+hq+l = y$. Upon substitution of these quantities for x and y in the equation $ax^2+bx+c = y^2$, the result is

$$a(d^2p^2+2dfp+f^2+e^2q^2)+bdp+bf+c+(2adep+2afe+be)q$$
$$= g^2p^2+2gpl+h^2q^2+l^2+(2ghp+2hl)q.$$

In this equation write ap^2+bp+c in place of q^2, and we have

$$(ad^2+a^2e^2-g^2-ah^2)p^2+(2adf+bd-2ql+abe^2-bh^2)p$$
$$+(af^2+bf-l^2-ch^2+ace^2+c)+(2adep+2afe+be-2ghp-2hl)q = 0.$$

Now equating separately to zero the terms

$$ad^2+a^2e^2-g^2-ah^2 = 2adf+bd+abe^2-2gl-bh^2 =$$
$$af^2+bf-l^2-ch^2+ace^2+c = 2ade-2gh = 2afe+be-2hl = 0,$$

we have that $\frac{ade}{h} = g$ and $l = \frac{2afe+be}{2h}$; in the first equation substitute $\frac{ade}{h} = g$, whence $d^2h^2+ae^2h^2-ad^2e^2-h^4 = -(h^2-d^2)(h^2-ae^2) = 0$. But h^2-ae^2 cannot be 0, unless a is a square. Therefore $h^2-d^2 = 0$, and it follows that $h = \pm d$. Now writing the values d, ae, and $\frac{2afe+be}{2d}$

for h, g, and l respectively in the second equation, we will have

$$2adf+bd+abe^2-2ae\left(\frac{2afe+be}{2d}\right)-bd^2 = -\frac{1}{d}(d^2-ae^2)(bd-b-2af),$$

whence $f = \frac{bd-b}{2a}$; writing these quantities for their values in the equation $af^2+bf-l^2-ch^2+ace^2+c = 0$ gives $\frac{1}{4a}(b^2-4ac)(d^2-ae^2-1) = 0$. We find the corresponding integer values of the quantities d and e in the equation $d^2 = ae^2+1$, which we may call α and π respectively; then

$$x = dp+eq+f = p\alpha+q\pi+\frac{b}{2a}(\alpha-1), \text{ and } y = ap\pi+q\alpha+\frac{b\pi}{2}.$$

Therefore the successive values of the unknown quantity x will be

$$p,\ p\alpha+q\pi+\frac{b(\alpha-1}{2a},\ 2p\alpha^2+2\pi\alpha q-p+\frac{b(\alpha^2-1)}{a},\ \text{etc.}$$

(Let E, F, and G be three successive values in this sequence, then

$$2\alpha F-E+b\cdot\frac{\alpha-1}{a} = G.)$$

Successive values of the unknown quantity y turn out to be

$$q,\ ap\pi+\alpha q+\frac{b\pi}{2},\ 2a\pi\alpha p+2\alpha^2 q-q+b\pi\alpha,\ \text{etc.}$$

(If three successive terms of this sequence are P, Q, and R, then $R = 2\alpha Q-P$.)

The equation $v^2+(A+Bx)v+Cx^2+Dx+E = 0$ may easily be transformed into the form $y^2 = ax^2+bx+c$, or even $y^2 = ax^2+d$; nothing more is required than to remove the second term of the given equation.

Note 1. The coefficient a should not be a square; if it is the given equation has an infinite number of solutions.

Note 2. In resolutions of this sort, the greater the number of independent unknown quantities there are, the more general will be the resolution.

Note 3. If r and s are corresponding values of x and y in the equation $ax^2+1 = y^2$, then $\frac{s}{r}$ is approximately equal to $\sqrt{a}$; this approximation is much better when the magnitudes of r and s are large: but this remark is more relevant to infinite series.

Case 2. (1) In a given algebraic equation which is free of fractions and surds, let m be its highest degree in all the unknown quantities x, y, z, ... In this equation,

for x write $au^n + bu^{n-1} + cu^{n-2} + \ldots$,
" y " $a'u^{n'} + b'u^{n'-1} + c'u^{n'-2} + \ldots$,
" z " $a''u^{n''}+b''u^{n''-1}+c''u^{n''-2}+\ldots$,

and so on, the result being an equation

$$Au^3 + Bu^{s-1} + Cu^{s-2} + \ldots = 0.$$

Now equate its coefficients A, B, C, ... to 0, and for each of the unknown quantities a, b, c, ... a', b', c', ... a'', b'', c'', ... in the resulting equations $A = 0$, $b = 0$, $C = 0$, ... find if possible a rational value. Then it follows that an infinite number of rational or integral values may be found for the unknown quantities x, y, z, ..., merely by assuming for u all integral or rational quantities.

In this case the number of unknown quantities a, b, c, ... a', b', c', ... a'', b'', c'', ... necessarily exceeds by two, three, four, or more, the number (s+1) of equations $A = 0$, $B = 0$, $C = 0$, etc.

The lower the indices n, n', n'', ... are chosen, the more general will be the values of the quantities x, y, z, ... which this method finds.

Whether or not the suggested strategy will find a resolution depends in large part on the number of unknown quantities and their dimensions in the given equation.

(2) For x, y, z, ... we may assume quantities of the form

$$\frac{au^n+bu^{n-1}+\ldots}{pu^m+qu^{m-1}+\ldots},\quad \frac{a'un'+b'u^{n'-1}+\ldots}{p'u^{m'}+q'u^{m'-1}+\ldots},\text{ etc.,}$$

which, upon substitution in their respective places in the given equation and all the terms of the resulting equation being reduced to a common denominator, gives an equation whose numerator is

$$Au^s + Bu^{s-1} + Cu^{s-2} + \ldots = 0.$$

Now set $A = 0$, $B = 0$, $C = 0$, ... as before, whence the corresponding values of the unknown quantities a, b, ... p, q, ... a', b', ... p', q', ... etc. are found, and so an infinity of corresponding values of the unknown quantities x, y, z, ..., contained in the given equation, may be found.

(3) The resolution will be still more general, if for the quantities x, y, z, ... in the given equation we substitute quantities of the form

$$au^n + (b+cw+\ldots)u^{n-1} + (d+ew+\ldots)u^{n-2} + \ldots,$$

etc., or any rational function of the letters u, w, U, etc. (where u, w, U, etc. denote any arbitrary numbers or rational quantities which may be assigned), and reduce all the terms of the resulting equation to a common denominator. Then equate to zero the numerator of each term (arranged by their degrees in all of the unknown quantities u, w, U, etc.) and find corresponding rational values of each of the unknown quantities a, b, ... p, q, ... a', b', ... p', q', ... etc. in the resulting equations.

The same principle may be used to locate several solutions to a given equation involving some particular irrationality, or to several equations in several unknowns whose solutions are to be integers, rational numbers, or irrationalities of prescribed types.

Problem LX

> *Given any number expressed in the common notation of Indian origin, as a sum of multiples of unity and the numbers* $10, 10^2, 10^3$, *etc., or any other similar positional notation, to transform this into another represented by multiples of unity and of the numbers* $\alpha, \alpha^2, \alpha^3, \alpha^4$, *etc.*

Divide the number N by α and call the quotient N' and the remainder R, i.e. $N/\alpha = N' + \frac{R}{\alpha}$; likewise divide N' by α, calling the quotient N'' and the remainder R'; then divide N'' by α and let the quotient be N''' and the remainder R''. Continue in this manner until a quotient is found which is less than α, call this quotient $N^{(m)}$. Then the required representation will be $N^{(m)} \cdots R''''R'''R''R'R$.

Example. To transform the number $37987 = N$ from the common notation into a positional representation based on unity and the powers $6, 6^2, 6^3$, etc. The operation will be

$6331 = N'$	$1055 = N''$	$175 = N'''$	$29 = N''''$	$4 = N'''''$
6)37987	6)6331	6)1055	6)175	6)29
36	6	6	12	24
19	33	45	55	5th rem=5=R''''
18	30	42	54	
18	31	35	4th rem=1=R'''	
18	30	30		
7	2nd rem=1=R'	3rd rem=5=R''		
6				
1st rem=1=R				

whence the required notation will be $N'''''R''''R'''R''R'R = 451511$. Exactly the same methods are used for addition, subtraction, multiplication, division, extraction of roots, etc., of numbers or fractions designated in this fashion, as they are for the same operations in the common numeric notation.

On the properties of integers

Theorem 46

(1) Let $N = a \cdot b \cdot c \cdot d \cdots$; now suppose a number M is divided by a, and let the quotient be M' and the remainder p; further let M' be divided by b, with quotient M'' and remainder q, i.e. let the division proceed as

$$\begin{array}{r|c|c|c|l} & b & c & d & \\ a\ M & M' & M'' & M''' & \cdots; \\ \hline p & q & r & s & \end{array}$$

then the remainder on division of M by N will be $p+aq+abr+abcs+\ldots$.

(2) If the remainders, when the numbers a, b, c, d, e, ... are divided by α, are respectively π, ρ, σ, τ, ..., then the remainder when the number $ma+nb+rc+sd+te+\ldots$ is divided by α will be the same as the remainder when the number $m\pi+n\rho+r\sigma+s\tau+\ldots$ is divided by α. [11]

(3) Let the quantity $a^m b^n c^r d^s \cdots$ be any number where a, b, c, d, ... are primes, no two of them equal; then the number of divisors of this number will be $(m+1)(n+1)(r+1)(s+1)\cdots$.

In fact they will be a, a^2, a^3, ... a^m; b, ab, a^2b, a^3b, ... a^mb; b^2, ab^2, a^2b^2, a^3b^2, ... a^mb^2; b^3, ab^3, a^3b^3, a^3b^3, ... a^mb^3; ... b^n, ab^n, a^nb^n, a^3b^n, ... a^mb^n; ac, bc; a^2c, abc, b^2c; ... ac^2, bc^2, ...; etc. Hence the sum of all the divisors will be equal to the product $(1+a+a^2+a^3+\ldots+a^m)$ $(1+b+b^2+b^3+\ldots+b^n)$ $(1+c+c^2+c^3+\ldots+c^r)$ $(1+d+d^2+d^3+\ldots+d^s)$ $(\cdots)$, and the number of divisors is evidently $(m+1)(n+1)(r+1)(s+1)\cdots$.

Corollary 3.1. Hence given the prime divisors of a quantity, and their powers, we can easily deduce the total number of that quantity's divisors.

Corollary 3.2. Hence we can find an infinity of numbers having a given number of divisors, as well as the smallest number having the given number of divisors. Choose a, b, c, d, ..., the smallest mutually prime numbers, and indices m, n, r, s, ..., such that $(m+1)(n+1)(r+1)(s+1)\cdots$ = the given number of divisors, and $a^m b^n c^r d^s \cdots$ will be the least such required number.

Corollary 3.3. If and only if the number of divisors is odd, the number is a square: for if the number of divisors is odd, then all of $m+1$, $n+1$, $r+1$, $s+1$, ... will be odd numbers, and so all of m, n, r, s, ...

are even numbers, therefore the quantity $a^m b^n c^r d^s \cdots$ is a square. [12]

(4) If 2^n-1 is a prime number, then $2^{n-1}(2^n-1)$ is a *perfect* number: the sum of its divisors, by the preceding theorem, is $(1+2+4+8+\ldots+2^{n-1} = 2^n-1 = A)(1+2^n-1 = 2^n = B) = AB = 2^n(2^n-1)$; from this we subtract the number itself, $2^{n-1}(2^n-1)$ and we are left with $2^{n-1}(2^n-1)$. Therefore the sum of all divisors of this number, except the number itself, is equal to this number.

(5) Let N and M be *amicable* numbers, i.e. $N = SM-M$ and $M = SN-N$, where SM and SN denote the sums of the divisors of M and N respectively. Let $2^n x$ and $2^n yz$ be two amicable numbers, where x, y, and z are prime numbers. The sum of the divisors of the first number (less the number itself) will be $2^{n+1}-1+(2^n-1)x = 2^n yz$, and of the second, $2^{n+1} - 1+(2^{n-1}-1)(y+z)+(2^n-1)yz = 2^n x$; from the first equation we have $x = \dfrac{2^n yz-2^{n+1}+1}{2^n-1}$, and then from the second, $z = \dfrac{(2^n-1)y+2^{n+1}-1}{y-2^n+1} = 2^n-1+\dfrac{2^{2n}}{y-2^n+1}$; so y should be chosen to be a prime such that $y-2^n+1$ is a divisor of 2^{2n}, and z and x also remain primes.

This method leads to the discovery of the amicable numbers 284 and 220, or 18416 and 17296, etc. The same strategy serves to discover an infinity of other amicable numbers, when in place of the factors 2^n in each of the expressions we choose any arbitrary numbers, and for x, y, and z we may choose composite numbers instead of primes.

Let a and b be given numbers, and let z, y, x, $\ldots$ and α, β, γ, $\ldots$ be the required primes, such that $axyz\cdots = N$ and $b\alpha\beta\gamma\cdots = M$ are amicable numbers. Let m and n be the sums of the divisors of a and b respectively. Then $m(x+1)(y+1)(z+1)\cdots-N = b\alpha\beta\gamma\cdots$, and $n(\alpha+1)(\beta+1)(\gamma+1)\cdots-M = axyz\cdots$.

It is easy to show that by this method (*viz.*, by assuming quantities whose factorization is known, and using equations to express the required relationships) an infinity of similar propositions may be deduced. Such would be the method of finding different numbers, the sums of whose divisors are all the same (e.g. if the number itself is counted among its divisors, then 6 and 11; 10 and 17; 14, 15, and 23; etc. will be numbers of this type), or numbers having some other relation among themselves — or numbers can be found, the sums of whose divisors have some prescribed relation to the numbers themselves.

(6) Let there be a number $a^m b^n c^r d^s \cdots = P$, the sum of whose divisors will be

$$\frac{a^{m+1}-1}{a-1} \cdot \frac{b^{n+1}-1}{b-1} \cdot \frac{c^{r+1}-1}{c-1} \cdot \frac{d^{s+1}-1}{d-1} \cdots = H;$$

multiply P by a number $Q = a^\alpha b^\beta c^\gamma d^\delta \cdots$, where α, β, γ, δ, ... are integers. Then the sum of the divisors of the number PQ will be

$$\frac{a^{m+\alpha+1}-1}{a-1} \cdot \frac{b^{n+\beta+1}-1}{b-1} \cdot \frac{c^{r+\gamma+1}-1}{c-1} \cdot \frac{d^{s+\delta+1}-1}{d-1} \cdots = K.$$

When the sum K is divided by the sum H, the quotient is

$$\frac{a^{m+\alpha+1}-1}{a^{m+1}-1} \cdot \frac{b^{n+\beta+1}-1}{b^{n+1}-1} \cdot \frac{c^{r+\gamma+1}-1}{c^{r+1}-1} \cdot \frac{d^{s+\delta+1}-1}{d^{s+1}-1} \cdots.$$

If m, n, r, ... are much larger than α, β, γ, ..., the quotient is very close to $a^\alpha b^\beta c^\gamma d^\delta$.

If α, β, γ, δ, ... are mutually prime and prime with respect to H, and if $N = \alpha\beta\gamma\delta \cdots H$, and the sum of the divisors of H is L, then the sum of the divisors of the number N will be $(\alpha+1)(\beta+1)(\gamma+1)(\delta+1) \cdots L$.

(7.1) Let $A = 1 \cdot 2 \cdot 3 \cdot 4 \cdot 5 \cdot 6 \cdot 7 \cdot 8 \cdots (l-2)(l-1)$. If this is multiplied by l, then the sum of the divisors of the resulting quantity has, to the sum of the divisors of A, a ratio not exceeding $(l+1) : 1$, and not less than $l : 1$. If the number h is less than l, then the sum of the divisors of lA will be greater than the sum of the divisors of hA; and so on.

(7.2) If $N = 1 \cdot 2 \cdot 3 \cdot 4 \cdot 5 \cdot 6 \cdot \cdots \cdot l$, then N has a ratio to the sum of its divisors which is less than the comparable ratio for any smaller number.

(8.1) Let $a^x b^z = A$, and let $(x+1)(z+1)$ be maximized given that a and b are fixed quantities. Then $a^{x+1} = b^{z+1}$.

For we may take the fluxion of the quantity $a^x b^z = A$, which will be

$$a^x b^z(\dot{x} \log a - \dot{z} \log b) = 0,$$

whence $\dot{z} = \dfrac{\log a}{\log b} \cdot \dot{x}$. But the fluxion of the product $(x+1)(z+1)$ will be zero, since by hypothesis this product itself is maximized, i.e. $x\dot{z}-z\dot{x}-\dot{x}+\dot{z} = 0$. Now in this equation for z write its previously computed value, and the result is $(x+1)\log a = (z+1)\log b$, and in consequence $a^{x+1} = b^{z+1}$.
Q.E.D.

(8.2) If there are more factors, i.e. $a^x b^z c^y d^v \cdots = A$, and once again we stipulate that $(x+1)(z+1)(y+1)(v+1) \cdots$ be maximized, we will have

$$(x{+}1) : (z{+}1) :: \log b : \log a;$$
$$(x{+}1) : (y{+}1) :: \log c : \log a;$$
$$(x{+}1) : (v{+}1) :: \log d : \log a;$$

and so on. In consequence the quantities $x{+}1$, $z{+}1$, $y{+}1$, $v{+}1$, ... will be in inverse ratio as the quantities $\log a$, $\log b$, $\log c$, $\log d$, ..., whence

$$a^{x+1} = b^{z+1} = c^{y+1} = d^{v+1} = \cdots.$$

(9.1) Under the same conditions we can show that the product

$$\frac{a^{x+1}-1}{a-1} \cdot \frac{b^{z+1}-1}{b-1} \frac{c^{y+1}-1}{c-1} \cdot \frac{d^{v+1}-1}{d-1} \cdots$$

is maximized when $a^{x+1} = b^{z+1} = c^{y+1} = d^{v+1} = \cdots$.

(9.2) Let a, b, c, d, ... be primes, and let x, y, z, v, ... be integers. Then $(x{+}1)(z{+}1)(y{+}1)(v{+}1)\cdots$ will be the number of divisors, and

$$\frac{a^{x+1}-1}{a-1} \cdot \frac{b^{z+1}-1}{b-1} \cdot \frac{c^{y+1}-1}{c-1} \cdot \frac{d^{v+1}-1}{d-1} \cdots$$

will be the sum of the divisors, of the quantity A. In this case however, we may not in general have a^{x+1}, b^{z+1}, c^{y+1}, d^{v+1}, ... strictly equal to each other; but the number and sum of the divisors will still be very large when these quantities are nearly equal to each other.

(9.3) If the numbers a, b, c, d, ... are not given, still the number and sum of the divisors will be larger when a, b, c, d, ... are nearly equal to each other, and where there are many quantities a, b, c, d, ...

(10) $$S(n) = S(n{-}1) + S(n{-}2) - S(n{-}5) - S(n{-}7) + S(n{-}12) + S(n{-}15) - S(n{-}22) - S(n{-}26) + S(n{-}35) + S(n{-}40) - \ldots$$

where $S(n)$, etc. denote the sum of the divisors of n, etc. The signs $+$ and $-$ alternate by pairs of terms.

This formula contains every number of the form $\frac{3z^2 \pm z}{2}$, for z any integer.

The series terminates when the argument of the term $S(x)$ becomes negative or zero; if it becomes exactly zero, for this last term write n.

Proof. Let $z = xS(1){+}x^2S(2){+}x^3S(3){+}x^4S(4){+}x^5S(5){+}\ldots$, whence

$$z = 1(x{+}x^2{+}x^3{+}x^4{+}x^5{+}\ldots){+}2(x^2{+}x^4{+}x^6{+}x^8{+}\ldots){+}3(x^3{+}x^6{+}x^9{+}x^{12}{+}\ldots)$$
$$+4(x^4{+}x^8{+}x^{12}{+}\ldots) \quad + 5(x^5{+}x^{10}{+}\ldots) \quad + \ldots$$
$$= \frac{1x}{1-x} + \frac{2x^2}{1-x^2} + \frac{3x^3}{1-x^3} + \frac{4x^4}{1-x^4} + \ldots,$$

whence $$\frac{z\dot{x}}{x} = \frac{\dot{x}}{1-x} + \frac{2x\dot{x}}{1-x^2} + \frac{3x^2\dot{x}}{1-x^3} + \ldots,$$

whose fluent will be $-\int \frac{z\dot{x}}{x} = \log(1-x) + \log(1-x^2) + \log(1-x^3) + \ldots$
$= \log\left[(1-x)((1-x^2)(1-x^3)(1-x^4)\cdots\right]$
$= \log(1-x-x^2+x^5+x^7-x^{12}-\ldots)$

whence $z = \dfrac{x+2x^2-5x^5-7x^7+12x^{12}+15x^15-\ldots}{1-x-x^2+x^5+x^7-x^{12}-x^{15}-\ldots}$
$= xS(1)+x^2S(2)+x^3S(3)+x^4S(4)+\ldots,$

from which equations it follows that

$S(1) = 1,\ S(2) = S(1) + 2,\ S(3) = S(2) + S(1),$
$S(4) = S(3) + S(2),\ S(5) = S(4) + S(3) - 5$, etc., and in general
$S(n) = S(n-1) + S(n-2) - S(n-5) - S(n-7) + \ldots$

Note 1. Let the equation be

$$(x-1)(x^2-1)(x^3-1)(x^4-1)(x^5-1)\cdots(x^n-1)$$
$$=x^h-px^{h-1}+qx^{h-2}-rx^{h-3}+sx^{h-4}-\ldots$$
$$=x^h-x^{h-1}-x^{h-2}+x^{h-5}+x^{h-7}-x^{h-12}-x^{h-15}+x^{h-22}$$
$$+x^{h-26}-x^{h-35}-x^{h-40}+x^{h-51}+\ldots\pm x^{h-n}\pm\ldots = A = 0.$$

The + and − signs alternate in pairs up to the term x^{h-n}. Coefficients of the terms up to x^{h-n} are +1, −1, or 0; they will be +1 for terms x^{h-v} where $v = \frac{3z^2\pm z}{2}$ and z is an even number, and −1 for $v = \frac{3z^2\pm z}{2}$ and z is an odd number; in all other cases they will be 0.

The numbers 1, 2, 5, 7, 12, 15, 22, 26, 35, 40, ..., which are to be subtracted from the index h, are obtained as sums of the numbers 1, 1, 3, 2, 5, 3, 7, 4, 9, 5, 11, 6, ..., which are derived by merging together the two sequences 1, 3, 5, 7, 9, 11, ... and 1, 2, 3, 4, 5, 6, 7, ...

Note 2. The sum of the mth powers of the roots of $A = 0$ will be $S(m)$, where $S(m)$ is the sum of the divisors of m, provided $m \leq n$.

Corollary 10.1. By this rule we observe that $S(m) =$
$pS(m-1)-qS(m-2)+rS(m-3)-sS(m-4)+\ldots$
$= S(m-1)+S(m-2)-S(m-5)-S(m-7)+S(m-12)+S(m-15)-S(m-22)$
$-S(m-26)+\ldots$, which is the property of divisors given by Euler.

Corollary 10.2. If in the given equation

$$S(m) = S(m-1)+S(m-2)-S(m-5)-S(m-7)+\ldots$$

we substitute the values $S(m-2)+S(m-3)-S(m-6)-S(m-8)+\ldots,$
$S(m-3)+S(m-4)-S(m-7)-S(m-9)+\ldots,$
etc. for $S(m-1)$, $S(m-2)$, etc. respectively, we can derive an expression for $S(m)$ in terms of the divisors of numbers less than $m-1$, or

recursively less than $m-2$, and so on.

The same method, carried out to its conclusion, may be applied to certain subsequent propositions.

Corollary 10.3. Here, by Corollary 2 of Problem 3, either 0, or $\pm 1 \cdot 2 \cdot 3 \cdot 4 \cdots m$, will be the value of

$$\frac{m-1}{2}S(2) + m(m-1)\frac{m-2}{3} - m(m-1)(m-2)\frac{m-3}{4}S(4) + \dots$$
$$+m(m-1)\frac{m-2}{2} \cdot \frac{m-3}{2^2}(S(2))^2 - \dots,$$

in which equation the sum of the divisors of the number m is expressed by the sums of the divisors of the lesser numbers $m-1$, $m-2$, $m-3$, ... and their powers. If v is an even number, then $\pm 1 \cdot 2 \cdot 3 \cdots m$ has the same sign as the coefficient; if odd, then it has the contrary sign; and if the coefficient itself is zero then the product vanishes.

Note 3. Let H be the number of ways by which a number m can be expressed as the sum of any two of the numbers 1, 2, 3, 4, ... $m-2$, $m-1$; H' the number of ways to express m as the sum of three of these numbers; H'', H''', H'''', etc., the number of ways m may be expressed as the sum of 4, 5, 6, etc. of these numbers respectively. Then $1-H+H'-H''+H'''-\dots = \pm 1$ or 0: if $m = \dfrac{3z^2 \pm z}{2}$, it will be $+1$ if z is odd and -1 if z is even; and it will be 0 in all other cases.

Note 4. Let the equation be

$$(x-1)(x^2-1)(x^3-1)(x^5-1)(x^7-1)(x^{11}-1)(x^{13}-1)\dots(x^n-1)$$
$$= x^{h'}-x^{h'-1}-x^{h'-2}+x^{h'-4}+x^{h'-8}-x^{h'-10}-x^{h'-11}+x^{h'-12}+x^{h'-16}$$
$$-x^{h'-17}-x^{h'-19}+x^{h'-21}-x^{h'-23}+2x^{h'-24}-x^{h'-26}-x^{h'-27}$$
$$+x^{h'-28}-\dots = A' = 0;$$

then the sum of the mth powers of the roots of the equation $A' = 0$ will be $S'(M)$, where $S'(M)$ denotes the sum of all prime divisors of m for $m \leq n$.

Corollary 10.4. By the prescribed rule,

$$S'(M) = S"(M-1)+S'(M-2)-S'(M-4)-S'(M-8)+S'(M-10)+S'(M-11)$$
$$-S'(M-12)-S'(M-16)+S'(M-17)+S'(M-19)-S'(M-20)+S'(M-23)$$
$$-2S'(M-24)+S'(M-26)+S'(M-27)-S'(M-28)+S'(M-29)-\dots.$$

If in this or the preceding similar series, the index in $S(m-r)$ or $S'(M-r)$ vanishes, i.e. it becomes $S(0)$ or $S'(0)$, use r in place of the S or S' expression.

Corollary 10.5. Let L be the coefficient of $x^{h'-m}$; then by the aforementioned Corollary to Problem 3,

$$\pm 1\cdot 2\cdot 3\cdots m\times L = 1-m\cdot\frac{m-1}{2}S'(2)+m(m-1)\frac{m-2}{3}S'(3)\pm\ldots,$$

an equation which expresses a relation between the prime divisors of the numbers 1, 2, 3, 4, ... m and their powers.

Corollary 10.6. The coefficient $\pm L$ is equal to the difference between the number of ways the number m can be constructed as a sum of the primes 1, 2, 3, 5, 7, 11, 13, 17, ..., and the number of ways m can be so constructed without including the number 2.

Note 5. Let the equation be $(x^\alpha-1)(x^\beta-1)(x^\gamma-1)(x^\delta-1)\cdots = x^h - px^{h-1} + qx^{h-2} - \ldots = 0$, then the mth power sum of the roots of this equation will be the sum of all the divisors of m which are to be found among the numbers α, β, γ, δ, ...

The coefficient of x^{h-m} will be the difference between two ways of expressing m as the sum of numbers $\alpha, \beta, \gamma, \delta, \ldots$, one of which includes an odd number of even integers, the other an even number of them, or no even integers at all.

This last principle can be extended to equations whose factors have the form $x^\alpha\pm k$, and from the lower order power sums and coefficients the coefficients of the higher terms can be determined. (The same can be accomplished, of course, using only the coefficients.)

(11) Let a, b, c, d, e, f, ... be primes. Then the ratio of the number of integers which are not divisible by these primes, to the number of integers which are so divisible, will be as $P : 1-P$, where

$$\begin{aligned}P = 1 &- \left(\tfrac{1}{a}+\tfrac{1}{b}+\tfrac{1}{c}+\tfrac{1}{d}+\tfrac{1}{d}+\ldots\right) + \left(\tfrac{1}{ab}+\tfrac{1}{ac}+\tfrac{1}{bc}+\tfrac{1}{ad}+\tfrac{1}{bd}+\tfrac{1}{cd}+\tfrac{1}{ae}+\ldots\right)\\ &- \left(\tfrac{1}{abc}+\tfrac{1}{abd}+\tfrac{1}{acd}+\tfrac{1}{bcd}+\tfrac{1}{abe}+\ldots\right)+\left(\tfrac{1}{abcd}+\tfrac{1}{abce}+\tfrac{1}{acde}+\tfrac{1}{bcde}+\ldots\right)\\ &- \left(\tfrac{1}{abcde}+\tfrac{1}{abcdf}+\tfrac{1}{acdef}+\tfrac{1}{bcdef}+\ldots\right) + \left(\tfrac{1}{abcdef}+\ldots\right) - \ldots\end{aligned}$$

(12) Let a, b, c, d, e, ... be specifically the prime numbers 2, 3, 5, 7, 11, 13, ...; then stopping the sequence at a=2 we have $P = \frac{1}{2}$; with the first two terms, a=2 and b=3 we have $P' = \left(1-\frac{1}{2}\right)\cdot\frac{1}{3} = \frac{1}{6}$; for the series a=2, b=3, c=5 we have $P'' = \left(1-\left(\frac{1}{2}+\frac{1}{3}\right)+\frac{1}{2\cdot 3}\right)\cdot\frac{1}{5} = \frac{1}{15}$; for the case a=2, b=3, c=5, d=7,

$$P'' = \left[1-\left(\tfrac{1}{2}+\tfrac{1}{3}+\tfrac{1}{5}\right) + \left(\frac{1}{2\cdot 3}+\frac{1}{2\cdot 5}+\frac{1}{3\cdot 5}\right) - \frac{1}{2\cdot 3\cdot 5}\right]\cdot\frac{1}{7} = \frac{4}{105};$$

and similarly

$$P''' = \left[1-\left(\tfrac{1}{2}+\tfrac{1}{3}+\tfrac{1}{5}+\tfrac{1}{7}\right) + \left(\frac{1}{2\cdot 3}+\frac{1}{2\cdot 5}+\frac{1}{3\cdot 5}+\frac{1}{2\cdot 7}+\frac{1}{3\cdot 7}+\frac{1}{5\cdot 7}\right)\right.$$
$$\left.-\left(\frac{1}{2\cdot 3\cdot 5}+\frac{1}{2\cdot 3\cdot 7}+\frac{1}{2\cdot 5\cdot 7}+\frac{1}{3\cdot 5\cdot 7}\right)+\frac{1}{2\cdot 3\cdot 5\cdot 7}\right]\cdot\frac{1}{11} = \frac{26}{1155};$$

etc., therefore the infinite sum $P+P'+P''+P'''+\ldots$ is 1.

(13) Let N be a number whose factorization is $N = p^\alpha q^\beta r^\gamma s^\delta \cdots$, where p, q, r, s, ... denote prime numbers. Then the number of integers less than N and prime to N will be some integral function of $p\cdot q\cdot r\cdot s\cdots$, involving also an integer L which is prime to N. This proposition may easily be generalized further. [13]

(14) Let a number $a+b$ be given, and assume $10c+b = (10-\pi)m$, or $100c+b = (100-\pi)m$, etc., where a, b, c, π, and m are integers. then $a-10c+10m$ (or $a-100c+100m$, etc.) is divisible by π if the given number, $a+b$, is divisible by π.

For b write $(10-\pi)m-10c$, and $a+b$ becomes $a-10c+10m-\pi m$, whence the proposition is proved. If we write 7 for π, we have the method of Herr Krafft. [14]

(15) Let $\frac{f+e}{x}$ be an integer, m a divisor of f, and $m-x = n$. Then

$$\frac{\frac{nf}{m}+e}{x} = \frac{\frac{(m-x)f}{m}+e}{x} = \frac{f+e}{x} - \frac{f}{m}$$

will be an integer.

Example. (1) Let the number $fedcba$ be a multiple of the number x, where f, e, d, c, b, and a denote the respective decimal digits of the number. Say $10-x = n$: multiply n by f and add e, and append to the sum (denoted nf_e) the digits $dcba$, and the number nf_{edcba} will be a multiple of x, as shown by Fontenelle; thus also nfe_{dcba}, $nfed_{cba}$, etc. will be multiples of x. (2) Let $100-x = n$, then the numbers nf_{edcba}, nfe_{dcba}, $nfed_{cba}$, ... will be multiples of the number x, if the number $fedcba$ is a multiple of the number x. Thus if $fedcb = 41573$ is a multiple of $x = 7$, taking $10-x = 3$, then $nf_{edcb} = 3\times 4_{1573} = 13573$, $nfe_{dcb} = 3\times 41 = 12873$, etc. Likewise taking $100-x = n = 93$, $nf_{edcb} = 93\times 4_{1573} = 38773$, $nfe_{dcb} = 93\times 41_{573} = 38703$, etc. will be multiples of the number $x = 7$. The example is easily proved by substituting for n the values $10-x$, $100-x$, etc.

(16) Let a chosen number *fedcba* be divided by α, and assume $n = 10-\alpha$ or $n = 100-\alpha$, etc. If, for example, $n = 100-\alpha$, the division may be written as

$$\begin{array}{l} 100-n\,\overline{)fedcba(}\,fedc \\ \quad (fedc \\ \quad\quad -n\times fedc) \\ \quad \overline{+fedc\times n+ba} = \text{first remainder} \\ \quad\quad \text{etc.} \end{array}$$

Repeating this operation continuously allows us to converge on the true quotient.

Example. Let the number 5430856 be divided by the number 83. In this case $100-83 = 17 = n$, so the operation will be

$$\begin{array}{rcr} \text{first quotient} & = & 54308\,.\,56 \\ 17\times\frac{54308}{100}+56 & = & 9232\,.\,92 \\ 17\times\frac{9232}{100}+92 & = & 1570\,.\,92 \\ 17\times\frac{1570}{100}+36 & = & 267\,.\,26 \\ 17\times\frac{267}{100}+26 & = & 45\,.\,65 \\ 17\times\frac{45}{100}+65 & = & 8\,.\,30 \\ 17\times\frac{8}{100}+30 & = & 1\,.\,66 \\ 17\times\frac{1}{100}+66 & = & .\,83 \\ \hline \text{Quotient} & & 65432 \end{array}$$

This was the method of Johann Andreas Castelvetro.

Theorem 47

(1) Let x and z be any integers, and let a and b be two numbers having no common divisor. Then any number greater than $ab-a-b$ can be expressed as $ax+bz$.

(2) If n is any integer, then any number divisible by n that is greater than $nab-na-nb$ may be expressed by $nax+nbz$.

(3) Every integer is either a triangular number, or a sum of two or three triangular numbers, i.e., is composed of one, two, or three of the numbers 1, 3, 6, 10, etc.

(4) Every integer is either a pentagonal number, or the sum of two, three, four, or five pentagonal numbers, i.e. is composed of one, two, three, four, or five of the numbers 1, 4, 10, 20, 35, ...; likewise it is the sum of one, two, three, four, five, six, seven, or eight of the hexagonal numbers 1, 5, 15, 35, 70, 126, 210, etc.

(5) Every integer is a square, or the sum of two, three, or four squares.

(6) Every integer may be expressed as the sum of quantities of the form a^2-b^2 and a^2-b^2-2, where a and b are integers (including 0).

(7) Every number which is greater than any of the given numbers p, q, r, s (which are regarded as parameters, and have no common divisors) can be expressed by a quantity $pa^2+qb^2+rc^2+sd^2$, where a, b, c, d are any integers (including 0); or more generally, by

$$pa^2+ha+qb^2+kb+rc^2+lc+sd^2+md+\alpha ab+\beta ac+\gamma ad+\delta bc+\epsilon bd+\zeta cd$$

where h, k, l, m, α, β, γ, δ, ϵ, and ζ are also integers.

(8) Every integer can be written as the sum $(a^2 \pm ab+b^2)+(p^2 \pm pq+q^2)$, where a, b, q, and q are integers.

(9) Every integer is a cube or the sum of two, three, ... nine cubes; every integer is also the square of a square, or the sum of up to nineteen such; and so forth. Similar laws may be affirmed (*exceptis excipiendis*) for the correspondingly defined numbers of quantities of any like degree. [15]

(10) Every number $4n+2$, divisible by 2 but not by 4, can be composed as the sum of two or three squares.

(11.1) In general every integer (subject to certain limits arising from the given quantities) is the sum of a finite number of quantities of the form

$$ax^m+bx^n+cx^r+\ldots$$

or of some number of quantities of the forms

$$ax^m+bx^n+cx^r+\ldots, \quad ex^\alpha+fx^\beta+gx^\gamma+\ldots,$$

etc., where m, n, r, ..., α, β, γ, ..., a, b, c, ..., e, f, g, ... are all integers.

We must be careful not to apply these limits for the zero case. Often the above quantities do not begin with the value 1, and it is necessary to determine the limit above which they begin to apply. Also the quantities might be of forms such as $3x^4+6x^3+24$, where all terms are divisible by 3, and consequently cannot be used to construct any integer which is not divisible by 3.

(11.2) By a similar principle it is possible to identify all numbers which can be composed from quantities of a particular type — e.g., all numbers which can be expressed as the sum of terms $(ax+e)^2+(bx+f)^2+(cx+g)^2+(dx+h)^2$.

In all these cases the number may often be composed in different ways: e.g., there are numbers which may be composed in several ways as the sum of one, two, three, or four squares.

It is easy to add many similar propositions concerning fractional and irrational quantities.

Theorem 48

Let $N = a^2+mb^2$ and $M = p^2+mq^2$, where a, b, p, q, and m denote integers; then $N\times M = (ap+mbq)^2+m(aq-bp)^2 = (ap-mbq)^2+m(aq+bp)^2$.

Theorem 49

Let $N = a^2+rb^2$. Then N^{2m+1} and N^{2m+2} may be composed in $m+1$ different ways using quantities of the form p^2+rq^2.

These quantities may be derived as follows:

$$N^{2m} = 4ra^2b^2(a^2-rb^2)^{2m-2}+(a^2-rb^2)^2(a^2+rb^2)^{2m-2}, \qquad \text{or in general,}$$

$$N^{2m} = \frac{\left[(a+b\sqrt{-r})^{2l}+(a-b\sqrt{-r})^{2l}\right]^2+\left[(a+b\sqrt{-r})^{2l}+(a-b\sqrt{-r})^{2l}\right]^2(\sqrt{-1})^2}{4} \times (a^2+rb^2)^{2m-2l},$$

where l is any integer less than or equal to m. Likewise,

$$N^{2m+1} = ra^2b^2(a^2-rb^2)^{2m}+(a^2-rb^2)^2(a^2+rb^2)^{2m}, \qquad \text{or in general,}$$

$$N^{2m+1} = \frac{\left[(a+b\sqrt{-r})^{2l+1}+(a-b\sqrt{-r})^{2l+1}\right]^2+\left[(a+b\sqrt{-r})^{2l+1}+(a-b\sqrt{-r})^{2l+1}\right]^2(\sqrt{-1})^2}{4} \times (a^2+rb^2)^{2m-2l}.$$

This theorem may be demonstrated by appeal to the following easily deduced proposition, namely that

$$(a+b\sqrt{-r})^l = a^l+lb\sqrt{-r}a^{l-1}-\frac{l(l-1)}{2!}b^2ra^{l-2}-\frac{l(l-1)(l-2)}{3!}b^3r\sqrt{-r}a^{l-3}$$
$$+\frac{l(l-1)(l-2)(l-3)}{4!}b^4r^2a^{l-4}+\frac{l(l-1)(l-2)(l-3)(l-4)}{5!}b^5r^2\sqrt{-r}a^{l-5}$$
$$-\dots;$$

then

$$\frac{(a+b\sqrt{-r})^l+(a-b\sqrt{-r})^l}{2} = a^l-\frac{l(l-1)}{2!}b^2ra^{l-2}+\frac{l(l-1)(l-2)(l-3)}{4!}b^4r^2a^{l-4}-\dots$$
$$= P, \text{ say, and}$$

$$\frac{(a+b\sqrt{-r})^l+(a-b\sqrt{-r})^l}{2}\sqrt{-1}$$
$$= \sqrt{r}\left(lba^{l-1}-\frac{l(l-1)(l-2)}{3!}b^3ra^{l-3}+\frac{l(l-1)(l-2)(l-3)(l-4)}{5!}b^5r^2a^{l-5}-\dots\right)$$
$$= \sqrt{r}\cdot Q,$$

whence $P^2+rQ^2 = a^{2l}+lb^2ra^{2l-2}+\frac{l(l-1)}{2!}b^4r^2a^{2l-4}+\dots = (a^2+rb^2)^l$.

Corollary 1. Thus we see that, if the number N is composed in n distinct ways of such quantities, so also will its square be composed in n distinct ways; its cube and biquadrate will be so composed in $2n$ distinct ways; and in general its $(2m+1)^{\text{th}}$ and $(2m+2)^{\text{th}}$ powers will be so composed in $(m+1)n$ distinct ways, of quantities of the form p^2+rq^2. (In some of these cases, two or more of the values may be equal to each other.)

Corollary 2. Let N, R, S, T, ... be numbers which are composed (in n, r, s, t, ... distinct ways, respectively) of quantities of the above form $a^2 + mb^2$. Then the number $N^\alpha R^\beta S^\gamma T^\delta\cdots$ is composed (in $\frac{\alpha+1}{2}(\beta+1)(\gamma+1)(\delta+1)\cdots nrst\cdots$ distinct ways) of quantities p^2+mq^2. (If the factor $\alpha+1$ is odd, substitute for it $\alpha+2$, and likewise for the others.)

This is established by considering the last two theorems together.

Those cases are excepted in which two or more values are equal to each other.

Corollary 3. Thus we can find numbers which can be composed in n ways, for whatever value of n we desire, of quantities of the form p^2+mq^2. For let $2n = (\alpha+1)(\beta+1)(\gamma+1)\cdots$ if α, β, γ, δ, ... are all odd numbers, but use $\alpha+2$ instead of $\alpha+1$ if α is an even number, etc. Then choose numbers N, M, R, S, ..., each composed of only two squares, and $N^\alpha M^\beta R^\gamma S^\delta \cdots$ will be the required number. And the solution may be made much more general.

We may also deduce the number of ways by which a number can be composed of two squares: let $N = a \cdot b \cdot c \cdot d \cdots$, and assume that we know in how many ways the numbers a, b, c, d, ... can be composed of two squares. Then by this theorem we can deduce the number of ways by which N may be composed of two squares.

Corollary 4. (1) Let the number $a^2+mb^2 = A = p \cdot q \cdot r \cdot s \cdot t \cdots$, where p, q, r, s, t, ... are prime numbers and $p = \alpha^2+m\beta^2$; then $q \cdot r \cdot s \cdots = \gamma^2+m\delta^2$. But if A is a prime number, then the assertion $A = ma^2+b^2 = mc^2+d^2$ will be false. For if it were, then we could write

$$\begin{aligned}(a^2-c^2)A &= a^2(mc^2+d^2)-c^2(ma^2+b^2) = a^2d^2-b^2c^2\\ &= (ad+bc)(ad-bc).\end{aligned}$$

But if A is prime, it must be a divisor of either the factor $ad+bc$ or the factor $ad-bc$; however, $2A = ma^2+b^2+mc^2+d^2$

and
$$\begin{aligned}2A-2ad-2bc &= (m-1)a^2+(m-1)c^2+a^2+c^2+b^2+d^2-2ad-2bc\\ &= (m-1)(a^2+c^2)+(a-d)^2+(b-c)^2.\end{aligned}$$

But if m is not less than 1 these quantities are positive, and therefore $A>ad+bc$, so A cannot be contained in $ad+bc$ as a factor.

(2) Let $mp^2+q^2 = P$ be a prime divisor of the number $ma^2+b^2 = N$. Then the quotient $\dfrac{ma^2+b^2}{mp^2+q^2}$ will be of the same form.

For $p^2(ma^2+b^2)-a^2(mp^2+q^2) = p^2b^2-q^2q^2 = (pb+aq)(pb-aq)$ will be divisible by P, whence either $pb+aq$ or $pb-aq$ will be divisible by the prime number P, and hence

$$aq\pm bp = P = mrp^2+rq^2.$$

Hence $a = \dfrac{p(mrp\mp b)}{q} + rq$,

whence $\dfrac{p(mrp\mp b)}{q}$, and consequently $\dfrac{mrp\mp b}{q}$, is an integer. Suppose $mrp\pm b = \mp sq$, then

$$b = sq\pm mrp,\quad a = \mp sp+rq,$$

and $A = ma^2+b^2 = (mp^2+q^2)(mr^2+s^2)$,

consequently for the quotient,

$$\frac{ma^2+b^2}{mp^2+q^2} = mr^2+s^2.$$

Also, $m(a^2+mb^2) = m^2b^2+ma^2 = c^2+md^2$.

Theorem 50

Let $(a^2+b^2+c^2+d^2)(p^2+q^2+r^2+s^2) = x^2+y^2+z^2+v^2$, where a, b, c, d and p, q, r, s are integers; then x, y, z, v may also be integers, *viz.*, $x = ap+bq+cr+ds$, $y = aq-bp\pm cs\mp dr$, $z = ar\mp bs-cp\pm dq$, $v = as\pm br\mp cq-dp$.

Similar assertions apply to more quantities of the same sort.

Theorem 51

$$\begin{aligned}(a^2+ab+b^2)(c^2+cd+d^2) &= (ac+ad+bd)^2+(ac+ad+bd)(bc-ad)-(bc-ad)^2\\ &= (ac+bd+bc)^2+(ac+bd+bc)(ad-bc)+(ad-bc)^2\\ &= (ac+ad+bc)^2+(ac+ad+bc)(bd-ac)+(bd-ac)^2\\ &= (ad+bd+bc)^2+(ad+bd+bc)(ac-bd)+(ac-bd)^2\\ &= P.\end{aligned}$$

Corollary 1. Hence, if a quantity of the form $p^2\pm pq+q^2$ (where p and q are integers or zero) is multiplied by another quantity of the same form, the product will be yet another quantity of the same form.

Corollary 2. The product $(a^2+ab+b^2)(c^2+cd+d^2)$ is composed in two different ways of quantities of the form p^2+pq+q^2, and likewise in two different ways of quantities of the form p^2-pq+q^2 (where p and q are positive integers or zero), unless the differences $bc-ad$, $ad-bc$ are equal to $bd-ac$, $ac-bd$ in the stated or opposite order.

Corollary 3. Hence we see that if the numbers N and M are composed, in n and m ways respectively, of quantities of the form p^2+pq+q^2, the number $N\times M$ is composed in $2nm$ ways of quantities of the same form.

Every number which can be generated by the formula p^2+pq+q^2 can also be generated by the formula p^2-pq+q^2, and conversely every number which can be expressed as p^2-pq+q^2 can be expressed as p^2+pq+q^2; but expressions of the form p^2-pq+q^2 occur twice for each occurrence of the form p^2+pq+q^2. This is a consequence of the theorem which follows.

Corollary 4. Every divisor of a number of the form $p^2 \pm pq + q^2$, where p and q have no common divisors, will also be a number of the same form.

This corollary may be proved by a method similar to that discussed in Theorem 49, Corollary 4.

Corollary 5. If the number P has no square divisor a^2, and is not divisible by 3, and can be resolved in only one way as a number of the form p^2+pq+q^2, then P must be a prime number. Also by the same principle, if $A = a^2+mb^2$ has neither a divisor a^2 nor ma^2, and can be expressed as a^2+mb^2 in only one way, then A is a prime number.

Corollary 6. The sum of two cubes, a^3+b^3, has divisors which are divisors of the number $a+b$, and those contained in a number of the form $p^2 \pm pq+q^2$. In these instances a and b are prime to each other.

Corollary 7. The same may be applied to quantities of the form $a^2+mab+nb^2$; these may easily be reduced to quantities of the preceding form, as may many other quantities.

Theorem 52

(1) Let n, a, α, and m be integers. Then on division of the quantity

$$(na+\alpha)^m = n^m a^m + mn^{m-1}a^{m-1}\alpha + \frac{m(m-1)}{2}n^{m-2}a^{m-2}\alpha^2 + \ldots + mna\alpha^{m-1} + \alpha^m$$

by n, the remainder will be the same as the remainder upon division of the quantity a^m by n. If m is an odd number and α is negative, then if the remainder is l when α^m is divided by n, the remainder will be $n-l$ on division of $(na-\alpha)^m$ by n. And so also $(\alpha-na)^m$ has the same residue, with respect to n, as does α^m.

(2) Let n be prime, and let a be any integer. If the power a^n, divided by n, leaves the remainder l, then

$$(a+1)^n = a^n + na^{n-1} + \frac{n(n-1)}{2!}a^{n-2} + \frac{n(n-1)(n-2)}{3!}a^{n-3} + \ldots + \frac{n(n-1)}{2!}a^2 + na + 1$$

will have the same residue as the number a^n+1 when divided by n, and thus its residue will be $l+1$.

Corollary. Hence the remainder of a^n upon division by n will be the same as the remainder of a divided by n, and it follows that $\frac{a^n-a}{n}$ will be an integer: whence, if a is not a multiple of n, the observation that $\frac{a^n-a}{n} = a(\frac{a^{n-1}-1}{n})$ leads us to conclude that $\frac{a^{n-1}-1}{n}$ must be an integer.

Aliter. [16] Let n be a prime and a an integer. Then $a^{n-1}-1$ may always be divided by n, unless n divides a. In the case where n divides $a^{n-1}-1$, then a^n-a is divisible by n; and then further $(a+1)^n-a-1$ must be divisible by n. For

$$(a+1)^n = a^n+na^{n-1}+\frac{n(n-1)}{2}a^{n-2}+\ldots+na+1,$$

and every term of this series, except the first and last (provided that n is prime) is divisible by n. But a^n-a is divisible by n, and so then must be $(a+1)^n-a-1$. And so, if the proposition is true when a is some particular number, it must be true whatever the value of a.

Corollary 1. If n is prime, then $\frac{a^{n-1}-b^{n-1}}{n} = \frac{a^{n-1}-1}{n} - \frac{b^{n-1}-1}{n}$ will be an integer, if neither a nor b is divisible by n.

Corollary 2. Let m, n, p, q, ... be distinct primes, and let A be the least common multiple of the same numbers each diminished by unity, i.e. of $m-1$, $n-1$, $p-1$, $q-1$, etc. Then a^A, divided by $mnpq\cdots$, has a remainder of 1, unless a itself is divisible by one or more of the numbers m, n, p, q, etc.

For by the Theorem, with a, b, c, d, ... and α denoting integers, $a^{m-1} = mb+1$, whence the remainder, when the quantity

$$a^{(m-1)\alpha} = (mb+1)^\alpha$$
$$= m^\alpha b^\alpha+\alpha m^{\alpha-1}b^{\alpha-1}+\frac{\alpha(\alpha-1)}{2!}m^{\alpha-2}b^{\alpha-2}+\ldots+\frac{\alpha(\alpha-1)}{2!}m^2b^2+\alpha mb+1$$

is divided by m, will be 1. Now let $\alpha = n-1$, and the remainder from division by n of the quantity

$$a^{(m-1)(n-1)} = (mb+1)^{n-1} = ne+1$$

will be 1, unless $a^{m-1} = mb+1$ is divisible by n: therefore $a^{(m-1)(n-1)}-1$ is divisible by both m and n, and since m and n are mutually prime, it is divisible by mn; and so $a^{m-1)(n-1)}$ divided by mn leaves the remainder 1. And likewise the same method of reasoning applies to the rest of the factors $p-1$, $q-1$, etc.

Corollary 3. [17] Let $2n+1$ be a prime, and p an integer. Then

$\frac{2^n+1}{2n+1}$ will be an integer, if $n = 4p+1$ or $4p+2$, otherwise

$\frac{2^n-1}{2n+1}$ will be an integer. And if $n = 6p+2$ or $6p+3$ then

$\frac{3^n+1}{2n+1}$ will be an integer, otherwise $\frac{3^n-1}{2n+1}$ will be an integer.

For $a^{2n}-1$ is always divisible by $2n+1$, unless a is divisible by $2n+1$. But $a^{2n}-1 = (a^n+1)(a^n-1)$ is divisible by $2n+1$; and it is easy to determine which of the factors a^n+1 or a^n-1 actually contains $2n+1$ as a factor.

Corollary 4. Let p be the greatest common divisor of the quantities m and $2n$. Then a^p-b^p will be divisible by $2n+1$, if $2n+1$ is prime and a^m-b^m is divisible by $2n-1$. For the greatest common divisor of the quantities a^m-b^m and $a^{2n}-b^{2n}$ will be a^p-b^p. Therefore, if the prime number $2n+1$ divides both of the quantities a^m-b^m and $a^{2n}-b^{2n}$, it must divide their greatest common divisor, which is a^p-b^p.

This is easily verified from the operation discussed in Problem XLVII, i.e. by finding common divisors of the quantities a^m-b^m and $a^{2n}-b^{2n}$.

Corollary 5. Let α, β, γ, δ, ... be the remainders, respectively, upon division of the number A by the numbers n, m, r, s, t, ..., i.e., $nx+\alpha = A$, $my+\beta = A$, $rv+\gamma = A$, $sw+\delta = A$, etc., where x, y, v, w, ... denote integers. From the equation

$$nx+\alpha = my+\beta$$

we may, by Problem LII, find integral expressions X, Y, for x and y in terms of a parameter π. Then we equate the two numbers

$$nX+\alpha = rv+\gamma$$

and by the same technique find expressions X', V for the quantities X and v in terms of a parameter ρ. Then we equate the two numbers

$$nX'+\alpha = sw+\delta$$

and from the resulting equation we may express X' and w as integer quantities X'' and W, in terms of a parameter σ, and so on. Thus integer values of the quantities x, y, v, w, ... may be found in terms of a single parameter, the final one, such as σ.

Corollary 6. (1) The sum of two squares, having the formula a^2+b^2, cannot be divided by any prime number of the form $4n-1$ unless both roots a and b are divisible by $4n-1$.

For the difference $a^{4n-2}-b^{4n-2}$ can be divided by $4n-1$, whence $a^{4n-2}-b^{4n-2}+2b^{4n-2} = a^{4n-2}+b^{4n-2}$ is not divisible by $4n-1$. But $a^{4n-2}+b^{4n-2}$ is divisible by a^2+b^2, and therefore a^2+b^2 is not divisible by $4n-1$.

(2) By the same principle we may prove that $a^{2m}+b^{2m}$ has no odd divisors, except those of the form $2^{m+1}n+1$.

Corollary 7. $(c^n+ap)^{\frac{p-1}{n}} = c^{p-1}+\frac{p-1}{n}c^{\frac{p-1}{n}\cdot n-n}ap+\ldots$, when divided by p, a prime number, gives a remainder of 0 or 1.

(In this and the following Corollaries, α, a, c, and $\frac{p-1}{n}$ will denote integers.)

Corollary 8. Let $ab^n = c^n\pm\alpha p$, p a prime, then $a^{\frac{p-1}{n}}$ leaves a remainder of 0 or 1 on division by p. For $(c^n\pm\alpha p)^{\frac{p-1}{n}} = a^{\frac{p-1}{n}}b^{p-1}$, and both $(c^n\pm\alpha p)^{\frac{p-1}{n}}$ and b^{p-1}, on division by p, leave the remainder 1, whence $a^{\frac{p-1}{n}}-1$ is divisible by p. If $p = mn+1$, then a^m-1 will be divisible by p when ab^n-c^n is divisible by p.

Corollary 9. If a number of the form a^m-1 is divisible by a prime number $p = mn+1$, then we can choose numbers x and y such that ax^n-y^n shall be divisible by the prime number p; and further when x^{mn} and y^{mn} both leave the remainder 1 on division by p, the number $a^m x^{mn}-y^{mn}$ will be divisible by p, although evidently neither x nor y is divisible by p. Now $a^m x^{mn}-y^{mn} =$

$$(ax^n-y^n)(a^{m-1}x^{mn-n}+a^{m-2}x^{mn-2n}y^n+a^{m-3}x^{mn-3n}y^{2n}+\ldots+y^{mn-n}),$$

and if the first factor is not divisible by p, the other one is. In the second factor, $a^{m-1}x^{mn-n}+a^{m-2}x^{mn-2n}y^n+a^{m-3}x^{mn-3n}y^{2n}+\ldots+y^{mn-n}$, substitute respectively 1, 2, 3, 4, $\ldots$ $(p-n)(mn-n+1)$ for y, and find the differences of the resulting quantities in order to eliminate the quantities a and x, the final resulting difference being $1\cdot 2\cdot 3\cdot 4\cdot 5\cdots(mn-n)$. But since each of the resulting quantities is divisible by p, so must be the final one, $1\cdot 2\cdot 3\cdot 4\cdots(mn-n)$. But this product is not divisible by $mn+1$, so the latter factor is not always divisible by p; hence we have shown the existence of a case in which ax^n-y^n will be divisible by p.

Theorem 53

(1) Let the residues of division of the numbers A, B, C, $\ldots$ by the number a be α, β, γ, $\ldots$ respectively, i.e., let $A = ha+\alpha$, $B = ka+\beta$, $C = la+\gamma$, $\ldots$, where h, k, l, $\ldots$ are integers. Then the residue of division of the number $ABC = (ha+\alpha)(ka+\beta)(la+\gamma)\cdots$ by a will be the same as the remainder when the number $\alpha\beta\gamma\cdots$ is divided by a.

(2) Conversely, let the residues of the numbers A and AB, on division by a, be α and $\alpha\beta$ respectively, where α is prime to the number a; then the residue of B, on division by a, will be β. For suppose the residues to be α and γ, when A and B respectively are divided by a; then $\alpha\gamma$ will be the residue of division of the product AB by a. But by hypothesis this residue will be $\alpha\beta$. Therefore $\alpha\gamma = ma+\alpha\beta$, whence $\gamma = \frac{ma}{\alpha}+\beta$. But since α and a are mutually prime, and $\frac{ma}{\alpha}$ is an integer, α must divide m, so that the remainder $\gamma = \beta$.

Corollary 2.1. Let the residues of division of the numbers A, B, C, ... by n be 1 in all cases. Then the residue of division of the product $ABC \cdots$ by n will be 1.

Corollary 2.2. Let q be a prime number, and let the residue of division of a^q by p be 1; then a^q is the lowest power of a, whose remainder on division by p is 1, unless the number a itself leaves a residue of 1 on division by p. For if a^q is not the least power as stated, then suppose a^r is the lowest power of a which, divided by p, has a remainder of 1, and let the residues of division of the numbers a, a^2, a^3, ... a^r, by p, be respectively α, β, γ, δ, ..., 1. Then the residues of division of the numbers $a^{r+1} = a^r \cdot a$, a^{r+2}, a^{r+3}, ..., a^{2r} will also be, respectively, α, β, γ, δ, ..., 1; hence the residues of division of all the numbers a^r, a^{2r}, a^{3r}, a^{4r}, ... by p will be 1. But in this series of numbers, r, $2r$, $3r$, $4r$, ..., there is no prime except r itself, whence r and q are equal to each other.

(3) Let α, β, γ, δ, ... be the residues of division of the numbers A, B, C, D, ... by n. Then α^m, $\alpha\beta$, $\alpha\gamma$, $\alpha\delta$, $\beta\gamma$, $\beta\delta$, $\gamma\delta$, ..., $\alpha\beta\gamma$, $\alpha\beta\delta$, ..., and in general $\alpha^h\beta^k\gamma^l \cdots$, will be the residues of division of numbers of the form $A^hB^kC^l \cdots$ by n, where A, B, C, ... are given and h, k, l, ... are arbitrary integers, or h, k, l, ... are given and A, B, C, ... are arbitrary. Let π, ρ, σ, ... be numbers which are not residues of division of quantities of the said form, $A^hB^kC^l \cdots$, by a prime number n; then $\alpha\pi$, $\beta\pi$, $\gamma\pi$, ..., $\alpha\rho$, $\beta\rho$, $\gamma\rho$, ..., $\alpha\sigma$, $\beta\sigma$, $\gamma\sigma$, etc. are also non-residues.

Corollary 3.1. Hence the number of non-residues of division of the terms of the infinite sequence 1^m, 2^m, 3^m, 4^m, ..., or 1, a, a^2, a^3, a^4, ..., or in general of all numbers of the form $A^hB^kC^l \cdots$, by a prime number p, will be 0 or n, $2n$, $3n$, $4n$, or ..., where n is the number of residues arising from such divisions.

Corollary 3.2. Let p be a prime, and let a be the least power of a number q which leaves a remainder of 1 when divided by p. If q is less than $p-1$ it is also not greater than $\frac{p-1}{2}$; if q is less than $\frac{p-1}{2}$ it will not be greater than $\frac{p-1}{3}$; if less than $\frac{p-1}{3}$ it will not be greater than $\frac{p-1}{4}$; and so on.

Corollary 3.3. Let α, β, γ, δ, ... be the residues of division of the squares 1, 4, 9, 16, 25, ... by a prime number p, and let π be a non-residue. Then $\alpha\pi$, $\beta\pi$, $\gamma\pi$, $\delta\pi$, ... will all be non-residues, but the product of any two non-residues, such as $\alpha\pi \cdot \beta\pi = \alpha\beta \cdot \pi^2$, will be a residue, where α, β, and π^2 are residues.

(4) Let the residues of division of the numbers 1^m, 2^m, 3^m, 4^m, 5^m, ... by n be respectively α, β, γ, δ, ..., where m is an odd number. Then the residues will be α and $n-\alpha$, β and $n-\beta$, γ and $n-\gamma$, etc., or α and $-\alpha$, β and $-\beta$, γ and $-\gamma$, etc., in complementary pairs. The residues of division of numbers of the form $(kn-\alpha)^m$ by n will be α^m and $-\alpha^m$, etc.

Let α, β, γ, δ, ... be the residues of division of the numbers 1, a, a^2, a^3, ... (to infinity) by a prime p. Then if the complement of α occurs, the complements of all residues will occur. For if the two complements α and $-\alpha$ occur, then also $\frac{-\alpha}{\alpha} = -1$ occurs, and so the complements $-1\times\alpha = -\alpha$, $-1\times\beta = -\beta$, $-\gamma$, $-\delta$, etc.

If m is even and p is a prime, and one complementary pair α and $-\alpha$ occurs among the residues of division of the numbers 1^m, 2^m, 3^m, 4^m, 5^m, ... by the prime p, then the complement of each residue always occurs. The same assertion holds if one complementary pair occurs among the residues of division of $A^h B^k C^l \cdots$ by the prime number p, where A, B, C, ... are given and h, k, l, ... are any arbitrary numbers, or h, k, l, ... are given and A, B, C, ... are arbitrary numbers.

These are proved by the principles discussed in the preceding cases.

Theorem 54

(1) Let r be the greatest common divisor of the numbers m and $n-1$, where n is prime. Then $\frac{n-1}{r}+1$ will be the number of distinct residues

of division of the numbers 1^m, 2^m, 3^m, 4^m, 5^m, 6^m, ... (to infinity) by n, counting among these residues always 0 and 1.

The number of equal residues among the numbers 1^r, 2^r, 3^r, 4^r, ... $(n-1)^r$, divided by the prime number n, cannot be greater than r.

(2) Let

$$a^l b^k c^i \cdots \times B = a^\pi b^\rho c^\sigma \cdots \left(\frac{a^\pi b^\rho c^\sigma \cdots -1}{2}\right)\left(\frac{a^\pi b^\rho c^\sigma \cdots -2}{3}\right)\cdots\left(\frac{a^\pi b^\rho c^\sigma \cdots -h+1}{h}\right)$$

where a, b, c, ... are primes and B is prime to a, b, c, ...; and where π, ρ, σ, ... are given numbers and h is an integer less than $a^\pi b^\rho c^\sigma \cdots = L$. Now let l', k', i', ... be values of the quantities of the quantities l, k, i, ... (to which correspond the values h', h'', h''', ... of the quantity h) such that

$$\begin{aligned} l' &+ \pi'(a^\pi - h') = p \\ k' &+ \rho'(a^\rho - h'') = q \\ i' &+ \sigma'(a^\sigma - h''') = r, \end{aligned}$$

etc. are minimized (where π', ρ', σ', ... are given). Then the number of residues of division, by $a^p b^q c^r \cdots$, of the numbers 1^m, 2^m, 3^m, 4^m, 5^m, ... (where $m = a^\pi b^\rho c^\sigma \cdots$) cannot be greater than $a^{\pi'} b^{\rho'} c^{\sigma'} \cdots -1$. The same assertion holds for division of the stated numbers by $a^{\pi'} b^{\rho'} c^{\sigma'} \cdots$, provided that p', q', r', ... are not greater than p, q, r, ... respectively.

(3) Let the number $A = abcd \cdots$, where a, b, c, d, ... are mutually prime, and let the number of residues of division of a sequence by a, b, c, d, ... be respectively m, n, r, s, ... Then the number of residues of division by A cannot be greater than $mnrs \cdots$; frequently also some of these are doubly, triply, or with higher multiplicity equal to each other.

For let α and β be residues of division, respectively, by the quantities a and b, and let x and y be integers which are to be determined. Assume $ax+\alpha = by+\beta = H$, and let X and Y be the correspondent values derived for the quantities x and y. Then by Problem LII, $b\pi+X$ and $a\pi+Y$ (where π is any integer) will be general expressions for the quantities x and y; and consequently there is only one residue of division of the quantity H by the product ab, *viz.* $aX+\alpha = bY+\beta$, whence by comparing each of the m, n, ... residues with each other the proposition is established.

Hence there cannot be more than $\left(\frac{a+1}{2}\right)\left(\frac{b+1}{2}\right)\left(\frac{c+1}{2}\right)\cdots$ residues of division of all squares by the number A; and often not nearly that many will occur. Here 0 is always assumed to be one of the residues.

Corollary 3.1. Let α, β, γ, δ, ... be the residues of division of the infinite sequence 1^m, 2^m, 3^m, 4^m, 5^m, ... by some number n, and let a, b, c, d, e, ... be non-residues. Then the number h^m cannot be represented as $nr+a$ or $nr+b$, etc., where r is an integer.

Example 3.1.1. The residues of division of the squares 1, 4, 9, 16, ... by 4 are 0 and 1, hence 2 and 3 are non-residues, and thus $4r+2$ or $4r+3$ (and so also $4r-1$) cannot represent a square.

Example 3.1.2. The residues of division of the cubes 1, 8, 27, 64, ... by 7 are 1, 6, and 0, whence the non-residues are 2, 3, 4, and 5. Therefore neither $7r+2$, $7r+3$, $7r+4$, nor $7r+5$ can be equal to a cube.

Corollary 2. Let a', b', c', d', ... be numbers which are not the sums of any two of the residues α, β, γ, δ, ...; let a'', b'', c'', d'', ... be numbers which are not the sums of any three residues, and so on. Then no number of the form $nr+a'$, $nr+b'$, ... can be the sum of two m^{th} powers h^m+k^m; and likewise none of $nr+a''$, $nr+b''$, ... can be the sum of three numbers $h^m+k^m+l^m$.

Example 3.2.1. The residues of division of cubes by 7 are 0, 1, and 6; the sum of any two residues can be 0, 1, 2, 6, and 5(=6+6−7); so 3 and 4 are numbers which cannot be the sum of two such residues. It follows that $7r+3$ and $7r+4$ cannot be the sum of two positive cubes.

Example 3.2.2. The residues of division of 1^5, 2^5, 3^5, 4^5, ... by 11 are 0, 1, and 10, and so the sum of each two residues will be 0, 1, 10, 2, and 9, with 3, 4, 5, 6, 7, and 8 never being such a sum. Thus neither $11r+3$ nor $11r+4$, etc., can be equal to the sum of two fifth powers h^5+k^5.

Example 3.2.3. The sum of each three residues 0, 1, and 10 will be 0, 1, 2, 3, 8, 9, and 10; among these sums 4, 5, 6, and 7 do not appear. Therefore neither $11r+4$, $11r+5$, $11r+6$, nor $11r+7$ can be equal to the sum of three fifth powers $h^5+k^5+l^5$.

Example 3.2.4. The differences, as well as sums, of pairs of the residues 0, 1, and 10 will be 0, 1, 10, 2, and 9; of the residues 0, 1, 6 they will be 0, 1, 2, 6, and 5. Therefore no number of the form $11r+3$, $11r+4$, $11r+5$ $11r+6$, etc., can be equal to a sum or difference $h^5 \pm k^5$, and likewise no number of the form $7r+3$ or $7r+4$ can be equal to a number $h^3 \pm k^3$.

Corollary 3.3. In general, let α, β, γ, δ, ... be the residues of division of the numbers 1^m, 2^m, 3^m, 4^m, ... by n, and let α', β', γ', δ', ... be the residues of division of $1^{m'}$, $2^{m'}$, $3^{m'}$, $4^{m'}$, ... by n. Now if the number H cannot be expressed by any number of the form $A\alpha \pm B\beta \pm C\gamma \pm \ldots$, then $nr+H$ cannot be a number of the form $Ah^m \pm Bk^m \pm Cl^m \pm \ldots$; and if H' cannot be expressed by the formula $A\alpha \pm B\beta \pm C\gamma \pm \ldots \pm A'\alpha' \mp B'\beta' \pm C'\gamma' \pm \ldots$, then $nr+H'$ cannot be a number of the form $Ah^m \pm Bk^m \pm Cl^m \pm \ldots \pm A'h'^{m'} \pm B'k'^{m'} \pm C'l'^{m'} \pm \ldots$, where h, k, l, ..., h', k', l', ..., m, m', ..., A, B, C, ..., and A', B', C', ... are all integers.

Corollary 3.4. The complements α and $-\alpha$ do not both occur among the residues when the squares 1, 4, 9, 16, ... are divided by a prime of the form $4n-1$. Therefore the sum of two squares, a^2+b^2, is not divisible by $4n-1$ unless both a^2 and b^2 are multiples of $4n-1$. Hence the sum of two mutually prime squares cannot be divided by a composite number of the form $4n-1$, since each composite number $4n-1$ has among its prime divisors some of the same formula.

(4) Let a', b, c, d, e, ... be non-residues of division of the numbers 1, a, a^2, a^3, a^4, a^5, ... (to infinity) by n. Then $nr+a'$, $nr+b$, $nr+c$, $nr+d$, ... (where r is any integer) cannot be any power of the quantity a.

Let α, β, γ, δ, ... be the residues of division of the numbers a, a^2, a^3, ... by n, and let α', β', γ', δ', ... be residues of division of the numbers b, b^2, b^3, b^4, b^5, ... by n. Then if H' is any number which is not expressible by the formula $A\alpha \pm B\beta \pm C\gamma \pm \ldots \pm A'\alpha' \pm B'\beta' \pm C'\gamma' \pm \ldots$, $nr+H'$ can not be a number of the form $Aa^h \pm Ba^k \pm Ca^l \pm \ldots \pm A'b^{h'} \pm B'b^{k'} \pm C'b^{l'} \pm \ldots$ (where all of the above letters denote positive integers).

Corollary 4.1. Let m be an odd number. Then there exist two m^{th} powers, a^m and b^m, prime to n (where n is any prime) whose sum a^m+b^m is divisible by n. If m is an even number, then there exist two or three numbers of the form x^m, whose sum (a^m+b^m or $a^m+b^m+c^m$) is divisible by the given number, unless $n-1$ and $m-1$ have a common divisor greater than 2.

(5) By such principles we can often show the impossibility of assigning integral values to each of the unknown quantities in a given equation: i.e., both sides of the equation are divided by the same arbitrary integer, and by the principles here given we attempt to show that the two resulting residues can never be equal to each other, and

thus that there are no integer solutions to the given equations for the stated unknown quantities.

Example 5. Let the equation be $13x^4+2 = 7z^6+5y^{12}$.

Divide both sides of this equation by 13: the left side, on division by 13, leaves a remainder of 2. But the right side, $7z^6+5y^{12}$, divided by 13, cannot have a remainder of 2. For the power z^6 divided by 13 will have the remainder 0, 1, or 12, and consequently $7z^6$ divided by 13 will have the remainder 0, $7\times 1 = 7$, or 6. Similarly $5y^{12}$ divided by 13 will have the remainder 0 or 5. But the number 2 cannot be generated by addition of any of the numbers 0, 7, or 6, to either 0 or 5. Therefore the remainders, when both sides of the equation have been divided by 13, cannot be equal to each other.

It does not matter whether or not the divisor happens to be a coefficient of the given equation. A divisor which produces fewer possible residues, when divided into the equation, is preferable to other, often smaller, candidate divisors which would have a larger number of possible residues. The number of residues is readily derived from the previous Theorem.

Let α, β, γ, δ, ... be arbitrary numbers contained in a given expression A, with all of them, or none, as the case may be, divisible by p. For α, β, γ, δ, ... in A write any function involving its unknown quantities π, ρ, σ, ..., such that if π, ρ, σ, ... are integers, then α, β, γ, δ, ... will also be integers. Then the resulting quantities likewise will all be divisible by p, or p will divide none of them.

Problem LXI

To find expressions for the divisors of any given quantities.

Case 1. (1) Reduce the given quantities to others, the form of whose factors is known, and from this you may deduce the form of the divisors of the given quantities.

(2) Choose quantities of a given form multiplied together, resulting in quantities which admit divisors of a particular type. For instance, you may choose quantities of the form $4n+1$, say some specific ones $4x+1$, $4y+1$, $4z+1$, ..., where the letters x, y, z, ... denote integers. Their product will be a number of the same form.

The product $(4x+1)(4y\pm1)(4z\pm1)\cdots$, on the other hand, will have the desired form only if the number of factors of the form $4n-1$ is even; otherwise the product itself will be of the form $4n-1$. This can be verified by multiplying the factors together.

(3) If all the factors of some quantity are subsumed by a given formula, and another formula can be found which includes all quantities of the given formula, then all the factors of the above quantities are encompassed by the latter formula.

Example 1.1. Given quantities a^2+nb^2, where a and b are any mutually prime numbers, and n is a given number, we know that some of their divisors must in each case be of the form p^2+nq^2. This formula can be expressed otherwise as $nx+\alpha$, where x is an integer and α is a constant. Now if the quantity p^2+nq^2 is divided by n, the residue will be the same as if the square p^2 had been divided by n; and thus if α, β, γ, δ, ... are the residues of division of all squares by n, then the prime divisors of all numbers of the form a^2+nb^2 must be contained in the formulas $nx+\alpha$, $nx+\beta$, $nx+\gamma$, etc.

If n is a prime number, the number of the abovementioned formulas will be $\frac{n+1}{2}$, whereas the number of residues is only $\frac{n-1}{2}$.

Example 1.2. The number of prime divisors of the form $4nx+\alpha$, of the qbove quantity (a^2+nb^2), will be $n-1$. Quantities of this form, divided by $4n$, will have the same number of prime divisors.

But if n is a quantity of the form $4x+1$, then these $n-1$ expressions of the form $4nx+\alpha$ may invariably be reduced to $\frac{n-1}{2}$ numbers of the form $2nx+\alpha$. Finally, if $n=2$ then $4nx+\alpha$ will be $8x+1$ or $8x+3$, and the given equation will also have divisors 2 and n.

If $n = p\cdot q\cdot r\cdot s\cdots$, where p, q, r, s, ... denote odd primes, then the number of prime divisors of the form $4nx+\alpha$ will be $(p-1)(q-1)(r-1)\cdots$; if $n = 2\cdot p\cdot q\cdot r\cdots$ then the above number will be $2(p-1)(q-1)(r-1)\cdots$: in other words, the divisors will be 2, p, q, r, etc.

If $4nx+\alpha$ is the formula for some divisor of a quantity a^2+nb^2, then $4nx-\alpha$ cannot be the formula for any divisor of that quantity.

Every expression $4pqx+\alpha$, which generates divisors of the quantity a^2+pqb^2, also generates divisors of the quantity pa^2+qb^2, because $p(a^2+pqb^2) = pa^2+p^2qb^2 = pa^2+qc^2$.

If $4nx+a$ is an expression which generates divisors of quantities of the form a^2+nb^2, then every prime number $A = 4nx+\alpha$ is also expressible as a^2+nb^2.

Case 2. If $a^2-nb^2 = A$, where $n = pqrs\cdots$ or $2pqrs\cdots$ (and p, q, r, s, ... denote prime numbers) then the number of prime divisors of the form $4nx+\alpha$ will be — as in the preceding case — either $(p-1)(q-1)(r-1)\cdots$ or $2(p-1)(q-1)(r-1)\cdots$.

But in this case if $4nx+\alpha$ is the formula of its divisors, then $4nx-\alpha$ will also be the formula of its divisors.

If $n = 4x-1$, then the formulæ of the prime divisors may be expressed by the formula $2nx\pm\alpha$. These will also give the prime divisors 2, p, q, r, etc.

It is also easy to determine all the different values α may take (one of them is always 1).

If the number P has the formula $8n+1$ or $8n+3$, and cannot be resolved into the form $2a^2+b^2$, then it is not a prime number.

Corollary 2.1. If there exist no two numbers a, b such that n is the sum of a square a^2 and a triangular number $\frac{b^2+b}{2}$, then $8n+1$, which accordingly cannot be equal to $8a^2+4b^2+4b+1 = 2(4a^2)+(2b+1)^2 = 2p^2+q^2$, is not a prime.

Corollary 2.2. If n cannot be expressed by the formula $a^2+a+\frac{1}{2}(b^2+b)$, then $8n+3$ cannot be expressed by the formula

$$8a^2+8a+2+4b^2+4b+1 = 2(2a+1)^2+(2b+1)^2 = 2p^2+q^2,$$

and consequently is not prime.

Case 3. If n is a prime number, and is also prime to m, then every formula $na+\alpha$ generates a number of the form p^2+mq^2. For if the quantity p^2 is divided by m, the potential residues are, say, α, β, γ, δ, ...; if also mq^2 is divided by n, the same residues, each multiplied by the residue (say π) of division of m by n, result. If π happens to be one of the residues α, β, γ, ..., then the residue that results will also be one of them; otherwise all the numbers that were not among the aforesaid residues will be produced. Combining these two possibilities, we see that all residues less than n result.

Case 4. The form of all divisors of quantities a^2+ab+b^2, where a and b are mutually prime, will be p^2+pq+q^2. Hence a number of the form

$3x+2$, to name one instance, cannot be generated by the said formula.

Thus it is easy to find many expressions which cannot be squares. Find a formula for a quantity which cannot be a factor of a quantity a^2+nb^2, and from the formula for that quantity subtract nb^2. The result is a quantity which cannot be a square. For example, the quantity $4nx-1$ can never be a factor of a quantity a^2+nb^2, where $n=4v+1$; whence $(4nx-1)u$ (u being any integer) cannot be equal to a^2+nb^2, and thus (taking $b=1$), $4nxu-u-n$ cannot be a square. And likewise for other cases.

Theorem 55

(1) Let P be any integral function of the letters $x, y, z, \ldots$, and for x in P write $\pm x+ny+mz+\ldots+a$, where the letters $x, y, z, \ldots, n, m, \ldots$ and a denote whole numbers. Call the resulting integral function Q. Then the set of values of P and Q, resulting from the substitution of every integer for $x, y, z, \ldots$, will be the same.

Example 1. Let the function be x^2-xy+y^2, in which we substitute $x+y$ for y, with the result $x^2-x(x+y)+(x+y)^2 = x^2+xy+y^2$, whence all the values of the function x^2+xy+y^2, obtained from the substitution of every integer for x and for y, will be the same as all the values of the function x^2-xy+y^2, obtained from the substitution of every integer for the two letters there also.

An infinity of similar propositions may be deduced.

Example 2. Let $Ax^2-Bxy+Cy^2$ be the function, in which we shall write $z-ay$ in place of x. The result is $Az^2-(2Aa+B)zy+(Aa^2+Ba+C)y^2$. Therefore the numbers generated by these two formulæ are the same.

(2) Let $Ax = ap^n+bp^{n-1}+cp^{n-2}+dp^{n-3}+\ldots+P$, where the letters A, $a, b, c, d, \ldots, P$ denote positive or negative integers, and x and p denote any integers. Then if integer values (call them α and π) are given for x and p, another value of p can be found which is less than $A/2$.

For we may write $\pi\pm\mu A$ (where μ is a positive or negative whole number) in place of p in the given equation, the result being

$$Ax = a(\pi\pm\mu A)^n+b(\pi\pm\mu A)^{n-1}+c(\pi\pm\mu A)^{n-2}+\ldots+P.$$

Now if $a\pi^n+b\pi^{n-1}+c\pi^{n-2}+\ldots+P$ is divisible by A, then plainly the above quantity $a(\pi\pm\mu A)^n+b(\pi\pm\mu A)^{n-1}+c(\pi\pm\mu A)^{n-2}+\ldots+P$ is divisible by A, whatever may be the value of the quantity μ. Therefore the value of μ may be so chosen that $\pi \pm \mu A$ is not greater than $A/2$, and so we have displayed a value of the quantity x that is not greater than $A/2$.

Lemma

(1) Let a^2+b^2 be a square, with a and b mutually prime. Then $a = p^2-q^2$ and $b = 2pq$, and thus $a^2+b^2 = (p^2+q^2)^2$, with p and q also mutually prime. One of them must be even, and the other odd, otherwise a and b cannot be relatively prime.

(2) Let a^2-b^2 be a square, with a and b mutually prime. Then $a = p^2+q^2$ and $b = p^2-q^2$ or $b = 2pq$, where p and q are mutually prime, one of them even and the other odd.

Theorem 56

Case 1. If a given equation can be transformed into another similar to it, by rational substitution of the unknown quantities of the resulting equation for the unknown quantities of the given equation, then from one corresponding value among the unknown quantities of the resulting equation, which is similar to the given one, more values of the said substitution may be derived.

Case 2. If the values of the unknown quantities in the resulting equation are less than the values of the corresponding unknown quantities in the given equation, and by similar substitutions the resulting equation can be further depressed to another, whose unknown quantities are still smaller, and so on, then either

(*a*) from the small values of the given equation many others may be deduced; or

(*b*) the given equation admits no solutions all of whose corresponding values are integers.

Case 3. More generally, if by a reciprocal rational substitution — i.e., by a rational substitution of the unknowns of the resulting equation for the unknown quantities of the given equation, which could equally be stated as a rational substitution of the unknown quantities of the given equation for those of the resulting equation — the given equation can be transformed into another, then the corresponding values of the unknown quantities of the given equation can be rational if and only if the corresponding values of the deduced equation can also be rational.

Example 1. Let $x^4+ay^4 = z^2$, where a is an integer. Set $z = m^2+an^2$, whence $z^2 = (m^2-an^2)^2+a(2mn)^2$, and thus $y^2 = 2mn$, $x^2 = m^2-an^2$. Now let $x = p^2-aq^2$, whence $x^2 = (p^2+aq^2)^2-a(2pq)^2$, whence $m = p^2+aq^2$ and $n = 2pq$. On substituting these two values into the equation $y^2 = 2mn$ we get $y^2 = 4pq(p^2+aq^2)$. Then let $p = s^2$, $q = t^2$, $u^2 = p^2+aq^2$, giving rise to $y = 2stu$, and then $s^4+at^4 = u^2$, as in the given equation, and $x = s^4-at^2$, $y = 2stu$, $z = (s^4+at^4)+a(2s^2t^2)^2$.

Example 2. (1) The sum of two biquadrates $(\alpha^4+\beta^4)$ cannot be a square. First, if α and β have a common divisor, i.e., $\alpha=ca$, $\beta=cb$, then $\alpha^4+\beta^4 = c^4(a^4+b^4)$, and a^4+b^4, where a and b are mutually prime, will be a square if $\alpha^4+\beta^4$ is a square.

Then suppose a^4+b^4 is a square. In that case $a^2 = p^2-q^2$ and $b^2 = 2pq$, where p and q are mutually prime, one of them (say q) being even and the other (p) odd (otherwise p^2-q^2 could not be a square). But, since $2pq$ is a square both p and $2q$ must be squares. For p^2-q^2 to be a square, we would require that $p = m^2+n^2$, $q = 2mn$, where m and n are mutually prime numbers, again one even and one odd. But since $2q$ is a square, $4mn$ (and so mn) is a square, whence m and n will be squares. Therefore write $m = x^2$, $n = y^2$, so $p = m^2+n^2 = x^4+y^4$ will be a square. Hence if a^4+b^4 is a square, x^4+y^4 will be a square, where x and y are less than a and b; and likewise there will be two squares smaller than these, and so on by infinite descent, and so no biquadrates of this type can be assigned.

So the three numbers p, $2q$, and p^2-q^2, and consequently $2pq(p^2-q^2)$ as well as $2pq(p^2+q^2)$, cannot be squares.

(2) By the same reasoning it follows that a^4-b^4, and thus $ab(a^2-b^2)$, $2ab(a^2+b^2)$, and $2a^4+2b^4 = (a^2+b^2)^2+(a^2-b^2)^2$, cannot be squares, unless $a=b$.

Corollary 1. The sum $(a^2+b^2)^2+(a^2-b^2)^2 = 2a^2+2b^2$ cannot be a square, since $(a^2+b^2)^2-(a^2-b^2)^2$ is a square; otherwise

$$\begin{aligned}&((a^2+b^2)^2+(a^2-b^2)^2)((a^2+b^2)^2-(a^2-b^2)^2)\\&=(a^2+b^2)^4-(a^2-b^2)^4\\&=p^4-q^4\end{aligned}$$

would be squares.

Corollary 2. Neither can the difference $2a^4-2b^4$ be a square. For suppose a and b are prime numbers, and $2a^4-2b^4$ is a square: then a and b must both be odd numbers. But since

$$2(a-b)(a+b)(a^2+b^2)$$

is a square, the product

$$\left(\frac{a-b}{2}\right)\left(\frac{a+b}{2}\right)\left(\frac{a^2+b^2}{2}\right)$$

will be a square — and since its factors are prime to each other, each of them must be a square. Write $\frac{a-b}{2} = p^2$ and $\frac{a+b}{2} = q^2$, whence

$$a = p^2 + q^2, \quad b = q^2 - p^2,$$

whence
$$\frac{a^2+b^2}{2} = p^4 + q^4 .$$

But p^4+q^4 cannot be a square; therefore $2a^4-2b^4$ is not a square.

Corollary 3. $m(a^2-mb^2)(a^2+mb^2)$, where m, a, and b are mutually prime, cannot be a square; for if it were a square, then m, a^2-mb^2, and a^2+mb^2 would be squares. Therefore set $m = p^2$, thus $a^2-p^2b^2$ and $a^2+p^2b^2$, and in consequence $a^4-p^4b^4$, will be squares, which is absurd. Similarly we may show that it is not possible for $2m(a^2-mb^2)(a^2+mb^2)$ to be a square.

If m is not prime with respect to a, we may write $m=rs$ and $a=rc$, where r and s are mutually prime (i.e., we may assume that m is free of squared factors, for if it is not, the problem can be reduced accordingly). Then

$$r^5sc^4 - r^3s^3b^4 = ma^4 - m^3b^4$$

and
$$2r^5sc^4 - 2r^3s^3b^4$$

would both be squares. But if these numbers are squares, then their prime factors r, s, ... must be squares, which is easily seen (as in the previous case) to be absurd.

Corollary 4. The formulæ $ma^4+m^3b^4$ and $2ma^4+2m^3b^4$ cannot be squares. For if mp^2-mq^2 is to be a square, it is not possible that mp^2+mq^2 should be a square as well — otherwise we would have that $m^2(p^4-q^4)$ or $2m^2(p^4-q^4)$ were a square. Therefore write

$$mp^2 - mq^2 = \frac{(p-q)^2a^2}{b^2}, \text{ whence } mp+mq = \frac{a^2p-q^2q}{b^2},$$

and thus

$$q = \frac{p(a^2-mb^2)}{a^2+mb^2} .$$

Now let $p = a^2+mb^2$, whence $q = a^2-mb^2$, and

$$p^2 - \quad q^2 = 2a^4 + 2m^2b^2.$$

Therefore $m(p^2+q^2) = 2ma^4+2m^3b^4$, and from this also $2mp^2+2mq^2 = 4ma^4+4m^3b^4$, cannot be squares.

Corollary 5. No triangular number $\frac{x(x+1)}{2}$, except unity, can be a biquadrate. For x and $\frac{x+1}{2}$, or $\frac{x}{2}$ and $x+1$, are mutually prime, and thus both would have to be biquadrates. Thus, in the first case, suppose $x=p^4$ and $\frac{x+1}{2}=q^4$, then $p^4+1 = 2q^4$, whence $4q^4 = 2p^4+2$, which has already been shown to be impossible; and similarly for the second case.

Example 3. No number of the form a^4+2b^4 can be a square. Set $(a^2+\frac{m}{n}b^2)^2 = a^4+2b^4$, where (a, b) and (m, n) are pairs of mutually prime numbers. Then $\frac{b^2}{a^2} = \frac{2mn}{2n^2-m^2}$, which is, or after division by 2 becomes, an irreducible fraction. Supposing first that $2mn$ and $2n^2-m^2$ are mutually prime, it follows that $b^2 = 2mn$ and $a^2 = 2n^2-m^2$. In this case m is an odd number, and n is either even or odd: if even, $2n^2 = a^2+m^2$ cannot be true; in the latter case b^2 cannot be equal to $2mn$. Suppose that $2mn$ and $2n^2-m^2$ have the common divisor 2, and consequently that m is an even number. Suppose $m=2k$, then n will be an odd number, and

$$\frac{b^2}{a^2} = \frac{2kn}{n^2-2k^2},$$

where $2kn$ and n^2-2k^2 will be mutually prime numbers. Hence, because b and a are mutually prime, $b^2 = 2kn$ and $a^2 = n^2-2k^2$. But $2kn$ is not a square unless k is an even number, and thus n and $2k$ must both be squares. So write $n = c^2$, $2k = 4d^2$, where c is odd, and thus $a^2 = c^4-8d^4$. Now write

$$c^2 - 8d^4 = (c^2 - \frac{2p}{q}d^2)^2$$

and from this it follows that

$$q^2d^2 = pqc^2 - p^2d^2, \text{ or } \frac{d^2}{c^2} = \frac{pq}{p^2+2q^2},$$

where (c, d) and (p, q) are mutually prime. Assuming p is odd, $d^2=pq$ and $c^2=p^2+2q^2$, and thus p and q must necessarily be squares. Accordingly, let $p=x^2$ and $q=y^2$, and then $c^2=x^4+2y^4$. Whence, if a^4+2b^4 is a square, there will be a square x^4+2y^4, where x and y are less than a and b; and again two numbers less than these can be deduced by the same method, and so on. But if p is even, i.e., $p=2r$, whence $\frac{d^2}{c^2} = \frac{qr}{2r^2+q^2}$, we will have (because qr and $2r^2+q^2$ are mutually prime) $d^2=qr$, $c^2=2r^2+q^2$, and thus both q and r will be squares, i.e., $q=x^2$ and $r=y^2$, whence $c^2=2y^4+x^4$, and thus again if a^4+2b^4 is a square, there will be smaller one x^4+2y^4, and so on infinitely.

Corollary 6. If a^4+kb^4 cannot be a square, then $2k\alpha\beta^3y^4-2\alpha^3\beta x^4$ will never be able to generate a square.

Let $a^4 + kb^4 = (a^2 + \frac{m}{n}b^2)^2$, then $kn^2b^2 = 2mna^2 + mb^2$, and $\frac{b^2}{a^2} = \frac{2mn}{kn^2-m^2}$; but since a^4+kb^4 can never be a square, also $\frac{2mn}{kn^2-m^2}$ or $2mn(kn^2-m^2)$ will not be a square. Let $m = \alpha x^2$, $n = \beta y^2$, and the product becomes

$$2\alpha\beta(k\beta^2y^4-\alpha^2x^4), \text{ or } 2k\alpha\beta^3y^4-2\alpha^3\beta x^4,$$

which formulas cannot represent a square.

From this we can easily deduce that the formulas $\alpha^3\beta x^4+\alpha\beta^3y^4$, $\alpha^3\beta x^4-\alpha\beta^3y^4$, $\alpha^3\beta x^4+2\alpha\beta^3y^4$, $2\alpha^3\beta x^4-2\alpha\beta^3y^4$, $2\alpha^3\beta x^4+2\alpha\beta^3y^4$, and $2\alpha^3\beta x^4-4\alpha\beta^3y^4$ also cannot represent squares.

Example 4. No cube (except 8) is one less than a square, i.e., $\frac{a^3}{b^3}+1 = v^2$ cannot be true, and so neither is $a^3b + b^4 = b^4v^2 = u^2$.

In the first place, $b(a+b)(a^2-ab+b^2) = a^3b+b^4$ is a square only if $b(a+b) = a^2-ab+b^2$, and the solution to this, $a = 2b$, explains the one noted exception. Write $a+b - c$, then we have $cb(c^2-3cb+3b^2)$, which cannot be a square unless $c = 3b$, where c and b are mutually prime. Now consider two cases:

First, if c is a multiple of the number 3, *viz.* $c = 3d$, in this case the factors c and $c^2-3bc+3b^2$ have a common divisor, 3, and $cb(c^2-3cb+3b^2) = 9db(b^2-3bd+3d^2)$. Therefore, if the quantity $cb(c^2-3cb+3b^2)$ is a square, the quantity $db(b^2-3bd+3d^2)$, an expression of the same form but composed of smaller numbers, will also be a square, and so on through infinite descent.

Second, if c is not divisible by 3, then b, c, and $c^2-3cb+3b^2$ must all be squares, so write

$$c^2 - 3bc + 3b^2 = (\tfrac{m}{n}b-c)^2,$$

whence $\frac{b}{c} = \frac{3n^2-2mn}{3n^2-m^2} = \frac{2mn-3n^2}{m^2-3n^2}$,

the terms of which fraction must be prime to each other, unless they are divisible by 3. Then (1) if m is <u>not</u> divisible by 3, we will have $c = 3n^2-m^2$ or $c = m^2-3n^2$, and $b = 3n^2-2mn$ or $2mn-3n^2$. But $3n^2-m^2$ cannot be a square, so instead we suppose $c = m^2-3n^2 = (m-\frac{p}{q}n)^2$, whence $\frac{m}{n} = \frac{3q^2+p^2}{2pq}$, $\frac{b}{n} = \frac{2m}{n}-3 = \frac{3q^2-3pq+p^2}{pq}$; accordingly $pq(3q^2-3pq+p^2)$ must be a square, this expression being of the same form as the preceding one, with however smaller numbers, so once again by infinite descent we reach a contradiction. On the other hand, (2) if $m = 3k$, then will $\frac{b}{c} = \frac{n^2-2kn}{n^2-3k^2}$, whence $c = n^2-3k^2$ or $3k^2-n^2$, $b = n^2-2kn$ or $2kn-n^2$. But $3k^2-n^2$ cannot be a square, so we suppose instead that $c = n^2-3k^2 = (n-\frac{p}{q}k)^2$, from which $\frac{n}{k} = \frac{3q^2+p^2}{2pq}$, $\frac{b}{n^2} = 1-\frac{2k}{n} = \frac{p^2+3q^2-4pq}{3q^2+p^2}$; in consequence $(p^2+3q^2)(p-q)(p-3q) = tu(3t^2-3tu+u^2)$, which, having the same form as the expression $cb(c^2-3cb+3b^2)$ but being composed of smaller numbers, will also be a square; and so on.

Theorem 57

The last digit of every power a^{4m}, where a and m are integers, is 0, 1, 5, or 6; and of every square a^2 the last digit is 0, 1, 5, 6, 4, or 9. Every square has 0, 2, 4, 6, or 8 as its penultimate digit, unless the final digit of the root is 4 or 6, in which case the penultimate digit of the square will be odd. If the last digit is 5, then the penultimate digit of every even power will be 2, and of every odd power either 2 or 7.

If 0, 1, 2, 3, 4, 5, 6, 7, 8, 9 are respectively the final digits of the roots, then 0, 1, 2, 3, 4, 5, 6, 7, 8, 9 will be the final digits of the corresponding odd powers of index 1, 5, 9, 13, 17, ..., and 0, 1, 8, 7, 4, 5, 6, 3, 2, 9 will be the corresponding final digits for the odd powers of index 3, 7, 11, 15, 19, etc. Also 0, 1, 4, 9, 6, 5, 6, 9, 4, 1 will be respectively the corresponding final digits for the even powers of index 2, 6, 10, 14, ..., and 0, 1, 6, 1, 6, 5, 6, 1, 6, 1 will be the corresponding final digits for the even powers of index 4, 8, 12, 16, etc.

Corollary. There cannot be an arithmetic progression composed of four squares.

This Corollary is easily deduced, by addition, from the Theorem.

Problem LXII

Given an equation having two or more unknowns x, y, ..., *whose values are rational, to find an equation whose unknown quantity w bears any desired algebraic relation to the roots of the given equation for all of its unknowns.*

(1) If the relation is an algebraic one between the assumed quantity w and the correspondent values of the unknown quantities, then the problem is solved by reducing the several equations to one, to eliminate the unknown quantities.

(2) But if the algebraic relation is not only concerned with the corresponding rational roots of the unknown quantities, but involves either different rational values of the same unknown quantities, or rational non-corresponding roots of different unknown quantities, then we have to find a relation among those different roots of the unknown quantities, which can always be found when we have a general method to find one of the roots. Such a method, which in general discovers one root, will discover more if there are more to be found.

Transform the resultant equations into one, eliminating the unknowns, and so the problem is solved.

(3) But if no relation can be found among these different roots, then only by recourse to conjectures and trial and error may we hope to solve the problem.

Similar observations apply to determination of the sums of individual integer values of functions of the type under consideration, such that these sums shall have prescribed properties, as well as to the transformation of several equations of this sort.

Problem LXIII

Given any arbitrary expressions or numbers, to form from them quantities of a given type.

Choose some of these quantities, and from these assumed values the resulting quantities may be found. But if from the assumed quantities you cannot form quantities of the given type, then continue correcting and approximating, until you arrive at the desired values.

Example. Let the given quantities be the set of all cubes, and the type of the required quantity a square: specifically, required to find a cube, the sum of all of whose aliquot parts is a square. [18]

Start with a few small cubes, for each determining the sum of its divisors; and if it turns out that the sum of the divisors of the first, second, third, ... cube cannot be designated immediately as the product of even powers of primes (as can, for instance, the sum of the divisors of the cube $343=7^3$, which sum is $1+7+49+343 = 400 = 20\times20 = 2\times2\times2\times2\times5\times5 = 2^2\times2^2\times5^2$) we may be able to compose, by multiplication of two, three, or more of these cubes, a number — itself also a cube — the sum of whose divisors (which will be composed of the sums of divisors of the component cubes multiplied together) satisfies the constraint we have imposed. We apply a continuous correction process so as to match the unpaired prime divisors of the sum being developed, with companion primes of equal value which occur unpaired among the other aliquot sums, so as eventually to elicit a cube having the desired property.

Thus for the cube of the number 47, we can represent the sum of its divisors, $1+47+47^2+47^3 = 106080$, in terms of prime numbers, as $2^2\times2^2\times2\times3\times5\times13\times17$, which has the unpaired factors 2, 3, 5, 13, and 17. At the same time, the cube of 5 has the property that the prime factors of the sum of its divisors are 2^2, 3, and 13, of which the unpaired ones are 3 and 13. These, paired with the kindred unpaired factors 3 and 13, of the first term, leave only 2, 5, and 17 as unpaired primes. Next we may insert the cube of 13, the sum of whose divisors is $2^2\times5\times7\times17$, and so has companion primes for the unpqired primes 5 and 17 of the former sum, but now the remaining unpaired numbers are 2 and 7. Next we select the cube of 41, the sum of whose divisors will be $2^2\times3\times7\times29^2$, and omitting the newly paired factors we are left with solitary factors 2 and 3. Now the sum of the divisors of 11^3 is $2^2\times2\times3\times61$, whence the numbers 2 and 3 now have companions, but 61 is now unpaired. Finally, for 27^3 the sum of divisors is $2^2\times11^2\times61$, so this will supply a companion for the number 61, and the required cube root, from this process of repeated correction of preceding stages, will emerge as $47\times5\times13\times41\times11\times27$ — and thus an infinity of other solutions to this problem may be effected.

Similar strategies are appropriate for all problems of this sort.

Here it is appropriate to mention a few further properties of integers and of prime numbers.

(1) Every even number is the sum of two primes, and every odd number is either a prime or the sum of three primes. [19]

(2) The sum $\frac{1}{1}+\frac{1}{2}+\frac{1}{3}+\frac{1}{5}+\frac{1}{7}+\frac{1}{11}+\frac{1}{13}+\frac{1}{17}+\ldots$, taken to infinity, is larger than any finite quantity.

Corollary 2.1. The ratio of the infinite product $1\cdot2\cdot3\cdot5\cdot7\cdot11\cdot13\cdots$ to the infinite product $(1+1)(2+1)(3+1)(5+1)(7+1)(11+1)(13+1)\cdots$ is smaller than any finite ratio.

Corollary 2.1. Hence a number can always be found, the sum of whose divisors has to the number itself a ratio larger than any given ratio.

(3) There are no three prime numbers in arithmetic progression whose common difference is not divisible by the number 6, unless 3 is a term of the series — in which case there can be three terms, and not more, in the same series of prime numbers whose common difference is not divisible by 6. This excepts the two arithmetic series 1, 2, 3 and 1, 3, 5, 7.

There are no five prime numbers in arithmetic progression, unless their common difference is divisible by 30, or unless 5 is the first term of the series, in which case there can be five numbers and no more, which comprise such an arithmetic series. And thus in general there are no cases of 3, 5, 7, 11, 13, 17, ... prime numbers in an arithmetic progression, except when their common differences are divisible, respectively, by the numbers

$$1\cdot2\cdot3,$$
$$1\cdot2\cdot3\cdot5,$$
$$1\cdot2\cdot3\cdot5\cdot7,$$
$$1\cdot2\cdot3\cdot5\cdot7\cdot11,$$
$$1\cdot2\cdot3\cdot5\cdot7\cdot11\cdot13,$$
$$1\cdot2\cdot3\cdot5\cdot7\cdot11\cdot13\cdot17,$$

etc., unless 3, 5, 7, 11, 13, 17, ... are respectively the first term of each such series, in which case each series will have respectively only 3, 5,

7, 11, 13, 17, ... terms, whose common differences are not multiples of the indicated divisors. [20]

This property of prime numbers is easily demonstrated, and others may easily be drawn from the same well. It is easy to show from the preceding that, by repeated addition of the same number to a prime, a number divisible by 3, 5, ... is always produced, unless the number being added is itself divisible by 3, 5, etc. And numbers are often so constituted, in other ways, that by their constitution they are shown to have divisors, and consequently not to be prime.

(4) There are no arithmetic progressions, consisting exclusively of prime numbers, whose sum is a prime number.

(5) If n is a prime number, then $\frac{(n-1)!+1}{n}$ will be an integer, e.g.,

$$\frac{1\cdot 2+1}{3}=1,\quad \frac{1\cdot 2\cdot 3\cdot 4+1}{5}=5,\quad \frac{1\cdot 2\cdot 3\cdot 4\cdot 5\cdot 6+1}{7}=103,\quad \text{etc.}$$

This truly elegant property of prime numbers was discovered by a most excellent man, master of all matters mathematical, John Wilson, Esq.

But given that n is a divisor of such a quantity, an infinity of others can easily be deduced, of which the same number n is also a divisor — for example,

$$\frac{1\cdot 2\cdot 3\cdots(n-2)-1}{n},\ \frac{1\cdot 2^2\cdot 3\cdots(n-3)+1}{n},\ \frac{1\cdot 2^2\cdot 3^2\cdots(n-4)-1}{n},$$

$$\ldots,\ \frac{2^2\cdot 3^2\cdot 4^2\cdot 5^2\cdots\frac{(n-1)^2-1}{4}\pm 1}{n}$$

(where +1 is taken when $\frac{n-1}{2}$ is an even number, and -1 otherwise) are all integers.

Proofs of propositions of this sort will be very difficult, due to the lack of a concise notation to express primality. [21]

(6) Let
$$\begin{aligned} x^r = {} & x(x-1)(x-2)\cdots(x-r+1)\\ & +Px(x-1)(x-2)\cdots(x-r+2)\\ & +Qx(x-1)(x-2)\cdots(x-r+3)\\ & +Rx(x-1)(x-2)\cdots(x-r+4)\\ & +\ldots+Hx(x-1)\\ & +Ix; \end{aligned}$$

then

$$
\begin{aligned}
1^r+2^r+3^r+4^r+\ldots+x^r &= \frac{1}{r+1}(x+1)(x)(x-1)(x-2)\cdots(x-r+1)\\
&+\frac{P}{r}(x+1)(x)(x-1)(x-2)\cdots(x-r+2)\\
&+\frac{Q}{r-1}(x+1)(x)(x-1)(x-2)\cdots(x-r+3)\\
&+\frac{R}{r-2}(x+1)(x)(x-1)(x-2)\cdots(x-r+4)\\
&+\ldots+\frac{H}{3}(x+1)(x)(x-1)+\frac{I}{2}(x+1)(x)\\
&= S.
\end{aligned}
$$

Now let $A = 1+2+3+4+\ldots+(r-2)+(r-1)$,

$A' = 1+2+3+4+\ldots+(r-3)+(r-2)$,

$A'' = 1+2+3+4+\ldots+(r-4)+(r-3)$,

$A''' = 1+2+3+4+\ldots+(r-5)+(r-4)$, etc.;

further denote by B, B', B'', ... the sum of the product of each pair of the numbers $(1, 2, 3, 4, \ldots, r-2, r-1)$, $(1, 2, 3, 4, \ldots, r-3, r-2)$, $(1, 2, 3, 4, \ldots, r-4, r-3)$, respectively; and by C, C', C'', ..., D, D', D'', ..., E, E', E'', ..., etc. the sum of all products of three, four, etc. numbers from the above sets. Then $P = A$, $Q = PA'-B$, $R = QA''-PB'+C$, $S = RA'''-QB''+PC'-D$, and so on.

Corollary 6.1. Hence $1\cdot 2\cdot 3\cdot 4\cdots(r-1)(r)(r+1)\cdot S$ will always be divisible by $x(x+1)$.

Corollary 6.2. If $x+1$ is a prime number greater than $r+1$, then the above sum S will be divisible by $x+1$. If x is a prime number greater than $r+1$, then the sum S will be divisible by x.

Corollary 6.3. *First*, Let $x+1$ be a prime number, and let a, b, c, d, ... be integers whose sum is less than x. Further let α, β, γ, δ, ... represent the numbers $1, 2, 3, 4, 5, \ldots, x$. Then the sum of all individual values of the products

$$\alpha^a\beta^b\gamma^c\delta^d\cdots+\alpha^b\beta^a\gamma^c\delta^d\cdots+\alpha^a\beta^b\gamma^d\delta^c\cdots+\alpha^b\beta^a\gamma^d\delta^c\cdots+\ldots$$

will be divisible by $x+1$.

Second, if x is a prime number and $a+b+c+d+\ldots$ is less than $x-1$, then the stated sum will be divisible by x.

This follows from Problem III of Chapter 1; from it we can see that each term in the sum, by the method given in that Problem, is divisible by $x+1$ in the former case, and by x in the latter case.

(7) Let $a+b+c+d+\ldots = x$, and with a view to Corollary 1 of Problem III we see that the product of the x individual roots, i.e. $\alpha\cdot\beta\cdot\gamma\cdot\delta\cdots$ is $1\cdot 2\cdot 3\cdot 4\cdots(x-1)\cdot x$; by the preceding Article (where the significance of a, b, ..., α, β, ... is identical to their usage here) each term in the sum of such products will be divisible by $x+1$ and its powers (if $x+1$ is a prime number), except for the term which is the sum of the xth powers, i.e. $1^x+2^x+3^x+4^x+\ldots+(x-1)^x+x^x = T$. But by Theorem 52 each term 1^x, 2^x, 3^x, 4^x, ... on division by $x+1$ leaves a remainder of 1, so the sum T divided by $x+1$ leaves a remainder of x, and consequently $1\cdot 2\cdot 3\cdot 4\cdots(x-1)(x)+1$, divided by $x+1$, leaves a remainder of 0, if $x+1$ is a prime number.

Corollary. Hence the product

$$(z-1)(z-2)(z-3)\cdots(z-x)$$
$$= z^x-Az^{x-1}+Bz^{x-2}-Cz^{x-3}+\ldots\pm Pz^2\mp Qz\pm 1\cdot 2\cdot 3\cdot 4\cdots(x-1)\cdot x,$$

where z is an integer, will upon division by $x+1$ leave the remainder 0 or x. If z is not divisible by $x+1$, evidently one of the factors $z-1$, $z-2$, $z-3$, ... $z-x$ must be divisible by $x+1$, and thus the above remainder will be 0. But if z is divisible by $x+1$, then the residue of division of the above quantity by $x+1$ will be the same as the residue of division of the product $1\cdot 2\cdot 3\cdot 4\cdots(x-1)\cdot x$ by $x+1$.

In other words, we have just shown that the numbers A, B, C, D, ... P, Q are divisible by $x+1$; that z^x leaves the remainder 1 on division by $x+1$, if z is not divisible by $x+1$; and that the final product $1\cdot 2\cdot 3\cdot 4\cdots(x-1)\cdot x$ divided by $x+1$ leaves a remainder of x; whence, if z^x cannot be divided by $x+1$, the sum $z^x+1\cdot 2\cdot 3\cdot 4\cdots(x-1)\cdot x$ divided by $x+1$ leaves the remainder $1+x$, i.e. 0, and thus the Corollary is proved.

(8) if $1^r+2^r+3^r+4^r+\ldots+x^r = S$, where r is an odd number, then $1\cdot 2\cdot 3\cdot 4\cdots r\cdot(r+1)\cdot S$ is always divisible by $(x+1)^2(x^2)$.

Corollary. Let

$$1+2+3+4+\ldots+r-3 = A,$$
$$1+2+3+4+\ldots+r-4 = A',$$
$$1+2+3+4+\ldots+r-5 = A'',$$
$$1+2+3+4+\ldots+r-6 = A''',$$

etc.;

and let B, B', B'', B''', ... be the sums of the products of each pair of the numbers

1, 2, 3, ..., r−3,
1, 2, 3, ..., r−4,
1, 2, 3, ..., r−5,
etc.,

respectively, and likewise let C, C', C'', C''', ..., D, D', D'', D''', ..., E, E', E'', E''', ..., etc. be the sum of all products of 3, 4, 5, etc., of the above numbers. Further, let

$$\begin{aligned} S = (x{+}1)^2(x^2)\,[\quad &(x-1)(x-2)\cdots(x-r+3) \\ &+H(x-1)(x-2)\cdots(x-r+4) \\ &+I\ (x-1)(x-2)\cdots(x-r+5) \\ &+K\cdots \\ &+l\ (x-1)(x-2)+m(x-1)+n\,] \end{aligned}$$

whose increment is

$$\begin{aligned} (x^2)(x-1)^2\,[\quad &(x-1)(x-2)\cdots(x-r+2) \\ &+H(x-1)(x-2)\cdots(x-r+3) \\ &+I\ (x-1)(x-2)\cdots(x-r+4) \\ +\ldots &+l\ (x-1)(x-2)(x-3)+m(x-1)(x-2)+n(x-1)]. \end{aligned}$$

Then set $(r{+}1)A + r-3 = rH$,

$$((r{+}1)B{+}(r-3)A)-(rHA'{+}(r-4)H) = -(r-1)I,$$

$$((r{+}1)C{+}(r-3)B)-(rHB'{+}(r-4)HA'){+}((r-1)IA''{+}(r-5)I) = (r-2)K,$$

$$\begin{aligned} ((r{+}1)D{+}(r-3)C)-(rHC'{+}(r-4)HB'){+}((r-1)IB''{+}(r-5)IA'') \\ -((r-2)KA'''{+}(r-6)K) = -(r-3)L\,, \end{aligned}$$

etc.

Then

$$\begin{aligned} &(r-3) \quad \cdot 1\cdot 2\cdot 3\cdot 4\cdots(r-3) \\ &-(r-4)\cdot H\cdot 1\cdot 2\cdot 3\cdot 4\cdots(r-4) \\ &+(r-5)\cdot\ I\ \cdot 1\cdot 2\cdot 3\cdot 4\cdots(r-5) \\ &-(r-6)\cdot K\cdot 1\cdot 2\cdot 3\cdot 4\cdots(r-6) \\ &+(r-7)\cdot L\cdot 1\cdot 2\cdot 3\cdot 4\cdots(r-7) \\ &-\ldots \\ &\pm 3\cdot\ l\ \cdot 1\cdot 2\cdot 3 \\ &\mp 2\cdot m\cdot 1\cdot 2 \\ &\pm n \\ &= 0 \end{aligned}$$

provided that r is an odd number.

(9) If r is an even number, then $1\cdot 2\cdot 3\cdot 4\cdots r\cdot(r{+}1)\cdot S$ will always be divisible by $x(x{+}1)(2x{+}1)$.

Corollary 9.1. Let

$$1+2+3+4+\ldots+r-2=A,$$
$$1+2+3+4+\ldots+r-3=A',$$
$$1+2+3+4+\ldots+r-4=A'',$$
etc.;

and let B, B', B'', B''', ... be the sums of the products of each pair of the numbers

1, 2, 3, ..., r-2,
1, 2, 3, ..., r-3,
1, 2, 3, ..., r-4,
etc.,

respectively, and likewise let C, C', C'', C''', ..., D, D', D'', D''', ..., E, E', E'', E''', ..., etc. be the sum of all products of 3, 4, 5, etc., of the above numbers. Further, let

$$\begin{aligned} S=(2x+1)(x+1)(x)\,[\quad &(x-1)(x-2)\cdots(x-r+2)\\ &+H(x-1)(x-2)\cdots(x-r+3)\\ &+I\,(x-1)(x-2)\cdots(x-r+4)\\ &+K(x-1)(x-2)(x-3)\cdots(x-r+5)+\ldots\\ &+l\,(x-1)(x-2)+m(x-1)+n\,] \end{aligned}$$

Then set $(2(r+1)A+r-2)=2rH$,

$$(2(r+1)B+(r-2)A)-(2rHA'+(r-3)H)=-2(r-1)I,$$
$$(2(r+1)C+(r-2)B)-(2rHB'+(r-3)HA')+(2(r-1)IA''+(r-4)I)=2(r-2)K,$$
$$\begin{aligned}&(2(r+1)D+(r-2)C)-(2rHC'+(r-3)HB')+(2(r-1)IB''+(r-4)IA'')\\ &\qquad -(2(r-2)KA'''+(r-5)K)=-2(r-3)L\,,\end{aligned}$$

etc.

Then
$$\begin{aligned} &(r-2)\quad\cdot 1\cdot 2\cdot 3\cdot 4\cdots(r-2)\\ &-(r-3)\cdot H\cdot 1\cdot 2\cdot 3\cdot 4\cdots(r-3)\\ &+(r-4)\cdot I\cdot 1\cdot 2\cdot 3\cdot 4\cdots(r-4)\\ &-(r-5)\cdot K\cdot 1\cdot 2\cdot 3\cdot 4\cdots(r-5)\\ &+\ldots\\ &\pm 3\cdot l\cdot 1\cdot 2\cdot 3\\ &\mp 2\cdot m\cdot 1\cdot 2\\ &\pm n\\ &=0 \end{aligned}$$

provided that r is an even number.

Corollary 9.2. The same letters denoting the same quantities, if $a+b+c+d+\dots$ is less than x, and the sum $a+b+c+d+\dots$ is an odd number, and $x+1$ is prime, then the sum of all quantities of the form $\alpha^a\beta^b\gamma^c\delta^d\cdots+\alpha^b\beta^a\gamma^c\delta^d\cdots+\alpha^a\beta^b\gamma^d\delta^c\cdots+\dots$ will always be divisible by $(x+1)^2$.

Corollary 9.3. If the sum $a+b+c+d+\dots$ is an odd number less than $x-1$, and x is a prime number, then the preceding expression will always be divisible by x^2.

Corollary 9.4. If the sum $a+b+c+d+\dots$ is less than $2x$, with a, b, c, d, ... all even numbers, and if $2x+1$ is prime, then the preceding expression is divisible by $2x+1$.

(10). Let S be the least common multiple of the numbers 1, 2, 3, 4, 5, 6, 7, ..., M, $m+1$; let $P = 1^m+2^m+3^m+4^m+5^m+\dots+x^m$ and let $Q = 1^2+2^2+3^2+4^2+5^2+\dots+x^2$. Then, first, if m is an even number, SP is always divisible by Q; secondly, if m is an even number and $1+2+3+4+5+\dots+x = p$, $1^3+2^3+3^3+4^3+5^3+\dots+x^3 = p'$, then SP is divisible by p and p'.

In several similar cases it is easily shown that P itself is divisible in the first instance by Q, and in the latter by p and p'.

(11) Let α, β, γ, δ, ... be the roots of the equation $z^n-1 = 0$, and $a+b+c+d+\dots$ not greater than n. Then the sum of all terms

$$\alpha^a\beta^b\gamma^c\delta^d\cdots+\alpha^b+\beta^a+\gamma^c+\delta^d\cdots+\alpha^a\beta^c\gamma^b\delta^d\cdot+\dots = P$$

is divisible by no prime number greater than n. Further if the number of letters α, β, γ, δ, ... in each term is m, then P is divisible by no prime number greater than $m-1$, except for n or its prime divisors.

Theorem 58

(1) $$(a+b)b+a^2,$$

$$(a+b+c)c+(a+b)b+a^2$$

or $$[(a+b+c)c+(a+b)b+a^2]c+[(a+b)b+a^2]b+a^3,$$

$$(a+b+c+d)d+(a+b+c)c+(a+b)b+a^2$$

or $$\begin{aligned}&[(a+b+c+d)d+(a+b+c)c+(a+b)b+a^2]d\\ &+[(a+b+c)c+(a+b)b+a^2]c\\ &+[(a+b)b+a^2]b+a^3 \qquad = P\end{aligned}$$

or $$\begin{aligned}Pd&+\Big[[(a+b+c)c+(a+b)b+a^2]c+[(a+b)b+a^2]b+a^3\Big]c\\ &+\Big[[(a+b)b+a^2]b+a^3\Big]b+a^4,\end{aligned}$$

and $(a+b+c+d+e)e+(a+b+c+d)d+(a+b+c)c+(a+b)b+a^2 = P$

or $Pe+ [(a+b+c+d)d+(a+b+c)c+(a+b)b+a^2]d$
$[(a+b+c)c+(a+b)b+a^2]c$
$[(a+b)b+a^2]b+a^3 = Q$

or $Qe+\Big[[(a+b+c+d)d+(a+b+c)c+(a+b)b+a^2]d$
$+ [(a+b+c)c+(a+b)b+a^2]c+[(a+b)b+a^2]b+a^3\Big]d$
$+\Big[[(a+b+c)c+(a+b)b+a^2]c+[(a+b)b+a^2]b+a^3\Big]c$
$+\Big[[(a+b)b+a^2]b+a^3\Big]b+a^9$,

and so on to infinity,

are all identical functions of each of the quantities a, b, c, e, f, ..., and consequently will remain the same, however the order of the letters a, b, c, d, e, f, ... may be rearranged. Likewise the quantity

$$((((a+b)b+a^2)b+a^3)b+a^4)\cdots$$

is the same function of a and b in either order, and so on.

(2) Let F be a function in which the letters x, y, z, ... are symmetrically involved; and let $\phi(e, f, g, \ldots)$, $\phi(e, f, h, \ldots)$, $\phi(f, g, h, \ldots)$, ... be functions in which $(e, f, g, \ldots)$, $(e, f, h, \ldots)$, $(f, g, h, \ldots)$, ... are symmetrically involved. Further denote by $\pi(e)$, $\pi(f)$, $\pi(g)$, $\pi(h)$, ... the same functions of the letters e, f, g, h, etc. Then in the function F, if for the letters x, y, z, v, ... we write respectively $\pi(e)+\phi(f, g, h, \ldots)$, $\pi(f)+\phi(e, g, h, \ldots)$, $\pi(g)+\phi(e, f, h, \ldots)$, etc., there will result an expression which is symmetric in all of the letters e, f, g, h, etc.

Theorem 59

Let a, b, c, d, e, ... h, k, l, m, n be any arbitrary quantities; then
$ab(a+b) + (a+b)c(a+b+c) + (a+b+c)d(a+b+c+d)$
$+(a+b+c+d)e(a+b+c+d+e) + \ldots$
$+(a+b+c+d+e+\ldots+h)k(a+b+c+d+e+\ldots+h+k)$
$+(a+b+c+d+e+\ldots+h+k)l(a+b+c+d+e+\ldots+h+k+l)$
$+(a+b+c+d+e+\ldots+h+k+l)m(a+b+c+d+e+\ldots+h+k+l+m)$
$+(a+b+c+d+e+\ldots+h+k+l+m)n(a+b+c+d+e+\ldots+h+k+l+m+n)$

$$\begin{aligned}
&= nm(n+m) + (n+m)l(n+m+l)\\
&\quad +(n+m+l\quad)k(n+m+l+k)\\
&\quad +(n+m+l+k)h(n+m+l+k+h) + \ldots\\
&\quad +(n+m+l+k+h+\ldots+e)d(n+m+l+k+h+\ldots+e+d)\\
&\quad +(n+m+l+k+h+\ldots+e+d)c(n+m+l+k+h+\ldots+e+d+c)\\
&\quad +(n+m+l+k+h+\ldots+e+d+c)b(n+m+l+k+h+\ldots+e+d+c+b)\\
&\quad +(n+m+l+k+h+\ldots+e+d+c+b)a(n+m+l+k+h+\ldots+e+d+c+b+a)
\end{aligned}$$

This is demonstrated in Theorem 18 of my *Proprietates Algebraicarum Curvarum.*

Problem LXIV

(1) *Given an equation which is a function of the quantities* x, y, z, ..., *containing* n *coefficients which may be assigned any value; further given* n *corresponding values of each of the unknown quantities* x, y, z, ...; *to find the desired coefficients.*

If for the unknown quantities x, y, z, we write their n corresponding values, there result n equations in as many unknown quantities, which unknowns are in fact the n arbitrarily assumable coefficients, and so by solving these equations the desired coefficients may be found.

(2) If there are m equations in $n+m$ unknown quantities x, y, z, ..., and each of these equations contains r unknown coefficients which may be assumed at will, then from r given corresponding values of each of the unknown quantities x, y, z, ... the coefficients may be determined, by essentially the same method as in the previous case.

Problem LXV

(1) *To find a function which takes the values* S, S′, S″, S‴, ... *when* x $= \alpha$, y $= \beta$, z $= \gamma$, ...; *or* x $= \alpha'$, y $= \beta'$, z $= \gamma'$, ...; *or* x $= \alpha''$, y $= \beta''$, z $= \gamma''$, ...; *etc.*

We need several arbitrary functions of x, y, z, ... (call them H, K, L, ..., and A, B, C, D, ...). In the function H for the unknown quantities x, y, z, ... substitute α, β, γ ...; in K, L, ... substitute respectively

α', β', γ', ... and α'', β'', γ'', ..., etc., calling the respective results h, k, l, etc. In the function A for the unknowns x, y, z, ... substitute first α, β, γ, ..., calling the result a; then also in A substitute α', β', γ', ..., and afterwards α'', β'', γ'', ..., etc., and call these resulting quantities a', a'', a''', etc. Likewise in the functions B, C, D, ... substitute in succession the same sets of values, producing the quantities b, b', b'', ...; c, c', c'', ...; d, d', d'', ...; etc. The required function may then be written as

$$\frac{H(A-a')(A-a'')(A-a''')\cdots}{h(a-a')(a-a'')(a-a''')\cdots}\times S + \frac{K(B-b)(B-b'')(B-b''')\cdots}{k(b'-b)(b'-b'')(b'-b''')\cdots}\times S' + \frac{L(C-c)(C-c'')(C-c''')\cdots}{l(c''-c)(c''-c')(c''-c''')\cdots}\times S''+\ldots$$

Corollary. If it is required that the derived expression be the simplest possible, then H, K, L, ... should be regarded as invariable quantities, and for A, B, C, ... assume functions of x, y, z, ... of the form $ex+fy+gz+\ldots+d$, where the letters e, f, g, ..., and d also denote invariable quantities.

(2) Required a quantity which shall become S^a, S^b, S^c, S^d, ... when $x=\alpha$, $x=\beta$, $x=\gamma$, $x=\delta$, ... respectively: the required expression may be written

$$\frac{(x-\beta)(x-\gamma)\cdots}{(\alpha-\beta)(\alpha-\gamma)\cdots}\times S^a + \frac{(x-\alpha)(x-\gamma)\cdots}{(\beta-\alpha)(\beta-\gamma)\cdots}\times S^b + \frac{(x-\alpha)(x-\beta)\cdots}{(\gamma-\alpha)(\gamma-\beta)\cdots}\times S^c+\ldots$$

Scholium

This concludes our study of algebraic equations. The same results may also be applied to such other equations as can through substitutions be transformed into algebraic equations — e.g., if $x^{nv}+Ax^{(n-1)v}+Bx^{(n-2)v}+\ldots=0$, we can write z in place of x^v, and so the given exponential equation is transformed into an algebraic one, $z^n+Az^{n-1}+Bz^{n-2}+\ldots=0$. Let the roots of this equation be α, β, γ, δ, ...; then x^v will be equal to α, β, ..., so the quantity x^v will have n distinct values.

(1) If $e^v = a$, where e is a given positive quantity greater than 1, then if a is positive, v will be positive or negative according as a is or is not greater than 1. If a is negative, v will be a fraction with an even denominator and an odd numerator; again it will be positive or negative, depending on whether or not the magnitude of a is greater than 1. If e is positive and less than 1, then v will be negative or positive according as a is greater than +1 or less than −1, or lies between 1 and −1. If e is negative, with magnitude greater than 1, then v will be positive if the magnitude of a is greater than 1, otherwise v will be negative. If the magnitude of e is less than 1, then v will be positive if a lies between +1 and −1, and will be negative otherwise.

(2) Let $x^x = a$. Every quantity attains a maximum or minimum when its fluxion goes to zero; and the fluxion of the quantity x^x will be $x^x dx + x^x \log x dx$, which, equated to zero, reduces to $\log x = -1$. Thus we seek that quantity whose hyperbolic logarithm is −1, which will be $0.36\cdots = \frac{1}{2!} - \frac{1}{3!} + \frac{1}{4!} - \frac{1}{5!} + \ldots = \rho$.

Let $\rho^\rho = \pi$. Then *firstly*, if a is positive and greater than $\frac{1}{\pi}$, there will be one positive value of x; *secondly*, if a is between π and $\frac{1}{\pi}$, then there will be one value of x which is positive, and another which is negative; *thirdly*, if a is situated between 0 and π, then there will be only one value of x, and it will be negative. If a is negative and (1) greater in magnitude than $\frac{1}{\pi}$, then there will be a positive value of x in the form of a fraction with an even denominator and an odd numerator; if a is (2) between π and $\frac{1}{\pi}$, then x takes one negative and one positive value; and if a is (3) less than π, then x has only one value, which is negative.

In all of these cases there are an infinity of impossible roots as well.

And if there are several exponential quantities in the given equation, including exponentials of high degree, we must determine the number of values which may be taken by each particular exponential among all of them; and multiplying the numbers so obtained, we generally arrive at the number of values of the unknown quantity in the given equation. And thus for transforming equations of this sort, and of finding those of their roots that are impossible, positive, or negative, or integral or rational. For example let $x^x + \ldots = a$, where a is an integer; then if an integer solution exists for x it must be a divisor of a, and so on. Not all numbers can be composed from a finite set of numbers of the form x^x, e^x, etc., where x and e etc. denote integers. Successive terms are too large or too small by increasingly large ratios — thus for example $a^{x+1}-1$ is composed, at a minimum, of more than $(a-1)x$ terms of the type a^z. But it is not my intent to write about exponentials, as I now bring to a close these Meditations, which treat of transformations; of methods of finding impossible, positive, and negative roots; of constitutions, reductions, and resolutions; of eliminating irrational quantities from equations; and of properties of integer numbers and equations, and so on — in a way which is certainly more general than any other treatment of these subjects that has heretofore been published.

TRANSLATOR'S NOTES

[1] *Wassenaer, Jacob van.* Once in a letter to Mersenne (cited by Korteweg [1882]), Descartes remarked with surprise, "There is someone in Utrecht who understands my geometry." The person of whom he spoke was probably Jacob van Wassenaer, to whom Waring attributes the discovery of a method of finding linear factors of polynomials which now, having been extended to quadratic and higher degrees, is generally known as Newton's or the Newton-Kronecker method (cf. Knuth [1981, p. 431]).

A total of four published works by Wassenaer are listed by Bierens de Haan [1882], who speculates also that Wassenaer may have been the real person behind the pseudonym Julius Caesar van Padua; this speculation is dismissed as unlikely by Waard [1927], author of the only published biographical notice of Wassenaer. According to Waard, he was born in 1607, the son of Abel Wassenaer, professor of mathematics in the University of Leyden where Jacob studied in 1626 and 1627; Jacob became a surveyor for the Province of Utrecht. Records of the town of Utrecht, consulted for me by A. D. A. Monna of the University of Utrecht, show his marriage to Petronella van Schayck on 16 December 1641; he died in Utrecht on 8 December 1682, one of his survivors being a son (also named Jacob) who was surveyor and schoolmaster in Zalthommel, in the province of Guelders.

Wassenaer's published works are all related to a dispute with Jean Janz Stampioen "de Ionge" (i.e. "junior"), whom Bosmans [1927] identifies as the son, born in Rotterdam in 1610, of another man of the same name who was professor of mathematics at Rotterdam; the younger Stampioen succeeded his father in that position and in 1639 published a book which contained a flagrantly incorrect method of computing the cube root of a binomial expression, which was publicly attacked by Wassenaer. Stampioen, unaware of his own incompetence, challenged Wassenaer to a debate, to be arbitrated by other mathematicians, Golius and van Schooten, of the University of Leyden where the debate was to be held, Berlicom of Rotterdam (proposed by Stampioen because he believed Golius and Schooten already knew and would favor Wassenaer), and Schotanus of Utrecht. Each of the combatants posted a forfeit of 600 florins, with the understanding that the loser's money would be given to the poor of Leyden. The debate was in fact the first public challenge and defense of Descartes' mathematics (although according to Roth [1926] Stampioen had begun expressing anti-Cartesian sentiments as early as 1633); it was conducted in November and December of 1639, with some delays due to Wassenaer's ill health at the time, and it was not until May of 1640 that the poor were awarded Stampioen's 600 florins. It is clear that Descartes was closely involved in the affair, and a letter from Descartes to Wassenaer matches the preface of the latter's [1640] triumphant pamphlet, "Den On-Wissen Wis-Konstenaer I. I. Stampioenius ontdeckt". The English translator of Baillet's [1691] biography of Descartes sums up the affair thus:

> He [Rivet] it was that inform'd them concerning the so much talk't of wager about Mathematics, between young Wasseneur (*sic*) and Stampion, where Monsieur Des Cartes was present, drawn in by the indiscretion and ill-will of the latter; notwithstanding Wasseneur's Victory was very glorious; for our Philosopher, who went publicly for his Master, yet he managed this business but as a piece of Merriment, which was not worthy of Rivet's disturbance, nor the curiosity of the mathematicians of France.

Although there were many Wassenaers among the Dutch nobility then, Jacob was evidently not one of them. Whitrow [1971, p. 611] mistakenly gives his dates as 1610 - 1666, which are the dates of Johann Jakob van Wassenaer (*Nieuw Nederlandsch Biographisch Woordenboek*, Vol. II), who was not a mathematician; the confusion is easy to understand, as the biography of this Jacob van W. is in Vol. VII. As an amateur rather than a professor of mathematics he was not in a position to do much in the way of publishing his discoveries. It was Schooten [1649, pp. 323-335] who in fact first published Wassenaer's factorization strategy, indicating that the idea may have begun to develop even earlier in writings of Stifel. The only other published record of Wassenaer's mathematics is in correspondence with Oldenburg in 1668, in the early days of the Royal Society, which was passed to Wallis and is reproduced in his [1693, Vol. II, pp. 599-601] Latin *Algebra* under the heading "*De tribus Problematis è Belgio missis*".

[2] This paragraph very closely resembles §LI of Newton's *Arithmetica Universalis*, and the translation as I present it here is closely in keeping with Raphson's English version of Newton's work. I cannot refrain from adding an observation of my own: in essence, "had we but world enough and time", the search for factors of any degree of a given polynomial could be achieved simply by factoring the coefficients of its leading and constant terms, and then trying all polynomials whose leading and constant terms were taken from those factorizations, with a limit on the "height" of the intermediate coefficients based on the range of the roots of the given polynomial. However, the number of such trial divisors could be enormously large; by constructing sequences of numbers, of limited length ($n+1$ terms, where n is the degree of the proposed factor), and then rejecting any sequences that do not lead to an arithmetic progression, Newton has dramatically shrunk the search space of potential factors, although Knuth points out that for degrees above 5 even this method is very laborious compared to techniques based on factorization modulo several primes. Like Newton's rule for the detection of complex roots, this procedure was presented with no proof that it could never happen

to overlook a true factor; commentators such as Gravesande, who published the Leyden 1732 edition of the work, offered their own proofs, and a proof by Maguire is given in Wilder's [1769] annotated edition in English. A thesis by Willich [1738] claims that the method for quadratic divisors was actually first published by Hamburg arithmetician Heinrich Meisner in 1692.

[3] For another method, based on Euclid's algorithm for the greatest common divisor, see Knuth [1981, pp. 326-327]; also compare Waring's treatment in Problem LIII of two equations with three or more unknowns.

[4] In fact if one counts the unknowns in the general expressions for degrees 4 and 8, there are more unknowns than there will be equations. However, for degree 16 or higher the assertion is true. It is not immediately obvious what Waring means by the "most general resolution", but the ground rules can be inferred from the technique by which irrationalities will be eliminated. The basic method is to isolate the most deeply nested radical expression on one side of the = sign, with x and all other square roots and rational terms on the other side. Then squaring both sides produces an equation in which no radical expression is as deeply nested as before, and the process may be repeated, reducing the extent of radical nesting by one each time, until there are no radical expressions left. However, this assumes that there is only one expression with the deepest nesting of radicals, and after the first squaring, there must be only one expression with the second deepest nesting of radicals (although it may be multiplied by an expression which includes radicals at the third deepest nesting level, etc.) Thus in the form of the general solution for the 8th degree, the expression $\sqrt{b+n\sqrt{c}}$ occurs twice (with different multipliers m and r outside the outermost root sign, to be sure) — but there is no expression, say, $\sqrt{e+\sqrt{f}}$, because if there were, one of them could be reduced to less nesting in one step, but another squaring step would be required for the other, doubling the degree of the final equation. Following this reasoning, the proposed general solution of an equation of degree 16 would be written

$$\sqrt{a+m\sqrt{b+n\sqrt{c+o\sqrt{d}}+p\sqrt{d}}+q\sqrt{c+o\sqrt{d}}+r\sqrt{d}}$$
$$+s\sqrt{b+n\sqrt{c+o\sqrt{d}}+p\sqrt{d}}+t\sqrt{c+o\sqrt{d}}+u\sqrt{d}+e\,.$$

This expression has a total of 14 unknowns, which will contribute to the 16 coefficients of the resulting equation, giving us, as Waring predicts, an over-determined set of non-linear equations to solve in integers. This is, of course, a specific case of Hilbert's Tenth Problem, which Yuri Matiyasevič showed in 1970 had no general solution (cf. Davis [1973]). Nevertheless it would be interesting to carry the elimination of radicals to the limit for this case, because then we could equate each of the 16 resulting coefficient expressions to 1, as if the proposed equation were $x^{16}+x^{15}+\ldots+x^2+x+1 = 0$, whose roots are the imaginary 17th roots of unity — and although it was not known during Waring's lifetime, this case has the solution $x = \cos(\frac{2\pi}{17})+i\sin(\frac{2\pi}{17})$, where

$$\cos(\tfrac{2\pi}{17}) = -\tfrac{1}{16}+\tfrac{1}{16}\sqrt{17}+\tfrac{1}{16}\sqrt{34-2\sqrt{17}}$$
$$+\tfrac{1}{8}\sqrt{17+3\sqrt{17}-\sqrt{34-2\sqrt{17}}-2\sqrt{34+2\sqrt{17}}}$$

[Gauss 1801, art. 365]; and of course $\sin(\frac{2\pi}{17}) = \sqrt{1-\cos^2(\frac{2\pi}{17})}$.

[5] This Problem and the following one are discussed by Bashmakova [1959, pp. 445-452], who actually referred to the 1770 edition where they were Problems XXXVII and XXXVIII. To my knowledge, the few lines of text in her paper are the only case where any of Waring's writings have appeared in Russian translation.

[6] Actually, if $n = 3$, $4n = 3n+3$, and if n is 1 or 2 there are fewer equations than unknowns, so reduction of equations of degree 4, 8, or 12 must be regarded as special cases. As in Problem XLVI (see note [4]), this method is theoretically sufficient (although computationally difficult) to reduce the 16th–degree equation of the 17th roots of unity to a quartic; had Waring chosen to work out this particular example he could have preceded Gauss to the compass-and-straightedge division of the circle into 17 equal arcs.

[7] This complicated pattern resembles some patterns described by Sylvester [1854] as "anakolouthic" symmetric functions.

[8] Waring also discusses this topic in *Meditationes Analyticæ*, Ed. II (1785), pp. 136ff.

[9] The numbered paragraphs between Corollary 1 and Corollary 1.1, and Corollary 1.1 itself, are from Addenda pages A3.7–A3.8 (the last printed and most error-prone set). There these paragraphs were assigned to "Pag. 332"; here (the start of page 323 of the Latin text) seems more likely what Waring intended. (The corollaries were unnumbered in the text and in the addendum.) The following sub-heading, "Supplement to Problem 56", I inserted to distinguish the remaining text from the addendum.

[10] Paragraph (2) of Case 5 comes from Addenda page A3.8, where it was assigned to page 323, line 25 (as "5.2"); in fact the place of reference appears to be page 326, line 25.

[11] In the 1770 edition Waring attributed this discovery to Stifel. (At that time this material was presented as Problem XLVIII, not as a Theorem.)

[12] A simpler proof: if a is a factor of n, and $a<\sqrt{n}$, then there is factor $n/a = b>\sqrt{n}$. Thus to all divisors $a_i<\sqrt{n}$ correspond an equal number of divisors $b_i>\sqrt{n}$, and conversely. Then the total number of divisors is even unless $\sqrt{n}$ is an integer, i.e., if and only if n is a square, it has an odd number of divisors.

[13] This rather vague proposition comes from the addenda and replaces a much more explicit formula in the text.

[14] Paragraph (14) was the conclusion of Case 2, Problem XLVIII, in the 1770 edition; there, as here, Waring attributed the result to "domini Kiraast", evidently a typographical error for Georg Wolfgang Krafft, whose [1735] paper was what Waring had seen. (Waring referred to both Krafft and Vandermonde as *dominus* (literally "lord"), and since *Monsieur* and *Herr* both also translate to "lord", I have used them in this translation.) Interestingly, Krafft attributed the origin of the concept to a book by Johann Krafft [1591]. (Dickson [1952, vol. 1, p. 338] says Krafft had actually rediscovered a method originally presented to the Académie Parisienne by Pascal in 1654, and first published in 1665.)

[15] This is the famous "Waring's conjecture", which first appeared in the 1770 edition, a few months before Lagrange had proved that every number was the sum of four or fewer squares. Hardy and Wright [1938] give a fairly thorough historical review of attempts to prove and/or generalize the conjecture. They also point out that the series $2^k+\left\lfloor\left(\frac{3}{2}\right)^k\right\rfloor-2$, which generates the numbers 4, 9, 19 for $k=2$, $k=3$, $k=4$, was first studied (in a different context) by Euler. In 1909 Hilbert proved that $g(k)$, the number of kth powers needed to represent any integer n, was finite for all values of $k \geq 2$. Modern research notes that the upper limit on $g(k)$ only applies to a few relatively small summands; more interesting is a function typically referred to in the literature as $G(k)$ representing the largest number of kth powers needed to represent any number greater than some limit (which also depends on k). Jacobi [1851] had in fact observed that of all numbers less than 12000, only 23 and 239 required nine cubes, only a few required eight or seven, and that any number between 8043 and 12000 could be represented as the sum of six or fewer cubes. A survey of results for values of k from 4 to 9 is found in Stewart [1986]. Bohman and Fröberg [1981] present convincing numerical evidence that $G(3)=4$; Dickson [1928] had proved Waring's conjecture that $g(3)=9$. Ribenboim [1989] summarizes the results for $g(k)$ and $G(k)$ for several values of k, in the context of an even more difficult question which he calls the Waring-Goldbach problem: restrict the summands to k^{th} powers of primes, not of just any integers, now what is $G(k)$ or $g(k)$?

[16] *Aliter* is a Latin term which generally translates to "in other words" or "alternatively"; here Waring uses it as a sub-heading (essentially a converse) within this Corollary (as he did in the *Philosophical Transactions*, 1779), and so I have left it untranslated, as I do also for terms such as *Scholium* and expressions like *exceptis excipiendis.*

[17] Corollaries 3 through 9 as presented in this translation were numbered 4 through 10 in the Latin text; a line in the Addenda advises the reader to "dele cor. 3.", which I have done, and accordingly renumbered the rest of them. For the record, here

is the translated text of the deleted Corollary:

> Let n be a prime number; then every power $\frac{a^{n-1}}{n} \times \frac{a^{n-1}}{n} \times \frac{a^{n-1}}{n} \times \cdots = \frac{a^{m(n-1)}}{n^m}$, whose exponent is $m(n-1)$, leaves a remainder of 0 or 3 when divided by n^m.

[18] The first discussion of the problem was published by Wallis [1685] under the heading, "Discourse of Combinations, Alternations, and Aliquot Parts", where he states that the problem was first posed by Fermat as a challenge to Wallis "and the other English mathematicians".

[19] This is the famous "Goldbach conjecture", first stated in 1742 by Goldbach in a letter to Euler; according to Struik [1969, p. 49] Waring was the first to publish it.

[20] In recent years supercomputers have been used to search for new sequences of primes in arithmetic progression. Moran and Pritchard [1991] developed an alternative method using a network of small workstations, and after that paper was written they achieved a new record, 21 primes in an arithmetic progression with a common difference of 1419763024680, the smallest prime in the sequence being 142072321123. The record was electronically published on USENET news on 22 January 1991, and was first mentioned in print in a newspaper, *The Australian* (2 July 1991, p. 26: "Scientists tap downtime to break prime-number record").

[21] Martin Gardner [1981, p. 17] retells an apocryphal story from the history of mathematics, in which Gauss hears of Wilson's theorem in a conversation where the author (Waring) was quoted as having said that the theorem "would never be proved" because of the lack of suitable notation, whereupon Gauss replied that it was not notations, but notions, that were lacking, and proceeded to work out a proof in about twenty minutes. In fact Waring acknowledged Lagrange's "elegant proof" in his preface to this edition. The set of additional formulas, adapted from the model invented by Wilson, leaves no doubt in my mind that Waring also had full command of the requisite "notions".

APPENDIX

Doctoral Dissertation of Franz X. Mayer

Translated from the German
by Dennis Weeks, August, 1991

Edward Waring's "Meditationes Algebraicæ"

Inaugural Dissertation for the Degree of Doctor of Philosophy submitted to the Philosophical Faculty II of the University of Zürich by Franz Xaver Mayer Pater Augustin Ordinis Cisterciensis of Weingarten, Württemberg.

Dedicated to Professor Dr. R. Fueter

1923
Aug. Feyel, Überlingen (Bodensee).

Introduction

Every mathematician has heard of "Waring's Formulas" and "Waring's Problem", and every teacher, in his lectures on higher algebra, specifically in the treatment of symmetric functions, eventually speaks of the English mathematician Waring, who treated this subject matter in his "Meditationes algebraicæ".

This work, however, is written in Latin and copies are scarce, so the average middle-school teacher in the German-speaking world probably does not have access to it; and yet he might be interested to know what sort of problems are discussed there.

This, and the choice of a thesis topic, gave me the motivation to study Waring's "Meditationes algebraicæ", to characterize the individual chapters, to examine the historical foundations of the work, and to give examples of its expository style.

In consequence my work falls naturally into three parts:

I. Characterization of the individual chapters of the "Meditationes algebraicæ".

II. Its historical foundations.

III. Critical discussion of specific topics.

In the characterization of the individual chapters I have borrowed from my teachers the precept that it is not very interesting just to relate what the author has said, but much more interesting to discuss what he has not said, in order to be able to measure the progress of mathematics.

Waring himself has supplied the historical foundations of the "Meditationes algebraicæ" in the Preface of his work, with a definite English perspective, and confined to those historical sources which, in the middle of the 18th century, were available to him.

The reader would be mistaken, however, to assume that Waring confined himself to mentioning to which mathematicians the credit was due for solution of the problems which are found in his book: on the contrary he discusses many different problems in continuous functions and series, only in exceptional cases mentioning the name of the author.

The subject of the third section of this work will be an overview of how Waring treats and presents a particular mathematical problem.

My esteemed teacher, Prof. Dr. Rudolf Fueter of Zürich, gave me many varied and extremely helpful suggestions, for which I here express to him my heartfelt thanks.

Biographical Introduction

Waring, Edward, b. 1734, was the eldest son of John Waring, a prosperous landholder of Old Heath, near Shrewsbury. From the school in Shrewsbury Edward came to Magdalen College, Cambridge, where in 1757 he graduated with highest honors. He was already considered a prodigy in mathematical disciplines, and on April 24, 1758, was named a Fellow of the University.

About the same time in Cambridge the so-called "Hyson Club" was founded and Waring and Paley, among the "sharpest minds of the University", joined it.

Even before Waring had qualified for the M.A. degree, he was in 1760 named Lucasian Professor at Cambridge. Because of his youth his candidacy encountered opposition. In order to demonstrate his extraordinary ability, before the selection was made he published the first chapter of his "Miscellanea analytica". However this work was attacked, for unspecified faults, by William Powell. More controversial pamphlets followed. In this verbal battle he was ably seconded by his friend John Wilson of Peterhouse College, of whom we shall hear more at the end of Section II. Waring was the victor. He secured the professorship in question, which he held until his death.

In the same year by royal decree he was issued the required degree of M(agister) A(rtium), and through his publications he attracted renewed attention to Cambridge University. He himself gave no lectures: "The profound researches of Dr. Waring were," as Dr. Parr remarked, "not adapted to any form of communication by lectures". [1]

It is interesting that in 1767 Waring took up the study of medicine and began visiting the hospitals of London. Indeed, by 1770 we find him listed as a physician at Addenbroke Hospital in Cambridge, and some time later at St. Joes, Huntingdonshire. However, since he was extremely near-sighted, and of a shy disposition, ultimately he had to give up the medical profession. Fortunately his income as Lucasian Professor was substantial, and besides he owned a comfortable estate.

Because his hope was frustrated of securing a position at Magdalen for his brother Humphrey, eventually he left Magdalen for Trinity College.

In 1776 he married Mary Oswell and they settled near Shrewsbury, where his brother in law resided. However, the climate there was not to his wife's liking, so he returned to his country estate, Plealy, in Petersbury. [2] There he died on August 15, 1798.

To help characterize Waring we cite the following:

In a letter to the Royal Astronomer, Dr. Maskeleyne, Waring complained about the contention by Lalande, in his "Life of Condorcet",[1] that in the year 1764 no first-rate analyst existed in England, pointing out that his book of 1762 had been commended by d'Alembert, Euler, and Lagrange.[2]

Waring further claimed to have published "300–400 new theorems, more than any other English writer". But he was forced to admit that he "had heard of no reader in England, outside of Cambridge, who had taken the trouble to read and understand his works".[3]

This was chiefly ascribed to the circumstance that his writing style was awkward and obscure, and besides that "the classification and arrangement of his material was thoroughly incomprehensible".[4]

Waring was called a man "of pride and modesty united, to the highest degree possible in the human character, yet pride was the predominating trait".[5] In the opinion of Bishop Watson Waring had a "hard head" — in his old age he sank into a religious melancholia, that bordered on madness.

Dugald Stewart called Waring "one of the greatest analysts that England has up to now produced".[6]

Already in 1763 Waring was named a Fellow of the Royal Society of London, and in 1784 he was awarded its Medal of Merit. He was a fellow of the Royal Societies of Göttingen and Bologna.[7]

---

[1] Actually an English rendition of "Vie de Condorcet"; later Lalande wrote a "Necrologie des hommes célèbres de France".

[2] Monthly Mag. May 1799, pp. 306-310.

[3] Essay on Human Knowledge, pp. 114-115, 1794, privately circulated ms.

[4] Ball, Mathematics at Cambridge, pp. 99-113; in Dictionary of National Biography, London 1899, LIX p. 384.

[5] Living Authors, 1798, pp. 364-365.

[6] Elements of Philosophy of Human Mind, pt. ii chap. i, in Works edit. 1854, iv. 218.

[7] Dictionary of National Biography. London 1899 LIX. 383 ff.

Waring acquaints us with his works in the context of the preface of the third edition of his "Meditationes algebraicæ".

His first paper was sent to the Royal Society in 1757 and then in 1760, under the title "Miscellanea analytica", was printed and circulated among friends; then in 1762 it was published, with an appendix on the properties of algebraic curves, as "Miscellanea analytica de æquationibus algebraicis et curvarum proprietatibus".

The second edition of this work was prepared in 1768 and 1769, and was published in 1770 with the title "Meditationes algebraicæ" with the "proprietates algebraicarum curvarum" removed. The latter appeared separately in 1772.

The third edition of the "Meditationes algebraicæ" appeared in 1782, and here many of the methods that were mentioned briefly in the earlier editions were now greatly enlarged and generalized.

In my work all citations are from this third edition.

The "Dictionary of National Biography" cites two other writings by Waring in the English language, namely

> "On the Principle of translating Algebraic Quantities into Probable Relations and Annuities" (1792).
>
> "An Essay on the Principles of Human Knowledge" (1794).

Both are very rare. One copy of the first is found in the British Museum, while the latter was never actually published, but rather was circulated as a manuscript among friends.

Part I

Characterization of the Individual Chapters of the "Meditationes algebraicæ"

Chapter I

De methodo inveniendi æquationem, cuius radices sint quæcunque algebraica datæ æquationis vel datarum æquationum functio.

A method to find an equation whose roots are any algebraic function of the roots of the given equation or of the given equations.

In the first Chapter Waring covers the theory of *symmetric functions*, that is, of those integral rational functions which remain unaltered by any permutation of their variables. However Waring does not refer to them by this name, but merely as algebraic functions of the roots of a generalized equation.[1]

A characteristic method of Waring's is to establish, wherever possible, general principles and rules, and then to apply these to specific cases. So does he here.

He gives a general law for the mth power sum as a function of the roots of the given equation, in that he not only computes, like Newton, the power sums in ascending sequence, but presents a *general formula* and then explains how it is derived.

For example: let an equation be given, of degree n, arranged by decreasing exponents, with the signs of the coefficients alternating, required to find a new equation whose coefficients consist of the sums of the mth powers of the roots of the aforesaid equation.[2]

In this general form Waring has discussed the elementary symmetric functions. So also are both systems of formulas established, which reveal the relations between the power sums of the roots of an equation and its coefficients, and which, written in this general form, came directly to be known as *Waring's formulas.*[3] Waring also produced proofs for both formulas,[4] and did not content himself only, as Serret[5] claims, with simply proving the correctness of the formulas.

---

[1]Medit. algebr., Cantabrigiæ 1782, p. 1.

[2]loco citato; Problema I. p. 1.

[3]loco citato; p. 1 and p. 5.

[4]loco citato, p. 2-4 and p. 7-8.

[5]J. A. Serret, Cours d'Algèbre Sup. t. 1. Paris 1910, p. 449.

From the power sums Waring passes to the symmetric functions of the form

$$\sum \alpha^a\beta^b\gamma^c + \alpha^b\gamma^a\beta^c + \dots$$

and shows that every symmetric function of this form can be expressed as an integral function of the power sums, and thus also as an integral function of the coefficients of the given equation whose roots are α, β, γ, etc.[1]

But his speculative spirit went yet further and sought for an equation whose coefficients were symmetric functions of the roots of two or more equations,[2] and here also obtained positive results which no one before him had ever achieved.

However Waring had neither our summation symbol $\sum$, nor the factorial notation, at his disposal, so his formulas are very cumbersome and unsightly.

Be that as it may, even modern mathematicians must refer back to Waring for research of this sort.

[1]Problema III, p. 8.
[2]Problema V, p. 24.

Chapter II

De impossibilibus radicibus.

Imaginary roots.

Waring's research on the number of roots of an equation was based on the complex roots. In keeping with the usage of the time, he calls them "impossible" and recognizes as such $a+bi$ and $a-bi$; his notation is $a+\sqrt{-b^2}$ and $a-\sqrt{-b^2}$.[1] The two together he identifies as *corresponding roots*, which we would now call complex conjugate roots.

For the detection of complex roots Waring adduces extensive research, which however applies only in special cases, not founded on a generally applicable rule. So for instance in the cubic equation the cases which correctly find the true number of imaginary roots are to those that fail to find the correct number, in the ratio of $2:1$.[2] Elsewhere Waring transforms a general equation, with some powers not included and with alternating signs attached to the coefficients, into another whose roots are the squares of the differences of the roots of the original equation. When the resulting equation has coefficients of continually alternating signs, then the given equation has no complex roots; otherwise it has some.[3]

Under the same assumptions Waring gives a general method to find the positive and negative roots of a biquadratic[4] and of one of the fifth degree,[5] which are however of no practical use, all the more so because at this later time we can find much more easily the maximum number of positive and negative roots (through the theorems of Descartes and Rolle) and even (through Sturm's Theorem) the true number of real roots.

Of comparable insignificance is the method given by him to find the number of imaginary roots of the equation whose roots are equal to the squares of the differences of the roots of a given equation, knowing

[1] Theorema I, p. 36.
[2] Problem IX, p. 69.
[3] Problem XII, p. 82.
[4] Problem XII, p. 84.
[5] Problem XII, p. 85.

the number of real, positive, and negative roots of the given equation.[1]

Waring knows Descartes' rule of signs, and discusses it in the following terms: "Descartes' Rule of Signs may be stated thus: that a complete polynomial equation has as many positive roots as it has changes in sign, from + to - and from - to +, provided that the equation has no imaginary roots. For when any equation is multiplied by $x-a=0$, where a signifies a positive quantity, the number of sign changes will increase by at least 1; but if the given equation is multiplied by $x+a$, then between the first and penultimate term of the resulting equation there will be as many, or more by an even number, identities of sign as there were in the original equation. [3] If the resulting equation has just as many sign changes, from its first to its penultimate term, as were in the given equation, then the penultimate term of the resulting equation has the same sign as the final term of the given equation, hence the resulting equation will have one more sign progression appended, which is not to be found in the original equation. Thus in the resulting equation there will be an odd number of sign progressions in addition to those that were in the given equation."[2]

This statement is important, because here Waring reveals the deeper foundation of the rule of signs, which Gauss was to rediscover later on.

Details of this matter are in Part III of this work.

[1]Problem XVI, p. 104.

[2]Scholium, pp. 110 and 113.

Chapter III

De reductione et resolutione æquationum.

Reduction and resolution of equations.

By reduction *sensu latu* Waring comprehends every transformation of an equation that renders its solution easier; reduction *sensu strictu* means to him transformation to another equation of lower degree. Therefore reduction, to Waring, is not simply synonymous with decomposing a polynomial equation into factors.

The Law of Reduction, as stated by Waring, is "Equations may be reduced if they can be separated into two equal parts, such that the same roots may be extracted from either side".[1]

Waring also understands that the vanishing of the remainder (which is independent of the unknown x) is precisely the condition that two equations have a root in common. But only with the aid of determinants did it become possible to enunciate the rule by which the polynomial remainder sequences could easily be constructed.

Of the methods of solution of equations of high degree Waring knows only those that depend on transformations of the given equation; the resolvent-based methods were first made known by Lagrange.

In the transformation of equations Waring advances beyond Tschirnhausen's method, which he mentions by name.

But Waring knew more than just the way to remove the second and third terms from an equation of any degree by solving a quadratic equation: he could also, with a biquadratic equation, remove the second and fourth terms.[2] He believed that by this method an equation of any degree could be transformed into a *pure* equation, and stated the theorem that the reducing equation in general would be of degree at most $(n-1)!$, although he did not offer a proof of either of these assertions.

Later mathematicians have shown that with the aid of a quadratic and a cubic equation the second, third, and fourth terms of any equation can be made to vanish, but that Tschirnhausen and the English

---

[1] Theorem XIX, p. 138.

[2] Problem XXIII, p. 148.

mathematicians Waring and Jerrard overrated the importance of their methods, by which they believed they could remove all the intermediate terms in an equation.

Further since Waring knew nothing of group theory, he had no way of knowing that the simple equations he hoped to obtain could be solved by a chain of *cyclic equations* of prime degree, these in turn being solvable by radicals.

For the solution of the general equation of the fifth and sixth degree Waring produces three independent quadratic equations in three unknowns, and takes credit for being the first to notice that one of these unknowns could be arbitrarily specified.[1]

In this chapter Waring also discusses the type of equation we now know as *reciprocal*,[2] and for their solution we have a special method that is only justified by a deeper insight into the structure of such equations. Waring did not have this insight, so to solve reciprocal equations he applies another, more complicated, method. Here he follows his favorite paths, and establishes a generalized system of formulas.

Waring also knows the *binomial*[3] equations and their properties; but their trigonometric representation remained unknown to him.

As a special case of the binomial equation, Waring considers the equation

$$x^n - 1 = 0$$

and he uses this opportunity to introduce the study of the so-called *roots of unity*.[4] But he did not know of the distinction between primitive and imprimitive roots, and so of course he knew nothing of the elegant structure of the equations for inscribing regular polygons in circles.

On this subject Waring knows the following theorem:[5] "The sum of the mth powers of the roots of the equation $z^n - 1 = 0$ is 0, except when $n = m$, or when n is a factor of m, in which case the sum becomes n."

Proceeding beyond this theorem, Waring shows further that any symmetric function of the nth roots of unity,

$$\sum \alpha^m \beta^r \gamma^s \delta^t$$

[1]Problem XXII, p. 142.
[2]Problem XVII, p. 123.
[3]Problem XXVII, p. 157.
[4]Theorem XIV, p. 127.
[5]Lemma, p. 167.

is always equal to zero, except when $m+r+s+t$ is a multiple of n.[1]

As is his custom, Waring seeks to expand this theorem in various combinations of power sums, and to give a generalized formula for the method.[2]

Here it is worth mentioning that Waring does not apply the symmetric functions of the roots of unity to the rationalization of equations with irrational coefficients, but rather uses the method of corresponding coefficients.[3]

Finally we find in this chapter the development of the Binomial Theorem,[4] but without the symbol $\binom{n}{k}$.

The relations among the binomial coefficients are known to Waring only up to the point of computing the sum of all the odd coefficients, or of the first, fourth, seventh, tenth, etc.[5]

---

[1] l.c. p. 167.
[2] l.c. p. 168.
[3] Theorem XX, p. 168.
[4] Problem XXVII, p. 157.
[5] l.c. p. 159.

Chapter IV

De duabus vel pluribus æquationibus duas vel plures incognitas quantitates habentibus.

Two or more equations with two or more unknowns.

Waring knows that there is a specific, finite number of pairs of roots that simultaneously satisfy two given equations.

He also knows Bezout's Theorem: Two independent equations, in two unknowns with independent coefficients, of degrees m and n respectively, have at most $n \cdot m$ solutions.[1]

Minding's method, which permits calculation of the exact number of solutions in each special case, is naturally unknown to him.[2]

Waring then shows in this chapter how one of the two unknowns may be eliminated, either by finding common divisors,[3] or through cross-multiplying and dividing,[4] or through infinite series, which one inserts in the second equation and then multiplies the resulting values by each other,[5], or by multiplying the two equations by two arbitrary equations and then solving them by the method of corresponding coefficients.[6]

Through numerous examples, moreover, Waring comes to the conclusion that k equations in k unknowns can be transformed into a single equation in only one variable. And when he later finds that the degree of this equation is equal to the product of the degrees of the original k equations, he has thereby enunciated a theorem[7] that later came to be known as the Generalized Theorem of Bezout.

It is surprising that Waring, who is so versatile in his use of symmetric functions, did not in the elaboration of this topic hit upon such an elimination method for multiple equations as was later found by Poisson.[8]

---

[1]Theorem XXIV, p. 189.

[2]Crelle's Jour. Bd. 22, p. 178, 1841.

[3]Theorem XXIV, p. 189.

[4]Problem XXXI, p. 190.

[5]l.c. p. 208.

[6]l.c. p. 211.

[7]Problem XXXI, p. 204.

[8]Journal de l'École Polytechnique, cahier XI. t. X. IV. an IX. p. 199.

Moreover, since Waring did not know of the theory of *determinants*, its application to the elimination of a variable between two equations of arbitrary degree remained unknown to him also.

Subsequently in this chapter Waring introduces his investigation into *repeated roots* of multiple equations,[1] but was unable to get as far as the notion of the *discriminant*, which, as is well known, is a homogeneous integral function of the coefficients, of degree $2(n-1)$ and of weight $n(n-1)$. [4]

By the *transformation* of two equations Waring understands, as we do, the operation by which from the given equations one obtains another, whose roots have a specified algebraic relation to those of the given equation. In particular Waring shows how the coefficients of the transformed equation can be constructed by means of symmetric functions.[3]

Then Waring establishes *criteria* for the estimation of the real and imaginary roots of multiple equations. In order to investigate the imaginary roots, Waring arrives at a method by which he first determines whether there are any real roots in the given equations, and after he has removed these from the equations he looks for the imaginary roots in the reduced equation.[5]

How warmly Waring would have welcomed the intriguing *Cauchy Theorem*, whereby one can determine the number of roots, in any equation, whose real and imaginary parts lie between two respective sets of upper and lower bounds.

On page 233 Waring refers to his book *Proprietates algebraicarum curvarum* and reports that through his geometric interpretation he came to realize that if two, three, or more curved arcs, whose equations are given, pass through a point P, they define a double (or multiple) point, and that this result is the geometric equivalent of the property that two or more equations have a double (or multiple) common root.

Finally Waring gives a method for *calculating approximately* the real and imaginary roots; in the latter case $a+bi$ represents an approximation.

---

[1]Problem XXXIII, p. 216.

[3]Problem XXXVIII, p. 227.

[5]Theorem XXXI and XXXII, p. 229 ff.

Waring's method of approximation however requires too many assumptions to be of much practical use.

It seems that the "Regula Falsi" and Newton's method of approximation remained unknown to him.

Chapter V

De rationalibus et integris quantitatibus.

Rational and Integral Quantities.

As the title says, Waring seeks *integer* solutions of equations with one, two, or more variables, for example when he divides an equation, after the method of Descartes, by a binomial supposed to contain one of its roots;[1] or when he deduces an integer solution from the external form of the aforesaid equation;[2] or from the limits, within which one can see intuitively that such roots exist.[3] The elegant solution of the so-called *Diophantine Problem* in integers through solving a congruence is unfortunately unknown to him.

Furthermore in this chapter Waring introduces us to methods by which one can find *rational roots* in an equation with several variables,[4] and further, whether an equation can be solved by extraction of *square or cube roots*,[5] and how by approximation one can find the integral values of the unknown. But Waring himself refers to the latter process as trial and error, "continua tentamina".[6]

The method of converting a repeating decimal fraction to a simple fraction gives rise, for Waring, to a technique for the approximation of the roots of certain infinite series.[7]

The solution in integers of what Euler[8] incorrectly named "Pell's equation",

$$ax^2 + 1 = y^2$$

Waring correctly ascribes to Viscount Brounker.[9]

---

[1]Problem XLIV, p. 273. Waring ascribes this method to the Dutch mathematician Jacob van Wassenaer, when in the binomial $(x \pm a)$ one substitutes $a = 0, 1, -1, 10, -10$, etc.

[2]l.c. p. 277 No. 5.

[3]l.c. p. 278 No. 7.

[4]l.c. p. 274–277.

[5]Problem XLVI, p. 283; Problem XLVIII, p. 289.

[6]Problem LVII, p. 333.

[7]Problem LI, p. 302–308.

[8]Anleitung zur Algebra, Leipzig, p. 396

[9]Problem LIX, p. 336.

He also seeks integer solutions to the *exponential function* a^x and x^a.[1] However, the discovery of the remarkable property of the exponential function $e^{2\pi iz}$ — that its rational functions with rational coefficients resolve all equations whose coefficients are rational numbers, and that they form an Abelian group when z ranges over all rational numbers — was to remain for a later age.

After this Waring gives a method, different from the modern one, for the *transformation of numbers* from the decimal number system to another radix[2] and gives examples and constructive methods for *perfect*[3] and *amicable* numbers.[4]

But when Waring speaks of *remainders*[5] and *non-remainders*[6], he is referring to our modern concept of *congruences.* Quadratic residues and non-residues, however, were unknown to him.

Waring also foresaw the central place of prime number theory, e.g. when he showed that if

$$(a+b)^n \text{ and } (a-b)^n$$

have no common factor except 2^m, then a is relatively prime to b.[7]

In the course of this chapter additional number-theoretic theorems from *different authors* are introduced; for example, the approximation technique of *Wallis*, to convert an infinite decimal fraction into the equivalent rational fraction;[8] further Brounker's method of expressing a common fraction and its reciprocal as a continued fraction;[9] or to derive infinitely many integral solutions of an equation in several unknowns, if two or more such values are known; then the method of *Kiraast*,[5] which determines the divisibility by 7 of a composite number of the form $a+b$;[10] further the method of *J. A. Castelvetri* for the division of a number through an approximation process.[11]

---

[1]Problem LXIII, p. 380; Scholium, p. 387.

[2]Problem LX, p. 340

[3]Theorem XLVI, p. 342, Cor. 4.

[4]l.c. Cor. 5.

[5]Theorem LII, p. 355.

[6]Theorem LIII, p. 360

[7]Theorem XLIII, p. 287.

[8]Problem LI, p. 307.

[9]Theorem XLV, p. 308 ff.

[10]Theorem XLVI, p. 347.

[11]l.c. p. 348.

On page 349 Waring presents *eleven number-theoretic theorems*, all without proof. One of them is known as "Waring's Problem" and was recently proved by Hilbert. (Dr. Hilbert, "Beweis für die Darstellbarkeit der ganzen Zahlen durch eine feste Anzahl n^{ter} Potenzen. [Warings-Problem]").[1]

On page 379 Waring states the theorem: "Every *even* number is the sum of two prime numbers, and every *odd* number is a prime or the sum of three primes." This statement for *even* numbers is generally known as the *Goldbach conjecture*, but was first published by Waring. Goldbach in 1742 stated it briefly to Euler;[2] but their correspondence was not to be published until 1843.[3]

Waring also knew of the *lesser Fermat theorem*, but without using this name.[4] The general Fermat theorem for an arbitrary module P, we know, was first to be given by Euler.

The most famous number-theoretic theorem in Waring's work is *Wilson's Theorem*, which to be sure Waring states without congruential notation and without proof, and then adds: "The proof of theorems of this sort will therefore be difficult, since no notation has been found to express prime numbers."[5]

But we know that Lagrange succeeded in finding the proof of Wilson's theorem; he published it under the title, "Démonstration d'un théorème nouveau concernant les nombres premiers," Oeuvres, t. 3, Paris, Gauthier-Villars (1869), p. 425.

All these observations concern only *rational* number fields. But when Waring rejoiced so at Wilson's Theorem, what mental joy it would have given him, had he known modern number theory, which also deals with imaginary quadratic fields and produces an elegant linkage between two great mathematical disciplines, algebra and function theory.

The proof by Rudolf Fueter[6] is called to mind, that the elliptic modular function $j(z)$ does for the imaginary quadratic fields what the

---

[1]Göttinger Nachricten 1909, p. 236 ff.

[2]Euler, Werke, Bd. III, 2, p. 610.

[3]Corresp. math.(Fuß) I, p. 127, 135. St. Pétersbourg, 1842.

[4]Theorem LII, p. 355.

[5]Problem LXIII, p. 380.

[6]R. Fueter, "Abelsche Gleichungen in quadr.–imag. Zahlkörpern". Mathem. Annalen, Bd. 75, p. 177.

exponential function $e^{2\pi i z}$ does for the rational numbers, i.e., that one can solve all Abelian equations in an imaginary quadratic field through rational functions of $j(z)$ with coefficients taken from a quadratic field, provided only that z ranges over all imaginary quadratic fields.

Part II

The historical foundations of the "Meditationes algebraicæ"

Waring himself has given us the historical background to the "Meditationes algebraicæ" in the Præfatio (Foreword) of his work.

He identifies his objective as, first, that the discoverers of algebraic problems be given the credit that is due them, and further, that those who are thirsty for knowledge be at least partially satisfied.[1]

Waring offers us the historical foundations of his Meditationes algebraicæ in the form of a "History of Algebra" up to his time, naturally from the perspective of the English, and based on those historical sources which had appeared up to the middle of the 18th Century and were known to him.

Therefore it became my task also to validate Waring's historical statements, and eventually to correct and expand them.

As the first who wrote about algebra, Waring names Diophantus of Alexandria (325–409). His great work consisted of 30 books, of which however only the first six have survived. There one finds the solution in whole numbers of indeterminate linear equations with two unknowns, and a large number of solved problems which depend on linear and quadratic equations. Waring regards the methods of Diophantus as so clear and elegant that they have seldom been surpassed by more recent mathematicians.

When Waring asserts that *our number system*, with a zero, originated in India and was disseminated by the Arabs, he is in accord with more recent mathematical writers.

The *decimal system* was introduced by *Regiomontanus* (1436–1476), who fully understood its significance. While his first Table of Sines, like those of Johann von Gmunden and Georg von Peuerbach (1423–1461), still contained a mixture of sexagesimal and decimal systems, his third Table of Sines relies entirely on the decimal system. Also his Table of Tangents, the first in the Western world, is implemented with decimal notation.

Decimal fractions were introduced by *Petrus Ramus* (1515–1572) in his "Arithmetica" around 1560 in the following way: in the extraction of square or cube roots of numbers which are not perfect powers, one simply proceeds as if the number before him were an integer; then one will obtain numbers which contain integer and decimal parts.

---

[1] Præfatio p. 1.

But according to Wallis the same method was earlier given by the Cambridge student William Buclæus; Wallis read his "Arithmetic" as an addendum to the *Logic* of Seton in a Cambridge work published in 1631. Buclæus seems to have died around 1550 during the reign of Edward VI, who of course died in 1553.

Briggs' logarithms are not discussed in the body of the "Meditationes"; Waring mentions them only in the foreword.

Of the Arabian mathematicians Waring mentions only three, namely *Muhamed ben Mûsâ el Chowârezmî*, Abû Abdallâh (died c. 845), *Tâbit ben Quarra ben Merwân*, Abû'l-Hasah, el-Harrâmî (826–901), and *Omar ben Ibrahım el-Chaijamı*, Gijat ed-dın, Abu'l Fath[1] (945–1023), and remarks in passing that the Arabic manuscripts up to his time had not been studied closely enough.

Fortunately, today this is no longer true. Zürich Gymnasium professor *Dr. Heinrich Suter* has made an intensive study of the subject which he recounts in his work, "Die Mathematiker und Astronomen der Araber" with 528 of them, to which six are added in his "Nachträgen und Berichtungen".

With the help of the Indian number system and the positional values of digits, these Arabs perfected elementary arithmetic operations and also the beginnings of number theory, and they enriched algebra through the geometric solution of equations of the third and fourth degree (using conic sections): they built up plane and spherical trigonometry to a level that was not surpassed until Regiomontanus.

And certainly the best-known mathematician of the 13th Century, *Leonardo of Pisa*, erects his works on Arabian ground, having spent considerable time in the Orient and having acquired his mathematical knowledge from the Arabs. His German contemporary *Jordanus Nemorarius* was equally exposed to the Arabs in his schooling. [6]

Through these two men the mathematical science of the Arabs was conveyed to the Western world.

Leonardo's major works are his "Liber abaci" and his "Practica Geometriæ"; those of Jordanus were the "Algorithmus demonstratus",

[1]The orthography is that of H. Suter, "Die Mathematiker und Astronomen der Araber", in the Abhandlungen zur Geschichte der mathematischen Wissenschaften mit Einschluss ihren Anwendungen, X. Heft, Leipzig, Teubner 1900.

"De numeris datis", and "De triangulis". Especially the latter's "De numeris datis" was very influential in the development of algebra, although it was to be three centuries before an advance beyond his work would be made.

The remaining Italian mathematicians of the 16th Century will be discussed in a separate chapter.

The first to publish a *printed* work on algebra was, according to Waring, *Lukas de Burgo*, based on the fact that he recorded for posterity all that was known to that time regarding quadratic equations and their two roots.

The true founder of modern algebra is *Francis Vieta* (1540–1603). It was he who represented all quantities, whether known or unknown, by letters, and showed how all computations can be done with letters. Vieta was the first to concern himself with the numerical solution of equations of all degrees. He also showed simple ways to transform equations without knowing their roots, by setting $x+k$ or $x \cdot k$ in place of x. Vieta also understood how to raise a binomial to any higher dimension.

When Waring says that Vieta gave an elegant method to find the rational values of the unknown of a quadratic equation, he refers to Vieta's graphic method, by means of which the results of algebraic computation can be geometrically represented. It is well known that Vieta had a higher reputation as a geometer than as an algebraist.

Waring says of *Pascal* (1623–1662) only that he showed how to find triangular and pyramidal numbers.

Thomas Harriot (1568–1621) introduced the notation a^2, a^3, ...; he also reduced each equation to its null form and from this showed the true origin of higher equations, as well as the structure of coefficients. In finding the imaginary roots of a cubic equation he made use of the same principles which in their subsequent elaboration gave us Newton's Rule.

René Descartes (1596–1650) it was who used the last letters of the alphabet to represent unknowns, and the first letters for known quantities. He also found limits which cannot be exceeded by roots of a given equation. Further, he proved rigorously that each integer root is a divisor of the constant term of an equation; also he could find, through division of the given equation by $x-a$, whether a is or is not a root of the equation. Waring regards this method, due to its simplicity and ease of use, as in most cases preferable to that of Wassenaer, although many mathematical writers scarcely mention it.

Waring discusses Descartes' Rule of Signs in detail, as we shall do in the third part of this work.

According to Waring, *Gua* first proved Descartes' Rule of Signs through multiplication of the given equation by $x+a$ or $x-a$, to be sure only for the case where all roots are real. *Gua de Malves* (1714–1785), a Parisian clergyman, had a great predilection for mathematics.

Andreas von Segner (1704–1777) showed that in each equation there are at least as many sign changes among the coefficients as there are positive roots, and as many sign continuations as there are negative roots.

Continued fractions were first found by *Brounker* (1620–1684), as was the integral solution of the so-called Pell's Equation.

Of *Wallis* (1616–1703) Waring mentions only that he developed a method of reducing fractions to lowest terms. Of his major work, the "Arithmetica Infinitorum", nothing is said, perhaps because the infinite was a subject which Waring did not fully understand.

With regard to the power sums, according to Waring's credits, *Albert Girard* (1590–1622) first determined the first four sums and later *Newton* (1642–1727) established the general rule by which the power sums of higher degree are constructed from the known power sums of lower degree.

The generalized rule of composition, however, was already given by *Waring* in the first edition of his work, in the well-known formula. To be sure Waring remarks that Euler later published the same rule.

Waring was very proud of his discovery, as the following statements prove:

> "These rules were found out by me."[1]
>
> "No one earlier found any such root transformation...."[2]
>
> "No one before me computed the sums of any irrational or fractional function of the roots of an equation from the

---

[1]Præfatio, p. xxii.

[2]Præfatio, p. xxiii.

> coefficients. No one showed any method of finding combinations of the power sums of the individual roots, except Cramer, who derived by induction those of higher degree from those of lower degree, a method which because of its complexity will seldom be found useful in practice."[1]

Then Waring expands at considerable length upon the solution of general equations of higher degree by means of particular substitutions according to the methods of *Schooten* (1615–1660), *Hudde* (1624–1704), *Newton* (1642–1727), *Euler* (1707–1783), and *Bezout* (1730–1783). In this context priority disputes are mentioned, which are not of much interest however, since these pre-Lagrange mathematicians had not gotten beyond ad-hoc techniques.

Lagrange (1736–1813), through his "resolvents", took the first step towards the modern algebra based on the theory of permutation groups and symmetric forms.

One episode may be of interest, which Waring tells regarding his contemporaries.[2]

Already in 1763 Waring had sent a copy of the first edition of his work to "the renowned Euler", and in May 1770 he sent copies of the second edition to the "great mathematicians" d'Alembert, Bezout, Montucla, Lagrange, and Frisi. Whether any of them had received the volumes, he did not know; at least he had received a letter of acknowledgment from none of them; he makes an exception for the "renowned" Frisi,[3] whose distinguished work[4] he acknowledges receiving and whose friendship was to be dear to him for life.

When Waring comes to speak of the *number-theoretic* foundations of his work, he says, *verbatim*, "It is impossible for me to enumerate all the elegant solutions of such problems, which were discovered by celebrated men; in this Preface I will add only what has been discovered about factoring and remainders."

Already in the first edition of this work many number-theoretic theorems had been communicated. To these the following are added:

[1] Præfatio, pg. x.

[2] loco cit. p. xxi.

[3] Paolo Frisi (1728–1784), Professor of Mathematics in Milan.

[4] Operum tom. 1. Algebram et Geometriam analyticam continens.

To *Bachet de Meziriac* (1587–1638) is due the theorem, "Every number is the sum of four squares." — The first proof of this theorem was found by Euler for rational numbers, and then by Lagrange for integers.

Pierre de Fermat (1601–1663) found that every prime number of the form $4m+1$ is the sum of two squares, and that $a^{n-1}-1$ is always divisible by n, if n is prime and not a divisor of a.

Euler[1] proved the theorem, "It is impossible to find two cubes whose sum or difference is a cube."

In general Euler devoted the last chapter of his Algebra to the solution of problems in number theory.

More elegant discoveries in the factoring of numbers were made by *Euler*, *Lagrange*, *Beguelin*, and others.

Finally Waring mentions the "elegant proof" of Wilson's Theorem by Lagrange, which however Waring does not reproduce in his "Meditationes algebraicæ".

Since until now little has been known of Wilson as a person, a further word about this remarkable man will follow.

The editors of the "Nouvelle Correspondance Mathématique (Bruxelles)" went to considerable trouble to learn something more about the life of Wilson. Their first effort was fruitless, however, turning up a John Wilson who had already died in 1714, and therefore could not have written a pamphlet in defense of Waring in 1760.

The next effort however was crowned with success. The "Nouvelle Correspondance Mathématique" reported in 1876, pp. 110–114, on the source material found in England, which was amplified later through some notices in the "Bibliotheca mathematica" for 1902, p. 412, and 1903, p. 91, from which we can now construct the following biographical sketch.

John Wilson was born August 6, 1741, at The Howe, Applethwaite, and was heir to a small estate at Iroutbeck in Westmoreland. He studied at Peterhouse in Cambridge University. When he was still an undergraduate some considered him to be the most excellent algebraist at the

[1] Opera omnia. II. Teil, II. Abschnitt, cap. 15, §242.

University, and was called by Professor Waring one of the outstanding algebraists of the 18th Century. Yet Wilson seems to have been less concerned with the study of mathematics than with the law.

In 1761 he became "Senior Wrangler". Cambridge University in those days recognized three types of graduates. These were called, in ascending order, "Junior optimes, Senior optimes, Wranglers". The first wrangler was called "senior wrangler". This status was so highly rated in England that the name of its fortunate holder in the early days after the publication of the list of graduates was on every tongue. The newspapers would print his biography, and the illustrated magazines his portrait. (This honor was acquired, for example, by such well-known men as Stokes in 1841, William Thomson in 1845, Clerk Maxwell in 1854.)

Next Wilson became a "private tutor". When Paley, then in his third year of studies at the University, set himself the goal of becoming senior wrangler, Wilson was recommended to him as a private tutor (in student parlance, "coach".) Both were diligent at the task. But sometimes when Paley called on his private tutor, on the closed door he found the notice, "Gone fishing", a circumstance which annoyed Paley all the more because he himself was an avid fisherman.

In 1766 Wilson exchanged the classroom for the courtroom. He practiced as an attorney with the greatest success. In 1782 Wilson became a Fellow of the Royal Society. In the same year in Waring's "Meditationes algebraicæ" he is labelled "armiger", that is, one who has the right to bear arms, which may therefore be translated as the English "squire", a landed gentleman. On November 15, 1786, he was knighted by the King, and was thereby elevated into the nobility. He usually spent his holidays at his estate, Iroutbeck. There to his surprise he heard one day that one Lord Thurlow, with whom he was not acquainted, had nominated him for a position as "Judge of the Court of Common Pleas", an indication that Wilson was in great repute as a lawyer.

Waring calls him, in the fifth chapter of his "Meditationes algebraicæ", where he presents Wilson's Theorem, "vir clarissimus rerumque mathematicarum peritissimus" — a distinguished and in mathematics very expert man. And in the Preface he says of him "ab amicissimo et in omni mathesos parte versatissimo viro" — as his bosom friend and in all parts of mathematics a very versatile man.

Certainly Wilson seems to have had no scientific ambition; the only paper which he published, it seems, is a pamphlet defending his friend Waring's "Miscellanea analytica".

The theorem that bears Wilson's name, however, was to be published by Waring and so communicated to a wider public.

In 1788 Wilson was married to the daughter of Serjeant-at-law Adair, and this marriage was blessed with children.

Wilson died October 18, 1793, having attained the reputation of a great jurist and capable magistrate. What made him immortal in the eyes of posterity, however, was his *mathematical*, especially his number-theoretic knowledge.

Waring and the Italian mathematicians of the 16th Century.

Whereas algebra had previously concerned itself almost entirely with the solution of linear and quadratic equations, and the Arabs had presented the solution of equations of the third and fourth degree only in geometric representation, some Italian mathematicians succeeded in finding purely algebraic solutions of the cubic and biquadratic. Their names are

Cardan, Tartaglia, and Luigi Ferrari.

Cardan was born in Pavia on September 24, 1501, and died in Rome on September 21, 1576. On the title page of his works he is identified as a philosopher and medical doctor. He wrote in Latin.

The book which is especially under consideration in this work is "Artis magnæ seu de regulis algebraicis liber unus", which first appeared in 1545 in Nuremberg.

I was not able to examine this work; therefore my citations are from his Collected Works.[1]

Formally this work differs quite markedly from other similar writings: in the entire book there is not a single equation in the familiar form with unknowns x and y, and radical signs; rather the entire text is written in words, so that one must first cast the equation in the modern form, in order to understand the problem. For the radical sign Cardan used the shorthand ℞, and for plus and minus simply p̃, m̃. *Tpquad* means to him "tertia pars numerum quadratorum" (cf. Regula, cap. 17).

Generally Cardan presents first the *Demonstratio*, where by means of simple figures he illustrates the conditions of the verbal equation, and then deduces the *Regula* from the Demonstratio.

A further peculiarity is that he devotes an entire chapter to each of the equational forms

[1]Hieronymi Cardani Mediolanensis philosophi ac medici celeberrimi *Operum Tomus quartus*, quo continentur Arithmetica, Geometrica, Musica. Editio ut cæteris elegantior ita et accuratior. *Lugduni*. Sumptibus Joannis Antonii Hugetan et Marci Antonii MDCLXIII. cum privilegio regis.

$$\begin{array}{ll} x^3 = ax + b & x^3 = ax^2 + b \\ x^3 + ax = b & x^3 + ax^2 = b \\ x^3 + b = ax & x^3 + b = ax^2 \end{array}$$

and therefore handles each separately. Even the word *capitulum* means to him literally an algebraic equation with known specifications of form and symbols. The title of the seventh chapter attests to this: "De capitulorum transmutatione". This separate treatment of the different forms of equations explains why Cardan, like the Greeks, gives geometric proofs and wants to disallow negative roots.

Waring makes three assertions regarding Cardan:[1]

1. That Cardan ascribed the solution of a cubic equation to Scipione del Ferro.

2. That Cardan published Tartaglia's method of solution of a specific cubic equation, without the latter's consent, which method, to be sure, he himself improved and extended.

3. That Cardan showed that a cubic equation must have three roots, be they negative or positive.

Waring's *first* assertion, that Cardan ascribed to Scipione del Ferro the solution of a cubic equation, is corroborated by the following: in the first chapter of the *Ars magna*[2] Cardan says, "But in our time Scipione del Ferro found the solution to the equation *cube plus unknown equal to a number*,"

i.e., $$x^3 + ax = b\,,$$

"truly a beautiful and marvelous thing" (*rem sane pulchram et admirabilem*).

I want to mention further that Scipione del Ferro was also concerned with rationalizing the denominator of a fraction, a problem which doubtless at the beginning of the 16th Century must have seemed quite difficult.

With Waring's *second* assertion, that Cardan revealed Tartaglia's method of solving a cubic equation against the latter's will, we come right into the middle of the dispute, "*Tartalea contra Cardanum*",

[1]Medit. algebr., Præfatio, p. ii.

[2]Operum tom. IV, p. 222.

which we shall discuss in detail later, in order not to break up the continuity of this study.

For the time being let us make only the following observation: Cardan did not simply treat the equation form used by Tartaglia, in which the term involving x^2 was always lacking; he shows rather, how the cubic equation can be freed of its x^2 term by the substitution

$$x = y \pm a/3.$$

Waring's *third* assertion, that Cardan had found the existence of three roots of a cubic equation, is verified in the 18th chapter[1] by the following:

> *Regula*: Ex hoc patet, quod numerus quadratorum in his tribus exemplis:
>
> ex. 5.: Cubus et 10 res æquantur 6. quadratis p̃ 4
> $(x^3 + 10x = 6x^2 + 4)$
> ex. 6.: Sit cubus et 21 res æqualia 9. quadratis p̃ 5
> $(x^3 + 21x = 9x^2 + 5)$
> ex. 7.: Cubus et 26 res æquantur 12. quadratis p̃ 12
> $(x^3 + 26x = 12x^2 + 12)$
>
> in quibus æstimatio rei triplicatur, semper componitur ex tribus æstimationibus junctis simul, velut
> in quinto exemplo
> $2,\ 2+\sqrt{2},\ 2-\sqrt{2}$ componunt 6. numerum quadratorum;
> et in sexto exemplo
> $5,\ 2+\sqrt{3},\ 2-\sqrt{3}$ componunt 9. numerum quadratorum;
> et in septimo exemplo
> $2,\ 5+\sqrt{19},\ 5-\sqrt{19}$ componunt 12. numerum quadratorum;
> ideo duabus cognitis, tertia semper emergit, causa est cognita in initio huius libri. Et manifestum est, quod cum pervenimus ad res, quæ a cubo separantur, seu numerus rebus, seu cubo jungatur, semper emergunt tres æstimationes et causa dicta est superius ibidem, ubi de vera et ficta æstimatione locuti sumus.

[1]Operum tom. IV, p. 259

Cardan thus made a *double* assertion: first, that the given cubic equations have *three* roots, and then, that the *sum* of the three roots yields the coefficient of the quadratic term — a discovery that was of far-reaching importance for algebra.

Since Waring only credits Cardan with statements about roots of the cubic equation, it seems he must not have noticed that already in the first chapter of his "Ars magna" Cardan remarks that the equation $x^4+12 = 7x^2$ has *four* roots.[1]

That Cardan first of all refers to the four roots of the above equation appears noteworthy to me, as it suggests that perhaps it was through biquadratic equations of this sort that Cardan was led to his discovery that an equation may have more than two roots.

The first algebraic solution of the general equation of the *fourth degree* Cardan published in the 39th chapter of his "Ars magna", attributing the discovery to his pupil Ludovico Ferrari ("alia est regula ... Ludovici de Ferraris, qui eam me rogante invenit")[2]. There is no reason to doubt the veracity of this statement, and from the same chapter one may conclude with a high degree of confidence that Cardan was prompted to concern himself with this question by a work of *Giovanni da Coi.* Whether da Coi himself could solve problems leading to equations of the fourth degree, no one knows.

Of the *solution* of the biquadratic equation the following point is noteworthy: Cardan knew how to remove the quadratic term from a cubic equation, by setting the unknown x equal to a new unknown y, minus one-third of the quadratic term. In the biquadratic equation he logically makes a corresponding substitution in order to make the cubic term vanish. In its place in the equation

$$x^4 + 6x^3 = 64$$

he substitutes $$x = -\frac{\sqrt[3]{64}}{y}$$

and obtains $$y^4 + 6y = \sqrt[3]{64}$$

or in general $$y^4 + ay = \sqrt[3]{c}.$$

There follow calculations that are found nowhere in Tartaglia's writings, and so with confidence may be ascribed to Cardan. They are:

[1]Operum tom. IV, p. 223.

[2]loco citato p. 293.

(1) The "Regula aurea"[1] for computing approximate values of roots, a method found by no earlier Western mathematician:

We can most easily analyze Cardan's "Regula aurea" if we represent it *geometrically.*

Let the given equation be represented in the form $f(x) = k$, where $f(x)$ is a rational (integer or fractional) function of x with no term independent of x, hence the real roots of the equation are the abscissæ of the points of intersection of the curve $y = f(x)$ and the straignt line $y = k$. Suppose we know that one and only one root lies between a and $a+1$, then the "golden rule" says:

1. That if we connect the points on the curve whose abscissæ are a and $a+1$ by a straight line, the abscissa of the intersection of this straight line and the line $y = k$ is a first approximation x_1 to x.
2. That one can derive a better approximation by connecting the point on the curve whose abscissa is x, with one of the above points on the curve, by a straight line, calling x_2 the abscissa of the point where this line intersects the line $x = y$; whether we should connect to the point $x = a$ or $x = a+1$ depends on the sign of the quantity $f(x_1)-k$.
3. That by repeating this process we approach ever nearer to the value of x.

Geometrically then, the "Regula aurea" is easily stated; but *arithmetically* we may with Cantor[2] call it *untamed,* since it demands totally unnecessary arithmetic computations of the utmost complexity.

(2) Of importance to the future was Cardan's attempt to calculate with *negative radicands*, which previously had simply been called "impossible". These methods are found in Chapter 37, Rule 2, in the following wording:[3] "The second sort of 'false assumption' is this, a root of less than zero". I give an example: If one says,

[1] l. c. pp. 273–274.

[2] M. Cantor, Vorlesungen über Geschichte der Mathem. 2. Bd., p. 507.

[3] l. c. p. 287.

divide 10 into two parts whose product is 40, it is clear that this problem is impossible. But we may compute as follows:

$$(5+\sqrt{-15})(5-\sqrt{-15}) = 25-(-15) = 40.$$

Later these "impossible" roots, as Waring still calls them, were termed *imaginary*, and it is interesting that Cardan has already anticipated this terminology when he says: "Ideo imaginaberis": $\sqrt{-15}$.

It certainly appears that Cardan doubted the wisdom of introducing these difficult concepts, when he says:

> The external form of the above result is foreign to the nature of numbers and lines, but rather it is more nearly a quantity which might be called *sophistical*,[7] since we are unable to perform computations with it as we may with pure negative or other quantities, or even to investigate what sort of meaning it has.[1]

Also in practice Cardan considered this computation with imaginary quantities to be of no worth; but naturally he used such square roots to give an algebraic solution to the so-called "casus irreducibilis".

Cardan's "Ars magna" can be labelled a varied collection of old and new, of useful methods and wild suggestions.

Nicolo Tartaglia ("the stutterer") was born in Brescia in 1500 and died in Venice on December 14, 1557. Raised in poverty, through self-study he rose to the position of professor of mathematics. He taught in Verona (1521–1533), Brescia, and Venice.

Of Tartaglia's works, the "*Quæsiti et Inventioni*" is of particular importance to our study, and of this work especially *Book 9*, since it discusses algebraic problems, while the earlier books address more mathematico-physical problems.

"Quæsiti" is the name given to the individual chapters, as Tartaglia always proceeds from the statement of a particular question which was proposed to him by his mathematical colleagues.

[1] l. c. p. 287.

The edition of Tartaglia's works that I have consulted dates from the year 1554, *privately published* by the author and printed "at Venice by Nicolo de Bascarini at the instance and request of the Author, Nicolo Tartaglia, at his own expense".

The Quæsiti are presented in Italian, and in this first edition are printed on 128 double-sided sheets.

The 9th *Book* first considers problems that may be solved through square roots. The most interesting "Quæsiti", however, are Nos. 34–40. They are concerned with the solution of cubic equations, in which however the quadratic term is always lacking. At the same time these Quæsiti give us an insight into the literary quarrel between Tartaglia and Cardan, which Waring describes briefly (Præfatio, p. ii):

> The solution of the cubic equation Cardan ascribes to Scipione del Ferro, who however wished to keep his method secret. In the meantime Tartaglia discovered the same method "proprio Marte" and communicated it to Cardan after the latter had begged it of him and had promised to keep silent about it. Cardan however broke his promise and publicized Tartaglia's method, which to be sure he enlarged and augmented.

From the original works of the two authors, the quarrel falls out thus:

On March 25, 1534, in his house in Milan Cardan said to Tartaglia, "I swear to you by God's holy Gospels, and as a true gentleman, never to publish anything at all of your discovery, if you reveal it to me; but I promise you also and pledge by my true Christian faith, to write it down in a secret code, so that after my death no one may understand it. — If you want, you may believe me; if not, let us drop the matter."

Nicolo: "If I did not want to trust your many sworn assurances, I would surely acquire a reputation as a faithless one; but I have decided to ride to Vigevano to meet with His Excellency the Marquis,[8] which I have already put off for three days, and I am ready, upon my return, to show you everything of which I have spoken."

Messer Hieronymo: "Since you have decided, in any case, to ride to Vigevano to the Lord Marquis, I want to give you a letter of introduction, which you may deliver to His Excellency. But before you depart from here, I want you to show me the solution of this your equation, which you have promised me."

Nicolo: "I agree to that. But you should know that I have composed the solution in rhymes, in order by my many-faceted occupation to be able to remember it; for had I not taken this precaution, I would already have often forgotten the rule. And if my rhymes seem awkward, I worry little about that; it is enough for me that they serve my purpose to call the rule always back into my memory, so often do I forget it.

"This Chapter I will write in my own hand, so that you shall be sure that I communicate to you a right and good discovery."

Like the riddles of the Sphinx, Tartaglia now gave him the method of solution in verse, that he should see and not see, hear and not understand. At the end however, Tartaglia remarked roguishly that this rule was so clear that His Excellency himself would be able to understand it without further examples.

Messer Hieronymo: "Whether I understand it now or in the future does not matter; only go and on your return I will let you know if I have understood it."

Nicolo: "Your Excellency will remember not to break your given promise; for if you should unhappily be remiss in this matter, whether by publishing the method of solution, either in the book you have now in preparation, or in some other, even if you set my name under it and identify me as its only discoverer, I promise, yes I swear to you, that I will immediately publish three pages for every one of yours, things that you may not find pleasant."

Messer Hieronymo: "Do not doubt that I intend to keep my promise. Go and be completely at ease. Give this letter from me to my lord the Marquis."

Nicolo: "I bid you farewell."

Messer Hieronymo: "Godspeed."

In the following *Quæsito XXXV*, Messer Hieronymo says: "Regarding the question of your equation, unknown plus cube equal to a number, I thank you very much that you have given this to me, and I would not appear ungrateful to you. But I also confess my inability, that is, that I have not had enough insight, to be able to understand the problem, and therefore I beg you, by the love you bear for me and by the friendship which I hope will last all the days of our lives, that you will send me the solution of the following problem:

1 Cubo più 3 cose egual à 10:

and I hope that you will be made no less happy by sending it than I will be by receiving it. That is all. Christ preserve you from evil!

Milan, 9 April 1539. Hieronymo Cardano medico.

Sincerely yours."

Nicolo: "With regard to the said equation of the kind

Unknown plus cube equal to a number

I am greatly surprised that Your Excellency did not understand the rule, which I stated so clearly in the said chapter. But I think you have encountered something of the following sort:

"al terzo cubo dette cose netto"

— that is, you must have taken the third part of the cube of the unknown, when you should have taken the cube of one-third of the unknown; e.g. to solve the following equation, which it pleased Your Excellency to submit to me,

1 Cubo più 3 cose egual à 10

one must, I say, find two numbers (or quantities) whose difference is 10 (or whatever the number is) and the product of these two numbers must be just equal to 1, that is, the cube of the third part of the unknown. From both of these numbers or quantities you find one, either algebraically or by other means, which appears more convenient; i.e. the smaller will be $\sqrt{26}-5$ and the larger $\sqrt{26}+5$; then you must find the cube root of each of these two quantities; that of the smaller becomes $\sqrt[3]{\sqrt{26}-5}$ and that of the larger, $\sqrt[3]{\sqrt{26}+5}$; then you must subtract the smaller from the larger, and what remains is the value of the unknown.

"The said remainder, namely, will be what remains after taking the difference of the two cube roots, namely

$$\sqrt[3]{\sqrt{26}+5} - \sqrt[3]{\sqrt{26}-5},$$

and the unknown is just so large.

"This result is supported by the process of complete verification, i.e. if we raise the said quantity to the third power, and to this cube add three times the said quantity, this sum turns out to be exactly 10, as was supposed.

"But the soundness and correctness of this operation may also be proved geometrically, e.g. if it were given

1 cubo più 1 cosa egual à 11,

then one must try to find two numbers of quantities, one of which is 11 greater than the other, and their product must be $\frac{1}{27}$, that is, $\frac{1}{3^3}$. Therefore proceed as described above, so you will find that our unknown has the value

$$\sqrt[3]{\sqrt{30\tfrac{31}{108}}+5\tfrac{1}{2}} - \sqrt[3]{\sqrt{30\tfrac{31}{108}}-5\tfrac{1}{2}}.$$

No more. God protect you from evil!

Venice, April 23, 1539.
Remember your promise!"

In the following *Quæsito XXXVI* Cardano once more assures his dearest friend (amicissimo Tartaglia) of his gratitude and the promise he has given, and feels almost insulted that he is so little trusted.

In *Quæsito XXXVII* a former pupil of Tartaglia's, Maphio Poveiani of Bergamo, remarks that he has heard from a friend in Milan that Cardan is planning to publish another work about algebra, concerning certain types of equations that have recently been solved, and he suspects that it will be precisely those which Tartaglia had discussed with him.

Tartaglia was naturally upset by this news, and he spoke of his abhorrence should the rumor prove true.

Then in the following letter to Cardano, on August 7, 1539, he warns him of this and says,

> "I give you notice that, if you are not true to your promise to me, then surely I do not threaten (for such is not my way) but I assure you that you will get eightfold of that which I have already promised you."

But in this letter Cardan had also sent his friend Tartaglia a cubic equation whose solution gave him great difficulty, *viz.*

$$x^3 = 9x + 10$$

in which $(\frac{9}{3})^3 > (\frac{10}{2})$. This case came later to be known as the "casus irreducibilis".

From this Tartaglia could see that Cardan was diligently occupying himself with the solution of cubic equations. He now sought to trap him in his answers; he even reproached him that he could not once correctly work out $\sqrt[3]{24}$, which is very elementary, the sort of thing every mathematical pupil learns.

By way of answer, Cardan sent him on October 18, 1539, the solution of the equation $x^3 = 12x+20$ with the remark, "As for your regret that you ever showed me the solution, I am not excited about that; it is nothing against the promise I gave you."

Then six years later, in 1545, the "Ars magna" of the Milanese medical doctor and mathematical professor *Hieronymo Cardano* appeared in Nuremburg.

In the eleventh chapter of the book the solution of the cubic equation was *published* for the first time, with the following introduction:

> Scipione del Ferro over 30 years ago discovered this method and disclosed it to Antonio Maria di Fiori of Venice. As this man was once engaged in a scientific contest with Nicolo Tartaglia of Bologna, he gave thereby the impetus to Nicolo to find the solution for himself, which he then at our request disclosed to us, but without proof. With trust in this clue, we sought for a proof, which after overcoming many obstacles we now explain in the following manner.

The *Demonstratio* follows.

Eneström[1] investigates the following question in this regard: "Did Tartaglia derive his solution of the cubic equation from del Ferro?", and from his fundamental research he is led to the following conclusions:

"1. We have no specific basis from which to assume that del Ferro's solution agrees exactly with Tartaglia's."

"2. The hypothesis that Tartaglia borrowed even part of his solution from del Ferro is consequently unfounded."

In concluding this discussion I raise the further question: *Did Cardan obtain the solution of the cubic equation, which he published in the Ars magna, exclusively from the confidential communications of Tartaglia, under a seal of secrecy, or did he acquire the knowledge in some other way?*

[1] Bibl. mathem. 1906, 3. Folge. 7. Bd. pp. 42–43.

Luigi Ferrari, pupil and later colleague of Cardan, supported the latter claim, and took up a lance on his teacher's behalf. There were six debates (*Cartelli*)[1] between Ferrari and Tartaglia, which were written in Latin by Ferrari and answered by Tartaglia in Italian.

In the second Cartello, page 3, Ferrari relates the following episode:

> Five years ago Cardan travelled to Florence, and I was his companion. In Bologna we met with the ingenious and benevolent Annibale de Nave; he showed us a little book, written long before in the hand of his father-in-law, Scipione del Ferro, in which that discovery is elegantly and clearly stated. [9] I would not write this, were not Annibale still alive at this hour and available as a witness in this controversy; for I would not have it seem as if I, as is your custom, misrepresent that, as far as I am concerned. — Are still other witnesses required? Do you yourself not confess, in the last part of your book, yes, in that very section where you so shamelessly besmirch the name of Cardan, that your adversary Antonio di Fiori was already many years earlier in possession of this discovery? So you must needs take your refuge in the following unrhymed poetry:
>
> 'Although others before me have made this discovery, although its publication by Cardan has brought me not shame, but for a small thing considerable fame; although there are no grounds that my friend deserved to be slandered by me, still it has seemed good to me that he give me his sworn assurances, never to publish my secret.'
>
> But how will you prove the truth of this? Which witness, which tribunal, which sophism — for a true one you have not — do you want to bring up for this?

From this passage one may draw a *twofold* conclusion:

1. That Cardan gained insight regarding solution of the cubic equation from Scipione del Ferro.

[1] Raccolti, autografati e pubblicati da Enrico Giordani, Milano 1876.

2. That Cardan could very well have believed that his publishing would not be displeasing to Tartaglia, especially if he identified him as the discoverer.

In fact, had Tartaglia known beforehand that the "Ars magna" would in any case include a solution of the cubic equation, it is by no means disproved that he would have preferred to have his own solution used, thereby having his name connected with such an important discovery.

Zeuthen[1] identifies yet another reason why Tartaglia would be reluctant about Cardan's publication: *viz.*, that Tartaglia himself was unable to overcome the difficulties in solution of the cubic equation in the "casus irreducibilis".

After these inquiries we may *summarize* in the following way: If Cardan came to his knowledge of the solution of the cubic equation, not entirely through Tartaglia but by other means as well, and might under certain conditions have assumed Tartaglia's tacit consent, then *what Cardan did appears not quite so reprehensible* as it might seem at first glance. [10]

Although Tartaglia in his "Quæsiti et Inventione" treats mainly of the history of the discovery of the solution of the cubic equation, *he also solves other mathematical problems* there. For example, in Quæsito XXXX, seven years before *Stifel*, he published a procedure for extracting cube roots of binomials. [11] From a practical standpoint his method is certainly not very satisfactory, but as the *first* such publication it definitely deserves to be mentioned.

As *Tartaglia's accomplishments*, then, we must acknowledge that he

1. rediscovered by himself the solution of the equation $x^3+ax = b$, first found by Scipione del Ferro; and
2. added to this the solution of the equation in the form $x^3 = ax+b$ (in the casus reducibilis).

Set against these are *Cardan's accomplishments*, that he gave the solution of the *complete* cubic equation (also only in the casus reducibilis), whereas Tartaglia could solve only those cubic equations in which the quadratic term was lacking.

---

[1]Notes sur l'histoire des mathématiques par H. G. Zeuthen. II. Tartalea contra Cardanum; réplique relative à la question de priorité sur la résolution des équations cubiques. Kobenhavn 1893, pp. 303–330.

Part III

Analysis and Criticism of a theme of the "Meditationes algebraicæ"

Waring's Explanation of the Rule of Signs

The earliest beginnings of a Rule of Signs are found in a work by *Cardan*, "Ars magna Arithmeticæ seu Liber quadraginta Capitulorum et quadraginta Quæstionum", cap. 18, p. 323.[1]

Since the Latin text has many possible translations and is sometimes completely misunderstood, I propose to cite it verbatim:

"Septimum notandum est, quod com fuerint denominationes extremæ æquales extremis, semper æquatio erit una tantum, et casus possibilis quotquot fuerint denominationes: Cum vero denominationes intermediæ fuerint æquales extremis, tunc semper erunt plures æquationes in quæsito et casus poterit cum hoc etiam esse impossibilis."

By way of example,

$$x^2 = 3x + 10;$$
$$x^2 + 3x = 10;$$
$$x^3 = 3x^2 + 6;$$
$$x^3 + 3x^2 = 6;$$
$$x^3 = 4x + 10;$$
$$x^3 + 10x = 20;$$
$$x^3 + 3x^2 = 7x + 20;$$
$$x^3 = 3x^2 + 7x + 20;$$
$$x^4 + 3x^3 + 7x^2 = 10;$$
$$x^4 + 3x^3 + 7x^2 = 20x + 10.$$

From the examples advanced by Cardan to illustrate his proposition it is evident that at that time one wrote only positive coefficients, and thus it was a question of whether the outermost terms were on the same or opposite sides of the equals sign.

So now we understand the above proposition of Cardan's, which I translate as follows:

"If the outermost terms are like one another, there is always one and only one root, and we have the possible case before us. If on the other hand the intermediate terms are like the outermost, then there are more roots in the stated problem, and thereby the case may become impossible." Then Cardan continues:

"In omnibus enim his non potest verificari quæsitum, nisi in una certa quantitate et casus semper est possibilis."

---

[1]Opera, tom. IV, Lugduni 1663.

"In all these examples the problem can only be verified by a specific root value, and we have the possible case before us."

"Si vero dicat: $x^2 + 10 = 6x$; $x^3 + 10 = 6x^2$;

$$x^3 + 10 = 10x^2 + 3x;$$

$$x^4 + 3x^3 + 10 = 2x^2 + 5x$$

in omnibus talibus in quibus æquationes extremæ æquantur intermediis, casus potest verificari plus quam in una certa quantitate videlicet in duabus, vel etiam aliquando pluribus et quæsitum potest esse etiam impossibile sicut apparet, si quis dicat

1 cen. $\tilde{p}$. 10 æquatur 3 co. vel 1 cu. $\tilde{p}$. 12 æquatur 40 co.

et ita in reliquis, et hoc bene nota! Scias etiam quod aliquando non habent nisi unam æquationem tantum, sed hoc est per accidens, nam naturaliter habent aut plures æquationes aut nullam, quando casus est impossibilis."

"In all such examples, in which the outermost terms are equal to the terms between, the problem can be solved by more than one root value, *viz.* by two, or often also by several root values, and the stated problem may also be impossible, as manifestly in the following cases, when one says

$$x^2 + 10 = 3x \quad \text{or} \quad x^3 + 12 = 40x$$

and in similar cases. Observe also, that these equations sometimes have only *one* root; but that is only *accidental*; since ordinarily they have either several (real) roots or none, which is the casus impossibilis."

This statement by Cardan is found in a forgotten manuscript which was first published in 1663. But Descartes died in 1650. It may therefore remain as an open question whether Descartes in his Italian travels became acquainted with Cardan's observation. *In any case Descartes was the first to enunciate a proper rule of signs.* It is found in his work, "La Géometrie, Livre troisième", page 57.[1]

Eneström somewhere reminds us that the original of the "Géometrie" was written in French, and was subsequently translated into Latin. Since the French original has now been re-issued by A. Hermann, Paris (1886) and is easily accessible to every mathematician, in scientific works the French original, not the Latin translation, should be cited. The passage of importance to us reads:

[1]Edition Paris, 1886.

"On connoît aussi de ceci combien il peut y avoir de vraies racines et combien de fausses en chaque équation: à savoir il y en peut avoir autant de vraies que le signe + et − s'y trouvent de fois être changés et autant de fausses qu'il s'y trouve de fois deux signes + ou deux signes − que s'entre-suivant. Comme en la dernière [$x^4 - 4x^3 - 19x^2 + 106x - 120 = 0$] à cause qu'après $+x^4$ il y a $-4x^3$ qui est un changement du signe + en −, et après $-19x^2$ il y a $+106x$ et après $+106x$ il y a -120 qui sont encore deux autres changement, on connoît qu'il y a trois vraies racines; et une fausse, à cause que les deux signes de $-4x^3$ et $-19x^2$ s'entre-suivent."

In modern terminology of *sign changes* and *sign progressions* Descartes' theorem thus becomes:

"There can be as many positive roots (in each equation) as there are sign changes, and as many negative roots as there are sign progressions."

The English mathematician Wallis[1] translates this Rule of Signs into Latin as follows:

"Tot in unaquaque æquatione haberi posse radices veras, quot variationes reperiuntur signorum + et −, et tot falsas, quot vicibus ibidem deprehenduntur duo signa + vel duo signa −, quæ se invicem sequuntur."

Wallis then observes that this theorem in unrestricted form is false, and adduces the following example to show why:

If we multiply the equation	$x^4 + 6x^3 + 111x^2 + 1993x + 35878 = 0$
by	$x - 18 = 0$
we obtain	$x^5 - 12x^4 + 3x^3 - 5x^2 + 4x - 645804 = 0.$

According to Descartes' Rule of Signs the first equation has *four negative roots*, the second equation *one positive root*, but the third has *five positive roots*, which obviously cannot be true.

Only Wallis has here overlooked the little word *posse* (il peut), which expresses the maximum number of roots. And with this understanding, the theorem of Descartes states:

The aforesaid equation has *at most* five positive roots, but in fact it has only one, since the others are imaginary.

Accordingly it is not Descartes' theorem which is false, but the *interpretation* of *Wallis*, who did not fully understand that theorem.

[1]Wallis t. II. Algebra cap. XLI pp. 171-172, Oxoniæ 1699.

For his theorem on the maximum number of positive and negative roots Descartes gave *no rigorous proof*, no doubt because he did not know how to prove his theorem for the case of imaginary roots.

Waring[1] gives Descartes' Rule of Signs in the following way:

"Sit notissima Cartesii regula tot esse radices affirmativos, quot signorum in continua serie mutationes de + in − et − in +; si modo nullæ æquationis radices sint impossibiles."

"The Rule of Signs of Descartes may be generally understood as this: that a complete equation has as many positive roots as there are changes of sign from + to − and from − to +, provided only that the equation has no imaginary roots."

Comparing this with Descartes' works we conclude the following:

1. Waring has only translated into Latin the *first part* of Descartes' Rule of Signs, which speaks of the number of positive roots; the second part, which allows us to determine the number of negative roots, has been *omitted.*

2. Further Waring, like his countryman Wallis, has ignored the important phrase "il peut", which specifies the maximum number of real roots, which however is not realized in every equation. — However, Waring's translation "si modo nullæ æquationis radices sint impossibiles" is equivalent to "an equation, all of whose roots are real".

Then Waring's citation may be rephrased as follows:

"Generally the following is known as Descartes' Rule of Signs: 'A complete equation, all of whose roots are real, has as many positive roots as it has changes of sign' " — a statement which fully corresponds to the first part of Descartes' rule.

In the following excerpt, Waring first speaks of *sign sequences* (*signorum progressus*).

Through this investigation we come to the foundation, which was indeed first published by Waring (*hic primum asseritur*), of Descartes's Rule of Signs, and reads as follows:[2]

[1] Med. Alg. cap. II, p. 110.
[2] loco citato p. 110.

"Si enim data quævis æquatio multiplicetur per $x-a=0$, ubi a sit affirmativa quantitas, mutationum signorum numerus saltem augebitur per unitatem; si autem eadem æquatio multiplicetur per $x+a=0$, iidem saltem erunt signorum progressus de + in + vel − in − in nova, ac in data æquatione (cum uno adjuncto)[1] vel magis generaliter hic primum asseritur, si data quævis æquatio multiplicetur per $x-a$, tum singula mutatio signorum de + in − et − in + in data æquatione habet unam vel tres vel quinque, etc. correspondentes mutationes signorum de + in − et − in + in data æquatione: si vero multiplicetur per $x+a$, tum singulus continuus progressus de + in + et − in − in data æquatione habet unum vel tres vel quinque etc. continuos progressus de + in + et − in − in resultante æquatione."

"If any equation is multiplied by $x-a=0$, where a denotes a positive quantity, then the number of sign changes will increase by at least 1; but when the same equation is multiplied by $x+a=0$, then the sign progressions will be as many in the new equation as in the old, with one more added. Or in general, here may be stated the proposition, *given here for the first time*: If any such equation whatever is multiplied by $x-a$, then each sign change in the aforesaid equation has 1 or 3 or 5 corresponding *sign changes* in the resulting equation: but if the said equation be multiplied by $x+a$, then each *sign progression* in the said equation has 1 or 3 or 5 etc. sign progressions in the resulting equation."

Waring's proposition thus is equivalent to this:

If a given equation is multiplied by $\mathrm{x-a}=0$, *then to the individual sign changes in the given equation correspond an odd number of sign changes in the resulting equation.*

On the other hand, if the same equation is multiplied by $\mathrm{x+a}=0$, *then the individual sign progressions of the given equation each have an odd number of corresponding sign progressions in the resulting equation.*

Waring illustrates his proposition with the following examples:

I. Examples for Sign Changes

Let the equation[2] be given:

$$x^n+px^{n-1}+qx^{n-2}+\ldots+Px^{n-m}-Qx^{n-m-1}+\text{ etc.}=0$$

[1]vide Addenda et corrigenda pro pag. 110 l. 8 in fine.

[2]loco citato p. 110

where p, q, etc., P, and Q are positive quantities, and $+Px^{n-m}-Qx^{n-m-1}$ is the first sign change; if this equation is multiplied by $x-a$, then we obtain the resulting equation:

$$x^{n+1}+(p-a)x^n+(q-ap)x^{n-1}+\ldots-(Q+aP)x^{n-m}+\text{ etc.}=0.$$

But $-(Q+aP)$ is a negative quantity, thus there are in the resulting equation 1 or 3 or 5 etc. sign changes which correspond to the first sign change in the given equation.

Let the equation be given:

$$x^n+px^{n-1}+qx^{n-2}+\ldots+\alpha x^{n-r}-\beta x^{n-r-1}-\ldots$$
$$-\pi x^{n-s}+\rho x^{n-s-1}+\ldots+\phi x^{n-t}-\chi x^{n-t-1}\pm\text{ etc.}=0$$

where α, β, π, ρ, ϕ, χ denote positive quantities, and $+\alpha x^{n-4}-\beta x^{n-r-1}$ is any sign change from + to − in the given equation, and $-\pi x^{n-s}+\rho x^{n-s-1}$ is the first subsequent change from $-$ to $+$, which in turn is followed by $\phi x^{n-t}-\chi x^{n-t-1}$.

Now if this equation is multiplied by $x-a$, there results

$$x^{n+1}+(p-a)x^n+\ldots-(\beta+a\alpha)x^{n-r}-\ldots+(\rho+a\pi)x^{n-s}+\ldots$$
$$-(\chi+a\phi)x^{n-t}-\text{ etc.}=0.$$

In this equation the coefficient $-(\beta+a\alpha)$ is negative, $+(\rho+a\pi)$ is positive, and $-(\chi+a\phi)$ is negative. Therefore from the negative term $-(\beta+a\alpha)x^{n-r}$, to the positive term $+(\rho+a\pi)x^{n-s}$ there are one or more (3, 5, etc.) sign changes corresponding to the single sign change from the term $-\beta x^{n-r-1}$ to the term $+\rho x^{n-s-1}$ in the given equation; and likewise from the positive term $+(\rho+a\pi)x^{n-s}$ to the negative term $-(\chi+a\phi)x^{n-t}$ in the resulting equation there are 1 or 3 or 5 etc. sign changes which correspond to the single sign change from the term $+\rho x^{n-s-1}$ to the term $-\chi x^{n-t-1}$ in the given equation.

Therefore every individual sign change in the given equation has an odd number *more* of corresponding sign changes in the resulting equation; consequently there will be as many or more sign changes in the resulting equation

$$x^{n+1}+(p-a)x^n+(q-ap)x^{n-1}+\ldots+(M-aL)x-aM=0$$

from the first to the penultimate term $(M-aL)x$ as are to be found in the given equation, $x^n+px^{n-1}+qx^{n-2}+\ldots+Lx+M=0$.

Now if there are only as many sign changes from the first term to the penultimate, as the original equation had in all, then the penultimate term $(M-aL)x$, of the resulting equation, has the same sign as the last term, M, of the given equation. If the latter is the case, then from the penultimate term, $(M-aL)x$, to the last term, $-aM$, there is one sign change, namely, $+(M-aL)x-aM$; and in consequence the resulting equation has *an odd number more sign changes* than the given equation possesses.

II. Examples for Sign Progressions

Let the equation be given:

$$x^n-px^{n-1}+qx^{n-2}-rx^{n-3}+\ldots-Px^{n-m+1}+Qx^{n-m}+Rx^{n-m-1}+\ldots=0$$

where $p, q, r, \ldots P, Q, R, \ldots$ are positive quantities, and $+Qx^{n-m}+Rx^{n-m-1}$ is the first sign progression in the given equation.

Now if this equation is multiplied by $x+a$, the result is

$$x^{n+1}-(p-a)x^n+(q-ap)x^{n-1}-(r-aq)x^{n-2}+\ldots+(Q-aP)x^{n-m+1}+(R+aQ)x^{n-m}$$
$$+\text{ etc.}=0.$$

Now if, in the resulting equation up to the term $(Q-aP)x^{m-n+1}$, no sign progression occurs, and $p>a$, $q>ap$, $r>aq$, ..., $Q>aP$ and therefore $(Q-aP)$ has the same sign as Q, i.e., is positive, and $+(Q-aP)x^{n-m+1}+(R+aQ)x^{n-m}$ represents a sign progression, then there appear in the resulting equation *an odd number of sign progressions* to which the first progression in the given equation corresponds, i.e., is contained in just these terms.

By the same method, *mutatis mutandis*, one may also proceed to the following equation:[1]

$$x^n-px^{n-1}+qx^{n-2}-rx^{n-3}+\ldots+Px^{n-m+1}$$
$$-Qx^{n-m}-Rx^{n-m-1}\pm\ldots=0.$$

Here the first sign progression is from $-$ to $-$, that is, *an odd number of sign progressions* in the resulting equation correspond to the first sign progression in the original equation.

Let the equation be given[2]

$$x^n-px^{n-1}+qx^{n-2}-\ldots+\alpha x^{n-m+1}+\beta x^{n-m}-\gamma x^{n-m-1}$$
$$+\delta x^{n-m-2}-\ldots-\pi x^{n-r}+\rho x^{n-r-1}+\sigma x^{n-r-2}\pm\ldots+Lx+M=0,$$

in which $+\alpha x^{n-m+1}+\beta x^{n-m}$ represents a sign progression, and the next sign progression is $+\rho x^{n-r-1}+\sigma x^{n-r-2}$.

Now if this equation is multiplied by $x+a$, there results:

---

[1]loco citato, p. 112.

[2]loco citato, p. 112

$$x^{n+1} - (p-a)x^n + (q-ap)x^{n-1} - \dots$$
$$+(\beta+a\alpha)x^{n-m+1} - (\gamma-a\beta)x^{n-m} + (\delta-a\gamma)x^{n-m-1}$$
$$+(\rho-a\pi)x^{n-r} + (\sigma+a\rho)x^{n-r-1} \pm \text{etc.} = 0.$$

Since in the first equation α and β have the same sign, $(\beta+a\alpha)$ has the same sign as β; and the signs of the terms $+\beta x^{n-m} - \gamma x^{n-m-1} + \delta x^{n-m-2} - \dots - \pi x^{n-r} + \rho x^{n-r-1}$ in the given equation alternate continually, then by hypothesis $+\rho x^{n-r-1} + \sigma x^{n-r-2}$ constitutes the next sign progression that appears in the given equation.

Now if from the term $+(\beta+a\alpha)x^{n-m+1}$ to the term $+(\rho-a\pi)x^{n-r}$ in the resulting equation no sign progressions arise, then it must be true that $\gamma > a\beta$, $\delta > a\gamma$, $\rho > a\pi$, etc. Consequently $(\rho-a\pi)$ has the same sign as ρ, and since ρ and σ have the same sign, so also must $+(\rho-a\pi)$ and $+(\sigma+a\rho)$ have the same sign, and it follows that there is a sign progression in the resulting equation, namely $+(\rho-a\pi)x^{n-r} + (\sigma+a\rho)x^{n-r-1}$; hence there is one or more (3, 5, etc.) sign progressions, which correspond to any sign progression whatever in the given equation. And thus there are from the first to the penultimate term as many, or 2, 4, 6 (i.e. an even number) more sign progressions in the resulting equation, than were to be found in the given equation.

However, if there are just so many sign progressions from the first term, x^{n+1}, up to the penultimate term, $(M+aL)x$, in the resulting equation, then there will also be exactly as many sign progressions as appear in the given equation; consequently the penultimate term, $(M+aL)x$, of the resulting equation, has the same sign as the final term, M, of the given equation, and therefore the resulting equation has one more sign progression appended, namely $+(M+aL)x + aM$, than is found in the given equation.

From this reasoning it follows, therefore, that in the resulting equation *an odd number more sign progressions occur*, than in the original equation.

Corrolarium 1. Let the equation be given:[1]

$$x^n + px^{n-1} + qx^{n-2} + \dots - Px^{n-m} + Qx^{n-m-1} + Rx^{n-m-2} + \dots$$
$$\pm fx^{n-m-r} \pm gx^{n-m-r-1} \pm hx^{n-m-r-2} \pm kx^{n-m-r-3} \pm \dots$$
$$\pm \alpha x^{n-m-r-s} \pm \beta x^{n-m-r-s-1} \pm \gamma x^{n-m-r-s-2} \pm \dots$$
$$+\pi x^{n-m-r-s-t} - \rho x^{n-m-r-s-t-1} - \sigma x^{n-m-r-s-t-2} + \dots = 0$$

---

[1]loco citato, p. 113

in which $P, Q, R, \ldots; f, g, h, k, \ldots; \alpha, \beta, \gamma, \ldots; \pi, \rho, \sigma, \ldots$ are either positive or negative quantities; further let $-Px^{n-m}+Qx^{n-m-1}$ be any change of sign, and $\pi x^{n-m-r-s-t}-\rho x^{n-m-r-s-t-1}$ the next sign change; further let $g > af$, $ag > h$, $ah > k$, and so on up to β; then if the given equation is multiplied by $x-a$, there arise in the resulting equation,

$$x^{n+1} + (p-a)x^n + (q-ap)x^{n-1} + \ldots + (Q+aP)x^{n-m} + (R-aQ)x^{n-m-1} + \ldots + (g-af)x^{n-m-r} - (ag-h)x^{n-m-r-1} - (ah-k)x^{n-m-r-2} - \ldots - (a\alpha-\beta)x^{n-m-r-s} + (\gamma-a\beta)x^{n-m-r-s-1} \pm \text{etc.} = 0,$$

two sign changes more than in the original equation, namely $+(g-af)-(ag-h)$ and $-(a\alpha-\beta)+(\gamma-a\beta)$; and for each such case with respect to the given equation the axiom holds, that *in the resulting equation two additional sign changes occur*, which are not contained in the given equation.

Corrolarium 2. Assumed:[1]
$+qx^{n-m-1}+Rx^{n-m-2}$ to be any sign progression, and $-\rho x^{n-m-r-s-t-1}-\sigma x^{n-m-r-s-t-2}$ the next sign progression, the quantities $f, g, h, k, \ldots; \alpha, \beta, \gamma, \ldots$ alternating $+$ and $-$ signs; further assume that $g > af$, $ag > h$, $ah > k$, and so on up to the coefficient β, and $\gamma > a\beta$; then when the given equation is multiplied by $x-a$, in the resulting equation,

$$x^{n+1} + (p+a)x^n + (q+ap)x^{n-1} + \ldots + (Q-aP)x^{n-m} + (R+aQ)x^{n-m-1} + \ldots \pm (g-af)x^{n-m-r} \pm (ag-h)x^{n-m-r-1} \pm (ah-k)x^{n-m-r-2} \pm \ldots \mp (a\alpha-\beta)x^{n-m-r-s} \mp (\gamma-a\beta)x^{n-m-r-s-1} \mp \ldots - (\rho-a\pi)x^{n-m-r-s-t} - (\sigma+a\rho)x^{n-m-r-s-t-1} \pm \text{etc.} = 0,$$

there arise *two more sign progressions* than are found in the given equation, and for each such case with respect to the given equation the axiom holds, that *in the resulting equation two further sign progressions occur*, that are lacking in the original.

Conclusion: Thus it is easy to see:

If the above equation is multiplied by $x-a$, then the number of sign changes is *increased* by 1; but if the equation is multiplied by $x+a$, then the number of sign changes, augmented by 2, 4, 6, etc., is *diminished* by 1, and then the above equation accordingly must have 2, 4, 6, etc. imaginary roots.

[1] loco citato p. 114.

Compare the Latin original:[1]

"*Corrolarium*: Hinc facile patet, si modo ducatur data æquatio in $x-a$ vel $x+a$; et in priori casu numerus mutationum signorum de + in − et − in + *augeatur*; in posteriori vero casu *diminuatur* per unitatem a duobus, quatuor, sex etc. auctam; tum saltem duas, quatuor, sex etc. respective impossibiles radices habet data æquatio."

From these investigations we arrive at the following conclusion:

1. *Waring here truly explained the foundation of Descartes' Rule of Signs, which Gauss*[2] *was later the first to rediscover*, since to him Waring's "Meditationes algebraicæ" remained unknown. We have here, therefore, an analogy to the previously mentioned discovery of the solution of a cubic equation by Scipione del Ferro, and its later independent discovery by Tartaglia. [12]

2. *Waring explained Descartes' Rule of Signs in such a way that in it one can already see, with respect to the identification of complex roots, a correspondence to the Budan*[3]*-Fourier*[4] *Theorem.*

Afterword

In this work I hope to have explained *two topics*:

First, that Waring as a diligent compiler recorded the progress in the field of Algebra up to his time in a single book, and likewise himself advanced it, although in a form to which the requirements of contemporary methods and didactics no longer correspond.

Second, the reader will gain the impression that mathematics since Waring's time (the second half of the 18th Century) up to now, as regards higher algebra, number theory, and function theory, has made enormous advances by applying new methods to old problems.

---

[1] loco citato, p. 115.

[2] Gauss, Werke, t. III, p. 68

[3] Dés. Budan, Nouvelle méthode pour la résolution des equations numériques. Paris 1811.

[4] J. B. J. Fourier, Analyse des équations determinées, Livre 1. Paris 1831.

Curriculum vitæ

I, Franz Xaver Mayer, Pater Augustin, Ordinis Cisterciensis, was born on June 10, 1875, the son of Franz X. Mayer, master potter in Wurzach, Württemberg. After completing my public school studies in Weingarten, Württemberg, I was enrolled at the Cistercian Privat-Gymnasium in Mehrerau, Bregenz, Vorarlberg, where I took orders in 1895. I completed the final Gymnasium class at the Staatsgymnasium in Feldkirch, where I was certificated in 1897. I continued my philosophical and theological studies at the well-known academy in Mehrerau.

Initially planning to teach, due to poor health I switched to pastoral duties, spending 17 years in Switzerland. The last four years I was enrolled at the University of Zürich and attended lectures by Messrs. Amberg, Bützberger, Disteli, R. Fueter, Greinacher, Karrer, E. Meyer, Meißner, Ratnowsky, Speiser, Tank, and Wolfer.

On June 26, 1922, after taking the required examinations, I was awarded the Diploma for Advanced Education with Specialization in Mathematics, by the University of Zürich.[13]

To all of the above named gentlemen I owe a debt of enduring gratitude. But Professor Rudolf Fueter especially deserves my thanks, both for suggesting the present work and for his continuing interest in my efforts.

TRANSLATOR'S NOTES

[1] The quotation from Dr. (Samuel) Parr comes from Courtney's piece on Waring in the *Dictionary of National Biography.* I am unable to identify its original source (obviously Mayer also failed in this matter; he has even neglected to credit Courtney as the secondary source of the citation). In point of fact Colson, the previous Lucasian Professor, had also been a lecturer; although Waring succeeded him in the Lucasian Chair, Colson's instructional duties were assumed by "Mr. Wollaston" (*Gentleman's Magazine*, July 1760, p. 347) — presumably George Wollaston (1738–1826), who received his B.A. and M.A. degrees each one year after Waring. Waring did, however, involve himself with students at Cambridge, as Maseres [1800, 47] recounts:

> The outward pomp which is affected frequently in the higher departments in academic life, was no gratification to one whose habits were of a very opposite nature; and he was too much occupied in science, to attend to the intrigues of the university. There, in all questions of science, his word was the law; and at the annual examination of the candidates for the prize instituted by Dr. Smith, he appeared to the greatest advantage. The candidates were generally three or four of the best proficients in the mathematics at the previous annual examination for the bachelor's degree, who were employed from nine o'clock in the morning to ten at night, with the exception of two hours for dinner, and twenty minutes for tea, in answering *viva voce*, or writing down answers to the Professor's questions, from the first rudiments of philosophy, to the deepest parts of his own and Sir Isaac Newton's works. Perhaps no part of Europe affords an instance of so severe a process; and there was never any ground for suspecting the Professor of partiality. The zeal and judgment with which he performed this part of his office cannot be obliterated from the memory of those who passed through his fiery ordeal.

[2] Courtney (p. 841) and Nightingale [1813, 139] locate Plealy near Pontesbury (about eight miles from Shrewsbury), not Petersbury. Such a short distance would do little to improve the climate, unless Mrs. Waring's problem stemmed from industrial pollution which might have appeared in Shrewsbury about that time.

[3] Mayer has here omitted the corrigendum *cum uno adjuncto* (page A2.2), and in fact he appears to have hybridized lines 4–6 of page 110 with lines 15–29 of page 113. (The full Latin text, including the corrigendum, appears on page 430 in this translation.)

[4] In fact, Waring produces the discriminant of a single equation at several places in Chapter 2, although he refers to it only as the transformed equation whose roots are the squares of the differences of pairs of roots of the given equation.

[5] This is where Mayer was deceived by the typographical error in Waring's text, which I rectify by substituting *Krafft* for *Kiraast.*

[6] Crossley and Henry [1990] have recently done an English translation of a 13th-century manuscript, a Latin rendition of the original work by al-Khowarizmi (which was probably read by Leonardo, Jordanus, etc.)

[7] Cardan's *sophistica*, rendered as *sophistische* by Mayer, and as *sophistical* by me, is used in the sense of "something a sophist would contrive". Witmer [1968, p. 220] interprets this passage as "this truly is sophisticated" — which conveys the wrong impression unless we recall that in the English language (according to the Oxford English Dictionary, for instance) the use of "sophisticated" in its present favorable connotation began only about a century ago, and it probably was never used in that sense in Mediæval or Renaissance Latin. In essence Cardan was acknowledging that his reader had a right to be suspicious of this method of reasoning, although his attempt to justify it must have persuaded at least some of his colleagues of its legitimacy.

[8] According to Morley [1854, Vol. I, p. 246] this was the Marquis dal Vasto (or dal Guasto), whose surname was D'Avalos.

[9] Witmer [1968, p. x] reports that del Ferro's manuscript has been found at the University of Bologna. Maseres [1780] attempts to retrace the method of reasoning which led del Ferro to his famous discovery. It is a very speculative piece and reflects more the thinking of mathematical laymen in eighteenth-century England than the Italian Renaissance, but makes interesting reading.

[10] Mayer's analysis of Tartaglia vs. Cardan covers in detail all the points that might come up in a court of law, but I believe he has inspected so many trees that he fails to perceive the forest of the dispute. (Boorstin's account of the evolution of scientific thinking provides some important insights in this regard.) Consider that the time of the dispute was essentially at the beginning of the Renaissance in Italy; but on the other hand, not everyone

who lived during the early years of the Renaissance was in fact a participant in the rebirth of knowledge and culture. Cardan was surely a "Renaissance man": to him, knowledge was to be shared and disseminated, and public recognition was a fitting reward for anyone who discovered something new in any area of research. To Tartaglia, however, knowledge was a treasure to be hoarded, not shared; the reward for knowledge was one's ability to defeat an opponent by answering any questions he could pose, and posing at least one question that he could not answer. Once the cubic solution was published by Cardan, Tartaglia could no longer use it in a scientific contest such as the one where he had been publicly humiliated by Fiori. In short, Tartaglia represented the Mediæval culture, which was rapidly being overwhelmed by the Renaissance.

However, Cardan was a Renaissance man, not a twentieth-century scientist. O'Malley's biography of Vesalius [*Dictionary of Scientific Biography*, vol. 14, p. 3] points out that we know his subject's date of birth only because a horoscope, done by Cardan, has been found; in fact Morley reports that Cardan's interest in astrology got him in trouble with both Church and State during his lifetime.

[11] According to Hutton [1815, p. 76] Giovanni da Coi may have had the method for cube roots of binomials as early as January, 1540, a century before the Wassenaer-Stampioen debate.

[12] A quibble: Gauss was reportedly unaware that Waring had already explained the foundations of Descartes' rule, although as mentioned in endnote 19 (p. 380) of Chapter V, he had been at least indirectly exposed to the *Meditationes Algebraicæ*; Tartaglia reproduced the solution of the cubic equation (possibly in a way that would now be called "reverse engineering") after the dismaying revelation that a rival mathematician could already solve it. Thus the analogy as perceived by Mayer does not really apply.

[13] The only additional biographical information is in *Cistercienser-Chronik*, 48. Jahrg. Nr. 570 (1 August 1936), which has Mayer's obituary notice on pp. 254–255.

Bibliography

Baillet, Adrien [1691] *La Vie de M. Des-Cartes.* Paris: D. Horthemels; reprint, Paris: La Table Ronde, 1946; English translation, *The life of Monsieur Des Cartes, containing the history of his philosophy and works; as also, the most remarkable things that befell him during the whole course of his life. Translated from the French, by S. R.* London: printed for R. Simpson, 1693, 286 pp.

Bashmakova, I. G. [1959] Ob odnom voprose teorii algebraicheskikh uravneniĭ v trudakh I. Nīutona i E. Varinga. *Istoriko-Matematicheskie Issledovaniīa* **12**: 431–456.

Bierens de Haan, David [1882] Bibliographie Néerlandaise Historico-Scientifique, etc. (continuazione K–Z). *Bulletino di Bibliografia e di Storia delle scienze matematiche e physiche* **XV**: 225 - 440.

Bohman, Jan and Carl-Erik Fröberg [1981] Numerical investigation of Waring's Problem for cubes. *BIT* **21**: 118–122.

Boorstin, Daniel J. [1983] *The Discoverers*, 1st edition. New York: Random House.

Borchardt, Carl Wilhelm [1847] Développements sur l'équation à l'aide de laquelle on détermine les inégalités séculaires du mouvement des planètes. *Journal de Mathématiques pures et appliquées* (Liouville's Journal) **XII**: 50–67 (Fev. 1847)

Bosmans, Henri [1927] À propos de la correspondance de Descartes avec Constantin Huygens récemment publiée par M. Léon Roth. L'auteur principal de l'Onwissen wiskonstenaer I. I. Stampioenius ontdeckt door Jacobus a Waessenaer (Leyde, 1640). *Revue des Questions Scientifiques*, **11**: 113–141.

Bowes, Robert [1888] Notes on the University Printers. Cambridge Antiquarian Society *Proceedings and Communications*, vol. V.

Brill, A. [1893] Ueber symmetrische Functionen von Variabelnpaaren. *Nachrichten von der Königlichen Gesellschaft der Wissenschaften zu Göttingen* 1893(20): 757–762.

Buchberger, Bruno [1965] *Ein Algorithmus zum Auffinden der Basiselemente des Restklassenringes nach einem nulldimensionalen Polynomideal.* Dissertation, Universität Innsbruck.

Buchberger, Bruno [1976] A theoretical basis for the reduction of polynomials to canonical forms. ACM *SIGSAM Bulletin* **10**(3): 19–29 (August 1976).

Campbell, George [1728] A method for determining the number of impossible roots in adfected æquations. *Philosophical Transactions of the Royal Society* **XXXV**: 515–531.

Campbell, George [1729] *Remarks on a Paper published by Mr.* Maclaurin, *in the Philosophical Transactions for the Month of* May, 1729. (printer and publisher not specified.)

Campbell, George [1733] The number of impossible roots in any æquation reduced to the Cartesian rule for finding the number of positive and negative roots. Article XI (pp. 96–98) in *The Philosophical Transactions (From the Year 1720, to the Year 1732) ABRIDGED, and Disposed under General Heads* By Mr. Reid and John Gray, A.M. F.R.S., Vol. VI, Part I: The Mathematical Papers. London: Printed for William Innys and Richard Manby, Printers to the Royal Society, at the West End of St. Paul's.

Cantor, M. [1907] *Vorlesungen über Geschichte der Mathematik*, III, *1668–1758* (zweite Auflage). Leipzig: B. G. Teubner.

Chalmers, A. [1812–1817] *The General Biographical Dictionary, containing an historical and critical account of the lives and writings of the most eminent persons in every nation; particularly the British and Irish; from the earliest to the present time. A new edition, revised and enlarged by Alexander Chalmers, F.S.A.* London: J. Nichols and Son; (etc.)

Courtney, William P. [1899] "Waring, Edward". *Dictionary of National Biography*, Vol. XX, pp. 840–842. London: Humphrey Milford (since 1917, Oxford University Press).

Crossley, John N. & Alan S. Henry [1990] Thus Spake al-Khwārizmī: A Translation of the Text of Cambridge University Library Ms. Ii.vi.5. *Historia Mathematica* **17**: 103–131.

Davis, Martin [1973] Hilbert's Tenth Problem is Unsolvable. *American Mathematical Monthly* **80**: 233–269.

DeMoivre, Abraham [1730] *Miscellanea analytica de seriebus et quadraturis*. Londini, excudebant J. Tonson & J. Watts, 1730.

Dickson, L. E. [1928] Simpler proofs of Waring's theorem on cubes with various generalizations. *Transactions of the American Mathematical Society* **30**: 1–30.

Dickson, L. E. [1952] *History of the Theory of Numbers*. 3 volumes. New York: Chelsea.

Dictionary of Scientific Biography. (Charles Coulston Gillespie, Editor in Chief). 16 volumes, New York: Scribner's, 1970–1980; supplement (vols. 17–18), 1990.

Englefield, H. [1793] *On the Determination of the Orbits of Comets, according to the Methods of Father Boscovich and M. De la Place. With new and complete Tables; and examples of the Calculation by both Methods.* By Sir Henry Englefield, Bart. F.R.S. and F.A.S. London: printed by Ritchie and Sammels for Peter Elmsly.

Gardner, Martin [1981] "Mathematical Games", *Scientific American*, Vol. 244, No. 2 (February 1981).

Gauss, Carl Friedrich [1801] *Disquisitiones Arithmeticæ*. Lipsiæ: Gerh. Fleischer, Jun. English translation by Arthur A. Clarke, New Haven and London: Yale University Press, 1966.

Gelfond, A. O. [1960] *Transcendental and Algebraic Numbers*. Translated from the first Russian edition by Leo F. Boron. New York: Dover.

Gerono, C. C. [1843] Note sur l'équation aux sommes des racines, prises deux à deux, d'une équation donnée. *Nouvelles Annales de Mathématiques* **2**: 128–132.

Gonggrÿp, B. [1915] Quelques théorèmes concernant la relation entre les zéros d'un polynome et ceux d'un polynome de degré inférieur. *Journal de Mathématiques Pures et Appliqués*, Sér. 7, **I**: 353 - 365.

Hardy, G. H. [1908] *A Course of Pure Mathematics.* Cambridge: at the University Press.

Hardy, G. H. [1943] *A Course of Pure Mathematics*, Seventh Edition. Cambridge: at the University Press.

Hardy, G. H. and E. M. Wright [1938] *An Introduction to the Theory of Numbers.* Oxford: Clarendon Press.

Hilbert, David [1909] Beweis für die Darstellbarkeit der ganzen Zahlen durch eine feste Anzahl n^{ter} Potenzen (Waringsche Problem). (mit einiger Veränderungen und Zusätzen abgedrückt aus dem Nachrichten der Ges. d. Wiss. zu Göttingen 1909, Sitzung vom 6. Februar.) *Mathematische Annalen* **67**: 281–300.

Hirsch, M. [1827] *Hirsch's Collection of Examples, Formulae, and Calculations of literal Calculus and Algebra.* Translated from the German, by the Rev. J. A. Ross, A. M. London: Black, Young, & Young (first German edition, Berlin: Duncker und Humblot, 1804.)

Hutton, Charles [1815] *A Philosophical and Mathematical Dictionary: containing an explanation of the terms, and an account of the several subjects,.... A new edition, with numerous additions and improvements.* London: printed for the Author; ...; and Simpkin and Marshall.

Jacobi, C. G. [1851] Über die Zusammensetzung der Zahlen aus ganzen positiven Cuben; nebst einer Tabelle für die kleinste Cubenanzahl, aus welcher jede Zahl bis 12000 zusammengesetzt werden kann. Von Herrn C. G. Jacobi, †weiland Professor zu Berlin. *Journal für die reine und angewandte Mathematik* **42**: 41 - 69.

Jones, A. [1986] (ed.) *Pappus of Alexandria: Book 7 of the Collection.* Part 1: Introduction, Text, and Translation. New York, Berlin: Springer–Verlag.

Junker, F. H. [1891] Die Relationen, welche zwischen den elementaren symmetrischen Functionen bestehen. *Mathematische Annalen* **38**: 91–114.

Junker, F. H. [1893] Ueber symmetrische Functionen von mehreren Reihen von Veränderlichen. *Mathematische Annalen* **43**: 225–270.

Junker, F. H. [1894] Die symmetrischen Functionen und die Relationen zwischen den Elementarfunctionen derselben. *Mathematische Annalen* **45**: 1–84.

Knuth, Donald E. [1981] *The Art of Computer Programming*, Vol. II, *Seminumerical Algorithms.* (second edition) Reading, MA: Addison–Wesley.

Kogge, Peter [1974] Parallel Solution of Recurrence Problems. *IBM Journal of Research and Development* **18**: 138–148.

Korteweg, D. J. [1888] Notes sur Constantijn Huygens considéré comme amateur des sciences exactes et sur ses relations avec Descartes. *Archives néerlandaises des Sciences exactes et naturelles* **XXII**: 422–466.

Krafft, Georg W. [1735] Observationes arithmeticæ de Septenario. *Commentarii Academiæ scientarum imperialis petropolitanæ* **7**: 41–45.

Krafft, Johann [1591] *Ein neues und wohlgegründtes Rechenbuch.* Ulm.

Lagrange, J. L. [1771] Demonstration d'un theorème d'arithmétique. *Nouveaux Mémoires de l'Académie de Berlin* **2**, 1773 (année 1771), pp. 125–137; *Oeuvres de Lagrange*, J. A. Serret, ed., vol. III, pp. 189–201, Paris: Gauthier-Villars, 1869; reprinted, Ann Arbor: University Microfilms, 1970.

Landen, J. [1755] *Mathematical Lucubrations: containing new improvements in various Branches of Mathematics.* London: J. Nourse.

Laplace, P. S. [1812] *Théorie analytique des probabilités.* Paris: V. Courcier.

MacDuffee, C. C. [1931] The discriminant matrices of a linear associative algebra. *Annals of Mathematics, 2nd Ser.* **32**: 60–66.

Maclaurin, C. [1726] A letter... concerning æquations with impossible roots. *Philosophical Transactions of the Royal Society* **XXXIV**: 104–113.

Maclaurin, C. [1729] A second letter... concerning the roots of equations, with the demonstration of other rules in algebra.... *Philosophical Transactions of the Royal Society* **XXXVI**: 59–96.

Maclaurin, C. [1730] *A defence of the letter published in the Philosophical Transactions for March and April 1729. Concerning the impossible roots of equations; in a letter from the author, to a friend at London.* (printer and publisher not specified).

Maseres, F. [1780] A Conjecture concerning the Method by which *Cardan's* Rules for resolving the Cubic Equation $x^3+qx = r$ in all cases (or in all magnitudes of the known quantities q and r) and the Cubic Equation $x^3-qx = r$ in the first Case of it (or when r is greater than $\frac{2q\sqrt{q}}{3\sqrt{3}}$, or $\frac{rr}{4}$ is greater than $\frac{q^3}{27}$) was probably discovered by *Scipio Ferreus*, of *Bononia*, or whoever else was the first Inventor of them. By *Francis Maseres,* Esq., F.R.S. Cursitor Baron of the Exchequer. *Philosophical Transactions of the Royal Society* **LXX**, Part I: 221–238.

(Maseres), F. [1800] Some Account of Doctor Waring, the late celebrated Mathematician. *Monthly Magazine* **11** (no. 55), February 1800: 46–49. (For justification of attribution to Maseres, see Translator's Note 11, Chapter 3, pp. 185–186.)

Meisner, Heinrich [1692] *Stern und Kern der Algebra.* Hamburg: Gleditsch (reprinted 1740).

Méray, C. [1867] Extension aux équations simultanées des formules de Newton pour le calcul des sommes de puissances semblables des racines des équations entières. *Annales scientifiques de l'École Normale supérieure* **IV**: 159–193.

Mills, S., (ed.) [1982] *The collected letters of Colin Maclaurin.* Nantwich, Cheshire: Shiva.

Mills, S. [1983] The controversy between Colin MacLaurin and George Campbell over complex roots, 1728–1729. *Archive for History of Exact Sciences* **28**: 149–164.

Moigno, F. N. M. [1840] Note sur la détermination du nombre des racines réelles ou imaginaires d'une équation numérique, comprises entre des limites données. Théorèmes de Rolle, de Budan ou de Fourier, de Descartes, de Sturm et de Cauchy. *Journal de Mathématiques pures et appliquées* (Liouville's Journal) **V** (Mar. 1840): 75–94.

Moran, A. & P. Pritchard [1991] The design of a background job on a local-area network. *Proceedings 14th Australian Computer Science Conference*, ed. G. Gupta, *Australian Computer Science Communications* **13** (1991): 17-1 – 17-11.

Morley, H. [854] *Jerome Cardan. The Life of Girolamo Cardano, of Milan, Physician.* London: Chapman & Hall.

Nangle, B. [1934] *The Monthly Review, first series, 1749 – 1789; indexes of contributors and articles.* Oxford: Clarendon Press.

Nangle, B. [1955] *The Monthly Review, second series, 1790 – 1815; indexes of contributors and articles.* Oxford: Clarendon Press.

Newton, Isaac [1707] *Arithmetica Universalis; sive De Compositione et Resolutione Arithmetica Liber.* Cui accessit Halleiana *Æquationum Radices Arithmeticæ inveniendi methodus.* In Usum juventutis Academiæ. (William Whiston, ed.) Cantabrigiæ, Typis Academicis, MDCCVII. (This is the edition cited by Waring.)

Newton, Isaac [1720] *Universal Arithmetick: or, a Treatise of Arithmetical Composition and Resolution*, to which is added, Dr. Halley's method of finding the roots of the Æquations arithmetically.

Translated from the Latin by the late Mr. Raphson, and revised and corrected by Mr. Cunn. London: Printed for J. Senex at the Globe in Salisbury Court, W. Taylor at the Ship, T. Warner at the Black Boy in Paternoster Row, and J. Osborn at the Oxford Arms in Lombard Street.

Newton, Isaac [1722] *Arithmetica Universalis; sive De Compositione et Resolutione Arithmetica Liber.* Editio Secunda, In qua multa immutantur et emendantur, nonnulla adduntur. Londini: Impensis Benj. & Sam. Tooke, Bibliopolarum, juxta Medii Templi Portam, in vulgo vocato Fleet-street, MDCCXXII.

Newton, Isaac [1728] *Universal Arithmetic...* – The Second Edition, very much corrected. London: Printed for J. Senex in Fleet Street, W. and J. Innys near S. Paul's, J. Osborn and T. Longman in Pater-noster Row.

Nichols, John [1812–1815] *Literary Anecdotes of the Eighteenth Century; comprizing biographic memoirs of William Bowyer, printer, F.S.A., ... ; and biographical anecdotes of a considerable number of eminent writers and ingenious artists.* 9 Volumes. London: printed for the author, by Nichols, son, and Bentley.

Nightingale, J. [1813] *A topographical and historical Description of the County of Salop*; .... London: Sherwood, Neely and Jones [etc.]

Pascal, Blaise [1665] De numeris multiplicibus. In *Traité du triangle arithmétique, avec quelques autres petits traitez sur la mesme matière*, Paris: G. Desprez; reprinted in *Oeuvres de Blaise Pascal, publiés suivant l'ordre chronologique avec documents complementaires, introductions et notes*, vol. 3, pp. 311 - 339, Paris: Hachette et cie., 1908 - 1921 (14 vols); or *Œuvres Complètes*, pp. 84–89, Paris: Éditions de Seuil, 1963.

Pinker, Aron [1973] On Waring's representation of symmetric polynomials in terms of the elementary symmetric polynomials. *FSC Journal of Mathematics Education* **6**: 34–42. (Frostburg State College, Maryland)

Poggendorff, J. C. [1863] *Biographisch–Litterarisches Handwörterbuch* .... Leipzig: Verlag von Johann Ambrosius Barth.

(Powell, W. S.) [1760*a*]. *Observations on the First Chapter of a Book called Miscellanea Analytica.* London, Printed for T. Merril, at Cambridge, MDCCLX.

(Powell, W. S.) [1760*b*]. *A Defence of the Observations on the First Chapter of a Book called Miscellanea Analytica.* London, Printed for T. Merril, at Cambridge, MDCCLX.

Ribenboim, Paolo [1989] *The Book of Prime Number Records*, 2nd edition. New York, Berlin: Springer–Verlag.

Riccati, Vincenzo [1771*a*] Esame del metodo, col quale il Chiarissimo Sig. EDMONDO VVARING ha pensato di aver ridotte l'Equazioni del quinto grado a quelle del terzo. *Nuova Raccolta d'Opusculi scientifici e filologici* (ed. A. Calogerà), Tomo XXI, art. XV. 10 pp., 1 fig.

Riccati, Vincenzo [1771*b*] Esame del metodo, col quale il Dottissimo Sig. EDMONDO VVARING ha giudicato d'aver ridotte l'Equazioni del sexto grado a quelle del terzo. *Nuova Raccolta d'Opusculi scientifici e filologici*, Tomo XXI, unnumbered article following art. XV. 14 pp., 1 fig.

Roth, Leon [1926] *The Correspondence of Descartes and Constantyn Huygens, 1635–1647, edited from manuscripts now in the Bibliothèque nationale, formerly in the possession of the late Harry Wilmot Buxton, F.R.A.S.* Oxford: Clarendon Press.

Rowe, R. Robinson [1971] (solution to Problem 19, February 1971) M.I.T. *Technology Review*, May 1971: 78–79.

Rubinstein, Z. [1972] On the property of the coincidence function associated with the logarithmic derivative of a polynomial. *Journal of Approximation Theory* 6(2): 1–5 (July 1972)

Ruffini, Paolo [1799] *Teoria generale delle equazioni in cui si demonstra impossibile la soluzione algebrica delle equazioni generali di grado superiore al quarto.* 2 vols. Bologna: Nella stamperia di S. T. d'Aquino. Reprinted in *Opere mathematiche di Paolo Ruffini, pubblicate sotto gli auspici dell'Unione matematica italiana a cura del Ettore Bortolotti.* Roma: Edizione Cremonese, 1953–1954.

Schooten, F. [1649] *Geometria à Renato Des Cartes Anno 1637 Gallicè edita ....* Lugduni Batavorum: Ioannes Maire, MDDXLIX (*sic*); Amstelodami: apud Ludovicum & Danielem Elzevirius, M.DC.LXIX.

Selmer, Ernst S. [1966] *Linear recurrence relations over finite fields* (mimeograph). Department of Mathematics, University of Bergen (Norway), 212 pp.

Simson, R. [1776] *Roberti Simson M.D.... Opera Quædam Reliquua.* Glasguæ: Robert and Andrew Foulis.

Stewart, Ian [1986] The Waring experience. *Nature* **323** (23 October 1986): 674.

Struik, D. J. [1969] *A Source Book in Mathematics, 1200 - 1800.* Cambridge, MA: Harvard University Press.

Sylvester, J. J. [1854] On some new theorems in arithmetic. *Philosophical Magazine* **VIII**: 187–190. Also in Baker, H. F. (ed.), *The collected Mathematical Papers of James Joseph Sylvester, F.R.S., D.C.L., LL.D., Sc.D.*, ..., Vol. II, pp. 40–43 (Cambridge, 1908).

Sylvester, J. J. [1864] Algebraical researches, containing a disquisition on Newton's rule for the discovery of imaginary roots.... *Philosophical Transactions of the Royal Society* **154**: 579–666. Also in Baker, H. F. (ed.), *The collected Mathematical Papers of James Joseph Sylvester* ..., Vol. II, pp. 376–479 (Cambridge, 1908).

Taylor, E. G. R. [1954] *The mathematical practitioners of Tudor and Stuart England.* Cambridge: for the Institute of Navigation, at the University Press.

Terquem, Olry [1843] ("Observation" appended to Gerono [1843]). *Nouvelles Annales de Mathématiques* **2**: 132–133.

Thomson, Thomas [1812] *History of the Royal Society: from its institution to the end of the Eighteenth Century.* Book II. London: printed for R. Baldwin.

Varnhorn, Mary C. [1939]. *Some properties of quartic functions of one variable* (dissertation). Washington, D. C.: Catholic University of America Press.

Waard, C. de [1927] "Wassenaer, Jacob van". *Nieuw Nederlandsch Biographisch Woordenboek*, vol. VII, cols. 1307–1308. Leiden: A. W. Sijthoff's Uitgevers-Maatschappij.

(Wales, William) [1778] (unsigned review of *Meditationes Analyticæ.*) *Monthly Review, or Literary Journal* **LIX**: 81–93 (August 1778).

Wallis, John [1685] *A Treatise of Algebra, both historical and practical....* London: printed by John Playford; Oxford: Richard Davis.

Wallis, John [1693] *Algebra* (Latin). in *Opera Mathematica*, Vol. II. Oxoniæ: E Theatro Sheldoniano, MDCXCIII.

Wallis, R. V. and P. J. [1986] *Biobibliography of British Mathematics and its Applications*, Part II: 1701–1760. Newcastle upon Tyne: PHIBB (Project for Historical Biobibliography); Letchworth, Herts: Epsilon Press.

Waring, Edward [1759]. *Miscellanea Analytica de Æquationibus Algebraicas, Curvarum Proprietatibus, Fluxionibus et Serierum Summatione.* Cantabrigiæ: typis academicis excudebat J. Bentham.

Waring, Edward [1760*a*] *A Reply to a Pamphlet entitled Observations on the first Chapter of a Book called Miscellanea Analytica.* Cambridge: Printed by J. Bentham, Printer to the University, for W. Thurlbourn & J. Woodyer in Cambridge.

Waring, Edward [1760*b*] *A letter to the Reverend Dr. Powell, ... in answer to his Observations on the First Chapter of a Book called Miscellanea Analytica, and his Defence of those Observations.* Cambridge: Printed by J. Bentham, Printer to the University, for W. Thurlbourn & J. Woodyer in Cambridge.

Waring, Edward [1762] *Miscellanea Analytica, de Æquationibus Algebraicis et Curvarum Proprietatibus.* Cantabrigiæ: Typis academicis excudebat J. Bentham.

Waring, Edward [1763] Problems by Edward Waring, M.A. and Lucasian Professor of Mathematics in the University of Cambridge, F.R.S. *Philosophical Transactions of the Royal Society* **LIII**: 294–298.

Waring, Edward [1770] *Meditationes Algebraicæ*, editio secunda. Cantabrigiæ: typis academicis excudebat J. Archdeacon, veneunt apud J. Woodyer, [etc.]

Waring, Edward [1772] *Proprietates Algebraicarum Curvarum*. Cantabrigiæ: typis academicis excudebat J. Archdeacon, veneunt apud J. Woodyer, [etc.]

Waring, Edward [1779] Problems concerning Interpolations. *Philosophical Transactions of the Royal Society* **LXIX**: 59–67.

Waring, Edward [1785] *Meditationes Analyticæ*, editio secunda. Cantabrigiæ: typis academicis excudebat J. Archdeacon, veneunt apud J. Nicholson.

Waring, Edward [1787] On finding the Values of Algebraical Quantities by converging Serieses, and demonstrating and extending Propositions given by *Pappus* and others. *Philosophical Transactions of the Royal Society* **LXXVII**: 71–83.

(Waring, Edward) [1794] (unsigned review of Englefield [1793]). *Monthly Review*, Series 2, XIV (June 1794): 155–158.

Waring, Edward [1799] Original letter of the late Dr. Waring to the Rev. Dr. Maskelyne, Astronomer Royal. *Monthly Magazine*, **9** (May 1799): 326–330.

Wassenaer, Jacob van [1640] *Den On-Wissen Wis-Kontenaer I. I. Stampioenius ontdeckt door sijne ongegronde Weddige ende Mislucte Solutien van sijne eygene questien. Midtsgaders eenen generalen regel om de Cubic-wortelen ende alle andere te trecken uyt twee-namighe ghetallen: dwelcke voor desen niet bekent en is geweest. Noch de Solutien van twee sware Geometrische questien door de Algebra dienstich om alle andere te leeren ontbinden door Jacobus Wassenaer Land-Meter tot Uytrecht.* Tot Leyden gedruct by Willem Christiaens, vor Johann Maire, anno 1640.

Weeks, D. [1974] Formal representations for algebraic numbers. ACM *SIGSAM Bulletin* **31** (August 1974): 91–95.

Weeks, D. [1991] The life and mathematics of George Campbell, F.R.S. *Historia Mathematica* (to appear)

Whitrow, Magda (ed.) [1971] *Isis Cumulative Bibliography; a bibliography of the history of science formed from Isis critical bibliographies 1–90, 1913–1965.* London: Mansell, in conjunction with the History of Science Society.

Wilder, Theaker [1769] *Universal Arithmetic: To which is added a Treatise upon the Measures of Ratios by James Maguire, A.M. The whole illustrated in a series of Notes by the Rev. Theaker Wilder, D.D., Senior Fellow of Trinity College, Dublin.* London: Printed for W. Johnston in Ludgate Street.

Willich, Michael Laurentius [1738] *Dissertationem mathematicam qua methodus Newtonianam reperiundi divisores æquationum illustrat atque demonstrat.* Gottingæ, apud Abram. Vandenhoeck Acad. Typographium.

(Wilson, John) [1760] *A Vindication of the Miscellanea Analytica, in Answer to a late Pamphlet entitled Observations, &c.* Cambridge, Printed by J. Bentham, Printer to the University, for T. & J. Merrill, in Cambridge, MDCCLX. (The only known copy in the United States is in the Rare Book Room at the Library of the University of Illinois, Urbana, where it is bound with Waring's *Meditationes Algebraicæ,* ed. III.)

Winkler, F., B. Buchberger, F. Lichtenberger, & H. Rolletschek [1985] Algorithm 628: an algorithm for constructing canonical bases of polynomial ideals. ACM *Transactions on Mathematical Software* **11**(1): 66–78.

Witmer, T. R. [1968] *The Great Art; or, the Rules of Algebra, by Girolamo Cardano.* Translated and edited by T. Richard Witmer, with a foreword by Oystein Ore. Cambridge, MA: M.I.T. Press.

Woepcke, F. [1850] Notice sur un manuscrit Arabe d'un traité d'algèbre par Aboul Fath Omar Ben Ibrahim Alkhayâmî, contenant la construction géométrique des équations cubiques. *Journal für die reine und angewandte Mathematik* (Crelle's Journal) **40**: 160 - 172.

Index